Sustainable Development Goals Series

The **Sustainable Development Goals Series** is Springer Nature's inaugural cross-imprint book series that addresses and supports the United Nations' seventeen Sustainable Development Goals. The series fosters comprehensive research focused on these global targets and endeavours to address some of society's greatest grand challenges. The SDGs are inherently multidisciplinary, and they bring people working across different fields together and working towards a common goal. In this spirit, the Sustainable Development Goals series is the first at Springer Nature to publish books under both the Springer and Palgrave Macmillan imprints, bringing the strengths of our imprints together.

The Sustainable Development Goals Series is organized into eighteen subseries: one subseries based around each of the seventeen respective Sustainable Development Goals, and an eighteenth subseries, "Connecting the Goals", which serves as a home for volumes addressing multiple goals or studying the SDGs as a whole. Each subseries is guided by an expert Subseries Advisor with years or decades of experience studying and addressing core components of their respective Goal.

The SDG Series has a remit as broad as the SDGs themselves, and contributions are welcome from scientists, academics, policymakers, and researchers working in fields related to any of the seventeen goals. If you are interested in contributing a monograph or curated volume to the series, please contact the Publishers: Zachary Romano [Springer; zachary.romano@springer.com] and Rachael Ballard [Palgrave Macmillan; rachael.ballard@palgrave.com].

Arthur Guseni Oliver Mutambara

Deploying Artificial Intelligence to Achieve the UN Sustainable Development Goals

Enablers, Drivers and Strategic Framework

Springer

Arthur Guseni Oliver Mutambara
Institute for the Future of Knowledge, University of Johannesburg
Johannesburg, South Africa

ISSN 2523-3084 ISSN 2523-3092 (electronic)
Sustainable Development Goals Series
ISBN 978-3-031-88422-1 ISBN 978-3-031-88423-8 (eBook)
https://doi.org/10.1007/978-3-031-88423-8

This Springer imprint is published by the registered company Springer Nature Switzerland AG
The registered company address is: Gewerbestrasse 11, 6330 Cham, Switzerland

This book is dedicated to technology-driven sustainable development as a basis for economic transformation.

Preface

The United Nations' Sustainable Development Goals (SDGs) are 17 interconnected goals designed to address the world's most pressing challenges by 2030. Adopted in 2015 as part of the 2030 Agenda for Sustainable Development, the SDGs cover a wide range of issues, including poverty eradication, quality education, gender equality, clean energy, climate action, and reduced inequalities. Each goal is supported by specific targets and indicators, providing a measurable framework for progress. The SDGs are universal, apply to all countries regardless of income level, emphasise the interconnectedness of global challenges, and thus require holistic solutions.

The SDGs are important because they provide a shared blueprint for global action and cooperation, uniting governments, businesses, and civil society to address critical challenges. By focusing on multidimensional goals, the SDGs encourage comprehensive strategies to tackle interconnected problems such as economic disparity, environmental degradation, and social inequities. They are vital for ensuring a sustainable future, reducing global inequality, and enhancing the quality of life for millions. Moreover, the SDGs emphasise leaving no one behind, targeting the most vulnerable populations, and ensuring progress benefits everyone.

However, there are challenges and complications. Progress towards achieving the SDGs has been mixed and uneven across different countries, regions, and goals. Indeed, significant advancements have been made in areas like poverty reduction, access to electricity, and improved healthcare. For example, global poverty rates declined steadily before the COVID-19 pandemic, and millions gained access to basic services such as clean water and education. Nevertheless, intractable problems are prevalent. Climate change continues to accelerate, inequality is rising, and many environmental targets, such as biodiversity conservation, are lagging behind. The COVID-19 pandemic and geopolitical conflicts, such as the unnecessary and avoidable wars in Ukraine, Gaza, and Sudan, have further slowed progress by disrupting economies, increasing inequalities, and redirecting resources away from SDG initiatives.

A significant barrier is inadequate funding. Achieving the SDGs requires trillions of dollars in investment annually, but many countries, especially low-income nations, lack the financial resources. Additionally, some regions' political instability, corruption, and poor governance hinder progress. The interconnected nature of the SDGs also means that failure in one area can impact others. For example, climate change exacerbates poverty and inequality, while conflicts and pandemics disrupt global efforts. Many countries' lack of data and monitoring capacity makes it challenging to track progress and identify effective solutions. The target date for attaining all 17 SDGs is 2030. It is a short five years away, yet according to the United Nations' 2024 SDG Report, only 17% of the SDG targets are currently on track to be achieved by 2030. Nearly half of the targets show minimal or moderate progress, while over a third are either stalled or regressing [132].

Indeed, the SDG agenda is characterised by an existential global crisis. So, what should be done? This book seeks to contribute to the resolution of this predicament. It aims to provide solutions to the challenges that have impeded the achievement of the SDGs by exploring broad and holistic interventions, technology-driven remedies, and, more specifically, the deployment of AI—the implementation of human intelligence in machines or systems programmed to perform tasks such as learning, reasoning, problem-solving, and decision-making. It is instructive to note that AI is just one tool. It is not a silver bullet. Achieving the SDGs by 2030 will require broader efforts and solutions than the use of AI systems.

Attaining the SDGs will depend on enhanced global cooperation, increased funding, improved infrastructure, and economic integration. Innovative solutions, such as leveraging technology and enhancing public-private partnerships, will be essential. The book proposes the acceleration of SDG progress by prioritising SDG implementation through stronger political commitment, integrated policies, and increased investment. There is a need to strengthen governance and policy frameworks. This ensures that resources are allocated efficiently, corruption is minimised, and accountability is upheld in implementing development programmes. Strong institutions, transparent decision-making, and inclusive policymaking are essential to ensure that progress towards the SDGs benefits everyone, particularly marginalised and vulnerable populations.

Countries must align their national strategies, policies, and budgets with SDGs, devising long-term plans to address interconnected challenges while ensuring inclusive and equitable policies. This includes enacting laws and regulations that promote equity, sustainability, and environmental conservation. Governments must introduce incentives for renewable energy adoption, enforce labour laws to reduce inequalities, and enhance land-use planning to protect biodiversity. There is a need to adopt innovations such as renewable energy for climate action and digital tools for education and healthcare. SDG progress can be driven by improving human capital through investments in education, healthcare, and social protection systems. Increasing funding for vocational training and public health campaigns can empower communities to address SDG-related challenges such as unemployment and public health crises.

International organisations, such as the United Nations and World Bank, must continue to support capacity-building and financial assistance for low-income countries. Achieving the SDGs requires bridging gaps between emerging or least industrialised economies and highly industrialised ones. There must be collaboration within the Global South and partnerships between the Global North and Global South. Global challenges like climate change, pandemics, and economic inequality require coordinated international efforts. Enhancing international cooperation and financing mechanisms is foundational to addressing funding gaps and sharing knowledge and resources for SDG implementation.

Highly industrialised countries must honour commitments to provide financial aid and technical support to emerging and least industrialised economies, mainly through mechanisms such as the Green Climate Fund and the Global Partnership for Education. There is a need to develop innovative climate financing models, including those based on carbon pricing such as carbon taxes, cap-and-trade systems, carbon markets, and green bonds. Fostering public-private partnerships can unlock investments in sustainable infrastructure, renewable energy, and other critical sectors. International trade policies should also be reformed to ensure fair market access for low-income countries. Strengthening multilateral institutions like the United Nations and World Bank is essential to coordinating global efforts, reducing systemic inequalities, and accelerating progress towards achieving the SDGs.

The arduous SDG journey towards 2030 will rise or fall on leadership. There is a need for visionary leadership at organisational, national, regional, continental, and global levels—characterised by the ability to create and articulate a clear, compelling future vision that inspires and motivates others to achieve the shared SDGs. It is essential for those driving the SDG agenda to have a unique blend of foresight, passion, and innovation, enabling them to see beyond the current reality and anticipate future trends and challenges. All this must be anchored by our shared common humanity and global interests, not narrow national, sectarian, or hegemonic interests. Unfortunately, events in the United States in early 2025 have signalled a shift from global collective leadership to isolationist, national interest-driven paradigms. US President Donald Trump signed executive orders to withdraw the United States from the Paris Agreement, the World Health Organisation, the UN Human Rights Council, and the United Nations Relief and Works Agency. He is also reviewing the country's role in the United Nations Educational, Scientific and Cultural Organization and has moved to dismantle the United States Agency for International Development. Achieving the SDGs in 2030 demands a different type of global leadership.

Technology (not necessarily Artificial Intelligence (AI)) is pivotal in advancing the SDGs by providing innovative solutions to global challenges. For instance, renewable energy technologies like solar panels, wind turbines, and hydropower systems are critical for achieving affordable and clean energy and combatting climate change. Similarly, advancements in water purification and sanitation systems directly address water and sanitation by ensuring access to safe drinking water and reducing waterborne diseases. Medical technologies, such as vaccines, portable diagnostic kits, and telemedicine platforms, are instrumental in advancing good health and well-being by improving healthcare access and disease prevention.

Desalination plants, water filtration systems, and wastewater recycling have significantly improved water security in arid and drought-prone regions. Drip irrigation schemes, bioengineered crops, and precision farming tools are transforming food production systems and reducing hunger. Emerging and least industrialised economies must move up the global value chains by implementing beneficiation and value addition. These two processes enhance the economic value of raw materials through local processing and manufacturing, generating higher revenues, creating jobs, and igniting sustainable development through the attainment of the SDGs.

The most significant contribution of this book is the use of AI to accelerate progress towards achieving the SDGs. Artificial Intelligence—the development of computer systems that can perform tasks typically requiring human intelligence, such as visual perception, speech recognition, decision-making, and language translation—is briefly introduced, and its unprecedented transformative nature is explained. The AI revolution is here. There were bold announcements on AI and its infrastructure in January 2025 from global leaders, such as UK Prime Minister Keir Starmer, then US President Joe Biden, and current US President Donald (immediately after his inauguration). Artificial Intelligence has become a key driver of competitiveness in every sector, and countries are unleashing massive investments in AI infrastructure. China's release, in the same month, of a ground-breaking open-source, low-cost, and less energy-intensive large language model called DeepSeek-R1, whose functionality is comparable to US offerings such as OpenAI's ChatGPT-4, Google's Gemini, and xAI's Grok 3, dramatises the equal-opportunity nature of the technology.

AI systems have the potential to provide innovative solutions to complex global challenges that impede the attainment of SDGs worldwide. For instance, AI-driven data analysis and predictive modelling can enhance decision-making processes across multiple SDG targets. In agriculture, AI systems can optimise crop yields by analysing weather patterns, soil quality, and pest activity, supporting the fight against hunger. These systems can guide farmers on when and where to plant crops, helping to reduce resource waste and adapt to climate-induced agricultural challenges. Similarly, AI-powered tools in supply chain management can minimise food loss and waste, ensuring that resources are used efficiently and equitably.

AI also plays a critical role in improving healthcare systems, contributing to good health and well-being. Machine Learning algorithms are used to detect diseases early, predict outbreaks, and personalise treatment plans based on patient data. For example, AI applications in medical imaging have proven highly accurate in diagnosing conditions like cancer and cardiovascular diseases. In addition, AI-driven tools enable the analysis of large-scale epidemiological data to predict the spread of infectious diseases, allowing governments to implement timely interventions. During the COVID-

19 pandemic, AI was leveraged to track the virus's progression, develop vaccines, and optimise healthcare delivery in overburdened systems.

Furthermore, AI can advance climate crisis mitigation by supporting climate modelling and environmental monitoring. AI-powered tools can analyse satellite imagery to monitor deforestation, track changes in land use, and measure carbon emissions. These insights help policymakers implement targeted conservation efforts and design effective climate adaptation strategies. Additionally, AI can optimise energy use by managing power grids, improving energy storage, and integrating renewable energy sources like wind and solar into the grid. By reducing energy waste and emissions, AI technologies contribute to the global transition towards a low-carbon economy.

For quality education, AI-powered adaptive learning platforms provide personalised education tailored to individual learning styles and needs, making quality education accessible to marginalised communities. AI also supports gender equality by identifying and addressing systemic biases in hiring processes and enabling women entrepreneurs to access financial services through AI-based credit scoring. Moreover, AI-driven financial inclusion initiatives, such as mobile banking and digital payment systems, empower underserved populations, advancing decent work and economic growth. By deploying AI thoughtfully and equitably, governments and organisations can harness its transformative power to address global inequalities and achieve the SDGs.

The book addresses AI safety, regulation, legislation, governance, risk mitigation, and carbon footprint while reviewing the AI semiconductor industry. It also assesses AI's potential negative impact on SDGs and examines AI deployment challenges. It emphasises aligning all national, regional, or continental plans with the AI-enabled SDG agenda. Summaries of AI-for-SDG experiences from six countries are outlined, and emerging best practices are harvested.

Details of enablers of AI deployment for SDGs are proposed and discussed—robust digital infrastructure and connectivity; awareness, education, and capacity building; regulations and ethical governance and accountability; guardrails; high-quality local data; financial resources and investment; beneficiation and value addition; research and development; and an accountable, capable, ethical developmental state. Similarly, critical mutually reinforcing drivers are presented: process efficiency and effectiveness; innovation and technological advancements; scalability and replicability; data-driven decision-making; addressing complex challenges; inclusivity and accessibility; partnerships and collaboration; private sector engagement; and policy support, governance frameworks, and global commitment.

Furthermore, it is essential to develop an AI ecosystem; embrace AI users' voices and insights; champion participatory approaches to AI design and deployment; incorporate diverse perspectives; and adopt feedback and iterative improvement mechanisms. There is efficacy in leveraging AI-enabled leapfrogging for SDGs, where emerging and least industrialised countries can bypass traditional stages of technological evolution and move directly to more advanced cutting-edge AI solutions. It is essential to embrace decoloniality in AI—a theoretical and practical framework aimed at dismantling the structures, knowledge systems, and power dynamics established during and after colonial rule, and likely to influence the essence and content of AI systems. In the same vein, it is imperative to democratise AI—making AI technologies, tools, knowledge, and opportunities accessible to a broader range of people, communities, and organisations beyond a privileged few.

Global governance for AI is vital. The key recommendations of the UN Secretary-General's 2024 AI Advisory Final Report are discussed. The strengths and flaws of this report are presented and explained. The principles of AI regulation/legislation and AI risk verticals are presented, while exemplary cases of AI legislation, such as the 2024 European Union AI Act, are reviewed, drawing lessons for other jurisdictions. However, the limitations of regulations and legislation as AI management tools are articulated, while the sociology of AI policy and adoption is also investigated.

While the book emphasises the need to embrace a broad range of enabling technologies, with a special focus on AI, it acknowledges the risks of technology-driven challenges such as digital imperialism and data colonialism, particularly in emerging and least industrialised economies. An incisive and robust case is made for decoloniality in AI on the SDG journey—a theoretical and practical framework aimed at dismantling the structures, knowledge systems, and power dynamics established during and after colonial rule and likely to influence the essence and content of AI systems. Furthermore, the book puts a premium on democratising AI in pursuit of the SDGs—making AI technologies, tools, knowledge, and opportunities accessible to a broader range of people, communities, organisations, countries, and beyond a privileged few individuals, institutions, and economies.

A key contribution of the book to AI adoption and thought leadership is the Strategic Framework for AI Deployment, which has six distinct but related components: Vision, Strategy, Policy, Governance, Legislation/Regulations, and Implementation Matrix (inclusive of Monitoring, Measurement, Evaluation and Feedback). In pursuit of the SDGs, every continent, regional bloc of states, country, organisation, or community must develop and adopt such a framework, where these structures dynamically influence each other. Within this context, the role of both regional and continental integration and political unity is articulated. The African Union's 2024 Continental AI Strategy is reviewed. Its strengths and weaknesses are discussed. The book provides details on deploying AI to achieve all 17 SDGs. Each goal is examined, its challenges are assessed, and detailed proposals for AI interventions to facilitate attainment are posited. AI adoption challenges and ethical considerations specific to the goal are discussed, and policy recommendations are proffered.

The potential future envisioned in the 2030 SDG agenda—a world free from poverty, hunger, and environmental degradation—is slowly becoming elusive, if not illusionary. That desired future—complete attainment of the SDGs—is not inevitable. It is contingent on immediate and transformative action. Political and business leaders, policymakers, academics, civil society activists, and ordinary citizens must reignite momentum towards the SDGs, ensuring that 2030 becomes a milestone of achievement rather than a moment of regret. Global cooperation, regional/continental integration, moving up global value chains, inclusive economic transformation, addressing the climate crisis, and use of advanced technology (in particular AI) can play a significant role in the arduous journey to 2030.

Of course, there is the danger that AI will widen global inequality. Left unchecked, AI can intensify global disparities by consolidating power and wealth in affluent nations while exploiting labour and resources in emerging and least industrialised countries. There is a real possibility that AI will entrench existing inequities, leading to heightened political instability, environmental degradation, and cultural dominance by a select few. This book seeks to mitigate these challenges. AI must serve as a transformative force for the collective good, benefiting the entire planet and all its inhabitants in an equitable manner. Harnessing this transformative technology to advance the SDGs in every country offers a strategic and practical starting point. The UN Secretary-General, António Guterres, is right: "We must never allow AI to stand for advancing inequality."

Johannesburg, South Africa — Arthur Guseni Oliver Mutambara

Acknowledgements

This book is a product of my research efforts within the Institute for the Future of Knowledge at the University of Johannesburg. The insights, examples, and illustrations were derived from collaborative research I have been involved in at various institutions in the United States, Europe, China, and South Africa. I greatly appreciate the comradeship, support, and suggestions provided by my colleagues in these countries.

I acknowledge all my research colleagues at the Massachusetts Institute of Technology (MIT), the University of Oxford, the Georgia Institute of Technology, the University of Pennsylvania, the University of Cape Town (UCT), the University of the Witwatersrand (Wits), and the University of Johannesburg (UJ), for reviewing the book material and providing insightful comments. I am deeply indebted to these academic research colleagues. Their questions, suggestions, and extensive feedback helped me make the material more readable.

In particular, I would like to thank Professors Edward Crawley (MIT), Chuchu Fan (MIT), David Autor (MIT), Gift Mugano, Durban University of Technology (DUT), and Koffi Kouakou (Wits) for their detailed remarks and suggestions. Jabulani Sibande, an IFK Research Assistant, assisted with recasting some diagrams and tables.

Contents

About the Author

Prof. Dr. Eng. Arthur G.O. Mutambara is Director and Full Professor at the University of Johannesburg's (UJ) Institute for the Future of Knowledge (IFK) in South Africa, where he also serves as Head of the Decentralised Artificial Intelligence and Control Systems (DAICS) Research Group. He is a world-renowned control systems specialist, roboticist, academic, author, and technology strategist. Professor Mutambara is a key African thought leader and presenter on the raging Artificial Intelligence revolution. He teaches Control Systems in two departments at UJ: Mechanical Engineering Science and Electrical and Electronic Engineering Science. Engineer Mutambara previously taught and conducted research in Control Systems, Robotics and Mechatronics in the Aeronautics and Astronautics Department at the Massachusetts Institute of Technology (MIT), and in the Mechanical Engineering Department at the Florida Agricultural and Mechanical University—Florida State University (FAMU-FSU), both in the United States.

He was also a Research Scientist at the National Aeronautics and Space Administration (NASA), a Visiting Professor at MIT, and a Visiting Research Professor at Carnegie Mellon University's Robotics Institute. In his academic research and teaching, he has received outstanding reviews and praise from students and peers worldwide. He is a Chartered Engineer, a Fellow of the Institution of Engineering and Technology (IET), a Professional Engineer, a Fellow of the Zimbabwe Institution of Engineers (ZIE), a Fellow of the Zimbabwe Academy of Sciences (ZAS), and a Senior Member of the Institute of Electrical and Electronics Engineers (IEEE).

Prof. Mutambara is a distinguished and prolific author of eight books. He has written three electrical engineering books widely used in undergraduate and graduate engineering (teaching and research) programmes in the United States, Europe, China, Japan, and Africa: *Decentralized Estimation and Control for Multisensor Systems (1998)*, *Design and Analysis of Control Systems (1999)*, and *Driving the Fourth Industrial Revolution (4IR): Design and Analysis of Control Systems (2024)*. He holds a PhD in Robotics and Mechatronics, and an MSc in Computer Engineering, both from the University of Oxford. Mutambara was there as a Rhodes Scholar from September 1991 to March 1995. He graduated with a BSc (Honours) in Electrical Engineering from the University of Zimbabwe in 1990. Beyond Engineering and Science, Professor Mutambara is a distinguished Pan-Africanist and a prolific public intellectual across the African continent and globally. He has conducted outstanding presentations on leadership, African unity, regional integration, management, and business strategy throughout the world, in addition to motivational speeches and seminars. He has authored three books on Thought Leadership: In Search of the Elusive Zimbabwean Dream Volume I: *The Formative Years and the Big Wide World (1983–2002)*; Volume II: *The Path to Power (2003–2009)*; Volume III: *Ideas & Solutions: Deputy Prime Minister and Beyond (2009–2023)*. He has also just published (2025) the ground-breaking research book: *Artificial Intelligence: A Driver of Inclusive Development and Shared Prosperity for the Global South*. In 2007, Mutambara was accorded the World Economic Forum (WEF) Young Global Leader status and subsequently attended WEF events from 2007 to 2013 in Davos (Switzerland), China, India, and Africa.

Professor Mutambara is the Former Deputy Prime Minister (DPM) of Zimbabwe. He was one of the three Principals who created and led the Government of National Unity (GNU) in Zimbabwe from 2009 to 2013. The other two were the late former Prime Minister Morgan Tsvangirai and the late former President Robert Mugabe. As the Deputy Prime Minister, his key functions included assisting the Prime Minister in policy formulation by the Cabinet and supervision of policy implementation by the Council of Ministers. He also specifically supervised the Ministries under the Infrastructure Cluster, such as Energy and Power Development, Transport and Infrastructure, Information Communication Technologies, Water Resources and Development, and Public Works. Professor Mutambara coordinated regional policy formulation within these sectors across Southern African Development Community (SADC) and Common Market for Eastern and Southern Africa (COMESA). Furthermore, DPM Mutambara drove three critical national efforts of the GNU: the development of a Shared National Vision, the Rebranding of Zimbabwe, and the formulation of a National Infrastructure Master Plan.

Professor Mutambara was also a management consultant with McKinsey & Company in the United States (Chicago Office) and the Director of Electronic Payments at Standard Bank in South Africa.

1 Introduction

1.1 Background: Why UN Sustainable Development Goals?

Sustainable development is essential because it aims to balance economic growth, social inclusion, and environmental protection to ensure a prosperous future for all. As the global population grows and resources become increasingly scarce, sustainable development offers a framework to address critical challenges such as poverty, inequality, and climate change. It ensures that the needs of the present are met without compromising the ability of future generations to meet their own needs. By integrating social, economic, and environmental goals, sustainable development promotes long-term resilience, equity, and the responsible use of natural resources [15, 121].

The UN Sustainable Development Goals (SDGs) are 17 interconnected goals designed to address the world's most pressing challenges by 2030. Adopted in 2015 as part of the 2030 Agenda for Sustainable Development, the SDGs cover a wide range of issues, including poverty eradication, quality education, gender equality, clean energy, climate action, and reduced inequalities. Each goal is supported by specific targets and indicators, providing a measurable framework for progress. The SDGs are universal, applying to all countries regardless of income level, and emphasise the interconnectedness of global challenges, requiring holistic solutions [110, 126].

The SDGs are important because they provide a shared blueprint for global action and cooperation, uniting governments, businesses, and civil society to address critical challenges. By focusing on multidimensional goals, the SDGs encourage comprehensive strategies to tackle interconnected problems such as economic disparity, environmental degradation, and social inequities. They are vital for ensuring a sustainable future, reducing global inequality, and enhancing the quality of life for millions. Moreover, the SDGs emphasise leaving no one behind, targeting the most vulnerable populations, and ensuring progress benefits everyone. Figure 1.1 shows the 17 SDGs.

1.2 Motivation for the Book

Progress towards achieving the SDGs has been mixed and uneven across different regions and goals. According to recent reports, significant advancements have been made in areas like poverty reduction, access to electricity, and improved healthcare. For example, global poverty rates declined steadily

A. G. O. Mutambara, *Deploying Artificial Intelligence to Achieve the UN Sustainable Development Goals*, Sustainable Development Goals Series,
https://doi.org/10.1007/978-3-031-88423-8_1

Fig. 1.1 UN Sustainable Development Goals [125]

before the COVID-19 pandemic, and millions gained access to basic services such as clean water and education. However, challenges remain. Climate change continues to accelerate, inequality is rising, and many environmental targets, such as biodiversity conservation, are lagging behind. The COVID-19 pandemic and geopolitical conflicts, such as the war in Ukraine, have further slowed progress by disrupting economies, increasing inequalities, and redirecting resources away from SDG initiatives.

Several factors have contributed to the slow progress of the SDGs. One of the most significant barriers is inadequate funding. Achieving the SDGs requires trillions of dollars in investment annually, but many countries, especially low-income nations, lack the financial resources to implement necessary initiatives. Additionally, some regions' political instability, corruption, and poor governance hinder progress. The interconnected nature of the SDGs also means that failure in one area can impact others. For example, climate change exacerbates poverty and inequality, while conflicts and pandemics disrupt global efforts. Finally, many countries' lack of data and monitoring capacity makes it challenging to track progress and identify effective solutions.

The target date for attaining all 17 SDGs is 2030. It is a short five years away, yet according to the United Nations' 2024 SDG Report, only 17% of the SDG targets are currently on track to be achieved by 2030. Nearly half of the targets are showing minimal or moderate progress, while over a third are either stalled or regressing. This is a crisis. Indeed, without immediate and innovative interventions, the vision for a sustainable and just world by 2030 could remain unfulfilled [132].

So, what is to be done? This book seeks to tackle this existential question. It aims to provide solutions to the challenges that have impeded the achievement of the SDGs. The starting point is an exploration of broad non-AI interventions, followed by a focus on general technology-driven remedies, and then proffering the deployment of AI as the key strategic approach for attaining all 17 SDGs.

1.3 Broad Interventions

Achieving the SDGs by 2030 is an ambitious and challenging objective. While some targets are within reach, many require urgent and accelerated action. Achieving the SDGs will depend on enhanced global cooperation, increased funding, and political commitment. Innovative solutions, such as leveraging technology and fostering public-private partnerships, will be essential. However, given the current trajectory and recent setbacks, including the pandemic and geopolitical conflicts, achieving all 17 goals by 2030 may not be realistic. Nevertheless, partial progress in key areas can still have transformative impacts on global development and quality of life.

The book proposes that to accelerate progress, countries must prioritise SDG implementation through stronger political commitment, integrated policies, and increased funding. Multilateral institutions, such as the United Nations and World Bank, must continue to support capacity building and financial assistance for low-income nations. Governments must align their national strategies with the SDGs, ensuring inclusive and equitable policies. Innovation and technology also play a critical role in addressing complex challenges, such as renewable energy for climate action and digital tools for education and healthcare. Most importantly, collective global action involving governments, businesses, and civil society must ensure no one is left behind.

Global cooperation is crucial because the SDGs address issues that transcend national boundaries, such as climate change, pandemics, and economic inequality. No country can tackle these challenges alone, making international partnerships essential for pooling resources, sharing knowledge, and coordinating efforts. Multilateral agreements, such as the Paris Agreement on climate change, exemplify the importance of collective action. Furthermore, achieving the SDGs requires bridging gaps between emerging or least industrialised and highly industrialised countries, addressing systemic inequalities, and enhancing mutual accountability. Sustained global cooperation ensures that progress is equitable, inclusive, and aligned with the overarching goal of sustainable development for all. However, all this will rise or fall on visionary leadership at organisational, national, regional, continental, and global levels. Political, business, and civil society leaders must create and articulate a compelling future vision that inspires and motivates countries and communities to achieve the SDGs.

Before exploring the possibility of deploying technology in general and AI in particular, it is imperative to flesh out and summarise the key broad non-AI interventions advanced in the book to enhance SDG achievement.

1.3.1 Strengthening Governance and Policy Frameworks

Strengthening governance and policy frameworks is one of the most critical non-AI interventions for improving SDG progress. Effective governance ensures that resources are allocated efficiently, corruption is minimised, and accountability is upheld in implementing development programmes. Countries must align their national policies and budgets with the SDGs, creating long-term strategies to address interconnected challenges. This includes enacting laws and regulations that promote equity, sustainability, and environmental conservation. For example, governments can introduce incentives for renewable energy adoption, enforce labour laws to reduce inequalities, and enhance land-use planning to protect biodiversity. Strong institutions, transparent decision-making, and inclusive policymaking are essential to ensure that progress towards the SDGs benefits everyone, particularly marginalised and vulnerable populations.

1.3.2 Investing in Human Capital Development

Another key intervention to accelerate SDG progress is improving human capital through investments in education, healthcare, and social protection systems. Education fosters long-term economic growth and reduces inequalities, particularly for girls and marginalised communities. Expanding access to quality healthcare services can significantly improve outcomes related to maternal and child health, disease prevention, and life expectancy. In addition, social protection programmes, such as cash transfers and unemployment benefits, help reduce poverty and enhance resilience to economic shocks. For instance, increasing funding for vocational training and public health campaigns can empower communities to address SDG-related challenges such as unemployment and public health crises. Governments can build a foundation for sustainable economic and social progress by prioritising human capital development.

1.3.3 Enhancing International Cooperation and Financing

Global challenges like climate change, pandemics, and economic inequality require coordinated international efforts. Enhancing international cooperation and financing mechanisms is crucial to addressing funding gaps and sharing knowledge and resources for SDG implementation. Developed countries must honour commitments to provide financial aid and technical support to developing nations, mainly through mechanisms such as the Green Climate Fund and the Global Partnership for Education. Fostering public-private partnerships can unlock investments in sustainable infrastructure, renewable energy, and other critical sectors. International trade policies should also be reformed to ensure fair market access for low-income countries. Strengthening multilateral institutions like the United Nations and World Bank is essential to coordinate global efforts, reduce systemic inequalities, and accelerate progress towards achieving the SDGs.

1.4 Leveraging Non-AI Technology

This book posits that technology—generally without necessarily focusing on AI—is pivotal in advancing the SDGs by providing innovative solutions to global challenges. For instance, renewable energy technologies like solar panels, wind turbines, and hydropower systems are critical for achieving affordable and clean energy and combatting climate change. These technologies reduce greenhouse gas emissions and bring electricity to remote and underserved areas, empowering communities and fostering economic development. Similarly, advancements in water purification and sanitation systems directly address water and sanitation by ensuring access to safe drinking water and reducing waterborne diseases. Technologies such as desalination plants, water filtration systems, and wastewater recycling have significantly improved water security in arid and drought-prone regions.

In the agricultural sector, non-AI innovations like drip irrigation, bioengineered crops, and precision farming tools are transforming food production systems and reducing hunger. These technologies optimise resource use, increase crop yields, and enhance food security, particularly in regions facing climate-induced agricultural challenges. Mobile connectivity and basic digital tools have also empowered rural farmers by providing access to weather forecasts, market prices, and best practices. Emerging and least industrialised economies must move up the global value chains through implementation of beneficiation and value addition. These two processes enhance the economic value of raw materials through local processing and manufacturing, generating higher revenues, creating jobs, and igniting sustainable development through attainment of the SDGs.

Non-AI medical technologies, such as vaccines, portable diagnostic kits, and telemedicine platforms, are instrumental in advancing good health and well-being by improving healthcare access and disease prevention. These innovations, coupled with global collaboration and investment in infrastructure, demonstrate the immense potential of non-AI technologies to create sustainable and equitable solutions across various SDG targets.

1.5 AI Deployment to Achieve SDGs

The most significant contribution of this book is the use of AI to accelerate progress towards achieving the SDGs. For a start, the AI deployment to attain these goals must be carried out in a strategic manner and framework where the vision is clear and the implementation is systematically planned and executed as illustrated in Fig. 1.2. This strategic structure must be replicated at organisational, national, regional, and continental levels. AI systems have the potential to provide innovative solutions to complex global challenges that impede the attainment of SDGs worldwide. For instance, AI-driven data analysis and predictive modelling can enhance decision-making processes across multiple SDG targets. In agriculture, AI systems can optimise crop yields by analysing weather patterns, soil quality, and pest activity, supporting the fight against hunger. These systems can guide farmers on when and where to plant crops, helping to reduce resource waste and adapt to climate-induced agricultural challenges. Similarly, AI-powered tools in supply chain management can minimise food loss and waste, ensuring that resources are used efficiently and equitably.

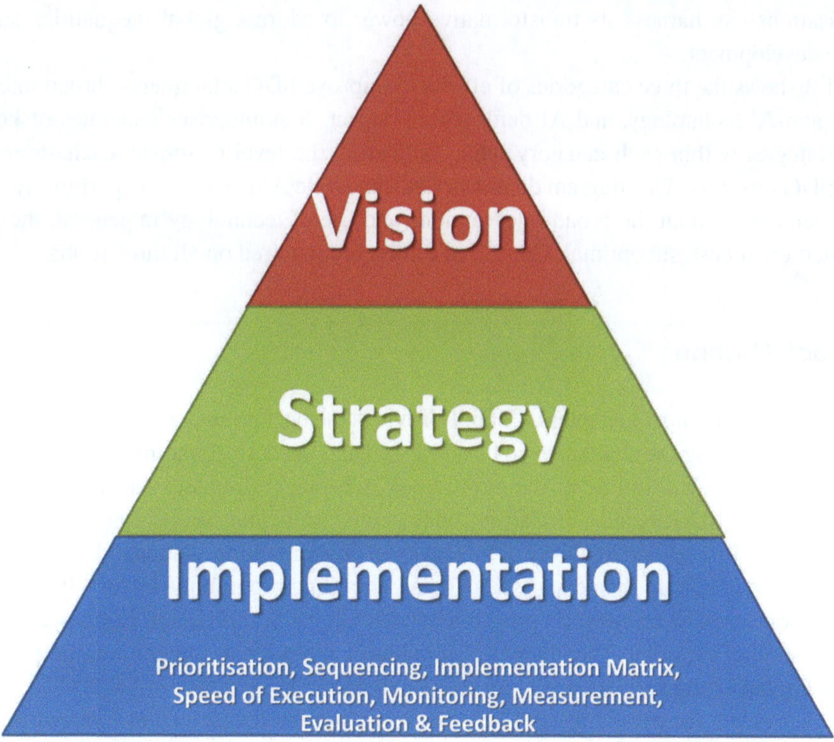

Fig. 1.2 Strategic framework for the deployment of AI to achieve the SDGs

AI also plays a critical role in improving healthcare systems, contributing to good health and well-being. Machine learning algorithms are used to detect diseases early, predict outbreaks, and personalise treatment plans based on patient data. For example, AI applications in medical imaging have proven highly accurate in diagnosing conditions like cancer and cardiovascular diseases. In addition, AI-driven tools enable the analysis of large-scale epidemiological data to predict the spread of infectious diseases, allowing governments to implement timely interventions. During the COVID-19 pandemic, AI was leveraged to track the virus's progression, develop vaccines, and optimise healthcare delivery in overburdened systems. Furthermore, AI can advance climate crisis mitigation by supporting climate modelling and environmental monitoring. AI-powered tools can analyse satellite imagery to monitor deforestation, track changes in land use, and measure carbon emissions. These insights help policymakers implement targeted conservation efforts and design effective climate adaptation strategies. Additionally, AI can optimise energy use by managing power grids, improving energy storage, and integrating renewable energy sources like wind and solar into the grid. By reducing energy waste and emissions, AI technologies contribute to the global transition towards a low-carbon economy.

Lastly, AI can enhance education, gender equality, and economic inclusion by bridging gaps in access to resources and opportunities. For quality education, AI-powered adaptive learning platforms provide personalised education tailored to individual learning styles and needs, making quality education accessible to marginalised communities. AI also supports gender equality by identifying and addressing systemic biases in hiring processes and enabling women entrepreneurs to access financial services through AI-based credit scoring. Moreover, AI-driven financial inclusion initiatives, such as mobile banking and digital payment systems, empower underserved populations, advancing decent work and economic growth. By deploying AI thoughtfully and equitably, governments and organisations can harness its transformative power to address global inequalities and achieve sustainable development.

Figure 1.3 shows the three categories of efforts to improve SDG attainment—broad interventions, leveraging non-AI technology, and AI deployment impact. It summarises examples of key actions, tools, or strategies within each category while indicating the level of impact each strategy has on achieving SDG progress. The diagram demonstrates that while AI presents an opportunity to enhance SDG achievement, without the broad interventions and use of technology in general, the AI impact will be muted or, at best, suboptimal. Concerted efforts are required on all three fronts.

1.6 Book Outline

This book is organised into 13 chapters. Chap. 1 provides the background and motivation for the book while clearly articulating the challenges to be addressed and the corresponding pathways to solutions. It also provides a detailed outline of the entire book. Chapter 2 discusses sustainable development to contextualise the SDG agenda. It also introduces environmental, social, and governance (ESG) concepts and their link to sustainability. Challenges of implementing sustainability are investigated and reviewed. The SDGs are then discussed in detail regarding programme history, linkages between SDGs, limitations of SDG framework, progress to date, Global South/Global North peculiarities, and emerging experiences/lessons. The chapter also explores the roles of integration, the private sector, and the UN Global Compact.

Chapter 3 presents the details of all 17 SDGs, including the measurement framework, examples of targets and indicators, and the monitoring and evaluation system. A review of essential non-AI interventions is carried out to hasten the attainment of the SDGs. Each of the 17 SDGs is discussed in terms of goal significance, targets, indicators, progress, and challenges. Chapter 4 presents an

Category	How It Works	Impact on SDG Progress
Broad Interventions	Strengthening governance and policy frameworks to ensure resource allocation, transparency, and equity.	High Impact
	Enhancing global cooperation and partnerships for funding, technical support, and knowledge sharing.	High Impact
	Investing in education, healthcare, and social protection to build human capital.	High Impact
	Mobilizing financial support through public-private partnerships and international aid.	Moderate Impact
	Aligning innovation and policies with SDG targets through integrated strategies.	Moderate Impact
Leveraging Non-AI Tech	Using renewable energy systems like solar and wind to reduce emissions and increase energy access.	High Impact
	Implementing water purification and sanitation technologies to enhance clean water availability.	High Impact
	Deploying smart irrigation and precision farming tools for agricultural productivity.	High Impact
	Advancing medical technologies like telemedicine, vaccines, and diagnostics for better healthcare.	Moderate Impact
	Adopting bioengineered crops to improve food security in climate-vulnerable areas.	Moderate Impact
AI Deployment Impact	Leveraging predictive analytics for data-driven decision-making and impact assessment.	High Impact
	Using AI for crop yield optimization, soil analysis, and climate adaptation in agriculture.	High Impact
	AI-powered disease detection, predictive modeling, and personalized healthcare systems.	High Impact
	Utilizing AI for climate modeling, satellite analysis, and risk management to mitigate climate crises.	Moderate Impact
	AI-driven adaptive learning platforms and virtual tutors to improve educational inclusivity.	Moderate Impact

Fig. 1.3 Three categories of interventions required to enhance SDG achievement

introduction to AI within the context of the 4IR. Key drivers of this revolution are also discussed, and classification, applications, and examples of AI systems are outlined. The issues of AI safety and the potential dangers of the technology are also discussed. There is also an introduction to AI regulation, governance, risk mitigation, carbon footprint, and a review of the AI semiconductor industry.

Chapter 5 tackles the deployment of AI to achieve the SDGs by presenting the potential role of the technology, examples of AI efficacy in SDGs, illustrative specific implementation cases, and AI interventions driving more than one SDG. Discussions cover the use of synthetic data, managing bias and inequality, ethics, data privacy and security, environmental impact, and the interplay between different AI interventions. The potential negative impact of AI on SDGs is also assessed, while AI deployment (for SDGs) challenges in the Global South are examined.

Chapter 6 reviews global experiences of countries deploying AI to achieve the SDGs. The starting point is a review of the worldwide state of play regarding SDG achievement, as detailed in the 2024 UN SDG Report. Global efforts to leverage AI for SDG attainment are then discussed. The special case of Africa is presented, and the corresponding implementation challenges are reviewed while refashioning and anchoring the continent's SDG strategies on AI. The need to align the efforts with other African initiatives, such as Agenda 2063, is posited. Summaries of AI-for-SDG experiences from six countries are outlined (United States of America, Singapore, Finland, China, Rwanda, United Arab Emirates, and South Africa), and emerging best practices are harvested.

Chapter 7 provides elaborate details of enablers of AI deployment in pursuit of SDGs. These include robust digital infrastructure and connectivity; awareness, education, and capacity building; regulations and ethical governance and accountability; guardrails; high-quality local data; financial resources and investment; beneficiation and value addition; research and development; and an accountable, capable, ethical developmental state.

Chapter 8 details the drivers of AI deployment in pursuit of SDGs. These mutually reinforcing critical success factors include visionary and transformational leadership; process efficiency and effectiveness; innovation and technological advancements; scalability and replicability; data-driven decision-making; addressing complex challenges; inclusivity and accessibility; partnerships and collaboration; private sector engagement; and policy support, governance frameworks, and global commitment. Further necessary igniters of the process include an AI ecosystem; embracing AI users' voices and insights; participatory approaches to AI design and deployment; incorporating diverse perspectives; and feedback mechanisms and iterative improvement.

Chapter 9 presents further critical success factors for deploying AI to achieve the SDGs. These include AI-enabled leapfrogging for SDGs, where emerging and least industrialised countries can bypass traditional stages of technological evolution and move directly to more advanced, AI cutting-edge solutions—skipping older, legacy systems and adopting more efficient AI technologies. Another indispensable tool is embracing decoloniality in AI—a theoretical and practical framework aimed at dismantling the structures, knowledge systems, and power dynamics established during and after colonial rule and likely to influence the essence and content of AI systems. The third aspect is democratising AI—the process of making AI technologies, tools, knowledge, and opportunities accessible to a broader range of people, communities, and organisations beyond a privileged few.

Chapter 10 discusses AI governance, legislation, regulations, sociology, and ethics with respect to the SDG journey. The reviewed matters include ethical guidelines, data privacy and protection, environmental impact, and mitigating inequality. In addition, accountability, international coordination, interoperability, standards, and cybersecurity regulations are discussed. The need for sector-specific AI regulations, regional and continental AI harmonisation, and global AI governance is presented. The key recommendations of the UN Secretary-General's 2024 AI Advisory Final Report are discussed. The strengths and flaws of this report are presented and explained. The chapter also outlines the principles of AI regulation and AI risk verticals while reviewing exemplary cases of

AI legislation, such as the 2024 European Union AI Act and India's AI-driven (2023) Aadhaar System. Lessons are drawn from these two pioneering legislative efforts. However, the limitations of regulations and legislation as AI management tools are articulated, while the sociology of AI policy and adoption is also investigated. There is also a discussion of ethics, privacy, and moral dilemmas in the AI-powered SDG journey.

Chapter 11 presents and discusses the strategic framework for AI deployment to achieve the SDGs—with six main components: vision, strategy, policy, governance, legislation/regulations, and implementation matrix (inclusive of monitoring, measurement, evaluation, and feedback). Every continent, regional bloc of states, country, organisation, and community must develop and adopt such a framework. Of course, these frameworks must dynamically influence each other. The chapter starts by presenting the adoption principles, objectives, and activities, and it discusses the infrastructure and investment requirements of the framework. The implementation and monitoring of pilot projects are discussed, and the role of global collaboration and partnerships is outlined. The role of both regional and continental integration and political unity is articulated. The African Union's 2024 Continental AI Strategy is reviewed. Its strengths and weaknesses are discussed.

Chapter 12 presents the details of deploying AI to achieve UN SDGs for each of the 17 goals. Each goal is examined, its challenges are assessed, and detailed proposals for AI interventions to facilitate attainment are posited. AI adoption challenges and ethical considerations specific to the goal are discussed, and policy recommendations are proffered.

Chapter 13 concludes the book by anchoring and summarising the key research contributions. Beyond SDG achievement, leveraging AI can lead to inclusive GDP growth. This is explained in this chapter. The key imperatives in deploying AI to achieve the SDGs are summarised. The fundamental question of whether the SDGs are achievable in 2030 is discussed. One of the significant challenges is resolving the risks that AI poses to attaining the SDGs. AI systems involve extensive computing facilities, substantial financial investments, consumption of a lot of energy, use of critical minerals, and generation of significant greenhouse gas emissions. This chapter addresses how this can be resolved. It also refers to the AI inequity between the Global North and the Global South. It advocates for undivided, razor-sharp attention, as the disparity undermines the latter's effectiveness in deploying AI to achieve the SDGs.

AI-driven advancements across various economic and social sectors are pivotal for achieving the SDGs as they create a synergistic effect that propels holistic development. Appendix A presents the details of two case studies of AI applications in agriculture and mining. Appendix B covers five case studies of AI applications in infrastructure, governance, and manufacturing, while Appendix C deals with five case studies of AI applications in financial services. Each case study identifies the sector problems, details the AI solutions, evaluates the impact of the AI intervention, and draws lessons for the deployment of AI to achieve the SDGs.

An extensive and comprehensive list of publications and sources used in authoring the book is presented in the **References**, while the **Index** lists the keywords, phrases and concepts to quickly direct the reader to relevant information or sections of the book.

1.7 Conclusion

The potential future envisioned in the 2030 SDG agenda—a world free from poverty, hunger, and environmental degradation—is slowly becoming elusive, if not illusionary. That desired future—complete attainment of the SDGs—is not inevitable. It is contingent on immediate and transformative action. Political and business leaders, policymakers, academics, civil society activists, and ordinary citizens must reignite momentum towards the SDGs, ensuring that 2030 becomes a milestone of achievement rather than a moment of regret. Global cooperation, regional/continental integration, moving up global value chains, inclusive economic transformation, addressing the climate crisis, and use of advanced technology (in particular AI) can play a significant role in the arduous journey to 2030. The emphasis is on bold but responsible AI development and deployment, ensuring that AI systems are grand and ambitious, but transparent, explainable, and aligned with human values. All national, regional, continental, and global stakeholders must understand and harness AI's potential to drive progress towards the SDGs.

Sustainable Development: Contextualising the SDGs

2

2.1 Introduction

Development is a multifaceted concept encompassing economic, social, political, human, and environmental aspects. It is often associated with progress in areas such as income, education, health, human rights, and industrialisation. The developmental objective is to improve a population's standard of living through factors such as wealth creation and distribution, social differentiation, industrial transformation, and economic growth. The starting point in exploring the deployment of AI to achieve the United Nations (UN) Sustainable Development Goals (SDGs) must be anchored by a thorough understanding of development, specifically sustainable development.

This chapter introduces the 17 UN SDGs within a broad context of sustainable development. The challenges, drivers, significance, and implementation of sustainability are fleshed out, while linkages are made to other frameworks, such as environmental, social, and governance (ESG). A brief history of the SDGs is outlined, and justification is provided for each SDG. Progress in achieving the SDGs is assessed, and limitations of the framework are discussed. Experiences from different regions of the world are compared and contrasted, while lessons are harvested. Broad interventions to resolve SDG implementation are proffered. The role of governments, the private sector, regional/continental bodies, and the UN Global Compact and the link between SDGs are investigated.

Indeed, before envisaging the deployment of AI to achieve the SDGs, it is instructive to contextualise the SDG agenda within the broad discussion of sustainable development. In a world increasingly threatened by the climate crisis, resource depletion, and social inequality, a holistic understanding of sustainability and the SDGs is essential to enhance cooperation and solidarity among all nations. Climate change and environmental damage respect no borders. A threat to one nation is a threat to the entire planet. There is no Planet B.

2.2 Types of Development

As already stated, there are five broad but related categories of development: economic, social, political, human, and environmental. It is prudent to outline these dimensions briefly. Economic development is a key aspect of general development. It speaks to growth in income and wealth,

A. G. O. Mutambara, *Deploying Artificial Intelligence to Achieve the UN Sustainable Development Goals*, Sustainable Development Goals Series,
https://doi.org/10.1007/978-3-031-88423-8_2

industrialisation and modernisation, and poverty reduction. Indeed, economic development often involves an increase in the standard of living, measured by income per capita, gross domestic product (GDP), or gross national income (GNI). The transition from agrarian-based economies to industrial and service-based ones is a key indicator of economic development. Although this transition is a measure of economic development, it is also a form of economic transformation—the continuous process of moving labour and other resources from lower to higher productivity sectors (structural change) and raising within-sector productivity growth. Of course, a key objective of economic development is to reduce poverty and ensure equitable resource distribution.

Social development is linked and interconnected to economics. This includes education, healthcare, gender equality, and social inclusion. Access to quality education is essential for development, empowering individuals and improving social mobility. Improvements in healthcare, including access to medical services and better health outcomes, are essential. Development involves ensuring equal opportunities for all genders, including access to education, employment, and political participation. There must be efforts to integrate marginalised communities and ensure that all members of society can participate fully in social and economic life.

Socio-economic development is pursued within a political context. Political development, a crucial component of overall development, involves good governance, democratic participation, and human rights. It often entails strengthening institutions and promoting transparency, accountability, and the rule of law. A key aspect of political development is ensuring that citizens have a voice in political processes and that their rights are protected. Upholding human rights and ensuring that all individuals can live free from oppression and discrimination is central to development.

Human development focuses on expanding people's opportunities and improving their quality of life, encompassing areas such as health, education, employment, and social well-being. It is centred on the idea of empowering individuals to lead productive, fulfilling lives by addressing inequalities and ensuring access to basic services and resources. Human development initiatives include improving healthcare systems, providing universal access to education, creating decent work opportunities, and fostering inclusive societies. By prioritising human rights and equity, these efforts contribute to reducing poverty, enhancing gender equality, and enabling people to participate fully in economic, social, and political life. Human development is not only a goal in itself but also a driver of sustainable growth, as healthier, better-educated, and empowered populations are more capable of contributing to and sustaining development outcomes.

Environmental development emphasises the sustainable management and use of natural resources to protect ecosystems and ensure environmental stability for future generations. It addresses critical challenges such as climate change, biodiversity loss, deforestation, and pollution, aiming to balance economic growth with ecological preservation. Environmental development strategies include transitioning to renewable energy sources, adopting sustainable agricultural practices, and promoting circular economies to reduce waste. It also involves protecting water, soil, and air quality and conserving natural habitats and wildlife. A healthy environment is essential for human development, as it underpins food security, public health, and economic livelihoods.

Countries have different histories and levels of the five types of development discussed in the preceding section. They are broadly classified into the Global North and the Global South. The rationale for and limitations of this categorisation and alternative classification are extensively covered in development literature. The Global South refers to a socio-economic and geopolitical concept rather than a precise geographical location. It describes emerging and least industrialised economies. The Global South broadly comprises Africa, Latin America, the Caribbean, Asia (excluding Israel, Japan, and South Korea), and Oceania (excluding Australia and New Zealand). On the other hand, the Global North—a grouping of the world's highly industrialised countries—includes North America, Europe, Israel, Japan, South Korea, Australia, and New Zealand. In this book, the expression "emerging and least industrialised countries/economies" is used interchangeably with the term Global South, while "highly industrialised countries/economies" constitute the Global North [27,76,86,107].

2.3 Sustainable Development

Sustainable development demands acute attention to the environment and climate change. It is an expansion of the notion of environmental development. It includes sustainability initiatives, resource management, and climate change adaptation and mitigation. Development today strongly emphasises sustainability, meaning that economic and social progress should not come at the expense of the environment or future generations. Effective management of natural resources, including water, land, and energy, is essential for sustainable development. Developing strategies to mitigate and adapt to the impacts of climate change is increasingly seen as a critical aspect of development. According to the UN, sustainable development is global progress "that seeks to meet the needs of the present without compromising the ability of future generations to meet their own needs". Put differently, sustainability refers to the practice of meeting current needs without compromising the ability of future generations to meet their own needs. It encompasses a broad range of activities and principles to promote the long-term health and stability of the environment, society, and economy. There are four key aspects to sustainability [15,42].

Environmental Sustainability
Efficient use and management of natural resources to prevent depletion include practices like water conservation, energy efficiency, and sustainable agriculture. Reducing emissions of pollutants into the air, water, and soil protect ecosystems and human health. Preserving diverse biological species and ecosystems maintain ecological balance and resilience. Mitigating climate change involves reducing greenhouse gas emissions and promoting renewable energy sources.

Social Sustainability
Social sustainability focuses on ensuring fair access to resources, opportunities, and benefits for all individuals and communities, regardless of their background or status; promoting healthy living conditions, access to healthcare, and overall well-being for all members of society; providing access to quality education and empowering individuals with the knowledge and skills needed for sustainable living; and strengthening communities through inclusive and participatory processes that foster social cohesion and resilience.

Economic Sustainability
Economic sustainability focuses on encouraging practices that reduce waste, promote recycling, and ensure products are made and consumed sustainably; building economies that can withstand and recover from crises while providing stable employment and economic growth; investing in technologies and practices that enhance productivity while minimising environmental impact; and ensuring that economic activities are conducted ethically, with fair wages, safe working conditions, and respect for workers' rights.

Governance Sustainability
Governance sustainability, often called sustainable governance, involves implementing and maintaining practices, policies, and frameworks that ensure an organisation's or society's operations are ethical, transparent, accountable, and sustainable over the long term. It is a crucial component of the broader concept of sustainability, including environmental and social dimensions. The key

aspects of governance sustainability include transparency, accountability, ethical conduct, stakeholder engagement, compliance, and regulation.

2.3.1 Importance of Sustainability

Sustainability is the path to a resilient future. It ensures that future generations have the resources and opportunities needed to thrive. Putting sustainability at the centre preserves natural habitats and biodiversity, reducing the impact of human activities on the planet. It enhances the quality of life for all individuals, promoting social justice and reducing inequalities. Sustainability supports long-term economic growth and resilience, creating sustainable livelihoods and reducing poverty.

Indeed, sustainability is no longer an abstract concept but a necessity for ensuring the well-being of current and future generations. It encompasses the responsible use of resources, preservation of the environment, and creation of systems that support long-term economic and social stability. In a world grappling with climate change, resource depletion, and growing inequalities, sustainability provides a framework for addressing these challenges holistically and ensuring that humanity thrives within the planet's limits.

At its core, sustainability is about balance—meeting the needs of today without compromising the ability of future generations to meet their own. This is particularly crucial in the face of environmental degradation. Deforestation, pollution, and the overconsumption of natural resources are straining ecosystems, leading to biodiversity loss and unpredictable climate patterns. Sustainable practices, such as transitioning to renewable energy, protecting forests, and promoting circular economies, are essential for mitigating these impacts and preserving the planet's health.

Sustainability also addresses pressing social issues by promoting equity, inclusion, and access to essential services. It calls for creating systems where everyone, regardless of socio-economic background, has access to clean water, nutritious food, education, and healthcare. By focusing on social sustainability, we build stronger communities that can adapt to challenges and support one another in times of need.

Economically, sustainability ensures that businesses and governments make decisions that consider long-term impacts rather than short-term gains. This approach fosters innovation, resilience, and efficiency, enabling economies to thrive in a rapidly changing world. Companies that embrace sustainable practices often gain a competitive edge, as consumers, investors, and employees increasingly prioritise ethics and environmental stewardship.

Ultimately, sustainability is vital because it acknowledges the interconnectedness of environmental, social, and economic systems. It is a call to action for individuals, communities, and organisations to work collaboratively to create a world that is equitable, prosperous, and environmentally sound. By embedding sustainability into the fabric of our lives, we can pave the way for a resilient future where people and the planet coexist harmoniously [61, 111].

2.3.2 Key Aspects of Sustainability

As already explained, there are four dimensions of sustainability—social, economic, environmental, and governance. The essential elements of sustainable development include a number of factors. Figure 2.1 illustrates the 5 Ps of sustainable development.

- Economic Prosperity: Sustainable development promotes inclusive economic growth, job creation, and poverty reduction while ensuring equitable access to resources and opportunities. It involves

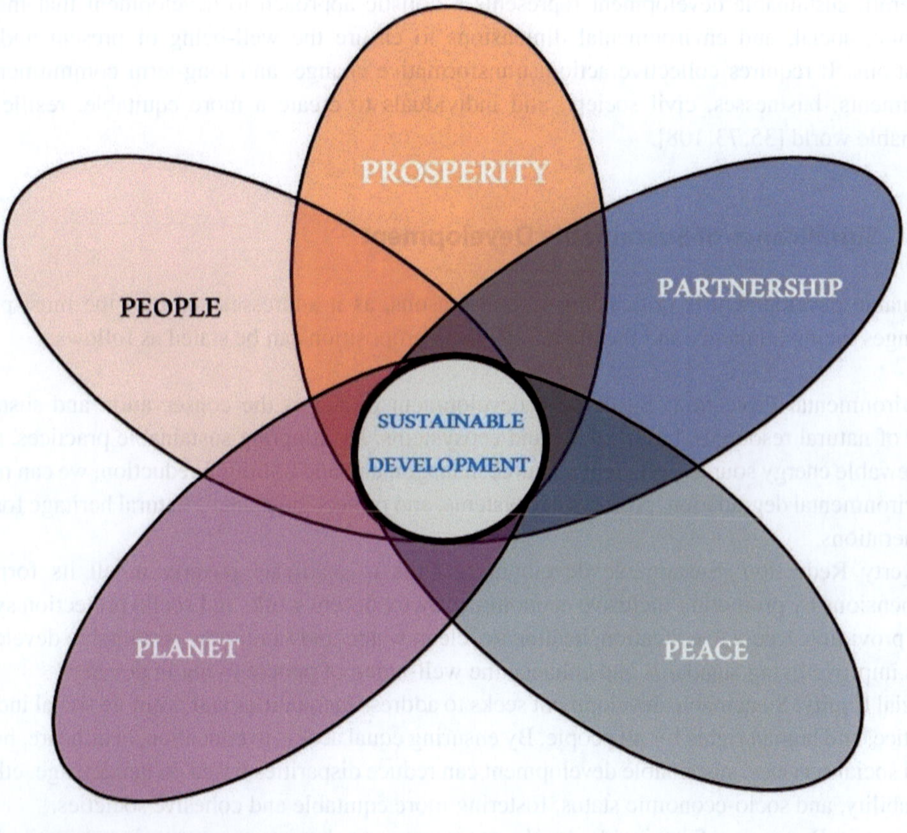

Fig. 2.1 The 5 Ps of sustainable development

fostering innovation, entrepreneurship, and investment in productive sectors that contribute to long-term prosperity and well-being.

- Social Inclusion: Sustainable development aims to ensure that all people, regardless of their gender, age, ethnicity, disability, or socio-economic status, have equal access to education, healthcare, housing, and social services. It seeks to address inequalities and promote social cohesion, diversity, and human rights.
- Environmental Sustainability: Sustainable development recognises the finite nature of natural resources and seeks to minimise environmental impacts while promoting conservation, biodiversity, and ecosystem resilience. It involves adopting sustainable resource management practices, reducing pollution and waste generation, and mitigating the impacts of climate change.
- Resilience and Adaptation: Sustainable development aims to build resilience to shocks and stresses, including natural disasters, climate variability, and economic downturns. It involves strengthening infrastructure, social safety nets, and community capacities to cope with and recover from adverse events while promoting adaptive strategies for long-term sustainability.
- Global Cooperation: Sustainable development requires international cooperation and partnership to address transboundary challenges, such as climate change, biodiversity loss, and global health pandemics. It involves sharing knowledge, resources, and best practices and promoting fair trade, technology transfer, and financial assistance to support sustainable development efforts worldwide.

Overall, sustainable development represents a holistic approach to development that integrates economic, social, and environmental dimensions to ensure the well-being of present and future generations. It requires collective action, transformative change, and long-term commitment from governments, businesses, civil society, and individuals to create a more equitable, resilient, and sustainable world [35, 73, 108].

2.3.3 Significance of Sustainable Development

Sustainable development is critical for several reasons, as it addresses some of the most pressing challenges facing humanity and the planet. Its value proposition can be stated as follows:

- Environmental Protection: Sustainable development promotes the conservation and sustainable use of natural resources, biodiversity, and ecosystems. By adopting sustainable practices, such as renewable energy sources, efficient resource management, and pollution reduction, we can mitigate environmental degradation, preserve ecosystems, and protect the planet's natural heritage for future generations.
- Poverty Reduction: Sustainable development aims to eradicate poverty in all its forms and dimensions by promoting inclusive economic growth, decent work, and social protection systems. By providing access to education, healthcare, clean water, and sanitation, sustainable development can improve living standards and enhance the well-being of people living in poverty.
- Social Equity: Sustainable development seeks to address inequalities and promote social inclusion, justice, and human rights for all people. By ensuring equal access to education, healthcare, housing, and social services, sustainable development can reduce disparities based on gender, age, ethnicity, disability, and socio-economic status, fostering more equitable and cohesive societies.
- Economic Prosperity: Sustainable development promotes long-term economic prosperity by fostering innovation, entrepreneurship, and investment in sustainable industries and infrastructure. By promoting job creation, income generation, and economic diversification, sustainable development can contribute to poverty reduction, economic stability, and shared prosperity.
- Climate Change Mitigation and Adaptation: Sustainable development plays a crucial role in addressing climate change by reducing greenhouse gas emissions, promoting renewable energy sources, and building climate-resilient infrastructure and communities. By taking action to mitigate climate change and adapt to its impacts, sustainable development can safeguard livelihoods, ecosystems, and economies from the adverse effects of climate change.
- Resilience and Security: Sustainable development helps build resilience to shocks and stresses, including natural disasters, conflicts, and economic crises. By strengthening social safety nets, community capacities, and institutional mechanisms, sustainable development can enhance resilience and promote stability, security, and peace.
- Interconnectedness and Global Cooperation: Sustainable development recognises the interconnectedness of global challenges and the need for collective action and global cooperation to address them. By fostering partnerships, knowledge sharing, and solidarity among countries and stakeholders, sustainable development can mobilise resources, leverage synergies, and accelerate progress towards common goals.

Overall, sustainable development is essential for ensuring the well-being of present and future generations, promoting environmental stewardship, social equity, economic prosperity, and resilience in the face of global challenges. It requires concerted efforts, transformative change, and long-term commitment from governments, businesses, civil society, and individuals to create a more sustainable and inclusive world [105, 106, 111].

2.3.4 Environmental, Social, and Governance (ESG)

An indispensable sustainability-anchored concept is environmental, social, and governance (ESG). This strategic framework refers to the three central factors used to measure the sustainability and societal impact of an investment in a company or business. These criteria help investors and stakeholders evaluate the potential long-term performance and ethical impact of their investments.

ESG investing, also known as sustainable investing or socially responsible investing (SRI), involves choosing investments based on ESG criteria. Investors use ESG ratings and reports to evaluate the sustainability and ethical impact of their investment choices, aiming to support companies that align with their values and contribute positively to society and the environment.

ESG criteria have emerged as a vital and strategic framework for evaluating organisations' sustainability and ethical impact, shaping how businesses operate in a rapidly changing world. Hence, it is prudent to present and discuss them in more detail. ESG encompasses three interconnected dimensions: the environmental responsibility to mitigate ecological harm, the social commitment to fostering equitable and inclusive practices, and the governance imperative to ensure ethical decision-making and accountability. Together, these pillars guide companies in balancing profitability with long-term value creation for stakeholders.

The environmental aspect of ESG focuses on reducing carbon footprints, managing natural resources responsibly, and addressing the challenges of climate change. Businesses are increasingly adopting renewable energy, improving energy efficiency, and investing in innovative technologies to mitigate their environmental impact. These practices are not only ethical imperatives but also strategies for resilience in an economy transitioning towards sustainability.

The social dimension emphasises corporate responsibility to employees, customers, and communities. It covers diversity and inclusion, workplace safety, labour rights, and community engagement. Companies that prioritise these factors cultivate trust and loyalty, which are essential for long-term success. Socially responsible practices also enable organisations to address systemic inequalities and contribute to societal well-being.

Governance, the third pillar, ensures that businesses operate transparently, ethically, and in compliance with regulatory standards. Strong governance frameworks prioritise accountability, diversity in leadership, and effective risk management. These principles build investor confidence and safeguard against scandals or mismanagement that can harm reputations and financial performance (Fig. 2.2).

Incorporating ESG principles is no longer a peripheral concern but a central tenet of modern business strategy. Investors, regulators, and consumers are increasingly demanding accountability from organisations, rewarding those that align with sustainable practices. ESG provides a road map for companies to not only minimise risks but also seize opportunities in a world that values ethical growth and environmental stewardship. By integrating ESG into core operations, businesses can drive positive change while securing their long-term viability in a rapidly evolving global landscape. Indeed, ESG criteria have emerged as a vital framework for evaluating the sustainability and ethical impact of organisations, shaping how businesses operate in a rapidly changing world [114, 138].

Fig. 2.2 The ESG criteria, principles, and strategy. (Image credit: iStock sarawuth702)

2.3.5 Implementing Sustainability

Implementing sustainability requires a multifaceted approach that integrates environmental, social, and economic dimensions into policies, practices, and everyday behaviours. Achieving a sustainable future is a collective endeavour that involves governments, businesses, communities, and individuals working together to align development goals with the planet's capacity to sustain them.

The first step towards sustainability is fostering awareness and education. People need to understand what sustainability means and why it is crucial. Public awareness campaigns, school curricula, and community workshops can equip individuals with the knowledge and skills to make sustainable choices. Educating communities about energy efficiency, waste reduction, and sustainable consumption habits empowers them to contribute meaningfully to sustainability efforts.

Governments play a critical role by creating supportive policies and regulations that prioritise sustainability. These policies may include incentives for renewable energy adoption, stricter emissions standards, or requirements for sustainable building practices. Governments must also lead by example, integrating sustainability into their operations and infrastructure development. At the international level, treaties and agreements, such as the Paris Agreement, promote global cooperation on pressing challenges like climate change.

Businesses are equally important in implementing sustainability by embedding ESG principles into their operations. This involves reducing waste, transitioning to renewable energy, and adopting circular economy models that recycle resources instead of discarding them. Companies can also engage in responsible sourcing, ensuring their supply chains uphold ethical labour practices and environmental standards. Transparent reporting on sustainability metrics not only builds consumer trust but also encourages continuous improvement.

Technology and innovation serve as powerful enablers of sustainability. Investments in clean energy technologies, such as solar and wind power, sustainable agriculture techniques, and waste-to-energy systems, make sustainable practices more efficient and scalable. Digital tools, such as artificial intelligence and data analytics, enhance decision-making by providing insights into resource management, climate patterns, and supply chain optimisation.

Community involvement and equitable participation are essential to the success of sustainability initiatives. Policymakers and organisations must include marginalised groups in decision-making processes to ensure that sustainability efforts are inclusive and fair. Programmes that prioritise equitable access to clean water, renewable energy, and education help bridge gaps and ensure that sustainability benefits everyone, not just a privileged few.

Individual action is a cornerstone of sustainability. Simple steps, such as conserving water, reducing single-use plastics, choosing sustainable products, and advocating for climate-friendly policies, contribute to the collective effort. When individuals adopt sustainable lifestyles, they create a ripple effect, influencing others to do the same.

Implementing sustainability is a dynamic, ongoing process that requires coordinated action and shared responsibility. By fostering education, enacting supportive policies, leveraging technology, and prioritising inclusivity, societies can transition to sustainable systems that ensure the well-being of both current and future generations.

Indeed, governments can create policies and regulations that promote sustainable practices, such as environmental protection laws, social welfare programmes, and economic incentives for sustainable businesses. Businesses can adopt sustainable practices in their operations, supply chains, and products, often guided by ESG criteria. As already indicated, ordinary people can contribute to sustainability by making eco-friendly choices, reducing waste, conserving energy, and supporting sustainable products and services. Communities can engage in local sustainability projects, such as urban gardening, renewable energy installations, and conservation efforts. By embracing sustainability, societies can work towards a balanced and equitable future where economic development, social inclusion, and environmental protection are harmonised.

It is instructive to note that sustainable development calls for concerted efforts to build an inclusive, sustainable, and resilient future for people and the planet. To achieve sustainable development, it is crucial to harmonise three core elements: economic growth, social inclusion, and environmental protection. These elements are interconnected, and all are crucial for the well-being of individuals and societies. Eradicating poverty in all its forms and dimensions is an indispensable requirement for sustainable development. To this end, sustainable, inclusive, and equitable economic growth must be promoted, creating greater opportunities for all, reducing inequalities, raising basic living standards, fostering equitable social development and inclusion, and promoting integrated and sustainable management of natural resources and ecosystems.

Sustainable development involves balancing economic, social, and environmental considerations to promote prosperity, equity, and resilience for all people while safeguarding the health of ecosystems and the planet. It aims to address the interconnected challenges of poverty, inequality, environmental degradation, and climate change, fostering inclusive and sustainable pathways to development [61, 111].

2.3.6 Challenges of Implementing Sustainability

While sustainability is a crucial goal for the future of humanity and the planet, its implementation faces significant challenges that complicate its adoption across societies, businesses, and governments. These challenges stem from a mix of economic, social, technological, and political factors, all of which require coordinated and innovative solutions to overcome.

One of the most pressing challenges is economic pressure. Transitioning to sustainable practices often requires substantial upfront investment, whether in renewable energy infrastructure, sustainable agriculture, or waste management systems. For businesses, the shift can mean higher operational costs, at least in the short term, which may deter companies that prioritise immediate profitability over long-term benefits. For developing countries, the challenge is even greater, as limited financial resources and competing priorities, such as poverty alleviation and infrastructure development, make sustainability seem less urgent.

Technological barriers also hinder sustainability efforts. While advancements in renewable energy, water purification, and waste recycling have made significant strides, the technologies needed to achieve widespread sustainability are not always accessible or affordable for all regions. Inadequate research and development in some sectors further delay progress, leaving critical gaps in the ability to implement sustainable practices efficiently.

On a social level, lack of awareness and resistance to change remain persistent issues. Many individuals and organisations are either unaware of the importance of sustainability or reluctant to alter established habits and systems. Cultural differences and varying levels of education exacerbate this challenge, creating disparities in how sustainability is perceived and prioritised worldwide. Without widespread understanding and support, it is difficult to build the collective momentum needed to drive meaningful change.

Political will and governance are perhaps the most significant obstacles to sustainability. Conflicting interests among stakeholders, including governments, corporations, and communities, often lead to fragmented policies and inconsistent enforcement of regulations. Short-term political cycles discourage long-term planning, as leaders may avoid sustainability initiatives that lack immediate results or involve unpopular decisions. Additionally, global cooperation is crucial for addressing cross-border challenges like climate change, but differences in national priorities and economic capacities complicate international agreements and coordinated actions.

Finally, the challenge of balancing trade-offs cannot be overlooked. Sustainability efforts often require compromises between environmental, social, and economic goals. For example, transitioning to renewable energy can disrupt industries reliant on fossil fuels, potentially leading to job losses and economic instability in affected regions. Managing these trade-offs requires careful planning and equitable solutions to ensure that no group is disproportionately burdened.

Despite these challenges, sustainability remains an essential goal. Overcoming these barriers demands innovative thinking, global collaboration, and a commitment to long-term well-being over short-term convenience. By addressing these challenges systematically, societies can move closer to achieving a sustainable future for all.

2.4 UN Sustainable Development Goals (SDGs)

2.4.1 An Introduction to the SDGs

The UN Sustainable Development Goals (SDGs) are 17 interconnected goals adopted by all UN Member States in 2015 as part of the 2030 Agenda for Sustainable Development. The SDGs aim to address global challenges and promote sustainable economic, social, and environmental development. The goals provide a comprehensive framework for achieving sustainability globally by 2030. These goals address many issues including poverty, hunger, health, education, gender equality, clean water, affordable energy, economic growth, and climate action.

The 2030 Agenda for Sustainable Development, adopted by all UN Member States in 2015, provides a shared blueprint for peace and prosperity for people and the planet, now and into the future. At its heart are the 17 SDGs, an urgent call for action by all developed and developing countries in a global partnership. They recognise that ending

Fig. 2.3 SDGs: Social, environmental, and economic classification [125]

poverty and other deprivations must go hand in hand with strategies that improve health and education, reduce inequality, and spur economic growth—all while tackling climate change and working to preserve our oceans and forests. As already explained, the 17 UN SDGs can be classified into social, environmental, and economic categories as illustrated in Fig. 2.3. Figure 2.4 illustrates the interdependence of SDGs through these three broad lenses of the environment, society, and economy [42, 74, 110, 126].

2.4.2 A Brief History of the SDGs

As already articulated, the UN SDGs were established to address global challenges such as poverty, inequality, environmental degradation, and peace. Their development is rooted in several key international efforts, starting with the Brundtland Report of 1987, formally titled "Our Common Future". It was a landmark publication by the United Nations' World Commission on Environment and Development (WCED), chaired by former Norwegian Prime Minister Gro Harlem Brundtland. The report aimed to address global environmental challenges while considering economic and social development needs. The report introduced and popularised the concept of sustainable development, defining it as: *Development that meets the needs of the present without compromising the ability of future generations to meet their own needs.* The SDG journey begins in earnest with the June 1992 Earth Summit in Rio de Janeiro, Brazil. Indeed, the SDGs build on decades of work by countries and the United Nations, including the UN Department of Economic and Social Affairs. At the Earth Summit of 1992, in Rio de Janeiro, more than 178 countries adopted Agenda 21, a comprehensive plan of action to build a global partnership for sustainable development to improve human lives and protect the environment. This conference marked a significant step towards international cooperation on environmental and developmental issues [15, 74, 110].

Then came the September 2000 Millennium Summit. World leaders adopted the Millennium Declaration, resulting in the Millennium Development Goals (MDGs). These eight goals targeted issues like extreme poverty and hunger, setting a 2015 deadline for achievement. Member states unanimously adopted the Millennium Declaration at this summit at the UN Headquarters in New

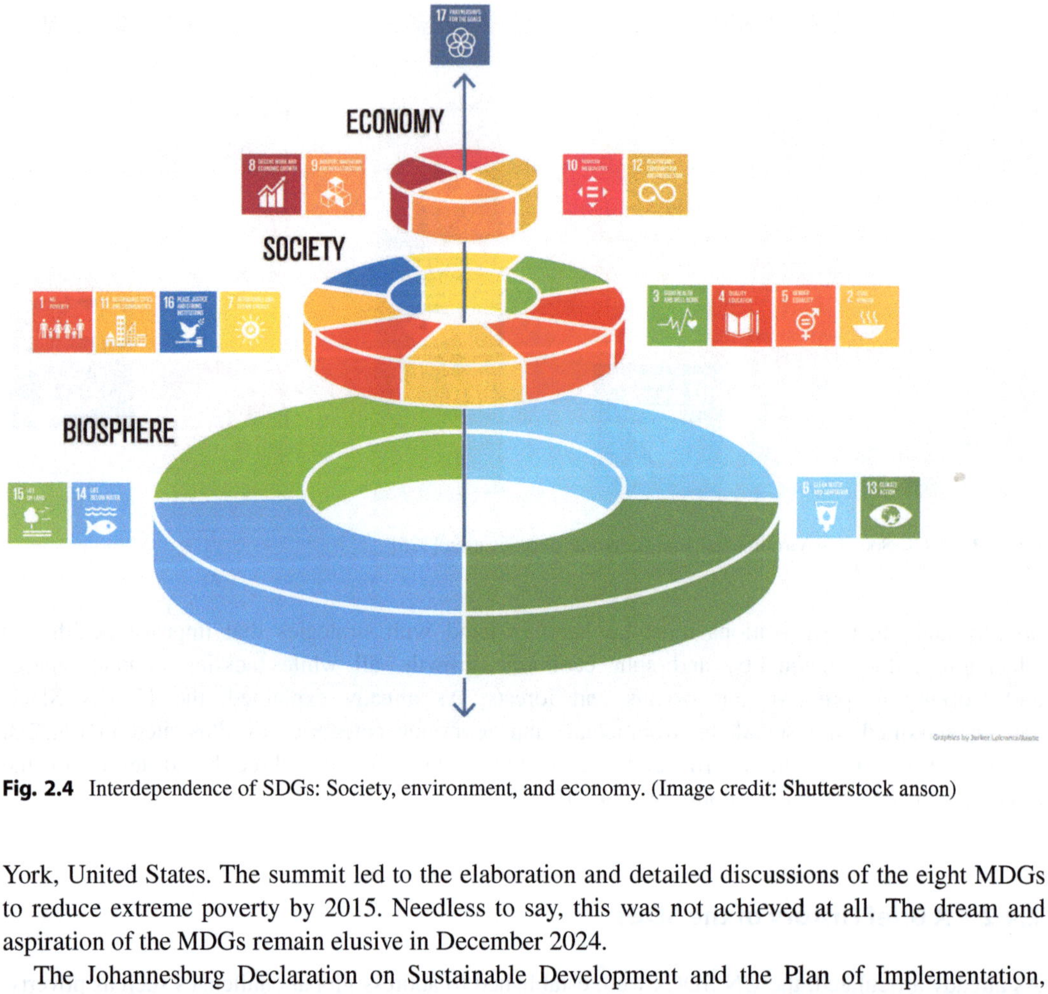

Fig. 2.4 Interdependence of SDGs: Society, environment, and economy. (Image credit: Shutterstock anson)

York, United States. The summit led to the elaboration and detailed discussions of the eight MDGs to reduce extreme poverty by 2015. Needless to say, this was not achieved at all. The dream and aspiration of the MDGs remain elusive in December 2024.

The Johannesburg Declaration on Sustainable Development and the Plan of Implementation, adopted at the 2002 World Summit on Sustainable Development in Durban, South Africa, reaffirmed the global community's commitments to poverty eradication and the environment. They built on Agenda 21 and the Millennium Declaration by emphasising multilateral partnerships. Commitment to the MDGs was reaffirmed in Durban.

At the June 2012 Rio+20 Conference, recognising the need for a more comprehensive framework, member states agreed to develop the SDGs, building on the MDGs and addressing broader sustainability challenges. Member states adopted the outcome document "The Future We Want", in which they decided, among other things, to launch a process to develop a set of SDGs to build on the MDGs and to establish the UN High-level Political Forum on Sustainable Development. The Rio +20 outcome also contained other measures for implementing sustainable development, including mandates for future programmes of work in development financing and small island developing states. Clearly, the Rio+20 conference aimed to secure renewed political commitment to sustainable development, assess progress since the 1992 Earth Summit, and discuss new challenges and opportunities.

Learning from and Leveraging the MDGs

The successes and failures on the journey to achieve the MDGs must provide instructive lessons for the SDG agenda. As stated earlier, the MDGs were established in 2000 by the United Nations and set ambitious targets to address global challenges such as poverty, hunger, education, and health. However, many of these goals were not fully achieved due to a range of structural and implementation challenges. One key reason was the lack of adequate funding and resources, particularly in emerging and least industrialised countries. Many economies relied heavily on international aid, which was often inconsistent, insufficient, or tied to conditions that hindered progress. Furthermore, the MDGs lacked mechanisms for ensuring accountability, making it difficult to track and enforce commitments. This was exacerbated by the global economic crises during the MDG period, which diverted attention and resources away from development initiatives, leaving the most vulnerable populations without the support needed to achieve significant improvements.

Another major factor was the one-size-fits-all approach of the MDGs, which failed to consider the diverse socio-economic, political, and cultural contexts of different countries. The goals were overly generalised, with insufficient emphasis on regional and local realities. As a result, countries faced difficulties in tailoring interventions to their specific needs. Additionally, conflict and political instability in many regions undermined efforts to implement development programs, while environmental challenges such as climate change and natural disasters further derailed progress. The MDGs also did not adequately address systemic issues like inequality and governance, which are critical for sustainable development. These shortcomings highlighted the need for more inclusive, adaptable, and integrated approaches, lessons which informed the subsequent SDG agenda [36, 48, 110].

Building on the MDGs and the outcomes of the Rio+20 Conference, discussions began on formulating a new set of sustainable development goals to succeed the MDGs after 2015. In 2013, the UN General Assembly established an Open Working Group (OWG) on SDGs to develop a proposal for the post-2015 development agenda. The OWG on SDGs, composed of representatives from UN member states, civil society, and other stakeholders, worked to identify and negotiate a set of SDGs and targets. Between March 2013 and July 2014, the OWG held multiple sessions and proposed 17 goals and 169 targets. Figure 2.5 summarises the history and ambition of the SDGs from the Brundtland Report of 1987 to the 2030 target year.

Adoption of the SDG Agenda

In January 2015, the General Assembly began negotiating the post-2015 development agenda. The process culminated in the subsequent adoption of the 2030 Agenda for Sustainable Development, with 17 SDGs at its core, at the UN Sustainable Development Summit in September 2015. At this 2015 Adoption of the 2030 Agenda, all UN Member States adopted the 2030 Agenda for Sustainable Development, which included 17 SDGs and 169 targets. This agenda serves as a shared blueprint for peace and prosperity for people and the planet. The SDGs are designed to be universal, apply to all countries, and emphasise the interconnectedness of social, economic, and environmental sustainability. They aim to leave no one behind, promoting inclusive progress globally.

Indeed, 2015 was a landmark year for multilateralism and international policy shaping, with the adoption of several major agreements: The Sendai Framework for Disaster Risk Reduction (March 2015), Addis Ababa Action Agenda on Financing for Development (July 2015), and Transforming our World: The 2030 Agenda for Sustainable Development with its 17 SDGs at the UN Sustainable Development Summit in New York in September 2015. Furthermore, there was the Paris Agreement on climate change (December 2015). The annual High-level Political Forum on Sustainable Development serves as the central UN platform for the follow-up and review of the SDGs.

The SDGs officially took effect on January 1, 2016, and set a universal agenda for achieving sustainable development by 2030. Following the adoption of the goals, efforts shifted towards implementing and monitoring progress towards the goals. Countries were encouraged to develop national strategies, plans, and policies aligned with the SDGs. At the same time, international

```
1987: Brundtland Report                | 1992: Earth Summit (Rio)
|                                      |
| Defines sustainable development      | Adoption of Agenda 21, a global   |
| as meeting the needs of the          | framework for sustainable
| present without compromising the     | development. Focuses on
| ability of future generations.       | deforestation, biodiversity loss,
|                                      | and climate change.
+--------------------------------------+----------------------------------
2000: Millennium Development           | 2012: Rio+20 Conference
| Goals (MDGs)                         |
|                                      | Agreement to develop goals
| UN adopts 8 goals to address         | integrating environmental, social,
| poverty, hunger, education,          | and economic sustainability,
| gender equality, and health,         | paving the way for SDGs.
| with a target of 2015.               |
+--------------------------------------+----------------------------------
2015: Adoption of SDGs                 | 2020: Decade of Action
|                                      |
| 17 SDGs with 169 targets adopted     | Global initiative to accelerate
| by the UN General Assembly as        | SDG progress with an emphasis on
| part of the 2030 Agenda for          | urgency and scaling up efforts.
| Sustainable Development.             |
+--------------------------------------+----------------------------------
2030: Target Year for the SDGs
|                                      |
| Aim to transform the world by        |
| ending poverty, protecting the       |
| planet, and ensuring prosperity      |
| for all.                             |
```

Fig. 2.5 A summarised history of the SDGs

organisations and stakeholders collaborated to support implementation efforts, mobilise resources, and monitor progress through data collection, reporting, and review mechanisms.

As of December 2024, the Division for Sustainable Development Goals (DSDG) in the UN Department of Economic and Social Affairs (UNDESA) provides substantive support and capacity building for the SDGs and their related thematic issues, including water, energy, climate, oceans, urbanisation, transport, science and technology, the Global Sustainable Development Report (GSDR), partnerships, and Small Island Developing States. DSDG plays a key role in the evaluation of the UN systemwide implementation of the 2030 Agenda and in advocacy and outreach activities relating to the SDGs. In order to make the 2030 Agenda a reality, broad ownership of the SDGs must translate into a strong commitment by all stakeholders to implement the global goals. DSDG aims to help facilitate this engagement.

Indeed, the adoption of the SDGs marked a significant milestone in global efforts to address sustainable development challenges and build a more equitable, resilient, and sustainable future for all. The SDGs provide a universal framework and shared vision for collective action to achieve a wide range of economic, social, and environmental objectives by 2030.

2.4.3 Justification for Each SDG

The goals were established through a comprehensive, inclusive process to address the world's most pressing challenges. But why these 17? What is the rationale and justification for each goal? The establishment process has been articulated in the history of the goals. However, it is prudent to restate some aspects of the process. As already stated, the journey began at the UN Conference on Sustainable Development in Rio de Janeiro, where member states agreed to develop a set of universal goals to address environmental, political, and economic challenges. An Open Working Group (OWG) (2013–2014) comprising representatives from 70 countries was formed to draft the goals. The OWG extensively consulted with various stakeholders, including civil society, academia, and the private sector. This was followed by the intergovernmental negotiations in 2015, where the OWG's proposal was discussed, leading to the refinement and finalisation of the goals. The UN General Assembly unanimously adopted the 2030 Agenda for Sustainable Development, encompassing the 17 SDGs and 169 targets. Each goal was justified as follows:

- SDG 1 (No Poverty): Eradicating poverty in all its forms is essential for sustainable development and ensuring that all individuals can live with dignity.
- SDG 2 (Zero Hunger): Ending hunger and achieving food security are fundamental to improving nutrition and promoting sustainable agriculture.
- SDG 3 (Good Health and Well-being): Ensuring healthy lives and promoting well-being at all ages is vital for sustainable development.
- SDG 4 (Quality Education): Inclusive and equitable quality education is crucial for enabling upward socio-economic mobility and breaking the cycle of poverty.
- SDG 5 (Gender Equality): Achieving gender equality and empowering all women and girls is essential for sustainable development and economic growth.
- SDG 6 (Clean Water and Sanitation): Access to safe water and sanitation is a basic human right and a critical component of health and development.
- SDG 7 (Affordable and Clean Energy): Ensuring access to affordable, reliable, sustainable, and modern energy is crucial for economic development and environmental sustainability.
- SDG 8 (Decent Work and Economic Growth): Promoting sustained, inclusive, and sustainable economic growth and full and productive employment is essential for reducing poverty.
- SDG 9 (Industry, Innovation, and Infrastructure): Building resilient infrastructure, promoting inclusive and sustainable industrialisation, and fostering innovation are key drivers of economic growth and development.
- SDG 10 (Reduced Inequalities): Reducing inequality within and among countries is necessary for social cohesion and sustainable development.

- SDG 11 (Sustainable Cities and Communities): Making cities inclusive, safe, resilient, and sustainable is essential as urbanisation continues to rise globally.
- SDG 12 (Responsible Consumption and Production): Ensuring sustainable consumption and production patterns is vital for reducing environmental degradation and resource depletion.
- SDG 13 (Climate Action): Taking urgent action to combat climate change and its impacts is critical for the survival of ecosystems and human societies.
- SDG 14 (Life Below Water): Conserving and sustainably using oceans, seas, and marine resources is essential for global biodiversity and livelihoods.
- SDG 15 (Life on Land): Protecting, restoring, and promoting sustainable use of terrestrial ecosystems is crucial for biodiversity and combatting desertification.
- SDG 16 (Peace, Justice, and Strong Institutions): Promoting peaceful and inclusive societies, providing access to justice, and building effective institutions are foundational for sustainable development.
- SDG 17 (Partnerships for the Goals): Strengthening global partnerships and mobilising resources are necessary to successfully implement all other goals.

Each goal was formulated to address specific global challenges, with the overarching aim of achieving a sustainable and equitable future for all by 2030 [126].

2.4.4 Linkages Between the SDGs

These 17 UN SDGs are intricately interconnected, with progress in one area often influencing outcomes in others. Understanding these relationships is crucial for developing integrated strategies that maximise synergies and minimise trade-offs.

- SDG 1 (No Poverty) and SDG 2 (Zero Hunger): Reducing poverty enhances access to nutritious food, while improved food security contributes to poverty alleviation.
- SDG 3 (Good Health and Well-being) and SDG 6 (Clean Water and Sanitation): Access to clean water and sanitation is fundamental for preventing diseases and promoting overall health.
- SDG 4 (Quality Education) and SDG 5 (Gender Equality): Providing quality education to all genders promotes gender equality, empowering individuals and communities.
- SDG 7 (Affordable and Clean Energy) and SDG 13 (Climate Action): Transitioning to clean energy sources reduces greenhouse gas emissions, mitigating climate change impacts.
- SDG 8 (Decent Work and Economic Growth) and SDG 9 (Industry, Innovation, and Infrastructure): Investments in infrastructure and innovation drive economic growth and create employment opportunities.
- SDG 10 (Reduced Inequalities) and SDG 16 (Peace, Justice, and Strong Institutions): Promoting inclusive societies and strong institutions helps reduce inequalities and fosters social cohesion.
- SDG 11 (Sustainable Cities and Communities) and SDG 12 (Responsible Consumption and Production): Developing sustainable urban areas encourages responsible consumption and production patterns.
- SDG 14 (Life Below Water) and SDG 15 (Life on Land): Protecting marine and terrestrial ecosystems ensures biodiversity and the provision of essential ecosystem services.
- SDG 17 (Partnerships for the Goals): Global partnerships are essential for mobilising resources and expertise to achieve all other SDGs.

Fig. 2.6 Interconnections and linkages among SDGs

Visual representations, such as the network diagram shown in Fig. 2.6, effectively illustrate these interconnections, highlighting how progress in one goal can influence others. For instance, the International Science Council's report, "A Guide to SDG Interactions: From Science to Implementation", provides detailed analyses of these relationships. Understanding these interdependencies is vital for policymakers, organisations, and stakeholders to design holistic approaches that leverage synergies and address potential conflicts among the SDGs [50, 91].

2.4.5 Limitations of the SDG Framework

The UN Sustainable Development Goals provide a comprehensive framework for addressing global challenges. The objective is to address global challenges and promote sustainable development in three related areas—economic, social, and environmental dimensions. However, the SDG formulation and approach have several limitations. The point of departure is the framework's non-binding nature. The SDGs are voluntary and lack enforcement mechanisms to ensure compliance. This absence of accountability can lead to inconsistent implementation across countries. Furthermore, achieving the SDGs requires substantial financial resources. The global economic downturns, exacerbated by events like the COVID-19 pandemic, have widened the funding gap, hindering progress. Thus, a shortage of funding has been a significant constraint.

The 17 goals' interconnectedness and interlinked nature necessitate coordinated efforts across sectors. Implementing them in an integrated manner is challenging, often leading to fragmented approaches. Hence, complexity and integration challenges characterise the achievement of the SDGs. There are also data and monitoring issues. Reliable data is crucial for tracking progress. Many countries face data collection and analysis challenges, resulting in gaps that impede effective monitoring and evaluation. Last but not least, there is a lack of awareness and engagement. Awareness of the SDGs remains limited in many regions, leading to insufficient community engagement and support for initiatives to achieve these goals.

A more fundamental conceptualisation challenge is the very notion of sustainable development. Taken in a limited sense, it can become an end in itself, which is counterproductive. Sustainable development must lead to economic transformation—a comprehensive process of structural change within an economy that enhances its productivity, competitiveness, and overall development. It involves shifting resources, such as labour and capital, from low-productivity sectors, like subsistence agriculture, to higher-productivity sectors, such as manufacturing, technology, and services. This process leads to diversification, modernisation, and value addition within the economy, fostering sustainable growth and improved living standards. All SDGs must specifically target the attainment of economic transformation. As currently formulated by the UN, they don't.

Economic transformation is particularly critical for developing regions like Africa and the rest of the Global South. It enables countries to break free from dependency on natural resources or primary commodities and move up the global economic value chains—towards diversified, sustainable economies that can better meet the needs of their populations. By fostering structural change, economic transformation lays the foundation for long-term growth and development. In most emerging and least industrialised countries, there is a need to decolonise and democratise the economies explicitly. The SDG framework does not speak to this foundational matter. The Global South's ambition for inclusive development and shared prosperity is not explicitly acknowledged in the formulation of the SDGs.

Nevertheless, there is an overlap between the SDG agenda and key aspects of economic transformation, such as industrialisation, the adoption of advanced technologies, the development of infrastructure, and increased investment in education and skills training. Both frameworks encompass changes in policies and institutions to support innovation, entrepreneurship, and inclusivity. Economic transformation seeks to create jobs, reduce poverty, and build resilience to external shocks by making economies more dynamic and competitive in the global market. Indeed, this aligns with the SDG ambition. However, understanding the strengths and weaknesses of the UN SDG framework is imperative. Addressing the limitations is essential for achieving and effectively realising the SDGs by 2030 worldwide.

2.5 UN SDGs: The Prognosis

The UN SDGs, established in 2015 with a target completion date of 2030, represent an ambitious framework that addresses global challenges such as poverty, inequality, and environmental degradation and promotes peace and justice. Assessing the achievement of the SDGs is a complex and ongoing process. As of December 2024, progress towards the SDGs varied significantly across regions and goals. Some goals have seen notable progress, while others lag behind. Assessments indicate that achieving all 17 SDGs by the 2030 deadline is increasingly unlikely. Indeed, the SDGs face significant challenges that make their achievement by 2030 a tall order. One of the primary obstacles is the persistent inequality within and between countries, exacerbated by global crises such as the COVID-19 pandemic, economic slowdowns, and conflicts. These events have reversed progress in areas such as poverty reduction, education, and healthcare, diverting resources and attention away from long-term goals. Many low-income and developing nations lack the financial resources and institutional capacity to implement the necessary changes. In contrast, high-income nations often fall short in meeting their commitments to provide aid or reduce their environmental impact. The gap between rhetoric and action, particularly in addressing climate change and sustainable development financing, has left critical SDG targets underfunded and unmet.

Additionally, the interconnected nature of the SDGs means progress in one area often depends on advancements in others. For instance, achieving universal education requires improvements in gender equality, infrastructure, and healthcare. However, fragmented efforts and competing priorities among governments, businesses, and international organisations create inefficiencies and contradictions. The lack of a binding mechanism to hold stakeholders accountable for their commitments further weakens global efforts. Climate change, in particular, threatens to derail progress across all SDGs, as its impacts disproportionately affect vulnerable communities and undermine gains in agriculture, water security, and public health. With just a few years left until 2030, the combined effects of financial, political, and systemic challenges make the ambitious vision of the SDGs increasingly difficult to realise.

The 2024 Sustainable Development Goals Report reveals that nearly half of the 17 targets are showing minimal or moderate progress, while over a third are stalled or regressing. This stagnation is attributed to factors such as the lingering effects of the COVID-19 pandemic, escalating conflicts, geopolitical tensions, and worsening climate chaos. A substantial funding gap of $4.2 trillion has been identified, hindering progress across various goals. This shortfall is exacerbated by reductions in official development assistance from major global funders, including the United States, United Kingdom, Germany, and France. Furthermore, there are environmental and health setbacks. The world is falling behind on critical goals related to the environment, health, and hunger. None of the 17 goals are on course to be met by 2030, with most targets showing limited or a reversal of progress.

Indeed, there are varying degrees of progress observed across different goals. As of December 2024, certain SDGs have seen more advancement, while others face significant hurdles. SDG 3 (Good Health and Well-being) has experienced significant progress. There have been improvements in reducing global child mortality and preventing HIV infections. Energy and mobile broadband access have also increased, improving health outcomes. Similarly, SDG 7 (Affordable and Clean Energy) has moved. Advancements in renewable energy technologies and increased investments have led to greater access to clean energy sources, particularly in developing regions.

SDG 1 (No Poverty) has stalled. The COVID-19 pandemic and escalating conflicts have reversed progress, pushing an additional 23 million people into extreme poverty in 2022 compared to 2019. Similarly, achieving SDG 2 (Zero Hunger) has not progressed. Over 100 million more people were suffering from hunger in 2022 compared to 2019, indicating a regression in efforts to eradicate hunger. SDG 13 (Climate Action) is an ambitious aspiration characterised by many challenges. The world experienced the warmest year on record in 2023, with global temperatures dangerously close to the 1.5 °C limit set by the Paris Agreement, highlighting insufficient progress in combatting climate change.

As for SDG 14 (Life Below Water), marine biodiversity continues to decline due to overfishing, pollution, and climate change, with limited progress in protecting ocean ecosystems. With respect to SDG 15 (Life on Land), challenges of deforestation and biodiversity loss persist, with efforts to protect terrestrial ecosystems falling short of targets. Indeed, while there have been strides in areas such as health and clean energy, significant challenges remain in eradicating poverty and hunger and addressing environmental sustainability. Accelerated and coordinated global efforts are essential to overcome these obstacles and achieve the SDGs by the 2030 deadline.

2.5.1 Exploring the Challenges

As explained in the preceding section, achieving the UN SDGs is challenging for several reasons. These include the goals' complexity and interrelatedness, structural and systemic constraints, and contextual factors. It is instructive to explore and explain these key reasons why achieving the SDGs by 2030 is difficult.

- Interconnected Nature of the Goals: The SDGs are interconnected and interdependent, meaning progress in one goal often requires progress in others. For example, achieving good health and well-being (SDG 3) may depend on addressing poverty (SDG 1), ensuring access to clean water and sanitation (SDG 6), and promoting gender equality (SDG 5). This interconnectivity requires integrated and holistic approaches to development, which can be challenging to implement.
- Complexity of Global Challenges: The SDGs address complex and multifaceted global challenges such as poverty, inequality, climate change, and environmental degradation. These challenges are deeply rooted in social, economic, political, and environmental factors, making them difficult to address comprehensively and sustainably.
- Resource Constraints: Implementing the SDGs requires significant financial, human, and technological resources. Many countries, particularly those in the Global South, face resource constraints, including limited funding, infrastructure, and capacity, which hinder their ability to invest in sustainable development initiatives.
- Inequality and Marginalisation: Inequality and marginalisation based on factors such as income, gender, ethnicity, disability, and geography pose significant barriers to achieving the SDGs. Marginalised populations often face disproportionate challenges in accessing education, healthcare, employment, and other essential services, exacerbating disparities and hindering progress towards the goals.
- Conflicts and Instability: Conflict, violence, and political instability undermine efforts to achieve the SDGs by disrupting social cohesion, damaging infrastructure, displacing populations, and diverting resources from development priorities. Fragile and conflict-affected countries face challenges in advancing sustainable development and achieving peace, justice, and strong institutions (SDG 16).
- Environmental Degradation and Climate Change: Environmental degradation, including deforestation, pollution, loss of biodiversity, and climate change, poses significant threats to sustainable development. Addressing these challenges requires transformative changes in production and consumption patterns, transitioning to renewable energy sources, and promoting sustainable resource management practices.
- Governance and Institutional Capacity: Weak governance structures, corruption, lack of transparency, and inadequate institutional capacity undermine efforts to implement and monitor progress towards the SDGs. Strengthening governance and institutions at the national, regional, and global levels is essential for effective policy implementation and accountability.
- Globalisation and Interdependence: Globalisation has led to increased interconnectedness and interdependence among countries and regions, but it has also exposed vulnerabilities and risks. Economic globalisation, trade imbalances, financial instability, and market volatility can impact countries' ability to achieve sustainable development goals, particularly in times of crisis.

COVID 19

In addition to the constraints above, the COVID-19 pandemic presented significant challenges to SDG implementation and progress. The pandemic has had a profound impact on the progress towards achieving the SDGs across the globe. Some of the key ways in which COVID-19 has affected SDG achievement include the following:

- Economic Impact: The pandemic triggered an unprecedented global economic downturn, leading to job losses, income inequality, and increased poverty rates. This has impeded progress towards SDG 1 (No Poverty) and SDG 8 (Decent Work and Economic Growth).
- Healthcare Systems: COVID-19 strained healthcare systems worldwide, diverting resources and attention away from other essential health services such as vaccinations, maternal health, and non-communicable disease prevention. This setback affects progress towards SDG 3 (Good Health and Well-being).
- Education Disruption: School closures and disruptions to education systems have affected millions of children and young people, exacerbating existing disparities in access to quality education. This poses challenges to achieving SDG 4 (Quality Education).
- Food Security: Disruptions in food supply chains, loss of income, and economic hardship have exacerbated food insecurity and malnutrition, particularly in vulnerable communities. This hinders progress towards SDG 2 (Zero Hunger).
- Gender Equality: The pandemic has disproportionately affected women and girls, exacerbating existing gender inequalities. Increased caregiving responsibilities, economic vulnerabilities, and disruptions to essential services have hindered progress towards SDG 5 (Gender Equality).
- Climate and Environment: While the pandemic temporarily reduced greenhouse gas emissions due to lockdown measures, long-term recovery efforts may prioritise economic growth over environmental sustainability, posing challenges to achieving SDG 13 (Climate Action) and other environmental goals.
- Global Partnerships: International cooperation and development funding have been strained due to the diversion of resources to pandemic response efforts. This affects progress towards SDG 17 (Partnerships for the Goals).

The COVID-19 pandemic has highlighted existing inequalities and vulnerabilities, making achieving the SDGs by 2030 more challenging. However, it has also underscored the importance of resilience, innovation, technology, and collaboration in addressing global challenges and building back better for a more sustainable and equitable future.

2.5.2 Global South-Global North Differences

The Global North refers to a grouping of economically advanced countries primarily located in the Northern Hemisphere. This category generally includes countries in Western Europe, North America (the United States and Canada), parts of Asia (such as Japan and South Korea), and parts of Oceania (such as Australia and New Zealand). High levels of industrialisation, wealth, technological advancement, and political stability often characterise these countries. In contrast, the Global South is a grouping of economically less developed countries primarily located in the Southern Hemisphere, including most economies in Africa, Latin America, Asia and Oceania. These economies are generally characterised by significantly lower levels of industrialisation, wealth, technological advancement, and political stability than the rest of the world. Figure 2.7 shows the geographical location of Global South and Global North countries. The stark socio-economic and political disparities and differences between the Global South and the Global North directly influence the relevance, urgency, nature of efforts, and success rate with respect to specific UN SDGs [86].

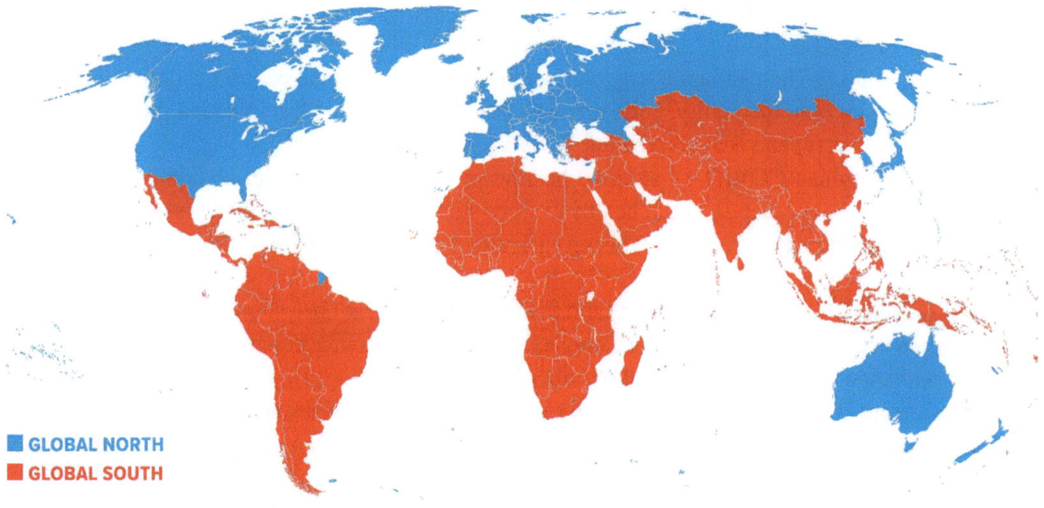

Fig. 2.7 Geographical distribution of Global South and Global North countries

2.5.3 Global South Imperatives

The UN SDGs are universally significant; however, certain goals are particularly important for the Global South due to specific regional challenges:

- SDG 1 (No Poverty): Many countries in the Global South experience high poverty rates, making poverty eradication a critical priority.
- SDG 2 (Zero Hunger): Food insecurity and malnutrition are prevalent in several Global South regions, necessitating focused efforts to eliminate hunger.
- SDG 3 (Good Health and Well-being): Addressing health disparities, combatting infectious diseases, and improving healthcare infrastructure are essential for enhancing well-being.
- SDG 4 (Quality Education): Access to quality education is vital for economic development and social mobility in the Global South.
- SDG 6 (Clean Water and Sanitation): Access to safe drinking water and sanitation facilities is crucial for public health and development.
- SDG 8 (Decent Work and Economic Growth): Promoting sustainable economic growth and employment opportunities is key to reducing poverty and inequality.
- SDG 9 (Industry, Innovation, and Infrastructure): Developing resilient infrastructure and fostering innovation are essential for industrialisation and economic progress.
- SDG 10 (Reduced Inequalities): Addressing income and social inequalities is important for inclusive development.
- SDG 13 (Climate Action): Climate change impacts often disproportionately affect the Global South, making climate action imperative.
- SDG 17 (Partnerships for the Goals): Strengthening global partnerships is crucial for resource mobilisation and knowledge sharing to achieve the SDGs.

Focusing on these goals can significantly contribute to sustainable development and improved quality of life in the Global South [68, 69, 103, 123, 150].

2.5.4 Global North Imperatives

As already explained, all 17 SDGs are globally relevant. Nevertheless, certain goals hold particular significance for countries in the Global North due to their specific socio-economic contexts and environmental responsibilities:

- SDG 7 (Affordable and Clean Energy): Transitioning to renewable energy sources is crucial for the Global North to reduce carbon emissions and combat climate change. Investments in clean energy technologies and infrastructure are pivotal in this transition.
- SDG 12 (Responsible Consumption and Production): High consumption patterns in the Global North necessitate sustainable production and consumption practices to minimise environmental impact and promote resource efficiency.
- SDG 13 (Climate Action): As major contributors to greenhouse gas emissions, countries in the Global North have a responsibility to lead in climate action initiatives, including emission reductions and support for global climate policies.
- SDG 9 (Industry, Innovation, and Infrastructure): Advancing sustainable industrialisation and fostering innovation are essential for maintaining economic growth and competitiveness in the Global North.
- SDG 11 (Sustainable Cities and Communities): Urbanisation challenges in the Global North require the development of sustainable cities that prioritise environmental sustainability, social inclusion, and resilience.
- SDG 17 (Partnerships for the Goals): The Global North plays a critical role in fostering international partnerships, providing financial resources, and sharing technological expertise to support global SDG implementation.

Focusing on these goals enables the Global North to address its unique challenges and responsibilities, contributing to the overall success of the SDGs worldwide [65, 71, 140].

Assessing progress towards the 17 UN SDGs reveals notable disparities between the Global North and Global South. Figures 2.8 and 2.9 provide a comparative overview of the journey to the attainment of these goals. These disparities underscore the need for tailored strategies and international cooperation to accelerate progress in the Global South, ensuring inclusive and equitable advancement towards the SDGs.

2.5.5 Unique Challenges in the Global South

Achieving the UN SDGs in emerging and least industrialised countries, such as those from Africa and Asia, faces numerous challenges stemming from a combination of structural, economic, social, and environmental factors. Here are some reasons why it is difficult to achieve the SDGs in the Global South [86]:

- Poverty and Inequality: Many countries in developing countries struggle with high levels of poverty and inequality, which hinder progress towards the SDGs. Limited access to basic services, resources, and opportunities exacerbates socio-economic disparities and impedes efforts to eradicate poverty and reduce inequalities (SDGs 1, 10).
- Weak Governance and Institutions: Weak governance structures, corruption, and ineffective institutions undermine the implementation of policies and programmes aimed at achieving the

SDG	Global North Progress	Global South Progress
1. No Poverty	Significant reduction in poverty rates; social safety nets are well-established.	Persistent high poverty levels; economic challenges and limited social protection systems.
2. Zero Hunger	Low levels of hunger; food security is generally stable.	High prevalence of hunger and malnutrition; food insecurity remains a critical issue.
3. Good Health and Well-being	Advanced healthcare systems; high life expectancy and low child mortality rates.	Limited healthcare access; higher rates of infectious diseases and maternal mortality.
4. Quality Education	High literacy rates; widespread access to quality education.	Lower literacy rates; challenges in educational access and quality.
5. Gender Equality	Progress in gender equality; remaining gaps in leadership roles and wage parity.	Significant gender disparities; cultural and structural barriers persist.
6. Clean Water and Sanitation	Universal access to clean water and sanitation facilities.	Limited access; challenges in infrastructure and resource management.
7. Affordable and Clean Energy	High energy access; ongoing transition to renewable energy sources.	Limited energy access; reliance on non-renewable sources; emerging renewable initiatives.
8. Decent Work and Economic Growth	Stable economic growth; low unemployment rates; diverse job markets.	Economic instability; high unemployment and underemployment; informal labor markets.
9. Industry, Innovation, and Infrastructure	Advanced infrastructure; strong emphasis on innovation and industrialization.	Underdeveloped infrastructure; limited industrial growth; emerging innovation sectors.

Fig. 2.8 Attainment of SDG 1 to 9: A Global North-Global South comparison

SDG	Global North Progress	Global South Progress
10. Reduced Inequalities	Moderate success in reducing income inequality; social policies in place.	Persistent income and social inequalities; limited access to resources and opportunities.
11. Sustainable Cities and Communities	Advanced urban planning; focus on sustainability and resilience.	Rapid urbanization challenges; inadequate infrastructure and services.
12. Responsible Consumption and Production	Initiatives promoting sustainable practices; higher resource consumption per capita.	Emerging efforts toward sustainability; challenges with waste management and resource efficiency.
13. Climate Action	Commitments to emission reductions; technological advancements in clean energy.	Vulnerable to climate impacts; limited resources for mitigation and adaptation.
14. Life Below Water	Policies for marine conservation; issues with overfishing and pollution.	Dependence on marine resources; challenges in enforcing conservation measures.
15. Life on Land	Conservation programs; issues with habitat loss due to development.	High biodiversity; threats from deforestation and land degradation.
16. Peace, Justice, and Strong Institutions	Stable institutions; challenges with political polarization and social unrest.	Issues with governance, corruption, and conflict; efforts toward institutional strengthening.
17. Partnerships for the Goals	Active in international cooperation; providing development assistance.	Recipients of aid; increasing participation in global partnerships.

Fig. 2.9 Attainment of SDG 10 to 17: A Global North-Global South comparison

SDGs. Lack of accountability, transparency, and rule of law hinder progress in areas such as sustainable development, peace, and justice (SDGs 16, 17).
- Limited Access to Education and Healthcare: Inadequate access to quality education and healthcare services in the Global South hampers efforts to improve health outcomes, promote gender equality, and reduce child mortality (SDGs 3, 4, 5). Limited investment in human capital development perpetuates cycles of poverty and underdevelopment.
- Environmental Degradation and Climate Vulnerability: Vulnerability to environmental risks, such as climate change, natural disasters, and resource depletion, poses significant challenges to sustainable development in the Global South (SDGs 13, 14, 15). Limited adaptive capacity and resilience exacerbate the impact of environmental shocks and disasters on livelihoods and ecosystems.
- Economic Vulnerability and Dependency: Economic vulnerabilities, including reliance on primary commodities, volatile markets, and external shocks, expose countries in the Global South to risks of economic instability and stagnation (SDG 8). Limited economic diversification and structural transformation hinder efforts to promote inclusive and sustainable economic growth.
- Conflict and Instability: Conflict, political instability, and insecurity in regions of the Global South disrupt development efforts and exacerbate humanitarian crises (SDG 16). Armed conflict, displacement, and insecurity undermine progress in achieving SDGs related to peace, justice, and strong institutions.
- Digital Divide and Technological Gaps: Limited access to digital technologies, internet connectivity, and ICT infrastructure in rural and remote areas of the Global South exacerbates the digital divide and hampers efforts to harness the potential of technology for sustainable development (SDG 9).
- Population Growth and Urbanisation: Rapid population growth, unplanned urbanisation, and informal settlements in many countries of the Global South strain infrastructure, services, and resources, exacerbating challenges related to poverty, inequality, and environmental sustainability (SDGs 1, 11).
- Dependency on Aid and External Assistance: Reliance on foreign aid and external assistance for development financing can undermine national ownership, sustainability, and accountability in achieving the SDGs. Aid dependency may also perpetuate unequal power dynamics and priorities in global development partnerships (SDG 17).

Addressing these challenges requires holistic and integrated approaches that prioritise poverty reduction, inclusive growth, social equity, environmental sustainability, and peacebuilding. It also requires global solidarity, partnership (within and outside the Global South), and commitment to supporting countries in the Global South on their development journeys [51, 69, 70, 103].

2.6 Resolving the Challenges and Forging Ahead

Several organisations, including the UN and various research institutes, track progress towards the SDGs using a range of indicators and metrics. These assessments typically include data on poverty reduction, access to education and healthcare, environmental conservation, gender equality, economic growth, and other relevant factors. While it's challenging to provide a definitive measure of how much of the SDGs have been achieved overall, ongoing efforts in monitoring progress, implementing policies, mobilising resources, and fostering international cooperation are crucial for advancing towards the 2030 Agenda's ambitious targets.

Despite the challenges articulated in previous sections, the SDGs remain a vital framework for global development. However, addressing all the constraints delaying their achievement is non-negotiable and must be done. It requires sustained political commitment, international cooperation, innovative approaches, and inclusive partnerships involving governments, businesses, civil society, academia, and international organisations. While achieving the SDGs by 2030 is ambitious, concerted efforts and collective action can help overcome these challenges and advance progress towards a more sustainable and equitable future.

To enhance progress, a number of actions are essential, starting with enhanced international cooperation. Strengthening partnerships among governments, private sectors, and civil society is crucial to mobilising resources and sharing best practices. Innovative financing mechanisms must be devised. Exploring alternative funding sources, including private investments and public-private partnerships, can help bridge the financial gap. Most significant of all is leveraging transformative technologies. Utilising advancements such as AI and data analytics can provide innovative solutions to complex development challenges.

While fully realising the SDGs by 2030 appears increasingly challenging, concerted and accelerated efforts can still lead to substantial progress, improving the quality of life for many and fostering a more sustainable and equitable world. Achieving the UN SDGs by 2030 requires significantly accelerating current efforts. A number of strategic actions are essential to effectively advance towards these goals [9, 44, 80, 93].

Enhanced International Cooperation and Partnerships
There is a need to strengthen multilateral collaboration. This can be achieved by fostering deeper cooperation among nations, international organisations, and stakeholders to address global challenges collectively. This includes sharing knowledge, resources, and best practices to implement effective solutions. Effective engagement of the private sector is instructive. This is achievable by encouraging businesses to align their operations with the SDGs, promoting sustainable practices and investments that contribute to development objectives. Public-private partnerships can mobilise additional resources and expertise.

Innovative Financing Mechanisms
Governments and international donors should enhance funding for SDG-related initiatives, particularly in developing countries. Addressing the estimated \$4.2 trillion funding gap is crucial for progress. To supplement public funding, private investments must be leveraged—develop financial instruments that attract private capital towards sustainable projects, such as green bonds and impact investing.

Adoption of Science and Technology
Countries must implement advanced technologies, including artificial intelligence and data analytics, to develop innovative solutions for complex development challenges. For example, AI can optimise resource management and enhance service delivery. (Using AI to drive the attainment of SDGs is the motive force behind this book.) Promoting research and development is pivotal. Countries must invest in research to generate new knowledge and technologies to drive sustainable development, ensuring that scientific advancements are accessible and applicable globally.

Policy Reforms and Institutional Strengthening
Implementing effective policies and strengthening institutions is necessary. Governments should enact and enforce policies that promote sustainability, social inclusion, and economic growth, ensuring alignment with SDG targets. They must also enhance institutions' capacity to effectively

implement and monitor SDG-related programmes, ensuring transparency, accountability, and good governance.

Community Engagement and Inclusivity
Communities must be involved in decision-making processes, ensuring that development initiatives are culturally appropriate and address local needs. More importantly, inequalities must be addressed by focusing on reducing disparities and implementing inclusive policies that benefit marginalised and vulnerable populations, ensuring that no one is left behind.

Monitoring and Accountability
There is a case for establishing robust monitoring systems by developing comprehensive frameworks to track progress towards SDG targets, utilising reliable data to inform policy adjustments and resource allocation. Accountability is essential—holding stakeholders accountable for their commitments to the SDGs, promoting transparency and regular reporting on progress and challenges.

By adopting these strategies, the global community can accelerate progress towards achieving the SDGs, fostering a more sustainable, equitable, and prosperous world for all.

2.6.1 SDG Achievement: Exemplary Economies

As of December 2024, progress towards achieving the Sustainable Development Goals (SDGs) varies greatly among countries. As already discussed, the SDGs consist of 17 goals adopted by the UN Member States in 2015, aiming to address global challenges such as poverty, inequality, climate change, environmental degradation, peace, and justice by 2030. It is difficult to authoritatively state which countries have achieved the most SDGs because progress can be subjective and vary across indicators and goals. However, some countries are making significant strides in certain areas. Nordic countries (such as Sweden, Norway, Finland, and Denmark) are often cited as leaders in sustainability, gender equality, and social welfare, making notable progress across multiple SDGs. European countries like Germany, Switzerland, and the Netherlands are also known for their efforts in sustainable development, innovation, and social inclusion. New Zealand is often highlighted for its strong commitment to environmental conservation and sustainability [65, 68, 71, 140].

Costa Rica stands out in environmental sustainability, having made substantial progress in areas like renewable energy and biodiversity conservation. Rwanda has shown remarkable progress in areas such as poverty reduction, healthcare, and gender equality since the devastating genocide in 1994. Bhutan is recognised for prioritising gross national happiness (GNH) over gross domestic product (GDP) and implementing policies aimed at sustainable development and environmental conservation. Uruguay has made significant strides in education, healthcare, and renewable energy, aiming for inclusive and sustainable development.

However, it's essential to recognise that all countries face unique challenges and have different starting points regarding SDG achievement. Progress also fluctuates over time due to various factors such as political stability, economic growth, policy implementation, and external shocks like natural disasters or global pandemics. Therefore, assessing which countries have achieved the most SDGs requires a nuanced understanding of each country's context and progress across multiple indicators.

2.6.2 Lessons from the Global North

Highly industrialised countries, primarily in Europe, North America, and parts of Asia, offer several lessons on achieving the UN SDGs. These advanced economies have made notable progress in achieving certain UN SDGs, particularly in areas such as quality education, access to clean water and sanitation, and economic growth. However, they face significant challenges in addressing others, including climate action, responsible consumption and production, and reducing inequalities within their borders. High-income nations often lead in technological innovation and funding for sustainable development but struggle to curb greenhouse gas emissions, overconsumption, and waste generation. Furthermore, many developed countries have yet to meet their commitments to provide financial and technical support to developing nations, impeding global progress on the SDGs. Rising inequality, ageing populations, and social disparities in health and education outcomes within these nations highlight the need for more inclusive and equitable policies. While some progress has been made, the overall pace and scale of efforts in developed countries are insufficient to fully achieve the SDGs by 2030. Nevertheless, these highly industrialised countries have some key lessons on how to achieve SDGs.

- Integrated Policies and Approaches: Many countries in the Global North have demonstrated the importance of integrating SDGs into national policies and strategies. Instead of treating each goal in isolation, they develop comprehensive approaches that address multiple interlinked goals simultaneously. This holistic approach ensures that progress in one area does not come at the expense of another.
- Investment in Education and Innovation: Education is crucial for achieving sustainable development. Countries in the Global North have invested heavily in education, research, and innovation, fostering a skilled workforce and technological advancements that contribute to sustainable development. Prioritising education and innovation enables societies to tackle complex challenges effectively.
- Strong Institutions and Governance: Effective governance structures and strong institutions are essential for implementing and monitoring progress towards the SDGs. The rule of law, transparency, accountability, and good governance practices are fundamental principles that underpin sustainable development efforts. Countries in the Global North often exemplify these principles through robust regulatory frameworks and public institutions.
- Partnerships and Collaboration: Achieving the SDGs requires collaboration among governments, civil society, businesses, academia, and international organisations. The Global North often engages in partnerships with various stakeholders to leverage expertise, resources, and networks. Collaborative initiatives facilitate knowledge sharing, innovation, and collective action towards common goals.
- Environmental Conservation and Sustainable Practices: Many countries in the Global North prioritise environmental conservation and sustainable practices. They implement policies and initiatives to reduce greenhouse gas emissions, promote renewable energy, conserve natural resources, and protect biodiversity. These efforts demonstrate the importance of environmental sustainability for long-term prosperity and well-being.
- Social Safety Nets and Welfare Systems: Social safety nets and welfare systems play a crucial role in reducing poverty, promoting social inclusion, and addressing inequality. Countries in the Global North often have robust social welfare systems that provide healthcare, education, housing, unemployment benefits, and other essential services to their citizens. These systems help ensure a basic standard of living and contribute to overall societal well-being.

- Long-Term Vision and Commitment: Achieving sustainable development requires long-term vision, commitment, and continuity of efforts across political cycles. Countries in the Global North demonstrate the importance of sustained commitment to SDGs despite changes in leadership or economic conditions. Setting clear goals, targets, and timelines facilitates progress monitoring and accountability.

While these lessons are valuable, it's essential to acknowledge that every country faces unique challenges and contexts. Therefore, approaches to achieving the SDGs must be tailored to local circumstances, priorities, and capacities. Learning from the experiences of the Global North can provide insights and inspiration for countries worldwide striving to advance sustainable development [65, 68, 71, 140].

2.6.3 Role of Regional and Continental Bodies

Regional and continental integration, buttressed and anchored by political unity and solidarity, is foundational to the attainment of the UN SDGs. Such entities as the Southern African Development Community (SADC), Economic Community of West African States (ECOWAS), African Union (AU), and European Union (EU) play pivotal roles in advancing the UN SDGs.

Policy coordination and harmonisation constitute a vital area of these institutions. A bloc like SADC can serve to align regional policies with the SDGs, focusing on areas like economic integration, infrastructure development, and environmental sustainability. For instance, SADC's Green Economy Strategy and Action Plan emphasises sustainable development practices. ECOWAS can develop regional strategies to address challenges such as poverty, health, and education, ensuring member states' policies are consistent with SDG objectives. The AU, a continental body, implements Agenda 2063, a strategic framework that complements the SDGs, aiming for inclusive growth and sustainable development across Africa. The EU integrates the SDGs into its policies, promoting sustainable development both within Europe and through its external actions. The EU's cooperation with SADC exemplifies this commitment.

Two essential roles of regional and continental institutions are resource mobilisation and technical assistance through partnerships such as SADC and EU collaboration: The EU has provided over 415 million to support SADC's regional integration process, focusing on economic integration, peace and security, and natural resource management. AU initiatives can be used to mobilise resources for continental projects that align with the SDGs, such as infrastructure development and health programmes.

ECOWAS monitors member states' progress towards the SDGs, providing data and analysis to inform policy decisions and ensure accountability. The EU regularly assesses its progress on the SDGs, publishing reports and setting benchmarks to guide member states. SADC facilitates training and workshops to enhance member states' capacities in areas like governance, environmental management, and economic development. The AU promotes knowledge exchange through platforms like the African Peer Review Mechanism, encouraging best practices in governance and development.

Another set of roles includes advocacy and partnerships. SADC and the EU engage in regular political dialogues to discuss progress and challenges related to the SDGs, fostering mutual understanding and cooperation. The AU advocates for Africa's interests in global forums, ensuring that the continent's priorities are reflected in international development agendas.

Through these multifaceted roles, regional and continental bodies act as catalysts, coordinators, and supporters in the collective effort to achieve the SDGs, ensuring that development initiatives are tailored to regional contexts and effectively implemented [3].

2.6.4 Role of the Private Sector

The private sector plays a pivotal role in advancing the UN SDGs by leveraging its resources, innovation, and global reach. Businesses drive economic development by creating jobs and fostering inclusive growth, directly impacting SDG 8 (Decent Work and Economic Growth). Companies can stimulate economic activity by adopting sustainable practices while ensuring social and environmental responsibility. The private sector invests in research and development, leading to technological advancements that address challenges such as clean energy (SDG 7) and climate action (SDG 13). Innovations in these areas are crucial for sustainable development.

Implementing responsible production and consumption patterns (SDG 12) helps reduce environmental impact. Companies adopting sustainable supply chains and resource-efficient processes contribute to environmental sustainability. Private investments are essential to bridge the funding gap for SDG-related projects. By aligning investment strategies with SDG objectives, businesses can support initiatives that promote sustainable development. Furthermore, public-private partnerships provide an efficacious platform for the business community's involvement. Collaborations between businesses, governments, and civil society enhance resource mobilisation and expertise sharing, vital for achieving SDG 17 (Partnerships for the Goals). Such partnerships facilitate comprehensive approaches to complex global challenges.

By integrating the SDGs into its core strategies, the private sector contributes to global development and benefits from sustainable business practices that can lead to long-term profitability and resilience.

2.6.5 The Link Between SDGs and ESG

ESG was discussed in Sect. 2.3.4 as encompassing three central factors used to measure the sustainability and societal impact of an investment in a company or business. ESG criteria help investors and stakeholders evaluate the potential long-term performance and ethical impact of their investments. The link between the SDGs and ESG principles lies in their shared vision of fostering a more sustainable, equitable, and resilient world. Both frameworks provide guidance for addressing global challenges, emphasising the interconnectedness of economic growth, social well-being, and environmental stewardship. While the SDGs are a global agenda set by the United Nations to address 17 specific goals by 2030, ESG serves as a corporate framework that evaluates an organisation's practices and impact in environmental, social, and governance dimensions. Together, they create a synergistic pathway for sustainable development. Figure 2.10 illustrates the connection of ESG to the sustainability of the Earth through the attainment of the SDGs.

The environmental aspect of ESG directly aligns with SDGs focused on climate action (SDG 13), clean energy (SDG 7), and life on land and water (SDGs 14 and 15). Companies adopting ESG principles often reduce carbon emissions, manage resources efficiently, and protect biodiversity—actions that contribute to achieving these goals. For instance, transitioning to renewable energy supports the SDGs' climate and energy targets while enhancing a company's ESG performance.

The social dimension of ESG aligns with SDGs addressing poverty (SDG 1), education (SDG 4), gender equality (SDG 5), and decent work (SDG 8). ESG principles encourage businesses to prioritise equitable labour practices, workplace diversity, and community engagement. Organisations that promote employee well-being, ensure safe working conditions, and invest in local communities directly contribute to these SDGs while building trust and loyalty among stakeholders.

The governance pillar intersects with SDGs related to peace, justice, and strong institutions (SDG 16). ESG emphasises transparency, accountability, and ethical decision-making in corporate gover-

Fig. 2.10 ESG leads to sustainability of the Earth through the attainment of the SDGs. (Image credit: iStock pcess609)

nance, which align with the SDGs' aim to build inclusive, responsible, and sustainable institutions. By fostering good governance, companies can create systems that support anti-corruption, equitable leadership, and effective risk management.

In practice, ESG provides a framework for companies to operationalise the SDGs. For example, businesses adopting ESG metrics often report their contributions to global sustainability efforts, such as reducing inequalities or mitigating environmental harm, aligning corporate strategy with the SDGs. This integration helps organisations track their progress while attracting investors, customers, and employees who value sustainability and ethical impact.

Ultimately, the relationship between SDGs and ESG highlights the complementary roles of global policy and corporate responsibility in driving systemic change. Together, they encourage businesses, governments, and communities to collaborate in building a sustainable future that benefits both people and the planet [114, 138].

2.6.6 UN Global Compact: Encouraging Corporate Responsibility

The UN Global Compact is a voluntary initiative that encourages businesses worldwide to adopt sustainable and socially responsible policies. Launched in 2000, it provides a framework for companies to align their strategies and operations with ten universally accepted principles in the areas of human rights, labour, environment, and anti-corruption. By participating in the Global Compact, businesses commit to upholding these principles while contributing to broader societal goals, including the UN SDGs. This global adoption of and cooperation for sustainability is illustrated in Fig. 2.11.

At its core, the UN Global Compact seeks to harness the power of the private sector to address global challenges. In the realm of human rights, companies are urged to support and respect the

Fig. 2.11 Worldwide corporate adoption of sustainable and socially responsible policies. (Image credit: iStock cobalt88)

rights of all individuals, ensuring that their operations do not contribute to human rights abuses. The labour principles emphasise fair working conditions, the abolition of forced and child labour, and the promotion of workplace equality. These commitments foster social justice and equity, both within companies and across their supply chains.

The environmental principles encourage businesses to adopt practices that minimise ecological harm, promote resource efficiency, and support innovation in sustainability. Companies are also expected to engage in environmentally friendly practices, such as reducing greenhouse gas emissions, conserving water, and adopting circular economy models. The principle of anti-corruption calls for ethical business conduct, emphasising transparency, accountability, and the rejection of bribery and other forms of corruption.

The Global Compact is not just a set of principles but a call to action. Participating companies are required to report their progress annually through Communication on Progress (COP) reports, fostering transparency and accountability. This process also provides a platform for knowledge sharing and collaboration, as businesses learn from each other's experiences and best practices.

By joining the UN Global Compact, businesses can align their operations with global priorities while enhancing their reputation, building trust with stakeholders, and fostering long-term success. The initiative highlights the critical role of the private sector in shaping a sustainable and equitable future, proving that profitability and responsibility can go hand in hand. Through collective action and commitment to the Global Compact, businesses worldwide can become agents of positive change, addressing societal challenges while creating value for both people and the planet. Indeed, the UN Global Compact provides a framework for corporate responsibility and agency on the journey towards the attainment of the UN SDGs by 2030 [44, 114, 138].

2.6.7 WHO and SDGs

The World Health Organization (WHO) plays a pivotal role in achieving the SDGs by providing leadership and fostering partnerships critical for a coordinated global health response. As the primary international health authority, WHO sets norms and standards, promotes evidence-based practices, and offers technical assistance to countries. This enables them to enhance their healthcare capacities essential for achieving SDG 3, which focuses on Good Health and Well-being. Furthermore, WHO's advocacy for integrating health in all policies underscores the interconnectedness of health with other SDGs, such as those targeting poverty, gender equality, clean water, and climate action.

In addition to its standard-setting and support roles, WHO is crucial in monitoring and evaluating global health trends to assess progress towards health-related SDGs. This monitoring helps identify gaps where efforts and resources need redirection to meet set targets effectively. WHO also plays a key role in managing health emergencies, mitigating their impact on health systems, and ensuring sustained progress towards the SDGs. Through these multifaceted efforts, WHO supports countries in improving health outcomes, which is fundamental to the success of the SDG agenda.

On 20 January 2025, immediately after his inauguration, US President Donald Trump signed an executive order to withdraw the United States from the WHO. This move could lead to the organisation losing a fifth of its budget. Clearly, the withdrawal of the United States from the WHO poses significant challenges to the SDGs, particularly those related to global health. Since the United States is one of the largest contributors to WHO, this action will undermine the WHO's capacity to provide essential services, such as public health initiatives, disease prevention, and emergency responsiveness worldwide. This reduction in funding and support will potentially jeopardise the progress towards achieving universal health coverage (SDG 3) and impact other interlinked goals such as poverty reduction and quality education. The absence of a major global leader in the WHO also affects collaborative efforts to tackle cross-border health issues like pandemics and climate change-related health problems. All this highlights the interconnectedness and the vulnerability of global health governance (and, by extension, overall SDG progress) to geopolitical shifts.

2.6.8 USAID, Funding of the UN, and SDGs

On 4 February 2025, US President Donald Trump signed an executive order withdrawing the United States from the UN Human Rights Council (UNHRC) and the United Nations Relief and Works Agency (UNRWA) while reviewing its role in the United Nations Educational, Scientific and Cultural Organization (UNESCO). Trump has also moved to dismantle the US Agency for International Development (USAID). All these actions meant to reduce US financial contribution to the UN and cut international aid will dramatically affect the SDG journey globally.

Indeed, reducing US funding for key UN agencies such as the UNHRC, UNRWA, and UNESCO could significantly affect the global SDG agenda. These agencies play vital roles in promoting human rights, providing education and relief to refugees, and fostering international cooperation in education, science, and culture—elements that are fundamental to several SDGs. A decrease in US contributions could strain the resources of these organisations, leading to scaled-back programmes and reduced support for the vulnerable populations they serve. For instance, reduced funding for UNRWA might worsen the conditions for millions of Palestinian refugees, impacting goals related to poverty reduction (SDG 1), education (SDG 4), and reducing inequalities (SDG 10). Similarly, diminished support for UNESCO could stifle international collaboration in science, education, and cultural preservation, adversely affecting goals that promote inclusive and equitable quality education (SDG 4), gender equality (SDG 5), and the fostering of innovation and resilient infrastructure (SDG 9). Overall, the funding cuts could hinder targeted initiatives and broader efforts to foster international peace and stability, which are essential for the successful achievement of the SDGs globally.

Furthermore, dismantling USAID, a major provider of international development assistance, would severely undermine global efforts to achieve the SDGs. Without USAID's support, many programmes focused on enhancing healthcare, promoting education, and alleviating poverty in underdeveloped regions would face significant funding shortages, stalling progress. Furthermore, the elimination of USAID would weaken global partnerships that are essential for coordinated international responses to challenges such as climate change, infectious diseases, and humanitarian crises, directly impacting the overall effectiveness and reach of the SDG agenda.

2.7 Conclusion

Understanding sustainability and the UN SDGs is crucial because these frameworks provide a road map for addressing the world's most pressing challenges. The SDGs encompass interconnected objectives that aim to eradicate poverty, reduce inequality, promote health and education, protect the planet, and foster economic growth, ensuring no one is left behind. By grasping the principles of sustainability, individuals, organisations, and governments can make informed decisions that balance economic, social, and environmental priorities. This understanding helps align actions with global efforts to build a resilient, equitable, and inclusive future, making it clear how everyday choices and policies contribute to long-term well-being.

Moreover, understanding the SDGs empowers stakeholders to identify gaps, drive innovation, and implement impactful and scalable solutions. For businesses, it offers opportunities to align operations with sustainable practices, enhancing their competitiveness and reputation while contributing to global progress. For individuals, it fosters a sense of responsibility and encourages sustainable lifestyles. At the policy level, understanding the SDGs helps governments prioritise resource allocation and international collaboration. More significantly, it is essential to comprehend and solve the challenges delaying SDG achievement while lessons and best practices are harvested and shared across the world. On a planet increasingly threatened by climate change, resource depletion, and social inequality, a comprehensive understanding of sustainability and the SDGs is essential to foster cooperation, drive meaningful action, and ensure a sustainable and just future for all. Global institutions such as the WHO, UNHRC, UNRWA, and UNESCO are foundational in the SDG journey and must be embraced and supported by all countries.

UN Sustainable Development Goals (SDGs) 3

3.1 Introduction

As discussed in Chap. 2, the 17 UN SDGs provide a comprehensive framework for global development, aiming to address a wide range of challenges by 2030. The UN SDGs are essential because they provide a comprehensive and unified framework to address the world's most pressing challenges. These 17 interconnected goals, adopted by the United Nations in 2015, aim to eradicate poverty, reduce inequality, combat climate change, and ensure peace and prosperity for all by 2030. Unlike fragmented approaches to development, the SDGs recognise that global issues are interdependent, requiring holistic solutions. For example, improving education (SDG 4) can lead to better health outcomes (SDG 3), gender equality (SDG 5), and economic growth (SDG 8). The SDGs encourage collective action to create a sustainable and equitable world by enhancing collaboration among nations, organisations, and communities.

The importance of the SDGs also lies in their ability to inspire action across sectors and levels of society. Governments, businesses, and individuals are encouraged to align their activities with these goals, ensuring that development efforts are inclusive and environmentally responsible. The SDGs emphasise the urgency of tackling climate change (SDG 13), protecting ecosystems (SDG 14 and SDG 15), and transitioning to sustainable energy (SDG 7), recognising that a healthy planet is critical for human survival. At the same time, they address social and economic challenges, such as ending hunger (SDG 2), promoting decent work (SDG 8), and building resilient infrastructure (SDG 9). By setting measurable targets and fostering accountability, the SDGs provide a clear road map for a better future, making them a vital tool for driving meaningful and transformative change worldwide. In this chapter, it is prudent to start with a summarised version of the nature of the 17 UN SDGs. Figure 3.1 shows these ambitious goals for humanity [110, 126].

A. G. O. Mutambara, *Deploying Artificial Intelligence to Achieve the UN Sustainable Development Goals*, Sustainable Development Goals Series,
https://doi.org/10.1007/978-3-031-88423-8_3

Fig. 3.1 The 17 UN SDGs [125]

3.2 Summary of the UN SDGs

As discussed earlier, the SDGs can be summarised as follows [126]:

1. **No Poverty**: Eradicate extreme poverty for all people everywhere, currently measured as living on less than US$1.25 a day.
2. **Zero Hunger**: End hunger, achieve food security and improved nutrition, and promote sustainable agriculture.
3. **Good Health and Well-being**: Ensure healthy lives and promote well-being for all at all ages.
4. **Quality Education**: Ensure inclusive and equitable quality education and promote lifelong learning opportunities for all.
5. **Gender Equality**: Achieve gender equality and empower all women and girls.
6. **Clean Water and Sanitation**: Ensure availability and sustainable management of water and sanitation for all.
7. **Affordable and Clean Energy**: Ensure access to affordable, reliable, sustainable, and modern energy for all.
8. **Decent Work and Economic Growth**: Promote sustained, inclusive, and sustainable economic growth, full and productive employment, and decent work for all.

9. **Industry, Innovation, and Infrastructure**: Build resilient infrastructure, promote inclusive and sustainable industrialisation, and foster innovation.
10. **Reduced Inequalities**: Reduce inequality within and among countries.
11. **Sustainable Cities and Communities**: Make cities and human settlements inclusive, safe, resilient, and sustainable.
12. **Responsible Consumption and Production**: Ensure sustainable consumption and production patterns.
13. **Climate Action**: Take urgent action to combat climate change and its impacts.
14. **Life Below Water**: Conserve and sustainably use the oceans, seas, and marine resources for sustainable development.
15. **Life on Land**: Protect, restore, and promote sustainable use of terrestrial ecosystems, manage forests sustainably, combat desertification, and halt biodiversity loss.
16. **Peace, Justice, and Strong Institutions**: Promote peaceful and inclusive societies, provide access to justice for all, and build effective, accountable institutions at all levels.
17. **Partnerships for the Goals**: Strengthen the means of implementation and revitalise the global partnership for sustainable development.

3.3 Classification of SDGs

As discussed in Chap. 2, these 17 SDGs can be classified into three interdependent categories—social, environmental, and economic. There are nine social, three environmental, and five economic goals. These are placed in these three overarching groups to ensure a holistic approach to global development. This classification reflects the interconnected nature of the challenges humanity faces. The nine social goals focus on addressing issues such as poverty, hunger, education, health, gender equality, and inequality, emphasising the need for inclusive human development. Social goals are foundational as they aim to improve the quality of life, ensure equitable access to resources, and build resilient societies. Without progress in these areas, efforts to achieve environmental and economic goals may falter due to weak social structures.

The three environmental goals emphasise protecting the planet by addressing climate change, preserving biodiversity, and ensuring sustainable use of ecosystems. These goals are essential because environmental degradation directly impacts human well-being and economic stability. For instance, unchecked climate change exacerbates social inequalities by disproportionately affecting vulnerable populations and undermining economic productivity through natural disasters. At the same time, conserving ecosystems is critical for food security, water availability, and the overall health of communities, directly linking environmental goals to social and economic objectives.

The five economic goals focus on fostering sustainable growth, decent work, innovation, infrastructure development, and reducing inequalities. Economic stability is a driving force behind achieving social and environmental progress. For example, robust economies provide the resources needed for investments in education, healthcare, and green technologies. However, economic growth must be balanced to avoid exacerbating environmental harm and social inequalities. The three categories—social, environmental, and economic—are deeply interconnected, with progress in one area often enabling advancements in the others. Sustainable development requires synergy between these dimensions, as failure in one can cascade into setbacks across all three. For example, unchecked economic activities may lead to environmental harm, which in turn undermines social welfare, highlighting the need for integrated strategies. Figures 3.2, 3.3, and 3.4 show the nine social, three environmental, and five economic SDGs, respectively.

Fig. 3.2 Social SDGs [125]

Fig. 3.3 Environment SDGs [125]

3.4 SDG Measurement Framework

If you cannot measure it, you cannot manage it. The achievement of the SDGs must be systematically monitored and measured. Hence, the SDG framework is accompanied by specific targets and indicators that provide a framework for measuring progress towards achieving each goal. Indeed, targets and indicators are essential tools for tracking progress towards achieving the SDGs. This is because they provide measurable benchmarks that translate broad goals into actionable objectives.

Fig. 3.4 Economic SDGs [125]

Targets specify the desired outcomes for each goal, such as halving global poverty by 2030 or ensuring universal access to clean water, offering a clear direction for governments, organisations, and communities. Indicators, in turn, provide the data points necessary to assess progress towards these targets, such as the percentage of the population living below the poverty line or the proportion of people with access to safe drinking water.

Breaking down the SDGs into concrete, quantifiable elements, targets, and indicators enables stakeholders to monitor progress, identify gaps, and make evidence-based decisions. These parameters ensure accountability and transparency, highlighting areas where additional resources or interventions are needed. They also allow for comparisons across countries and regions, fostering global collaboration and learning. Without targets and indicators, the ambition of the SDGs could become abstract and difficult to measure, undermining efforts to achieve meaningful and impactful change.

3.4.1 Targets

As discussed, each of the 17 SDGs is associated with a set of targets that outline specific objectives to be achieved by 2030. Targets are designed to be ambitious, measurable, and time bound, providing a clear road map for action. For example, SDG 1 (No Poverty) includes the target to "eradicate extreme poverty for all people everywhere, currently measured as people living on less than $1.25 a day. There are a total of 169 targets associated with the 17 Sustainable Development Goals (SDGs) adopted by the UN in 2015. Each SDG has a set of specific targets that outline the objectives to be achieved by 2030. These targets provide a clear road map for action and guide efforts towards sustainable development across various sectors and dimensions.

The 169 targets cover a wide range of areas, including poverty eradication, education, health, gender equality, environmental sustainability, climate action, and peace and justice. They address key challenges facing humanity and aim to promote inclusive, equitable, and sustainable development for present and future generations. It is important to note that each target is associated with one or more indicators, which are specific, measurable parameters used to track progress towards achieving the targets. The indicators provide data for monitoring and assessing progress over time, helping to identify trends, gaps, and areas requiring additional attention and resources.

Overall, the 169 targets associated with the SDGs provide a comprehensive framework for addressing global challenges and advancing sustainable development priorities at the local, national, regional, and global levels [51, 70, 80, 97, 128].

3.4.2 Indicators

Indicators are specific, measurable parameters used to track progress towards achieving each target. Indicators help assess trends, identify challenges, and inform policy decisions. The United Nations has established a comprehensive framework of indicators to monitor progress towards the 17 SDGs. This framework comprises 232 unique indicators, each corresponding to specific targets within the goals. These indicators serve as measurable benchmarks, enabling countries and organisations to assess advancements and identify areas requiring attention. Each target is associated with one or more indicators, which provide data for tracking progress over time. Each SDG has a set of indicators that provide specific, measurable parameters for tracking progress towards the corresponding targets.

The global indicator framework comprises a diverse range of indicators covering various aspects of sustainable development, including poverty eradication, education, health, gender equality, environmental sustainability, climate action, and peace and justice. These indicators help assess trends, identify challenges, and inform policy decisions at the local, national, regional, and global levels. The detailed list of all these indicators is in the official document provided by the UN Statistics Division [129]. This publication outlines each indicator associated with the respective SDG targets, offering a structured approach to tracking global development efforts. By utilising these indicators, stakeholders can ensure a data-driven and transparent process in measuring SDG progress worldwide.

It's important to note that the global indicator framework is periodically reviewed and updated to ensure its relevance and effectiveness in monitoring progress towards the SDGs. National statistical offices, international organisations, and other stakeholders play a key role in collecting data, analysing trends, and reporting on SDG indicators to assess progress and inform policy action. The Inter-Agency and Expert Group developed the current global indicator framework on SDGs.

Indicators were agreed upon (including refinements on several indicators) at the 48th session of the UN Statistical Commission held in March 2017. As has already been stated, the global indicator list is contained in the Report of the Inter-Agency and Expert Group on Sustainable Development Goal Indicators [129]. The list includes 232 indicators on which general agreement has been reached. The revised global list of SDG indicators consists of 244 indicators. However, since nine indicators repeat under two or three different targets, the total number of individual indicators on the list remains 232 [51, 70, 80, 97, 128].

3.4.3 Examples of Targets and Indicators

The SDG monitoring and measurement framework can be illustrated by considering targets and indicators across different SDGs as follows:

SDG Target	Indicator
SDG 1.1: By 2030, eradicate extreme poverty for all people everywhere, measured as people living on less than $1.90 a day.	Indicator 1.1.1: Proportion of the population living below the international poverty line, by sex, age, employment status, and geographic location.
SDG 3.2: By 2030, end preventable deaths of newborns and children under 5 years of age.	Indicator 3.2.1: Under-5 mortality rate. Indicator 3.2.2: Neonatal mortality rate.
SDG 4.1: By 2030, ensure that all girls and boys complete free, equitable, and quality primary and secondary education.	Indicator 4.1.1: Proportion of children and young people achieving at least a minimum proficiency level in reading and mathematics, by sex.
SDG 13.2: Integrate climate change measures into national policies, strategies, and planning.	Indicator 13.2.1: Number of countries with nationally determined contributions, long-term strategies, national adaptation plans, and other national climate instruments.

Fig. 3.5 Four SDG targets and their corresponding indicators

SDG 2 (Zero Hunger)

- Target 2.1: By 2030, end hunger and ensure access by all people, in particular, the poor and people in vulnerable situations, including infants, to safe, nutritious, and sufficient food all year round.
- Indicator 2.1.1: Prevalence of undernourishment (percentage of population).

SDG 4 (Quality Education)

- Target 4.1: By 2030, ensure that all girls and boys complete free, equitable, and quality primary and secondary education leading to relevant and effective learning outcomes.
- Indicator 4.1.1: Proportion of children and young people achieving proficiency in reading and mathematics.

SDG 7 (Affordable and Clean Energy)

- Target 7.2: By 2030, increase substantially the share of renewable energy in the global energy mix.
- Indicator 7.2.1: Renewable energy share in the total final energy consumption.

SDG 13 (Climate Action)

- Target 13.1: Strengthen resilience and adaptive capacity to climate-related hazards and natural disasters in all countries.
- Indicator 13.1.1: Number of deaths, missing persons, and directly affected persons attributed to disasters per 100,000 population.

These examples illustrate how targets provide specific goals to be achieved, while indicators provide measurable parameters for tracking progress towards those goals. Figure 3.5 highlights four specific targets and the indicators used to measure progress towards achieving those targets [80, 128].

3.5 SDG Monitoring and Evaluation

Monitoring SDG progress requires collecting data on the SDG indicators and analysing trends over time to assess whether countries are on track to achieve the targets by 2030. As discussed in the preceding section, the achievement of the 17 UN SDGs is measured through a framework of indicators and data collection mechanisms at the global, regional, national, and local levels. It is instructive to flesh out the monitoring framework and its several components.

1. Indicators: As explained in Sect. 3.4.2, indicators are specific, measurable parameters used to track progress towards each of the 17 SDGs and their associated targets. These indicators help assess trends, identify challenges, and inform policy decisions. The UN Statistical Commission, in collaboration with relevant international organisations and experts, has developed a global indicator framework comprising 232 indicators to monitor the SDGs.
2. Data Collection: Data collection is essential for monitoring progress towards the SDGs. Countries collect data on SDG indicators through various sources, including national statistical offices, surveys, administrative records, censuses, and research studies. Improving data quality, coverage, and disaggregation helps provide a more accurate picture of progress and identify disparities among different population groups.
3. Reporting Mechanisms: Countries report on their progress towards the SDGs through voluntary national reviews (VNRs) submitted to the UN High-level Political Forum on Sustainable Development (HLPF). VNRs allow countries to share experiences, lessons learned, and challenges encountered in implementing the SDGs. The HLPF also reviews thematic and regional progress reports prepared by international organisations, UN agencies, and other stakeholders.
4. Global Reports and Assessments: The UN periodically releases reports and assessments tracking global progress towards the SDGs. These reports analyse trends, highlight success stories, and identify areas requiring additional attention and resources. The UN Secretary-General's annual SDG Progress Report provides an overview of progress towards the goals, drawing on data and analyses from various sources.
5. Regional and Thematic Reviews: Regional organisations, such as the European Union, African Union, and Organisation of American States, conduct regional reviews of progress towards the SDGs. These reviews assess regional trends, challenges, opportunities, and regional cooperation efforts to advance sustainable development. Thematic reviews focus on specific SDGs, cross-cutting issues, or vulnerable populations, providing in-depth analysis and recommendations.
6. Independent Assessments and Research Studies: Independent assessments and research studies conducted by academia, think tanks, civil society organisations, and other stakeholders contribute to monitoring progress towards the SDGs. These studies provide valuable insights into the drivers of change, policy effectiveness, and areas requiring policy intervention.
7. Stakeholder Engagement and Participation: Stakeholder engagement and participation are integral to the monitoring and review process of the SDGs. Governments, civil society organisations, academia, the private sector, and international organisations collaborate to collect data, analyse trends, and assess progress towards the goals. Participatory approaches ensure transparency, inclusivity, and ownership of the SDG agenda.

Overall, monitoring progress towards the SDGs requires a coordinated and multi-stakeholder approach, leveraging data, indicators, reporting mechanisms, and stakeholder engagement to assess progress, identify gaps, and accelerate action towards achieving the goals by 2030. Progress towards the SDGs varies across countries, regions, and goals, and it's influenced by a wide range of factors such as economic development, policy implementation, technological advancements, and social and environmental factors [51, 70, 80, 97, 128].

3.6 Critical Non-AI Interventions

Progress towards the SDGs varies across countries, regions, and goals. Various factors, such as economic development, policy implementation, technological advancements, and social and environmental concerns, influence it. Before considering the potential role of AI in pursuing the goals, it is prudent to articulate critical non-AI interventions that have been used to hasten their attainment and measurement. AI is just one technology-based effort. Over the past 10 years, SDG progress has been assessed through traditional methods of data collection, analysis, and evaluation of policy implementation.

- Global Monitoring and Reporting: The United Nations periodically releases reports and assessments tracking progress towards the SDGs at the global, regional, and national levels. These reports analyse indicators and trends related to poverty, hunger, health, education, gender equality, environmental sustainability, and other dimensions of sustainable development.
- Country-Level Assessments: Many countries have developed their own frameworks, strategies, and indicators for monitoring progress towards the SDGs. National governments, along with civil society organisations, academia, and other stakeholders, assess progress, identify challenges, and develop policies and programs to accelerate progress towards the SDGs.
- Data Collection and Measurement: Data collection and measurement play a crucial role in assessing progress towards the SDGs. Countries collect data on various indicators related to the SDGs, including demographic, social, economic, and environmental data. Efforts to improve data quality, coverage, and disaggregation help provide a more accurate picture of progress and identify disparities among different population groups.
- Policy Implementation and Interventions: Governments, international organisations, and other stakeholders implement policies, programs, and interventions aimed at achieving the SDGs. These interventions may include investments in infrastructure, healthcare, education, social protection, environmental conservation, and poverty reduction.
- Multilateral Cooperation and Partnerships: Achieving the SDGs requires collaboration and partnerships among governments, international organisations, civil society, the private sector, academia, and other stakeholders. Multilateral cooperation helps mobilise resources, share knowledge and expertise, and leverage synergies to accelerate progress towards the goals.
- Challenges and Limitations: Despite progress in some areas, significant challenges remain in achieving the SDGs. These challenges include persistent poverty, hunger, inequality, gender disparities, environmental degradation, climate change, conflict, and humanitarian crises. Addressing these challenges requires sustained political commitment, resources, and innovative approaches.
- Monitoring and Accountability Mechanisms: Monitoring progress towards the SDGs is essential for accountability and transparency. Countries establish monitoring and reporting mechanisms to track progress, assess performance, and identify areas requiring additional attention and resources. International organisations, such as the United Nations, provide technical assistance and support to countries in monitoring and reporting on the SDGs.

Overall, progress towards achieving the SDGs, without considering AI, is multifaceted and influenced by many factors. While AI can accelerate progress in certain areas by providing new tools and insights, achieving the SDGs ultimately requires concerted efforts, resources, and partnerships across sectors and stakeholders. Indeed, the journey towards 2030 SDG victory has been driven by traditional but effective interventions and measured by techniques such as data collection, reporting, monitoring, and evaluation. Will the deployment of AI dramatically drive the attainment and improve the measurement of SDGs? How can the proliferation of AI systems and the ubiquitous nature of the technology expedite the journey to victory in 2030? Before engaging these questions, it is prudent to present the details of each of the 17 SDGs.

3.7 Details of the 17 UN SDGs

The UN SDGs are interconnected, recognising that progress in one area can influence outcomes in others, and they collectively aim to achieve a better and more sustainable future for all. It is instructive to explore the 17 SDGs in detail, outlining each goal's key targets. For each SDG, some indicators of its targets are presented. As indicated earlier, the complete detailed list of all 231 unique indicators (for all 17 SDGs) and their exact definitions are in the official document provided by the UN Statistics Division [125, 126, 129].

3.8 SDG 1 (No Poverty)

SDG 1 (No Poverty), established by the United Nations in 2015, aims to "end poverty in all its forms everywhere". This goal underscores the global commitment to eradicate extreme poverty and address the multifaceted dimensions of poverty by 2030. As of available data in December 2024, several countries have a majority of their populations living below the international poverty line of $2.15 per day (2017 PPP). This threshold is used to measure extreme poverty globally. Figure 3.6 highlights these countries.

These figures are based on the World Bank's Poverty and Inequality Platform, which provides comprehensive data on global poverty levels. It's important to note that these percentages are subject to change as new data becomes available and as countries implement policies aimed at poverty reduction.

3.8.1 Key Targets of SDG 1

- Eradicate Extreme Poverty: By 2030, eliminate extreme poverty for all individuals worldwide, defined as living on less than US$2.15 per day.
- Reduce Poverty by Half: By 2030, reduce by at least half the proportion of men, women, and children of all ages living in poverty in all its dimensions according to national definitions.
- Implement Social Protection Systems: Establish nationally appropriate social protection systems and measures for all, including social protection floors, and achieve substantial coverage of the poor and vulnerable by 2030.
- Ensure Equal Rights to Economic Resources: By 2030, ensure that all individuals, particularly the poor and vulnerable, have equal rights to economic resources, access to basic services, and ownership and control over land and other forms of property, inheritance, natural resources, appropriate new technology, and financial services, including microfinance.
- Build Resilience to Disasters: By 2030, enhance the resilience of the poor and those in vulnerable situations, reducing their exposure and vulnerability to climate-related extreme events and other economic, social, and environmental shocks and disasters.

Country	Percentage of Population Below $2.15/day
Burundi	79.0%
Democratic Republic of the Congo	77.2%
Madagascar	77.6%
Malawi	70.3%
Mozambique	62.9%
Central African Republic	61.3%
Guinea-Bissau	61.0%
Niger	60.8%
Liberia	50.9%
Chad	50.5%

Fig. 3.6 Countries with most of their people living below the international poverty line

3.8.2 Some Indicators of SDG 1 Targets

- Proportion of population living below the international poverty line
- Proportion of population covered by social protection systems
- Direct economic loss attributed to disasters
- Proportion of population living in poverty in all dimensions

3.8.3 Importance of SDG 1

Eradicating poverty is fundamental to achieving sustainable development. Poverty is linked to various issues, including hunger, poor health, lack of education, and social exclusion. Addressing poverty contributes to economic growth, social stability, and environmental sustainability.

Poverty rates vary significantly across countries, reflecting diverse economic conditions and measurement standards. Figure 3.7 highlights selected countries and their respective poverty rates, based on the available data in December 2024.

In the table, it is essential to note that the poverty rates are based on national poverty lines and may not be directly comparable across countries due to differences in measurement methodologies and economic contexts. These figures underscore the varying degrees of poverty worldwide and highlight the need for targeted policies and interventions to address the specific challenges faced by each state.

Country	Poverty Rate (%)	Year of Data
South Sudan	82.3	2024
Madagascar	70.7	2024
Guinea-Bissau	69.3	2024
Mozambique	63.7	2024
Nigeria	40.1	2024
India	21.9	2024
United States	11.8	2024
Canada	10.0	2024
United Kingdom	11.7	2024
Germany	9.7	2024

Fig. 3.7 Poverty rates in selected countries

3.8.4 Progress and Challenges

Significant progress has been made in reducing poverty globally; however, challenges remain. As of 2022, approximately 8.4% of the world's population, or about 670 million people, were living in extreme poverty. The COVID-19 pandemic has exacerbated poverty levels, reversing years of progress. If current trends continue, an estimated 7% of the global population could still be living in extreme poverty by 2030, with a significant concentration in Sub-Saharan Africa.

Achieving SDG 1 requires comprehensive strategies, including economic growth, social protection, access to essential services, and resilience-building measures to address the root causes of poverty and ensure no one is left behind.

3.9 SDG 2 (Zero Hunger)

Sustainable Development Goal 2 (SDG 2) aims to "end hunger, achieve food security and improved nutrition, and promote sustainable agriculture". This goal addresses the critical need to eliminate hunger and malnutrition while ensuring that food production systems are sustainable and resilient.

3.9.1 Key Targets of SDG 2

- End Hunger: By 2030, ensure access to safe, nutritious, and sufficient food for all people, particularly the poor and those in vulnerable situations, including infants.

- End All Forms of Malnutrition: By 2025, achieve internationally agreed targets on stunting and wasting in children under five years of age, and address the nutritional needs of adolescent girls, pregnant and lactating women, and older persons.
- Double Agricultural Productivity and Incomes: By 2030, double the agricultural productivity and incomes of small-scale food producers, especially women, indigenous peoples, family farmers, pastoralists, and fishers.
- Ensure Sustainable Food Production Systems: By 2030, implement resilient agricultural practices that increase productivity and production, help maintain ecosystems, and strengthen capacity for adaptation to climate change, extreme weather, drought, flooding, and other disasters.
- Maintain Genetic Diversity: By 2020, maintain the genetic diversity of seeds, cultivated plants, and farmed and domesticated animals and their related wild species.

3.9.2 Some Indicators of SDG 2 Targets

- Prevalence of undernourishment
- Prevalence of stunting and wasting in children under 5 years
- Agricultural productivity and income of small-scale farmers
- Proportion of agricultural area under sustainable practices

3.9.3 Importance of SDG 2

Achieving zero hunger is fundamental to sustainable development. Hunger and malnutrition impede economic growth, perpetuate poverty, and contribute to poor health outcomes. Sustainable agriculture ensures that food production can meet the needs of the present without compromising the ability of future generations to meet their own needs.

3.9.4 Progress and Challenges

Despite progress in reducing hunger, challenges remain. As of 2022, approximately 735 million people—9.2% of the world's population—were in a state of chronic hunger, an increase from 2019. Additionally, an estimated 2.4 billion people will face moderate to severe food insecurity in 2024. Factors such as conflict, climate change, and economic downturns have exacerbated food insecurity, highlighting the need for comprehensive strategies to achieve SDG 2.

Addressing SDG 2 requires coordinated efforts to enhance food production, improve access to nutritious food, and promote sustainable agricultural practices, ensuring that no one is left behind in the quest to end hunger.

3.10 SDG 3 (Good Health and Well-being)

UN SDG 3 aims to "ensure healthy lives and promote well-being for all at all ages". This goal addresses a broad spectrum of health-related challenges, emphasising the importance of universal health coverage, reducing mortality rates, and combatting various diseases.

3.10.1 Key Targets of SDG 3

- Reduce Maternal Mortality: By 2030, reduce the global maternal mortality ratio to less than 70 per 100,000 live births.
- End Preventable Deaths of Newborns and Children: By 2030, end preventable deaths of newborns and children under 5 years of age, aiming to reduce neonatal mortality to at least as low as 12 per 1000 live births and under-5 mortality to at least as low as 25 per 1000 live births.
- Combat Communicable Diseases: By 2030, end the epidemics of AIDS, tuberculosis, malaria, and neglected tropical diseases, and combat hepatitis, waterborne diseases, and other communicable diseases.
- Reduce Non-communicable Diseases: By 2030, reduce by one-third premature mortality from non-communicable diseases through prevention and treatment, and promote mental health and well-being.
- Strengthen Prevention and Treatment of Substance Abuse: Strengthen the prevention and treatment of substance abuse, including narcotic drug abuse and harmful use of alcohol.
- Halve Global Deaths from Road Traffic Accidents: By 2020, halve the number of global deaths and injuries from road traffic accidents.
- Ensure Universal Access to Sexual and Reproductive Health Services: By 2030, ensure universal access to sexual and reproductive healthcare services, including family planning, information, and education, and the integration of reproductive health into national strategies and programmes.
- Achieve Universal Health Coverage: Achieve universal health coverage, including financial risk protection, access to quality essential healthcare services, and access to safe, effective, quality, and affordable essential medicines and vaccines for all.
- Reduce Deaths from Hazardous Chemicals and Pollution: By 2030, substantially reduce the number of deaths and illnesses from hazardous chemicals and air, water, and soil pollution and contamination.

3.10.2 Some Indicators of SDG 3 Targets

- Maternal mortality ratio
- Under-5 mortality rate
- Incidence of major diseases (HIV, tuberculosis, malaria, etc.)
- Coverage of essential health services
- Mortality due to air pollution and unsafe water

3.10.3 Importance of SDG 3

Ensuring healthy lives and promoting well-being are essential to sustainable development. Health is a fundamental human right and a key indicator of sustainable development. Improved health outcomes contribute to economic growth, poverty reduction, and social stability.

3.10.4 Progress and Challenges

Significant progress has been made in improving global health; however, challenges remain. As of 2023, maternal mortality rates have declined, and child mortality rates have decreased. However, non-communicable diseases now account for 71% of all deaths globally, and mental health disorders are on the rise. The COVID-19 pandemic has further strained health systems, highlighting the need for resilient health infrastructure and universal health coverage.

Achieving SDG 3 requires comprehensive strategies, including strengthening health systems, ensuring equitable access to health services, and addressing the social determinants of health to promote well-being for all.

3.11 SDG 4 (Quality Education)

UN SDG 4 aims to "ensure inclusive and equitable quality education and promote lifelong learning opportunities for all". This goal emphasises the importance of accessible, high-quality education as a foundation for improving people's lives and achieving sustainable development.

3.11.1 Key Targets of SDG 4

- Free Primary and Secondary Education: By 2030, ensure that all girls and boys complete free, equitable, and quality primary and secondary education leading to relevant and effective learning outcomes.
- Early Childhood Development and Pre-primary Education: By 2030, ensure that all girls and boys have access to quality early childhood development, care, and pre-primary education so that they are ready for primary education.
- Equal Access to Technical, Vocational, and Higher Education: By 2030, ensure equal access for all women and men to affordable and quality technical, vocational, and tertiary education, including university.
- Relevant Skills for Employment: By 2030, substantially increase the number of youth and adults who have relevant skills, including technical and vocational skills, for employment, decent jobs, and entrepreneurship.
- Eliminate Gender Disparities in Education: By 2030, eliminate gender disparities in education and ensure equal access to all levels of education and vocational training for the vulnerable, including persons with disabilities, indigenous peoples, and children in vulnerable situations.
- Universal Literacy and Numeracy: By 2030, ensure that all youth and a substantial proportion of adults, both men and women, achieve literacy and numeracy.
- Education for Sustainable Development and Global Citizenship: By 2030, ensure that all learners acquire the knowledge and skills needed to promote sustainable development, including education for sustainable lifestyles, human rights, gender equality, promotion of a culture of peace and non-violence, global citizenship, and appreciation of cultural diversity.

3.11.2 Some Indicators of SDG 4 Targets

- Proportion of children achieving minimum proficiency in reading and math
- Participation rate in organised learning
- Number of trained teachers
- Equity in educational opportunities by gender, disability, and income

3.11.3 Importance of SDG 4

Education is a powerful tool for breaking the cycle of poverty, reducing inequalities, and fostering economic growth. It empowers individuals with the knowledge and skills necessary to participate fully in society and the economy. Quality education also promotes tolerance and understanding among people, contributing to more peaceful and inclusive societies.

3.11.4 Progress and Challenges

Significant progress has been made in increasing access to education, particularly at the primary level. However, challenges remain in ensuring quality education and equal access at all levels. As of 2020, about 258 million children and youth were out of school, and more than half of children and adolescents worldwide were not meeting minimum proficiency standards in reading and mathematics. The COVID-19 pandemic has further exacerbated educational disparities, with school closures affecting learning outcomes globally.

Achieving SDG 4 requires concerted efforts to improve educational infrastructure, train qualified teachers, provide learning materials, and address barriers to education, ensuring that all individuals have the opportunity to learn and thrive.

3.12 SDG 5 (Gender Equality)

UN SDG 5 aims to "achieve gender equality and empower all women and girls". This goal addresses the systemic barriers that have historically hindered women's and girls' full participation in various aspects of life, emphasising the importance of equal rights and opportunities.

3.12.1 Key Targets of SDG 5

- End Discrimination: Eliminate all forms of discrimination against women and girls everywhere.
- Eliminate Violence: Eradicate all forms of violence against women and girls in public and private spheres, including trafficking and sexual and other types of exploitation.
- Eliminate Harmful Practices: Abolish harmful practices such as child, early, and forced marriage and female genital mutilation.
- Recognise Unpaid Care and Domestic Work: Acknowledge and value unpaid care and domestic work through public services, infrastructure, social protection policies, and the promotion of shared responsibility within households.
- Ensure Participation and Leadership: Guarantee women's full and effective participation and equal opportunities for leadership at all levels of decision-making in political, economic, and public life.
- Ensure Access to Reproductive Health: Provide universal access to sexual and reproductive health and reproductive rights.
- Undertake Reforms for Economic Rights: Implement reforms to grant women equal rights to economic resources, including access to ownership and control over land, property, financial services, inheritance, and natural resources.
- Enhance Use of Technology: Promote the empowerment of women through technology, particularly information and communications technology.

- Adopt Policies for Gender Equality: Establish and strengthen policies and enforceable legislation to promote gender equality and empower all women and girls.

3.12.2 Some Indicators of SDG 5 Targets

- Proportion of women in leadership positions
- Prevalence of gender-based violence
- Proportion of women in decision-making roles
- Legal frameworks for gender equality

3.12.3 Importance of SDG 5

Gender equality is a fundamental human right and a cornerstone for a peaceful, prosperous, and sustainable world. Empowering women and girls leads to economic growth, social development, and improved health and education outcomes. Addressing gender disparities contributes to the achievement of all other Sustainable Development Goals.

3.12.4 Progress and Challenges

While progress has been made in areas such as education and political representation, significant challenges remain. As of 2023, women continue to face disparities in employment, wages, and leadership positions. Violence against women and girls persists globally, and harmful practices like child marriage and female genital mutilation remain prevalent in some regions. The COVID-19 pandemic has exacerbated existing inequalities, disproportionately impacting women's economic opportunities and increasing instances of domestic violence.

Achieving SDG 5 requires comprehensive efforts to dismantle discriminatory laws and practices, promote women's economic empowerment, ensure access to education and healthcare, and foster environments where women and girls can thrive equally alongside men and boys.

3.13 SDG 6 (Clean Water and Sanitation)

UN SDG 6 aims to "ensure availability and sustainable management of water and sanitation for all". This goal addresses the critical need for universal access to safe and affordable drinking water, adequate sanitation, and hygiene while emphasising the sustainable management of water resources.

3.13.1 Key Targets of SDG 6

- Safe and Affordable Drinking Water: By 2030, achieve universal and equitable access to safe and affordable drinking water for all.
- Adequate Sanitation and Hygiene: By 2030, achieve access to adequate and equitable sanitation and hygiene for all, and end open defecation, paying special attention to the needs of women and girls and those in vulnerable situations.

- Improve Water Quality: By 2030, improve water quality by reducing pollution, eliminating dumping, and minimising the release of hazardous chemicals and materials, halving the proportion of untreated wastewater, and substantially increasing recycling and safe reuse globally.
- Increase Water-Use Efficiency: By 2030, substantially increase water-use efficiency across all sectors, ensure sustainable withdrawals and supply of freshwater to address water scarcity, and substantially reduce the number of people suffering from water scarcity.
- Implement Integrated Water Resources Management: By 2030, implement integrated water resources management at all levels, including through transboundary cooperation as appropriate.
- Protect and Restore Water-Related Ecosystems: By 2020, protect and restore water-related ecosystems, including mountains, forests, wetlands, rivers, aquifers, and lakes.
- Expand International Cooperation and Capacity Building: By 2030, expand international cooperation and capacity-building support to developing countries in water- and sanitation-related activities and programmes, including water harvesting, desalination, water efficiency, wastewater treatment, recycling, and reuse technologies.
- Support and Strengthen Local Community Participation: Support and strengthen the participation of local communities in improving water and sanitation management.

3.13.2 Some Indicators of SDG 6 Targets

- Proportion of population with access to safely managed water and sanitation
- Level of water stress
- Proportion of bodies of water with good ambient quality
- Proportion of wastewater safely treated

3.13.3 Importance of SDG 6

Access to safe water and sanitation is fundamental to human health, dignity, and development. Water is essential not only for drinking but also for agriculture, industry, and energy production. Proper sanitation and hygiene are crucial for preventing diseases and ensuring well-being. Sustainable management of water resources is vital for maintaining ecosystems and biodiversity.

3.13.4 Progress and Challenges

Despite progress, significant challenges remain. As of 2022, 2.2 billion people still lacked safely managed drinking water, and 3.5 billion lacked safely managed sanitation. Water scarcity affects more than 40% of the global population, and this figure is projected to increase. Water pollution and inadequate infrastructure continue to hinder access to clean water and sanitation. Climate change exacerbates these challenges by altering water availability and increasing the frequency of extreme weather events.

Achieving SDG 6 requires concerted efforts to invest in infrastructure, promote sustainable water management practices, enhance international cooperation, and engage local communities in water and sanitation initiatives. Addressing these challenges is essential for health, economic development, and environmental sustainability.

3.14 SDG 7 (Affordable and Clean Energy)

UN SDG 7 aims to "ensure access to affordable, reliable, sustainable, and modern energy for all". This goal addresses the critical need for universal energy access, increased renewable energy usage, and improved energy efficiency to foster sustainable development and mitigate climate change.

3.14.1 Key Targets of SDG 7

- Universal Access to Modern Energy: By 2030, ensure universal access to affordable, reliable, and modern energy services.
- Increase Renewable Energy Share: By 2030, substantially increase the share of renewable energy in the global energy mix.
- Improve Energy Efficiency: By 2030, double the global rate of improvement in energy efficiency.
- Enhance International Cooperation: By 2030, enhance international cooperation to facilitate access to clean energy research and technology, including renewable energy, energy efficiency, and advanced and cleaner fossil-fuel technology, and promote investment in energy infrastructure and clean energy technology.
- Expand Infrastructure and Upgrade Technology: By 2030, expand infrastructure and upgrade technology for supplying modern and sustainable energy services for all in developing countries, particularly least developed countries, small island developing states, and landlocked developing countries, in accordance with their respective programmes of support.

3.14.2 Some Indicators of SDG 7 Targets

- Proportion of population with access to electricity
- Renewable energy share in the total energy mix
- Energy efficiency improvements
- Access to clean cooking solutions

3.14.3 Importance of SDG 7

Access to energy is fundamental for economic growth, social development, and environmental sustainability. Reliable and modern energy services are essential for healthcare, education, water supply, and industrialisation. Transitioning to renewable energy sources and improving energy efficiency are crucial for reducing greenhouse gas emissions and combatting climate change.

3.14.4 Progress and Challenges

Significant progress has been made in expanding access to electricity; however, challenges remain. As of 2021, 733 million people worldwide still lacked access to electricity, and 2.4 billion people relied on inefficient and polluting cooking systems. The share of renewable energy in the global energy mix has been increasing, but not at the pace required to meet the 2030 targets. Energy efficiency improvements have also been slower than needed. The COVID-19 pandemic has further impacted progress, highlighting the need for resilient and sustainable energy systems. Achieving SDG

7 requires concerted efforts to invest in renewable energy infrastructure, enhance energy efficiency, promote international cooperation, and implement policies that ensure equitable access to modern energy services for all. Addressing these challenges is essential for sustainable development and climate action.

3.15 SDG 8 (Decent Work and Economic Growth)

UN SDG 8 aims to "promote sustained, inclusive, and sustainable economic growth, full and productive employment, and decent work for all". This goal emphasises the importance of economic growth that benefits all segments of society, ensuring that employment opportunities are productive and provide fair wages and safe working conditions.

3.15.1 Key Targets of SDG 8

- Sustain Per Capita Economic Growth: Achieve higher levels of economic productivity through diversification, technological upgrading, and innovation, including a focus on high-value-added and labour-intensive sectors.
- Promote Development-Oriented Policies: Support productive activities, decent job creation, entrepreneurship, creativity, and innovation, and encourage the formalisation and growth of micro-, small-, and medium-sized enterprises (MSMEs), including through access to financial services.
- Improve Resource Efficiency: Enhance global resource efficiency in consumption and production and endeavour to decouple economic growth from environmental degradation.
- Achieve Full Employment and Decent Work: By 2030, achieve full and productive employment and decent work for all women and men, including young people and persons with disabilities, and equal pay for work of equal value.
- Reduce Youth Unemployment: By 2020, substantially reduce the proportion of youth not in employment, education, or training.
- Eradicate Forced Labour and Modern Slavery: Take immediate and effective measures to eradicate forced labour, end modern slavery and human trafficking, and secure the prohibition and elimination of the worst forms of child labour, including recruitment and use of child soldiers, and by 2025, end child labour in all its forms.
- Protect Labour Rights: Protect labour rights and promote safe and secure working environments for all workers, including migrant workers, particularly women migrants, and those in precarious employment.
- Promote Sustainable Tourism: By 2030, devise and implement policies to promote sustainable tourism that creates jobs and promotes local culture and products.
- Strengthen Financial Institutions: Strengthen the capacity of domestic financial institutions to encourage and expand access to banking, insurance, and financial services for all.

3.15.2 Some Indicators of SDG 8 Targets

- Annual growth rate of GDP per capita
- Proportion of informal employment
- Youth unemployment rates
- Fatal and non-fatal occupational injuries

3.15.3 Importance of SDG 8

Economic growth is a key driver of development, providing the resources needed to improve living standards, reduce poverty, and address social inequalities. However, for growth to be sustainable and inclusive, quality jobs must be created and benefits equitably distributed. Decent work opportunities contribute to social cohesion and political stability, fostering environments where individuals can thrive.

3.15.4 Progress and Challenges

While global unemployment rates have declined, challenges persist in ensuring that economic growth translates into decent work for all. As of 2022, many workers remain in informal or vulnerable employment, lacking social protection and labour rights. Youth unemployment remains a significant concern, with young people facing higher rates of joblessness compared to adults. The COVID-19 pandemic has further exacerbated these issues, leading to job losses and economic contractions worldwide.

Achieving SDG 8 requires comprehensive policies that promote economic diversification, support small- and medium-sized enterprises, protect labour rights, and ensure that economic growth benefits all members of society. Addressing these challenges is essential for building resilient economies and societies.

3.16 SDG 9 (Industry, Innovation, and Infrastructure)

UN SDG 9 aims to "build resilient infrastructure, promote inclusive and sustainable industrialisation, and foster innovation". This goal emphasises the importance of infrastructure development, industrial growth, and innovation as key drivers of economic development and social well-being.

3.16.1 Key Targets of SDG 9

- Develop Quality, Reliable, Sustainable, and Resilient Infrastructure: By 2030, develop infrastructure that supports economic development and human well-being, with a focus on affordable and equitable access for all.
- Promote Inclusive and Sustainable Industrialisation: By 2030, significantly raise industry's share of employment and gross domestic product, in line with national circumstances, and double its share in least developed countries.
- Increase Access to Financial Services and Markets: Increase the access of small-scale industrial and other enterprises, particularly in developing countries, to financial services, including affordable credit, and their integration into value chains and markets.
- Upgrade Infrastructure and Retrofit Industries: By 2030, upgrade infrastructure and retrofit industries to make them sustainable, with increased resource-use efficiency and greater adoption of clean and environmentally sound technologies and industrial processes.
- Enhance Research and Upgrade Technological Capabilities: By 2030, enhance scientific research and upgrade the technological capabilities of industrial sectors in all countries, particularly developing countries, encouraging innovation and substantially increasing the number of research and development workers per one million people.

- Facilitate Sustainable and Resilient Infrastructure Development: Facilitate sustainable and resilient infrastructure development in developing countries through enhanced financial, technological, and technical support to African countries, least developed countries, landlocked developing countries, and small island developing states.
- Support Domestic Technology Development and Industrial Diversification: Support domestic technology development, research, and innovation in developing countries, including by ensuring a conducive policy environment for industrial diversification and value addition to commodities.
- Increase Access to Information and Communications Technology: Significantly increase access to information and communications technology and strive to provide universal and affordable access to the internet in least developed countries by 2020.

3.16.2 Some Indicators of SDG 9 Targets

- Proportion of population with access to mobile networks
- Research and development expenditure
- Proportion of small-scale industries with loans
- CO2 emissions per unit of value added

3.16.3 Importance of SDG 9

Infrastructure, industrialisation, and innovation are crucial for economic growth, job creation, and improving living standards. Resilient infrastructure supports communities during disasters, sustainable industrialisation reduces environmental impact, and innovation drives progress and competitiveness. Together, these elements contribute to building inclusive societies and achieving sustainable development.

3.16.4 Progress and Challenges

While there has been progress in infrastructure development and industrial growth, challenges remain. As of 2023, many developing countries still face significant gaps in infrastructure, limiting economic opportunities and access to essential services. Small-scale industries often struggle with limited access to financial services and markets. Investment in research and development remains low in many regions, hindering innovation. The COVID-19 pandemic has further highlighted the need for resilient infrastructure and accelerated digital transformation.

Achieving SDG 9 requires concerted efforts to invest in sustainable infrastructure, promote inclusive industrial policies, support innovation ecosystems, and enhance international cooperation to build resilient and sustainable societies.

3.17 SDG 10 (Reduce Inequalities)

UN SDG 10 aims to "reduce inequality within and among countries". This goal addresses disparities in income, access to resources, and opportunities, emphasising the need for inclusive policies that promote social, economic, and political inclusion for all individuals, regardless of their background.

3.17.1 Key Targets of SDG 10

- Income Growth for the Bottom 40%: By 2030, progressively achieve and sustain income growth of the bottom 40% of the population at a rate higher than the national average.
- Social, Economic, and Political Inclusion: By 2030, empower and promote the social, economic, and political inclusion of all, irrespective of age, sex, disability, race, ethnicity, origin, religion, or economic or other status.
- Equal Opportunity and Reduced Inequalities of Outcome: Ensure equal opportunity and reduce inequalities of outcome, including by eliminating discriminatory laws, policies, and practices, while promoting appropriate legislation, policies, and actions in this regard.
- Fiscal, Wage, and Social Protection Policies: Adopt policies, especially fiscal, wage, and social protection policies, and progressively achieve greater equality.
- Regulation of Global Financial Markets: Improve the regulation and monitoring of global financial markets and institutions and strengthen the implementation of such regulations.
- Enhanced Representation for Developing Countries: Ensure enhanced representation and voice for developing countries in decision-making in global international economic and financial institutions to deliver more effective, credible, accountable, and legitimate institutions.
- Safe Migration and Mobility: Facilitate orderly, safe, regular, and responsible migration and mobility of people, including through the implementation of planned and well-managed migration policies.
- Special and Differential Treatment for Developing Countries: Implement the principle of special and differential treatment for developing countries, particularly least developed countries, per World Trade Organisation agreements.
- Encourage Financial Flows to Where the Need Is Greatest: Encourage official development assistance and financial flows, including foreign direct investment, to states where the need is greatest, in particular least developed countries, African countries, small island developing states, and landlocked developing countries, in accordance with their national plans and programmes.
- Reduce Transaction Costs for Migrant Remittances: By 2030, reduce to less than 3% the transaction costs of migrant remittances and eliminate remittance corridors with costs higher than 5%.

3.17.2 Some Indicators of SDG 10 Targets

- Income growth rates for the bottom 40% of the population
- Proportion of people experiencing discrimination
- Proportion of migrants in the population
- Financial flows to developing countries

3.17.3 Importance of SDG 10

Reducing inequality is essential for fostering social cohesion, economic stability, and sustainable development. High levels of inequality can lead to social unrest, hinder economic growth, and perpetuate cycles of poverty. Addressing disparities ensures that all individuals have the opportunity to contribute to and benefit from economic progress, leading to more resilient and inclusive societies.

3.17.4 Progress and Challenges

Significant challenges remain while some progress has been made in reducing income inequality within specific countries. As of 2023, income inequality has increased worldwide, and disparities in access to education, healthcare, and employment opportunities persist. The COVID-19 pandemic has exacerbated existing inequalities, disproportionately affecting marginalised and vulnerable populations.

Achieving SDG 10 requires comprehensive policies that address the root causes of inequality, including discriminatory practices, unequal access to resources, and imbalances in political representation. International cooperation, inclusive economic policies, and social protection measures are vital to creating equitable societies where all individuals can thrive.

3.18 SDG 11 (Sustainable Cities and Communities)

SDG 11, established by the United Nations in 2015, aims to "make cities and human settlements inclusive, safe, resilient, and sustainable". This goal addresses the challenges of urbanisation, emphasising the need for sustainable urban development that benefits all residents and minimises environmental impact.

3.18.1 Key Targets of SDG 11

- Adequate, Safe, and Affordable Housing: By 2030, ensure access for all to adequate, safe, and affordable housing and basic services and upgrade slums.
- Safe and Sustainable Transport Systems: By 2030, provide access to safe, affordable, accessible, and sustainable transport systems for all, improving road safety, notably by expanding public transport, with special attention to the needs of those in vulnerable situations, women, children, persons with disabilities, and older persons.
- Inclusive and Sustainable Urbanisation: By 2030, enhance inclusive and sustainable urbanisation and capacity for participatory, integrated, and sustainable human settlement planning and management in all countries.
- Protect Cultural and Natural Heritage: Strengthen efforts to protect and safeguard the world's cultural and natural heritage.
- Reduce Disaster Impact: By 2030, significantly reduce the number of deaths and the number of people affected, and substantially decrease the direct economic losses relative to the global gross domestic product caused by disasters, including water-related disasters, with a focus on protecting the poor and people in vulnerable situations.
- Reduce Environmental Impact of Cities: By 2030, reduce the adverse per capita environmental impact of cities, including by paying special attention to air quality and municipal and other waste management.
- Provide Access to Green and Public Spaces: By 2030, provide universal access to safe, inclusive, and accessible, green and public spaces, in particular for women and children, older persons, and persons with disabilities.
- Support Positive Economic, Social, and Environmental Links: Support positive economic, social, and environmental links between urban, peri-urban, and rural areas by strengthening national and regional development planning.

- Implement Integrated Policies: By 2020, substantially increase the number of cities and human settlements adopting and implementing integrated policies and plans towards inclusion, resource efficiency, mitigation and adaptation to climate change, and resilience to disasters, and develop and implement, in line with the Sendai Framework for Disaster Risk Reduction 2015–2030, holistic disaster risk management at all levels.
- Support Least Developed Countries: Support least developed countries, including through financial and technical assistance, in building sustainable and resilient buildings utilising local materials.

3.18.2 Some Indicators of SDG 11 Targets

- Proportion of urban population living in slums
- Access to public transportation
- Air quality and waste management in cities
- Number of deaths from disasters

3.18.3 Importance of SDG 11

With over half of the global population residing in urban areas—a figure projected to reach 70% by 2050—sustainable urban development is crucial. Cities are centres of economic growth and cultural exchange but also face challenges like pollution, inadequate infrastructure, and social inequality. Achieving SDG 11 ensures that urbanisation contributes positively to human well-being and environmental sustainability.

3.18.4 Progress and Challenges

Progress has been made in areas such as improving access to public transportation and reducing slum populations. However, challenges persist, including rising urban populations, insufficient public transport, and city expansion outpacing population growth. As of 2022, nearly 1.1 billion people lived in slums or slum-like conditions in urban areas, with an additional two billion expected to live in such conditions over the next 30 years. The growing number of slum populations is a manifestation of the housing crisis, a situation in many instances that has now been exacerbated by the impacts of the COVID-19 pandemic.

Achieving SDG 11 requires comprehensive urban planning, investment in infrastructure, and policies that promote inclusivity and environmental sustainability, ensuring that cities are livable for all residents.

3.19 SDG 12 (Responsible Consumption and Production)

UN SDG 12 aims to "ensure sustainable consumption and production patterns". This goal addresses the need to decouple economic growth from environmental degradation, promoting resource efficiency and sustainable lifestyles to safeguard the planet for future generations.

3.19.1 Key Targets of SDG 12

- Implement the 10-Year Framework of Programmes: Encourage all countries to take action, with developed countries leading, considering the development and capabilities of developing nations.
- Sustainable Management and Efficient Use of Natural Resources: By 2030, achieve sustainable management and efficient use of natural resources.
- Halve Per Capita Global Food Waste: By 2030, halve per capita global food waste at the retail and consumer levels and reduce food losses along production and supply chains, including post-harvest losses.
- Environmentally Sound Management of Chemicals and Wastes: By 2020, achieve environmentally sound management of chemicals and all wastes throughout their life cycle, significantly reducing their release to air, water, and soil.
- Substantially Reduce Waste Generation: By 2030, substantially reduce waste generation through prevention, reduction, recycling, and reuse.
- Encourage Companies to Adopt Sustainable Practices: Encourage companies, especially large and transnational ones, to adopt sustainable practices and integrate sustainability information into their reporting cycle.
- Promote Sustainable Public Procurement: Promote public procurement practices that are sustainable, in accordance with national policies and priorities.
- Ensure Information and Awareness for Sustainable Development: By 2030, ensure that people everywhere have the relevant information and awareness for sustainable development and lifestyles in harmony with nature.
- Support Developing Countries: Support developing countries to strengthen their scientific and technological capacity to move towards more sustainable patterns of consumption and production.
- Develop Tools to Monitor Sustainable Tourism: Develop and implement tools to monitor sustainable development impacts for sustainable tourism that creates jobs and promotes local culture and products.
- Rationalise Inefficient Fossil-Fuel Subsidies: Rationalise inefficient fossil-fuel subsidies that encourage wasteful consumption by removing market distortions, reflecting environmental impacts, and considering the specific needs of developing countries.

3.19.2 Some Indicators of SDG 12 Targets

- Food loss and waste per capita
- Material footprint and resource use efficiency
- Proportion of sustainable tourism activities
- Policies for sustainable consumption

3.19.3 Importance of SDG 12

Sustainable consumption and production are essential for reducing environmental degradation, conserving resources, and fostering economic growth. By adopting sustainable practices, societies can ensure that economic activities do not compromise the planet's ability to meet the needs of present and future generations.

3.19.4 Progress and Challenges

Significant challenges remain despite progress in areas like sustainable resource management and waste reduction. As of 2023, global material consumption continues to rise, and waste generation remains a critical issue. The COVID-19 pandemic has also impacted production and consumption patterns, highlighting the need for resilient and sustainable systems. Achieving SDG 12 requires concerted efforts from governments, businesses, and individuals to adopt sustainable practices, promote resource efficiency, and foster a culture of sustainability. This collective action is vital for ensuring the well-being of the planet and its inhabitants.

3.20 SDG 13 (Climate Action)

UN SDG 13 aims to "take urgent action to combat climate change and its impacts".

3.20.1 Key Targets of SDG 13

- Strengthen Resilience and Adaptive Capacity: Enhance the ability of all countries, particularly those most vulnerable, to adapt to climate-related hazards and natural disasters.
- Integrate Climate Change Measures: Incorporate climate change mitigation and adaptation strategies into national policies, strategies, and planning processes.
- Improve Education and Awareness: Elevate education, awareness, and institutional capacity concerning climate change mitigation, adaptation, impact reduction, and early warning systems.
- Implement Financial Commitments: Fulfil the financial commitments made by developed countries to mobilise US$100 billion annually by 2020 to support climate action in developing nations.
- Promote Mechanisms for Capacity Building: Encourage mechanisms that raise capacity for effective climate change-related planning and management in the least developed countries and small island developing states, focusing on marginalised communities.

3.20.2 Some Indicators of SDG 13 Targets

- Number of countries with climate strategies
- Total greenhouse gas emissions
- Disaster mortality rates due to climate-related hazards
- Financial resources for climate action

3.20.3 Importance of SDG 13

Climate change poses a significant threat to global development, disproportionately affecting the poorest and most vulnerable populations. Addressing climate change is essential for the successful implementation of all Sustainable Development Goals, as it impacts food security, water availability, health, and biodiversity.

3.20.4 Progress and Challenges

Despite international efforts, global greenhouse gas emissions continue to rise, leading to more frequent and severe weather events. The United Nations emphasises the need for accelerated action to meet the targets of SDG 13, including substantial reductions in emissions, increased investment in renewable energy, and enhanced resilience-building measures. Achieving SDG 13 requires collective action from governments, businesses, and civil society to implement sustainable practices, invest in green technologies, and promote policies that mitigate climate change and adapt to its impacts. However, it is prudent to note that the highly industrialised economies' commitment to the climate agenda is fickle. There is a history of failed promises and a lack of follow through on climate finance. In 2009, during the Copenhagen Climate Summit, developed nations pledged to mobilise a paltry US$100 billion annually by 2020 to assist developing countries in addressing climate change through mitigation and adaptation efforts. This commitment was reaffirmed in the 2015 Paris Agreement, emphasising the importance of financial support for vulnerable nations. However, this financial target was not met by the 2020 deadline. It wasn't until 2022 that (the inadequate) US$100 billion target was reportedly achieved, two years later than promised.

At the 29th United Nations Climate Change Conference (COP29) in Baku, Azerbaijan, in November 2024, significant progress was made in establishing a global carbon market and further financial promises made—a climate finance deal, committing developed countries (the main historical polluters) to provide US$300 billion annually by 2035 to assist developing nations (the primary victims of Loss and Damage) in addressing climate change impacts. However, there is a lot of scepticism about the sincerity of undertakings. On 20 January 2025, shortly after his inauguration, US President Donald Trump signed an executive order to withdraw the United States from the Paris Agreement, not for the first time but on a second occasion. Indeed, achieving SDG 13 (Climate Action) by 2030 is going to be a tall order.

3.21 SDG 14 (Life Below Water)

UN SDG 14 aims to "conserve and sustainably use the oceans, seas, and marine resources for sustainable development". This goal addresses the critical importance of oceans and marine ecosystems to human well-being, economic prosperity, and environmental health.

3.21.1 Key Targets of SDG 14

- Reduce Marine Pollution: By 2025, prevent and significantly reduce marine pollution of all kinds, particularly from land-based activities, including marine debris and nutrient pollution.
- Protect Marine and Coastal Ecosystems: By 2020, sustainably manage and protect marine and coastal ecosystems to avoid significant adverse impacts, including by strengthening their resilience and taking action for their restoration.
- Address Ocean Acidification: Minimise and address the impacts of ocean acidification, including through enhanced scientific cooperation at all levels.
- Regulate Harvesting and End Overfishing: By 2020, effectively regulate harvesting and end overfishing, illegal, unreported, and unregulated fishing, and destructive fishing practices, implementing science-based management plans to restore fish stocks.

- Conserve Coastal and Marine Areas: By 2020, conserve at least 10% of coastal and marine areas, consistent with national and international law and based on the best available scientific information.
- Prohibit Harmful Fisheries Subsidies: By 2020, prohibit certain forms of fisheries subsidies that contribute to overcapacity and overfishing, eliminate subsidies contributing to illegal, unreported, and unregulated fishing, and refrain from introducing new such subsidies.
- Increase Economic Benefits for Small Island Developing States and Least Developed Countries: By 2030, increase the economic benefits from the sustainable use of marine resources, including through sustainable management of fisheries, aquaculture, and tourism.
- Enhance Scientific Knowledge and Research Capacity: Increase scientific knowledge, develop research capacity, and transfer marine technology to improve ocean health and enhance marine biodiversity's contribution to developing countries.
- Support Small-Scale Fishers: Provide access for small-scale artisanal fishers to marine resources and markets.
- Implement International Law: Enhance the conservation and sustainable use of oceans and their resources by implementing international law as reflected in the UN Convention on the Law of the Sea.

3.21.2 Some Indicators of SDG 14 Targets

- Proportion of fish stocks within biologically sustainable levels
- Coverage of marine protected areas
- Levels of marine pollution
- Proportion of large marine ecosystems under sustainable management

3.21.3 Importance of SDG 14

Oceans cover over 70% of the Earth's surface and are vital for human survival, providing food, oxygen, and livelihoods for billions of people. They play a crucial role in regulating the global climate and are home to diverse ecosystems. However, oceans face significant threats from pollution, overfishing, and climate change, necessitating urgent action to preserve their health and resources.

3.21.4 Progress and Challenges

While there have been efforts to establish marine protected areas and reduce certain pollutants, challenges persist. As of 2023, overfishing remains a critical issue, with approximately one-third of global fish stocks overexploited. Marine pollution, particularly plastic debris, continues to threaten marine life and ecosystems. Ocean acidification, driven by increased carbon dioxide emissions, poses risks to marine biodiversity and fisheries. Achieving SDG 14 requires comprehensive international cooperation, effective policy implementation, and concerted efforts to address the multifaceted challenges facing the world's oceans. Protecting marine environments is essential for sustaining life on Earth and ensuring the well-being of future generations.

3.22 SDG 15 (Life on Land)

UN SDG 15 aims to "protect, restore, and promote sustainable use of terrestrial ecosystems, sustainably manage forests, combat desertification, and halt and reverse land degradation and halt biodiversity loss". This goal underscores the critical importance of biodiversity as humanity's life support system.

3.22.1 Key Targets of SDG 15

- Conserve and Restore Terrestrial and Freshwater Ecosystems: By 2020, ensure the conservation, restoration, and sustainable use of terrestrial and inland freshwater ecosystems and their services, particularly forests, wetlands, mountains, and drylands, in line with international agreements.
- Sustainable Management of Forests: By 2020, promote the implementation of sustainable management of all types of forests, halt deforestation, restore degraded forests, and substantially increase afforestation and reforestation globally.
- Combat Desertification and Restore Degraded Land: By 2030, combat desertification, restore degraded land and soil, including land affected by desertification, drought, and floods, and strive to achieve a land degradation-neutral world.
- Conserve Mountain Ecosystems: By 2030, ensure the conservation of mountain ecosystems, including their biodiversity, to enhance their capacity to provide benefits essential for sustainable development.
- Protect Biodiversity and Natural Habitats: Take urgent and significant action to reduce the degradation of natural habitats, halt the loss of biodiversity, and, by 2020, protect and prevent the extinction of threatened species.
- Promote Fair Sharing of Genetic Resources: Promote fair and equitable sharing of the benefits arising from the utilisation of genetic resources and promote appropriate access to such resources, as internationally agreed.
- Eliminate Poaching and Trafficking of Protected Species: Take urgent action to end poaching and trafficking of protected species of flora and fauna and address both demand and supply of illegal wildlife products.
- Prevent Invasive Alien Species: By 2020, measures were meant to be introduced to prevent the introduction and significantly reduce the impact of invasive alien species on land and water ecosystems and control or eradicate priority species.
- Integrate Ecosystem and Biodiversity Values: By 2020, integrate ecosystem and biodiversity values into national and local planning, development processes, poverty reduction strategies, and accounts.
- Mobilise Financial Resources: Mobilise and significantly increase financial resources from all sources to conserve and sustainably use biodiversity and ecosystems.
- Mobilise Resources for Sustainable Forest Management: Mobilise significant resources from all sources and at all levels to finance sustainable forest management and provide adequate incentives to developing countries to advance such management, including for conservation and reforestation.
- Combat Poaching and Trafficking: Enhance global support for efforts to combat poaching and trafficking of protected species, including by increasing the capacity of local communities to pursue sustainable livelihood opportunities.

3.22.2 Some Indicators of SDG 15 Targets

- Forest area as a proportion of total land area
- Proportion of key biodiversity areas protected
- Red List Index (species conservation status)
- Proportion of degraded land

3.22.3 Importance of SDG 15

Terrestrial ecosystems provide a series of goods, raw materials for construction and energy, food, and a series of ecosystem services including the capture of carbon, maintenance of soil quality, provision of habitat for biodiversity, maintenance of water quality, as well as regulation of water flow and erosion control. Maintaining those ecosystems greatly supports climate change mitigation and adaptation efforts.

3.22.4 Progress and Challenges

Despite some progress, significant challenges remain. Between 2015 and 2019, at least 100 million hectares of healthy and productive land were degraded every year, impacting the lives of 1.3 billion people. Agricultural expansion is the direct driver of almost 90% of deforestation. More than one million plants, animals, and other living things are facing the threat of extinction.

Achieving SDG 15 requires comprehensive efforts to protect and restore ecosystems, sustainably manage forests, combat desertification, and halt biodiversity loss. Addressing these challenges is essential for maintaining the planet's health and ensuring the well-being of all its inhabitants.

3.23 SDG 16 (Peace, Justice, and Strong Institutions)

UN SDG 16 aims to "promote peaceful and inclusive societies for sustainable development, provide access to justice for all, and build effective, accountable, and inclusive institutions at all levels". This goal emphasises the importance of peace, justice, and strong institutions as foundational elements for sustainable development.

3.23.1 Key Targets of SDG 16

- Reduce Violence: Significantly reduce all forms of violence and related death rates everywhere.
- End Abuse and Exploitation: End abuse, exploitation, trafficking, and all forms of violence against and torture of children.
- Promote the Rule of Law: Promote the rule of law at the national and international levels and ensure equal access to justice for all.
- Combat Organised Crime: By 2030, significantly reduce illicit financial and arms flows, strengthen the recovery and return of stolen assets, and combat all forms of organised crime.
- Reduce Corruption and Bribery: Substantially reduce corruption and bribery in all their forms.
- Develop Effective Institutions: Develop effective, accountable, and transparent institutions at all levels.

- Ensure Inclusive Decision-Making: Ensure responsive, inclusive, participatory, and representative decision-making at all levels.
- Strengthen Participation in Global Governance: Broaden and strengthen the participation of developing countries in the institutions of global governance.
- Provide Legal Identity for All: By 2030, provide legal identity for all, including birth registration.
- Ensure Public Access to Information: Ensure public access to information and protect fundamental freedoms per national legislation and international agreements.
- Strengthen National Institutions: Strengthen relevant national institutions, including through international cooperation, for building capacity at all levels to prevent violence and combat terrorism and crime.
- Promote Non-discriminatory Laws and Policies: Promote and enforce non-discriminatory laws and policies for sustainable development.

3.23.2 Some Indicators of SDG 16 Targets

- Homicide and conflict-related deaths per 100,000 people
- Proportion of people subjected to violence or bribery
- Proportion of children whose births are registered
- Public access to information and government accountability

3.23.3 Importance of SDG 16

Peaceful and inclusive societies, access to justice, and effective institutions are essential for sustainable development. They provide the foundation for economic growth, social stability, and the protection of human rights. Without peace, justice, and strong institutions, development efforts can be undermined, leading to instability and conflict.

3.23.4 Progress and Challenges

While there has been progress in some areas, significant challenges remain. As of 2023, many regions continue to experience high levels of violence, corruption, and weak institutions. The COVID-19 pandemic has further strained justice systems and exposed vulnerabilities in institutional effectiveness. Efforts to reduce violence, ensure justice for all, and build strong institutions are ongoing and require sustained commitment from all sectors of society. Achieving SDG 16 is crucial for creating societies where individuals can live free from fear and violence, have equal access to justice, and participate in decision-making processes that affect their lives. This goal is integral to the broader 2030 Agenda for Sustainable Development, as peace, justice, and strong institutions are foundational to achieving all other SDGs.

3.24 SDG 17 (Partnerships for the Goals)

UN SDG 17 aims to "strengthen the means of implementation and revitalise the Global Partnership for Sustainable Development". This goal emphasises the necessity of global cooperation and collaboration among governments, the private sector, civil society, and other stakeholders to achieve the 2030 Agenda for Sustainable Development.

3.24.1 Key Components of SDG 17

- Finance: Mobilise additional financial resources for developing countries from multiple sources. Assist developing countries in attaining long-term debt sustainability through coordinated policies aimed at fostering debt financing, debt relief, and debt restructuring.
- Technology: Enhance the development, transfer, dissemination, and diffusion of environmentally sound technologies to developing countries on favourable terms. Fully operationalise the technology bank and science, technology, and innovation capacity-building mechanisms for least developed countries.
- Capacity Building: Strengthen the capacity of developing countries to implement sustainable development plans and programmes.
- Trade: Promote a universal, rules-based, open, non-discriminatory, and equitable multilateral trading system under the World Trade Organisation. Significantly increase the exports of developing countries, aiming to double the least developed countries' share of global exports by 2020.
- Policy and Institutional Coherence: Enhance global macroeconomic stability and respect each country's policy space and leadership.
- Multi-Stakeholder Partnerships: Encourage and promote effective public, public-private, and civil society partnerships.
- Data, Monitoring, and Accountability: Enhance capacity-building support to developing countries to increase the availability of high-quality, timely, and reliable data.

3.24.2 Some Indicators of SDG 17 Targets

- Total official development assistance (ODA) as a proportion of gross national income
- Proportion of global exports by developing countries
- Internet access and digital technologies
- Financial resources mobilised for sustainable development

3.24.3 Importance of SDG 17

Achieving the ambitious targets set by the SDGs requires a revitalised and enhanced global partnership that brings together governments, civil society, the private sector, the UN system, and other actors. The ambition is to mobilise all available resources. SDG 17 underscores that the success of the 2030 Agenda depends on the active engagement and collaboration of all stakeholders at local, national, and global levels.

3.24.4 Progress and Challenges

While there have been efforts to strengthen global partnerships, challenges persist. As of 2023, many developing countries continue to face obstacles in accessing financial resources, technologies, and markets. The COVID-19 pandemic has further strained international cooperation and highlighted the need for resilient and inclusive partnerships. Addressing these challenges requires a renewed commitment to multilateralism, increased resource mobilisation, and the fostering of inclusive partnerships that prioritise the needs of the most vulnerable.

Achieving SDG 17 is crucial for achieving all other Sustainable Development Goals, as it provides the framework for the means of implementation, including finance, technology, capacity building, and trade, necessary to support sustainable development worldwide.

3.25 Conclusion

The UN SDGs are essential because they provide a comprehensive blueprint for addressing the most pressing challenges facing humanity and the planet. These 17 interconnected goals tackle issues such as poverty, hunger, health, education, gender equality, clean energy, climate change, and social justice, recognising that progress in one area impacts others. By setting measurable targets and fostering global collaboration, the SDGs create a unified framework that guides governments, businesses, and communities in working towards a sustainable future. They emphasise inclusivity, ensuring no one is left behind, while also promoting environmental stewardship to protect the planet for future generations. The SDGs are not just aspirational but a call to action to align economic growth with social equity and environmental sustainability, addressing systemic inequalities and creating a more resilient and equitable world.

Achieving the UN SDGs has proven challenging due to a combination of economic, social, political, and environmental factors. Many countries face resource constraints, limiting their ability to invest in critical areas such as education, healthcare, and infrastructure. Inequalities between and within nations further exacerbate disparities in progress, leaving marginalised populations disproportionately affected. Political instability and weak governance in some regions hinder the implementation of effective policies and the coordination necessary for large-scale change. Global crises, such as the COVID-19 pandemic, have disrupted progress, diverting resources and attention away from long-term development goals. Climate change, biodiversity loss, and environmental degradation also pose existential threats that require urgent, collective action but often face resistance due to economic interests or lack of international consensus. The complexity of the SDGs themselves, with their interconnected and sometimes competing priorities, adds another layer of difficulty, requiring systems-thinking approaches that are not always prioritised. These challenges highlight the need to accelerate progress towards the SDGs through stronger political will, global collaboration and partnerships, extensive use of technology, targeted research and development, innovative solutions, and effective monitoring and evaluation as the anchor. With the advent of the AI revolution, all these interventions can be enabled and driven by AI technology. Indeed, AI systems can be deployed to achieve the 17 UN SDGs by 2030. However, the daunting question is: How can this be accomplished?

Introduction to Artificial Intelligence (AI) 4

4.1 Introduction

Technology can be harnessed to enable and assist the attainment of the UN SDGs discussed in Chap. 3. This will involve a comprehensive approach that targets infrastructure, education, healthcare, governance, and economic development. For example, smart city technologies must be implemented to improve urban infrastructure, including efficient transportation systems, smart grids for electricity, and water management systems. Expanding broadband and mobile network coverage to remote and underserved areas is needed to improve connectivity. Academic institutions must use online education platforms to provide accessible and affordable education and skill development programmes and equip schools with digital tools and resources to enhance learning experiences and bridge the education gap.

Emerging economies must prioritise the implementation of telemedicine services to provide remote medical consultations and healthcare services, especially in rural areas. The use of electronic health records and health information systems will significantly improve patient care and streamline healthcare management, bringing hope for a brighter future in healthcare access.

High technology can power economic development, thus facilitating the achievement of the UN SDGs. When harnessed effectively, e-government platforms will streamline public services, reduce corruption, and improve citizen engagement, paving the way for more efficient and transparent governance. Data analytics will further inform policy decisions and improve the delivery of public services. Promoting mobile banking and fintech solutions to improve financial inclusion, especially for unbanked populations, is essential. Local businesses must also be supported by providing platforms for e-commerce and digital marketplaces to expand their reach.

Technology can be marshalled to address environmental sustainability and mitigate climate change challenges. Public and private sectors must invest in renewable energy technologies such as solar, wind, and hydropower to reduce reliance on fossil fuels and increase energy access. Smart waste management systems must be implemented to improve recycling rates and reduce environmental pollution. Countries must use technology for early warning systems to predict and respond to natural disasters, reducing their impact on vulnerable communities; they must also utilise GIS and other mapping technologies for crisis mapping and efficient disaster response and recovery efforts.

A. G. O. Mutambara, *Deploying Artificial Intelligence to Achieve the UN Sustainable Development Goals*, Sustainable Development Goals Series,
https://doi.org/10.1007/978-3-031-88423-8_4

Digital marketing and social media can be used to promote tourism, showcase unique cultural and natural attractions, and implement smart tourism solutions like mobile apps and virtual tours to enhance the visitor experience. Crowdsourcing platforms to fund community projects and initiatives, leveraging local and global support, and using social media to raise awareness about local issues, mobilise communities, and advocate for change are also viable options. There are illustrative case studies demonstrating technology's transformative and revolutionary impact in addressing developmental challenges leading to the attainment of the UN SDGs.

- M-Pesa in Kenya: A mobile money service that has transformed financial inclusion, allowing millions of people to perform financial transactions using their mobile phones.
- EcoCash in Zimbabwe: A mobile money service that has transformed financial inclusion, allowing millions of people to perform financial transactions using their mobile phones.
- FarmDrive: A Kenyan startup using data analytics to provide credit scores for smallholder farmers, improving their access to financing.
- Zipline in Rwanda and Ghana: A company using drones to deliver medical supplies to remote areas, improving healthcare access and response times.
- Andela: A technology company that trains software developers in Africa and connects them with global companies, addressing the tech talent gap.

By leveraging technology, the world can address many of its pressing challenges, fostering sustainable development and improving the quality of life for its people. With the advent of the AI-driven Fourth Industrial Revolution, most of the technology interventions discussed in the preceding sections have been enhanced by using AI systems. For example, precision agriculture has been attained by using IoT and AI technologies to optimise resource use, improve crop yields, and monitor soil health. Blockchain technology and AI systems have been implemented to enhance transparency and efficiency in industrial supply chains. Given the ubiquitous nature of AI applications across all sectors, countries must understand AI technology comprehensively. This is a precondition for the effective deployment of AI to achieve the UN SDGs.

4.2 A Primer in Artificial Intelligence

Artificial intelligence, including its global impact and consequences, must be appreciated and embraced within the Fourth Industrial Revolution (4IR) context. The 4IR is a global, all-embracing technological transformation that radically and fundamentally alters how individuals, institutions, businesses, communities, and countries operate, work, organise, relate, and innovate, primarily because of the changes' scale, scope, speed, and complexity. It is characterised by a range of new technologies that are fusing the physical, digital, and biological worlds, impacting all disciplines, economies, and industries and even redefining the very meaning of humanity. Consequently, this revolution demands an interdisciplinary, comprehensive, holistic, and integrated ecosystem approach involving all stakeholders across the globe, such as multilateral institutions, national governments, the public sector, civil society, the private sector, academia, and the media.

4.2.1 AI Drives the Fourth Industrial Revolution (4IR)

Under the 4IR, the physical world of industrial production is merging with the digital world of information. This revolution is characterised by connected supply chains, digitised industrial production processes, connected workers, machine components, materials, to-be-manufactured objects, and even logistics, with continuous data exchange. The 4IR requires new and different skills such as critical thinking, how to think, problem-solving, structured thinking, blended learning, interdisciplinary approaches, ecosystem thinking, emotional intelligence (EQ), cultural intelligence (CQ), judgement, acquisition of key competencies, negotiation, cognitive capacity, flexibility, knowledge production, knowledge management, and knowledge ownership.

The full scale of the 4IR started to manifest in around 2011. It was anchored by and built upon the Third Industrial Revolution (3IR) and its key drivers of Information and Communications Technology (ICT) and Electronics. The 4IR represents emerging technology breakthroughs in several fields, including big data, robotics, AI, advanced analytics, virtual and augmented reality, the metaverse, nanotechnology, high-performance computing, biotechnology, the Internet of Things (IoT), Internet of Everything (IoE), 3D printing, quantum computing, blockchain technology, renewable energy, battery energy storage systems (BESS), drones, driverless cars, and autonomous vehicles in general. These 4IR technologies have dramatically improved private and public institutions' efficiency, productivity, quality, impact, and scale. It is instructive to observe that most of the 4IR innovations are driven by AI. In fact, AI constitutes the DNA of the 4IR. It is the key driver of the 4IR [115, 116].

4.2.2 Key AI Definitions

Artificial intelligence refers to intelligence programmed into and demonstrated by machines or systems. This capability is in contrast to the natural intelligence manifested by humans and other animals. Put more succinctly, AI refers to the development of computer systems that can perform tasks that ordinarily require human intelligence, such as perception, reasoning, learning, decision-making, obstacle avoidance, and path planning. These AI computer programs can process and analyse data, learn from it, and make decisions based on that learning. Techniques used in AI include machine learning, deep learning, reinforcement learning, and neural networks.

Put differently, an AI system is a computer-based system designed to simulate human intelligence processes. It can perform tasks that typically require human intelligence, such as recognising patterns, learning from data, reasoning, decision-making, understanding natural language, and problem-solving. AI systems can vary in complexity and functionality, ranging from basic systems that automate specific tasks to advanced systems capable of self-learning and adapting to new situations. There are seven key components of an AI system [58, 85, 109]:

- Algorithms: These rules or instructions guide the system's decision-making and learning processes.
- Data: AI systems require large amounts of data to learn from and make predictions or decisions.
- Models: These are representations of knowledge or patterns derived from the data that the AI system uses to make inferences.
- Machine Learning: A subset of AI where systems learn from data through experience and improve their performance over time without being explicitly programmed for each task.
- Natural Language Processing: Enables AI systems to understand, interpret, and respond to human language.
- Computer Vision: Allows AI systems to interpret and make decisions based on visual input, such as recognising objects or faces.
- Autonomous Reasoning and Decision-Making: Some AI systems can make decisions independently based on input data, rules, or learned models.

AI has a wide range of applications, from speech recognition and natural language processing to image and video analysis, mechatronics, autonomous vehicles, and robotics. The technology is also used in various industries, such as healthcare, finance, education, and manufacturing, to improve efficiency, reduce costs, and develop new products and services. AI is currently transforming society in many ways as an effective tool in solving human challenges. It is still a rapidly evolving field with great opportunities and potential dangers. There are concerns about the ethical and social implications of AI, including issues related to privacy, bias, fake news/data/images, autonomous weapons, accountability, terrorism, and job displacement [85].

4.2.3 Examples of AI Applications

AI has a wide range of applications across various industries. Some of the key sectors with great opportunities to leverage this revolutionary technology include the following:

- Healthcare (disease diagnosis and prediction, personalised treatment plans, drug discovery and development, medical image analysis)
- Banking and Finance (fraud detection and prevention, algorithmic trading, customer service and chatbots, credit scoring and risk assessment)
- Agriculture (precision farming, crop monitoring and disease detection, autonomous farming equipment)
- Education (intelligent tutoring systems, adaptive learning platforms, grading and assessment automation, educational content creation)
- Manufacturing (predictive maintenance of machinery, quality control in production processes, supply chain optimisation, robot-assisted manufacturing)
- Energy Management (predictive maintenance for energy infrastructure, energy consumption optimisation, smart grid management)
- Retail (personalised shopping recommendations, inventory management and optimisation, supply chain management, customer service automation)
- Autonomous Vehicles (self-driving cars and drones, traffic optimisation, predictive maintenance for vehicles)

This non-exhaustive outline of AI applications demonstrates the ubiquitous and versatile nature of the impact of AI globally. The technology dramatically improves efficiency, decision-making, and innovation across all socio-economic domains. As the AI field evolves, new possibilities and applications are generated. For example, with advances in personalised AI and conversational AI, there is now a reduced cost of personal access to perfect information. There is an opportunity to have a perfect tutor in everyone's pocket, in their language and style. Potentially, everyone will be smarter, with a personal assistant at their disposal—a chief of staff for one's personal life.

4.3 Classification of AI Systems

Artificial intelligence systems can be placed in various categories in accordance with the nature or complexity of the cognitive capacity involved.

4.3.1 Traditional (Single-Task) Artificial Intelligence

Single-task AI systems are designed to perform a specific task or set of tasks rather than being able to perform a wide range of tasks. They are also called traditional AI systems. Single-task AI is typically developed using machine learning algorithms that are trained on a specific dataset to perform a particular function. Examples of single-task AI systems include speech recognition software, image recognition software, obstacle avoidance, path planning, and fraud detection systems. Most robots, drones, and driverless cars are driven by traditional AI systems. These highly specialised systems are typically designed to perform their task at a very high level of accuracy. Overall, single-task AI systems are highly effective for performing specific tasks, but they are not capable of the same level of flexibility and adaptability as more general-purpose AI systems.

4.3.2 Task-Based Versus Agentic Artificial Intelligence

Another classification is task-based AI versus agentic (agency-based) AI. Task-based AI models are designed to perform specific, well-defined tasks with precision and efficiency. These models excel in environments where the problem is narrow, and the inputs and outputs are clearly defined. For example, a task-based model might be trained to recognise objects in images, translate text between languages, or analyse sentiment in social media posts. Their strength lies in their ability to deliver highly accurate results within their specialised domain, often outperforming broader systems in terms of speed and reliability. However, their limitation is their lack of adaptability. They struggle to generalise beyond their trained tasks and require significant retraining or fine-tuning to handle new or slightly different problems.

In contrast, agentic AI models are presented with a goal instead of a task. They are built to exhibit autonomy and decision-making capabilities, mimicking human-like agency as they pursue the goal. Agentic models are not confined to a single task but are designed to operate in dynamic, unpredictable environments where they must perceive, reason, and act to achieve long-term goals. For instance, an agentic AI system might power a self-driving car, navigate complex traffic scenarios, or control a robot that learns to perform household chores. These systems often incorporate advanced techniques like reinforcement learning, memory networks, and planning algorithms, enabling them to adapt to new situations and improve over time. Their strength lies in their flexibility and ability to handle multistep, complex tasks that require strategic thinking and problem-solving. However, this adaptability comes at the cost of increased complexity, making them more challenging to design, train, and deploy.

The distinction between these two AI paradigms becomes particularly evident when considering their applications and limitations. Task-based models are ideal for industries where repetitive, specialised tasks are the norm, such as manufacturing, healthcare diagnostics, or customer service chatbots. They provide reliable, high-performance solutions that can be easily integrated into existing workflows. On the other hand, agentic models are better suited for environments that demand autonomy and adaptability, such as autonomous vehicles, robotics, or personalised AI assistants. These systems can navigate uncertainty, learn from experience, and make decisions in real time, making them invaluable in scenarios where predefined rules and static models fall short.

Ultimately, the choice between task-based and agentic AI models depends on the problem and the desired autonomy level. Task-based models offer precision and efficiency for well-defined tasks, while agentic models provide the flexibility and adaptability required for complex, dynamic environments. As AI continues to evolve, integrating these two paradigms—combining the reliability of task-based systems with the autonomy of agentic models—could pave the way for even more powerful and versatile AI solutions. This hybrid approach may be key to addressing some of the most challenging problems across industries, from healthcare and education to climate change.

4.3.3 Artificial General Intelligence

Definition

General-purpose AI systems are designed to perform a wide range of tasks. The ultimate and most advanced form of such capability is called artificial general intelligence (AGI). This is the intelligence of a machine that can learn, understand, and execute any intellectual task that a human being can. The idea is to have the ability to perform multiple complex tasks in a machine as opposed to specific task competence. Unlike narrow AI systems designed to perform specific functions, such as image recognition, language translation, obstacle avoidance, and path planning, AGI can reason, learn, and adapt to new situations while accomplishing multiple complex tasks in much the same way as a human being. The concept of AGI is sometimes associated with the idea of strong AI, which suggests that it is possible to create machines that are not only capable of mimicking human intelligence but are sentient, conscious, and self-aware, that is, machines that manifest intentionality, consciousness, and feelings. However, it is important to note that AGI and strong AI remain ostensibly areas of current active research—inconclusive work in progress. Whether and when such systems will be fully developed are open questions. Ethical considerations, religious beliefs, or fear of the unknown cannot be allowed to deter the pursuit of both AGI and strong AI. The research and experimentation must continue [13, 58, 63].

Applications and Implications

As already explained, AGI refers to highly autonomous systems that possess the ability to outperform humans at nearly any economically valuable work. AGI is a work in progress with lofty ambitions to develop a form of AI that has the capability to understand, learn, and apply knowledge across a wide range of tasks, similar to human intelligence. The development of AGI would have profound implications and applications across various aspects of society, as illustrated below:

- Automation of Complex Tasks: AGI could automate a broad spectrum of tasks across multiple industries, leading to increased efficiency and productivity.
- Scientific Discovery: AGI could accelerate scientific research by analysing vast datasets, identifying patterns, and generating hypotheses in fields such as medicine, physics, and materials science.
- Problem-Solving and Decision-Making: AGI systems could excel in complex problem-solving and decision-making scenarios, helping to address global challenges, such as climate change, poverty, and healthcare.
- Innovation and Creativity: AGI might contribute to innovation by proposing novel solutions, designs, and ideas that may not have been apparent to human researchers.
- Personal Assistance: AGI could provide advanced personal assistance, not only in routine tasks but also in complex decision-making processes, offering customised solutions based on individual preferences and needs.
- Education and Tutoring: AGI could revolutionise education by providing personalised and adaptive learning experiences, catering to the unique needs and learning styles of each student.
- Entertainment and Art: AGI could contribute to the creation of highly realistic and engaging entertainment experiences, including movies, music, literature, and virtual reality.
- Space Exploration: AGI systems could autonomously operate space probes and vehicles, analyse data from distant planets, and assist in planning and executing complex space missions.

- Ethical and Social Implications: The development of AGI raises ethical concerns related to its impact on employment, privacy, security, and the potential misuse of advanced technologies. Ensuring responsible development and use of AGI is crucial.
- Global Governance and Security: AGI development requires international collaboration to address potential risks and establish guidelines for responsible use. The control and security of AGI systems will be significant global concerns.
- Economic Impact: AGI has the potential to reshape economies by transforming industries, creating new job roles, and influencing the distribution of wealth. Preparing for these changes will be essential for societies to adapt.
- Human-AI Collaboration: AGI could collaborate with humans in various domains, enhancing human capabilities and augmenting our ability to address complex challenges.

It's important to note that the development of AGI also poses risks, including unintended consequences, biases, and the need for robust safety measures. Addressing these challenges requires careful consideration and responsible AI development practices. As already indicated, we are not there yet with respect to AGI. It is a work in progress. AGI development must focus on ethical considerations while using humanity's collective and shared interest as the central motivating and organising principle [63, 78].

4.3.4 Generative AI

Definition

This is a special category of AI that involves the use of algorithms and models to create new content or generate novel ideas. This can include the creation of images, music, text, or even entire virtual environments. Unlike traditional AI systems that are designed to perform specific tasks or solve specific problems, generative AI (GenAI) is designed to create something new that did not exist before. It uses complex algorithms and models trained on large amounts of data to create new content that is often indistinguishable from that created by humans. The goal is to be able to replicate or mimic the creative abilities of humans and produce new content that is original and novel. Several different techniques can be used to create generative systems, including machine learning, deep learning, and probabilistic modelling.

GenAI has many potential applications in fields such as art, music, and design, as well as in areas such as drug discovery and materials science. However, it also raises important ethical questions about the use of AI-generated content and the potential for misuse or abuse of these technologies [40, 102].

History and Examples

GenAI is not a particularly new concept. It has been around as a technology that refers to a class of artificial intelligence techniques that involves generating new content, such as images, text, or other data, based on patterns learned from existing examples. What has changed is the sophistication of GenAI because of the availability of massive amounts of data (for training) and immense advances in computational power. Historically, examples of GenAI have included Google Translate (a digital language translator in 2006 (digital translation from one language to another)), Siri (a digital assistant in 2011 (responds to users' spoken questions by speaking back to them through the device's speaker and presenting relevant information on the screen)), Amazon Alexa (a digital assistant in 2014 (receives voice commands, uploads them to Amazon servers for translation, and then delivers results in the form of audio, video, or text)), and the widespread cell phone application that tries to complete typed instructions or sentences.

Advanced Generative AI Systems

In terms of the broad development of GenAI, several progressively sophisticated systems have been launched, starting with the generative adversarial network (GAN) in 2014 [40, 102].

- A GAN is a type of neural network architecture that consists of a generator network and a discriminator network that are trained in tandem. The generator generates data, and the discriminator evaluates whether the generated data is real or fake. This adversarial training process creates realistic content, such as images or even deepfake videos.
- StyleGAN, developed by NVIDIA, is an extension of GANs specifically designed for generating high-quality and realistic images. StyleGAN allows for control over various aspects of image style, such as facial features. StyleGAN2 is an improved version that enhances the quality and diversity of generated images.
- OpenAI's GPT (generative pretrained transformer) series includes models GPT-3, GPT-3.5, GPT-4, and GPT-5. When it was launched in 2023, GPT-3 was one of the largest language models ever created. These models are pretrained on massive datasets and can generate coherent and contextually relevant text. GPT-3, GPT-3.5, and GPT-4 models have been used for various natural language processing tasks, including text completion, translation, and content generation. OpenAI's GPT-5 (to be released sometime in 2024) will be the world's most advanced large language model.
- BERT (Bidirectional Encoder Representations from Transformers), developed by Google, is another powerful language model based on transformer architecture. While not strictly a generative model, BERT is pretrained on large amounts of text data and can be fine-tuned for various natural language understanding tasks. It has been widely used in applications such as question-answering and sentiment analysis.
- DeepDream is a generative AI technique developed by Google that utilises convolutional neural networks (CNNs). It enhances and modifies images in a dreamlike or hallucinogenic manner by amplifying patterns detected by the neural network. This technique provides artistic and visually intriguing results.
- Pix2Pix is a type of conditional GAN that is trained to convert images from one domain to another. It can be used for tasks such as image-to-image translation, turning satellite images into maps, black-and-white photos into colour, and more.
- CycleGAN is another GAN variant designed for image-to-image translation without paired training examples. It can be used to transform images from one domain to another (e.g. turning photos of horses into zebras) without the need for labelled training data.
- BigGAN is a large-scale GAN developed by Google that is capable of generating high-resolution and diverse images. It has been trained on extensive datasets and can produce high-quality synthetic images across various categories.
- DALL-E is an AI model developed by OpenAI that generates images from textual descriptions. It can create novel images based on user-inputted textual prompts, demonstrating the ability to generate diverse and creative visual content. These examples showcase the versatility of generative AI across different domains, from image synthesis and transformation to natural language processing and content generation. It's important to note that while generative AI has many positive applications, it also raises ethical considerations, particularly in the context of deepfake technology and the potential misuse of AI-generated content [119].

4.3.5 Large Language Models

A large language model refers to a type of AI representation designed to understand and generate human-like text at a scale that involves a vast amount of parameters, which are the internal variables that the model learns during training. These models are typically based on deep learning architectures, with neural networks known as transformers being a common choice for natural language processing tasks. Large language models are the key innovation and driving technology behind GenAI.

The term "large" in "large language model" refers to the sheer number of parameters within the model. More parameters generally imply a greater capacity for the model to capture intricate patterns and nuances in the data it is trained on. Large language models are often pretrained on massive datasets containing diverse examples of human language from the internet. Once pretrained, these models can be fine-tuned for specific tasks or used as is to generate text or perform various language-related tasks.

Key Characteristics of Large Language Models

- Scale of Parameters: Large language models can have tens or hundreds of billions of parameters. The GPT-3 model, for example, has 175 billion parameters.
- Pretraining: These models are initially pretrained on a broad range of text data to learn the intricacies of language, grammar, context, and even factual knowledge. This pretraining phase allows them to capture general language patterns.
- Fine-Tuning: After pretraining, models can be fine-tuned on specific datasets or tasks to make them more applicable to particular use cases or industries. Fine-tuning involves training the model on a smaller, task-specific dataset.
- Transfer Learning: Large language models leverage the concept of transfer learning. They learn general language features during pretraining and can then transfer this knowledge to specific tasks during fine-tuning, making them versatile and capable across various applications.
- Natural Language Generation: These models can generate human-like text in response to prompts or inputs. They excel at tasks such as text completion, summarisation, translation, question-answering, and conversation.
- Contextual Understanding: Large language models exhibit a degree of contextual understanding, allowing them to generate contextually relevant responses based on the input they receive.

Prominent examples of large language models include GPT-3 (including GPT-3.5, GPT-4, and GPT-5), developed by OpenAI, and BERT (Bidirectional Encoder Representations from Transformers), developed by Google. These models have demonstrated state-of-the-art performance in various natural language processing benchmarks and applications. Other competitive large language models are Meta Llama, Anthropic Claude, xAI Grok, Alibaba Cloud, Mistral 3B, and Cohere Command. Figure 4.1 shows the top 10 LLMs in 2024 before the dramatic and disruptive entry of DeepSeek-R1 in January 2025.

While these AI models have shown remarkable capabilities, their deployment raises concerns related to ethical considerations, biases present in training data, environmental impact due to extensive computation requirements, and the potential for misuse. It is prudent to emphasise that traditional large language models are expensive and energy-intensive because they require significant compute resources to process and learn from enormous datasets during training. Additionally, the energy costs associated with powering high-performance computing systems for training and deploying these models contribute to their high operational expenses. Researchers and developers are actively working to address these challenges and improve the models' cost and energy efficiencies.

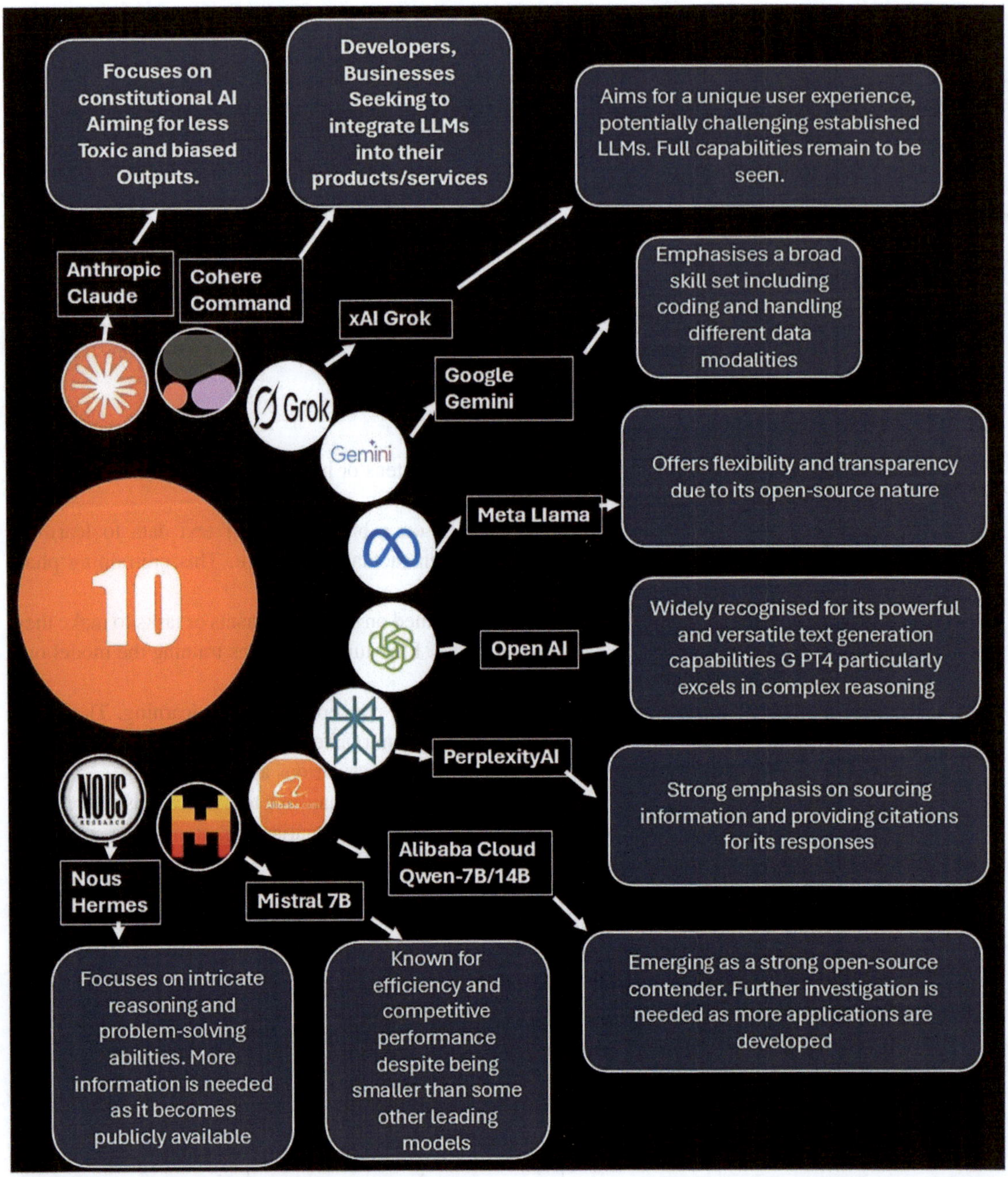

Fig. 4.1 Top 10 LLMs in 2024 [56]

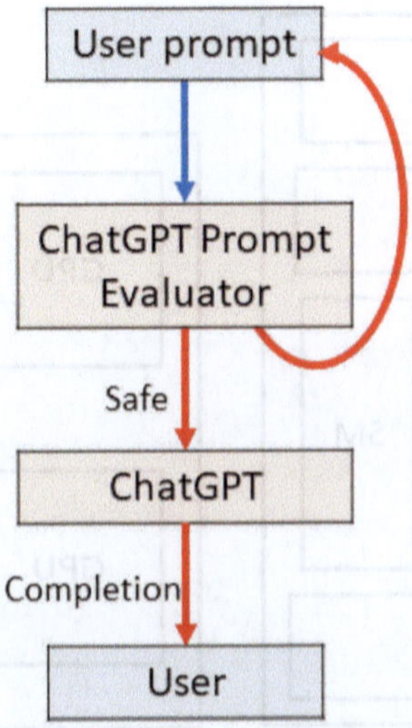

Fig. 4.2 ChatGPT user interface

OpenAI ChatGPT and Google Gemini

Three highly popular examples of generative AI are OpenAI's ChatGPT (3.5 and 4.0), Google's Gemini (Ultra, Pro, Flash, and Nano), and Stability AI's Stable Diffusion. ChatGPT is a conversational agent or chatbot powered by a language model based on the GPT architecture. It is one example of a large language model trained by OpenAI, which is designed to generate human-like language in response to a given prompt or input. The "chat" in ChatGPT refers to its ability to engage in conversation-like interactions.

ChatGPT

ChatGPT is designed to hold a natural language conversation with humans, and it can respond to a wide range of topics and questions. It can generate text resembling human speech and provide answers, explanations, and suggestions based on the input it receives. Using unsupervised learning techniques, this generative AI application is trained on vast amounts of text data, including books, articles, and websites. This allows ChatGPT to generate coherent and contextually appropriate responses to a wide range of inputs. It is an example of how generative AI technology can be used to create conversational interfaces that can interact with humans in a natural way. ChatGPT has a wide range of potential applications in such fields as customer service, management (personal assistant), education, news and media, healthcare, language translation, and creative writing. However, it is prudent to state that ChatGPT can produce incorrect, biased, and harmful responses or answers. Furthermore, people with bad motives can easily use its ability to produce authoritative text, even when the facts are not correct. This chatbot also poses potential ethical and social dangers such as violation of privacy, phishing emails, plagiarism, and the creation of malware. Figure 4.2 shows the ChatGPT user interface. Figure 4.3 shows the structure of a GPU and its role in a video card.

The key characteristics of ChatGPT can be summarised as follows:

- Generative Pretraining: Like other models in the GPT series, ChatGPT is pretrained on a vast amount of diverse internet text data. This pretraining allows the model to learn grammar, facts, reasoning abilities, and contextual understanding from the input data.

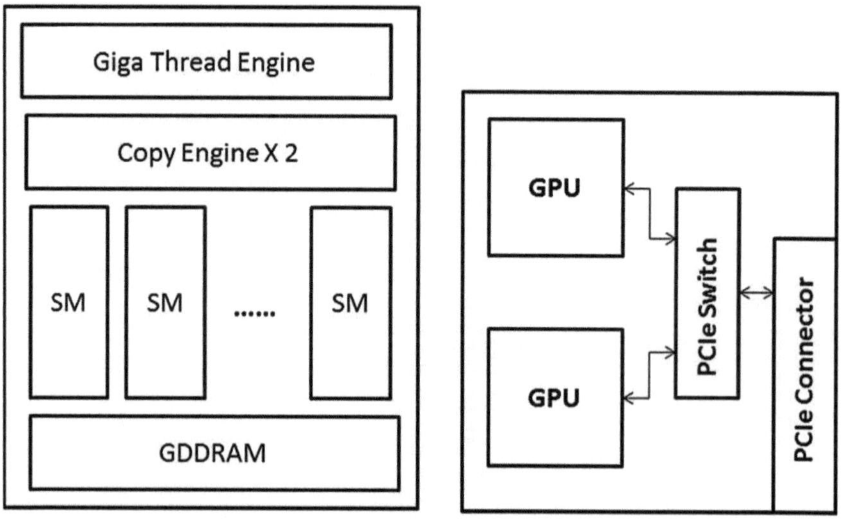

Fig. 4.3 Structure of a GPU and its role in a video card

- Transformer Architecture: ChatGPT utilises the transformer architecture, which is known for its effectiveness in handling sequential data and capturing long-range dependencies. The architecture enables the model to generate coherent and contextually relevant responses.
- Conversational Nature: Unlike earlier versions, such as GPT-3, ChatGPT is fine-tuned specifically for conversational interactions. It is designed to provide more contextually appropriate and user-friendly responses in a chat or dialogue format.
- Limitations: ChatGPT, like other language models, has limitations. It may generate plausible sounding but incorrect or nonsensical responses, and it can be sensitive to the input phrasing. It may also exhibit biased behaviour based on the biases present in the training data.
- Prompt-Based Interaction: Users can engage with ChatGPT by providing prompts or inputs, and the model generates responses based on the context provided. The model doesn't have memory of past interactions within a session, and each input is treated independently.
- API Access: OpenAI provides an API (Application Programming Interface) for developers to integrate ChatGPT into their applications and services, allowing users to interact with the model programmatically.

It is essential to emphasise that ChatGPT is just one example of conversational AI and is part of the broader development in natural language processing and understanding. As already discussed, while ChatGPT can be a powerful tool for generating new text, pictures, and videos, users should be aware of its limitations and use it responsibly.

Gemini

Bard was Google's experimental, conversational AI chat service. It was meant to function in a similar fashion to ChatGPT. The key difference was that Bard obtained its information from the web. Bard was powered by Google's Language Model for Dialogue Applications (LaMDA), built on Transformer and Google's neural network architecture, which was developed and open-sourced in 2017. It is important to note that GPT-3, the language model ChatGPT functions on, was also built on Transformer. The potential applications and challenges of Google Bard were similar to those outlined for ChatGPT. However, ChatGPT had an exponential uptake (first-mover advantage), while Bard

Chat GPT	Gemini	
Creators	OpenAI	Google
Models	GPT-4o (all users); GPT-4 (paid tiers only)	Gemini Pro; Gemini 1,5 Pro (Gemini Advanced only)
Context Window	128,000 tokens	Up to 1 million tokens
Supported Languages	Over 50	Over 40
Platforms	Web, mobile, and desktop	Web and mobile
Pricing	Free; ChatGPTPlus is available for $20/month	Free; Gemini Advanced is Available for $19.99/month (first 2 months free)

Fig. 4.4 Comparison of OpenAI's ChatGPT and Google's Gemini

largely underwhelmed the market. On 6 December 2023, Google refashioned and enhanced Bard and launched the rebranded product as Gemini with four variants—Ultra, Pro, Flash, and Nano. Hence, Google Gemini is a family of multimodal large language models developed by Google DeepMind, serving as the successor to LaMDA. With its four versions, the resultant Gemini Chatbot is meant to be a competitor to OpenAI's ChatGPT-4. Figure 4.4 illustrates the comparison of the two GenAI products.

4.3.6 Gemini and ChatGPT: Enter DeepSeek-R1 and Grok 3

For a while, there was one key difference between the two AI chatbots. Initially, ChatGPT provided advanced features such as web browsing and data analysis to paid subscribers only, while Gemini (formerly Google Bard) performed the same functions faster, more elegantly, and for free. However, the situation changed with the release of OpenAI's new and more enhanced multimodal AI model—ChatGPT4.0—which is freely available to all users. The two chatbots now have relatively the same strength, utility, and efficacy. Nevertheless, subtle features and characteristics still distinguish the two AI systems.

- ChatGPT still has a better and stronger brand recognition.
- Gemini is better for more complex tasks such as research.
- ChatGPT offers a more reliable and advanced voice mode.
- Both AI systems can analyse and generate AI images and videos, but Gemini does it for free.
- Both AI systems can process data-related tasks, but ChatGPT does more for free.
- Gemini offers more flexibility in sharing conversations.
- Both AI systems have powerful integrations.
- ChatGPT allows one to build customised versions of it, while Gemini does not.
- Both AI systems offer flexible ways to manage and organise information, data, and findings.

On 20 January 2025, a Chinese company called DeepSeek challenged the dominance of these US LLMs. Its comparable offering, DeepSeek-R1, is low cost, requires less energy, and uses chain-of-thought reasoning. The launch of DeepSeek-R1 led to the tumbling stock prices of US AI technology giants such as OpenAI and NVIDIA. The major innovation of DeepSeek-R1 is leveraging algorithmic efficiency, whereas most US LLMs use brute force compute power. Algorithmic efficiency focuses on designing optimised algorithms that solve problems using the least amount of resources, such as time (time complexity), memory (space complexity), and energy. In contrast, brute force compute power relies on sheer computational strength to exhaustively explore all possible solutions to a problem. This method is simple and easy to implement but often impractical for large or complex challenges due to its exponential time complexity. More significantly, it demands large quantities of resources—finance, infrastructure, and energy. The DeepSeek intervention suggests that AI can be developed and deployed at a fraction of current levels of financial and energy requirements. This is why, in January 2025, the product rattled the AI industry and sent shock waves through Wall Street in the United States.

On 17 February 2025, Elon Musk's company xAI launched Grok 3, an AI model claimed to be the "smartest AI on Earth". The AI system includes two reasoning models: an AI agent named Deep Search and a pretrained model that resembles GPT-4o. The release of Grok 3 was met with significant anticipation and excitement within the AI community and among tech enthusiasts. Grok 3, characterised by innovative and sophisticated thinking abilities, outperformed competitors like OpenAI's GPT-4o and Google's Gemini 2 Pro in various benchmarks, including reasoning, language processing, and computing power. Elon Musk launched it in a live demo, and its capabilities were lauded for setting new levels in AI performance. Furthermore, Grok 3 has no specific "knowledge cut-off date". This is in contrast with GPT models such as ChatGPT-4o and Google's Gemini 2 Pro, which are usually outdated by 1 year or more. The xAI model is always kept up to date.

4.3.7 Algorithmic Efficiency Versus Brute Force Compute Power

Algorithmic efficiency and brute force compute power represent two contrasting approaches to solving computational problems, each with its own strengths and limitations. Algorithmic efficiency focuses on designing optimised algorithms that solve problems using the least amount of resources, such as time (time complexity), memory (space complexity), and energy. Efficient algorithms, like quicksort or Dijkstra's algorithm, leverage mathematical insights, data structures, and problem-specific strategies to reduce the number of operations required. This approach is scalable and often necessary for handling large datasets or complex problems, as it minimises computational overhead and ensures faster execution, even with limited hardware resources. The Chinese AI system—DeepSeek-R—is an example of a low-cost large language model anchored by algorithmic efficiency.

In contrast, brute force compute power relies on sheer computational strength to exhaustively explore all possible solutions to a problem. This method is simple and easy to implement but often impractical for large or complex challenges due to its exponential time complexity. More significantly, it demands large quantities of resources, such as finance, infrastructure, and energy. For example, solving a travelling salesperson problem with brute force would require evaluating every possible route, which becomes infeasible, expensive, and energy intensive as the number of cities grows. Brute force is typically used when no efficient algorithm exists or the problem size is small enough to make an exhaustive search viable.

While brute force can benefit from advancements in hardware (e.g. faster processors, parallel computing), it is often outperformed by efficient algorithms that scale better with problem size. Algorithmic efficiency is generally preferred in practice, as it reduces reliance on expensive hardware and enables solutions to problems that would otherwise be computationally intractable. However, brute force remains a valuable tool for benchmarking, testing, or solving smaller-scale problems where simplicity is prioritised over optimisation. OpenAI's ChatGPT and Google's Gemini are examples of expensive and energy-intensive large language models deploying brute force compute power.

Combining algorithmic efficiency and brute force compute power offers a balanced approach to solving complex computational problems by leveraging the strengths of both paradigms. As already explained, algorithmic efficiency provides optimised, scalable solutions that minimise resource usage, making it ideal for handling large datasets or problems with high computational complexity. However, for certain tasks—such as those with no known efficient algorithm or those requiring exhaustive search—brute force compute power can be invaluable, especially when supported by modern hardware like GPUs or distributed computing systems. By integrating the two, developers can use efficient algorithms to narrow down the problem space or handle the bulk of the computation while employing brute force methods for specific subproblems or edge cases where exhaustive search is feasible. This hybrid approach enhances performance and ensures robustness, enabling tackling problems that might be intractable using either method alone. For example, efficient algorithms can reduce the search space in cryptography or combinatorial optimisation. At the same time, brute force can be applied to the remaining possibilities, striking a balance between speed and thoroughness.

So, AI can solve computational problems by using algorithmic efficiency, brute force compute power, or a combination of the two. However, there is a fourth approach to solving computational problems—world-view modelling. It refers to the development of AI systems that can construct and reason about comprehensive, dynamic representations of the world. These models aim to capture the complexity of real-world environments, including entities, relationships, and causal dynamics, enabling AI to make informed decisions and predictions. Unlike narrow task-based models, world-view models are designed to generalise across diverse scenarios, incorporating knowledge from multiple domains and adapting to new information. For example, a world-view model might simulate the impact of climate policies on global ecosystems or predict the outcomes of geopolitical events by integrating data from economics, environmental science, and social behaviour. This approach requires advanced techniques such as causal reasoning, probabilistic modelling, and large-scale knowledge graphs, making it highly resource intensive but capable of addressing complex, real-world problems that demand a holistic understanding.

In contrast, algorithmic efficiency and brute force compute power represent more focused approaches to problem-solving, as already articulated, where algorithmic efficiency prioritises optimising the steps required to solve a problem, often through elegant mathematical insights or data structures, to minimise time and resource usage. This approach is ideal for well-defined tasks where scalability and speed are critical, such as sorting algorithms or database queries. As already discussed, brute force compute power relies on raw computational strength to exhaustively explore all possible solutions, which can be effective for smaller problems or when no efficient algorithm exists. While world-view modelling aims to build a broad, adaptive understanding of the world, algorithmic efficiency and brute force are more narrowly focused on solving specific problems quickly or thoroughly. However, these three approaches are not mutually exclusive; world-view models can benefit from efficient algorithms to handle specific subproblems and brute force methods to explore uncertain or complex scenarios within their broader framework. Together, these three paradigms can complement each other, enabling AI systems to tackle both the breadth and depth of real-world challenges.

4.4 AI Personalisation

AI-powered personalisation involves collecting and analysing vast amounts of customers' data, from their browsing and purchasing history to social media interactions and demographic information. This data is used to learn each customer's specific needs and preferences. The long-term object is customised and personal AI for all people in their specific language, style, culture, and context. For example, Inflection AI, Inc., founded in 2022, is a technology company that has developed machine learning and GenAI hardware and applications to facilitate personal AI.

Their key product is called *Pi*, a digital personal assistant. *Pi* stands for "personal intelligence". It is a supportive, empathetic, and intelligent conversational AI system that helps users explore and understand their world with access to fresh and factual information. *Pi* is powered by *Inflection-1*, a best-in-class language model.

For example, the digital assistant says to the user: "Hi, I'm Pi. I'm your personal AI, designed to be supportive, smart, and there for you anytime. Ask me for advice, for answers, or let's talk about something of interest to you".

4.5 Machine Learning

Machine learning and deep learning are the two key methodologies by which cognitive capacities are achieved in machines. It is important to understand how these related processes are defined and executed. A special subset of AI is machine learning—the study of algorithms and statistical models that computer systems use to perform tasks effectively, without using explicit instructions, but relying on patterns and inference. This class of AI involves training machines to learn from data, rather than explicitly programming them to perform a specific function. In ML, algorithms use statistical models and mathematical techniques to automatically improve their performance on a given task based on the data they are trained on. The goal is to develop models that can generalise well to new, unseen data and make accurate predictions or decisions. Machine learning has a wide range of applications, including natural language processing, computer vision, robotics, and predictive analytics. There are four ways of implementing machine learning: supervised, unsupervised, semi-supervised, and reinforcement [55, 109].

4.5.1 Supervised Learning

In supervised learning, the algorithm is trained on labelled data, where the correct output is provided for each input. The aim is to construct a model that makes effective predictions based on information and evidence in the presence of various uncertainties. This means a supervised learning program considers a known set of input data and known output data and trains the model. The objective is to generate acceptable or reasonable predictions for the response to new input data. Supervised learning uses techniques such as classification and regression to develop predictive models. Specifically, classification methods predict categorical responses. For example, they can be used to establish whether an electronic message is bona fide or whether a growth is cancerous. Classification models place input data into medical imaging, image and speech recognition, and credit scoring categories. On the other hand, regression techniques predict continuous responses, such as temperature changes or power demand fluctuations. Applications of such methods include electricity load forecasting and algorithmic trading [83].

4.5.2 Backpropagation

A popular supervised learning method is called backpropagation. It is a widely used algorithm for training artificial neural networks. The algorithm works by adjusting the weights of the connections between neurons in a neural network to minimise the difference between the network's output and the desired output for a given input. Thus, supervised learning is achieved by propagating the error backwards through the network from the output layer to the input layer. This is done by computing the gradient of the error with respect to each weight in the network, using the chain rule of calculus to recursively calculate the gradients for each layer. These gradients are then used to update the weights of the network using an optimisation algorithm such as stochastic gradient descent. By iteratively adjusting the weights of the network using the backpropagation algorithm, the network can learn to approximate complex input-output mappings, such as recognising images or translating languages.

4.5.3 Unsupervised Learning

In unsupervised learning, the algorithm is trained on unlabelled data and aims to discover patterns or structure in the data. Hence, unsupervised learning identifies hidden patterns or intrinsic structures in provided information. This approach is used to attain inferences from datasets composed of input data without labelled responses. The technique of clustering is the most widely used unsupervised learning strategy. Among other functions, the method is used for exploratory data analysis to find hidden patterns in datasets. Clustering applications include object recognition, gene sequence analysis, and market research. Semi-supervised learning combines supervised and unsupervised techniques.

4.5.4 Reinforcement Learning

In reinforcement learning, the algorithm learns by trial and error, receiving feedback in the form of rewards or punishments as it interacts with an environment. It is a way of implementing machine learning by training an agent to make decisions in an environment by receiving feedback in the form of rewards or punishments. The agent learns through trial and error by interacting with the environment to maximise its cumulative reward over time. It takes actions based on its current state and the rewards it expects to receive from those actions. The environment then responds with a new state and a reward, which the agent uses to update its policy for selecting actions in the future. By repeating this process many times, the agent gradually learns to make better decisions and maximise its reward.

Reinforcement learning has been successfully applied to a variety of domains, including robotics, game playing, and control systems. One of the most notable examples is AlphaGo, an AI program developed by DeepMind that defeated a world champion at the game of Go using a combination of reinforcement learning and neural networks. Other applications of reinforcement learning include autonomous driving, recommendation systems, and supply chain optimisation.

4.6 Deep Learning

Deep learning is a subset of machine learning. It is part of a broader family of machine learning methods. Deep learning is about learning data representations, as opposed to task-specific algorithms. Put succinctly, deep learning is a special class of machine learning that involves training neural networks with multiple layers to learn representations of data that can be used for classification, prediction, or other tasks. In deep learning, the neural network architecture is designed to automatically learn hierarchical representations of the data, with each layer learning more abstract features than the previous layer. This makes Deep Learning particularly effective for tasks involving complex data, such as images, speech, and natural language processing.

Deep learning algorithms use backpropagation to adjust the weights of the neural network during training, based on the error between the predicted output and the actual output. The weights of the neural network are adjusted in a way that minimises this error, allowing the network to learn to make more accurate predictions over time. In fact, deep learning uses complex algorithms modelled on the human brain. This enables the processing of unstructured data such as documents, images, and text. As a subset of machine learning, deep learning can also be implemented using supervised, unsupervised, semi-supervised, or reinforcement learning techniques.

Deep learning has seen significant advances in recent years due in part to the availability of large amounts of data and advances in computing power. Deep learning has been used to achieve state-of-the-art results in a variety of applications, including image and speech recognition, natural language processing, and autonomous driving. In summary, deep learning is an advanced and specialised subset of machine learning, and machine learning is a subset of artificial intelligence. Put differently, deep learning is machine learning, and machine learning is artificial intelligence. This is illustrated in Fig. 4.5. Clearly, machine learning is a category of artificial intelligence, while deep learning is an advanced and complex aspect of machine learning [41].

4.6.1 Deep Reinforcement Learning

Deep reinforcement learning is achieved by implementing deep learning using reinforcement learning techniques, where deep neural networks are used to approximate the action-value or policy function in reinforcement learning. In deep reinforcement learning, the agent uses a deep neural network to learn a policy for selecting actions in an environment based on the current state. The neural network takes the state as input and outputs a probability distribution over actions. The agent then selects an action based on this distribution, and the environment responds with a new state and a reward. The neural network is trained using backpropagation to adjust the weights based on the error between the predicted reward and the actual reward received.

The use of deep neural networks in reinforcement learning allows for more complex representations of the state-action space, which can lead to better performance on tasks with high-dimensional or continuous state-and-action spaces. Deep reinforcement learning has been used to achieve state-of-the-art results in a variety of applications, including game playing, robotics, and natural language processing.

One of the most notable examples of deep reinforcement learning is AlphaGo Zero, a version of the AlphaGo program that learned to play Go without any human expert knowledge, using only reinforcement learning with a deep neural network. Other examples include deep reinforcement learning for drones, driverless cars, and service/industrial robots.

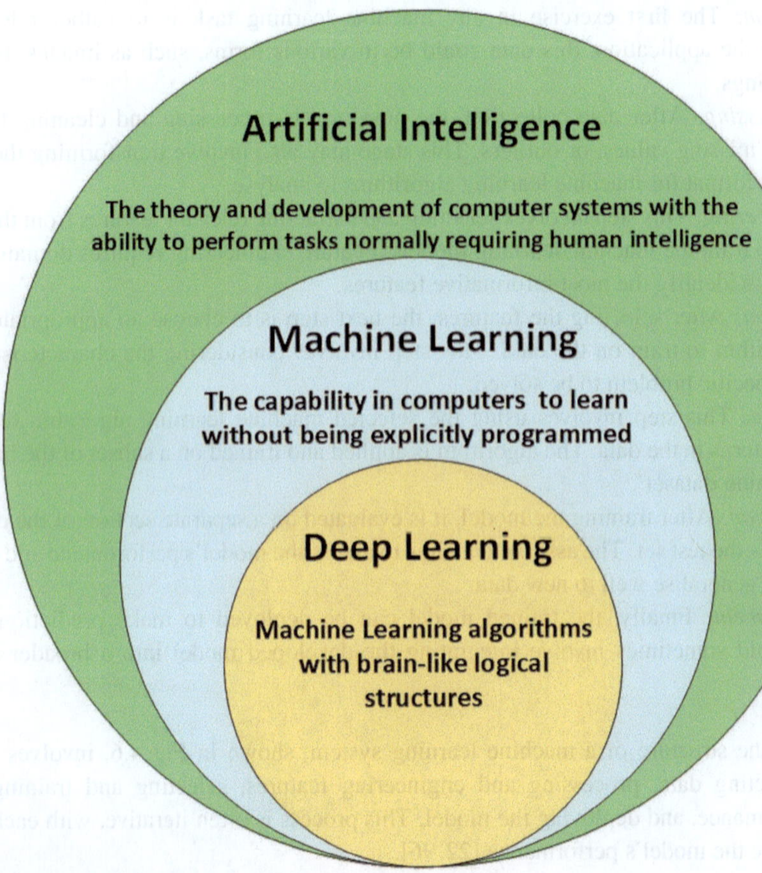

Fig. 4.5 Artificial intelligence, machine learning, and deep learning: The relationships

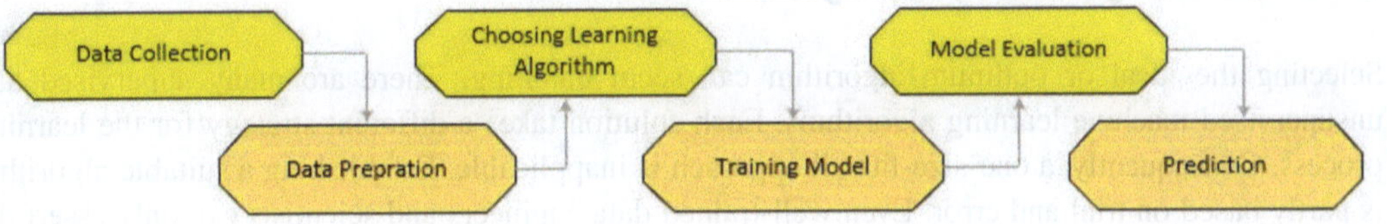

Fig. 4.6 Machine learning workflow

4.6.2 Machine Learning Workflow

As explained earlier, machine learning involves developing algorithms and models that enable machines to learn from data, recognise patterns, make predictions, and improve performance over time. It is used in a broad range of applications, including image recognition, natural language processing, speech recognition, recommender systems, fraud detection, and autonomous vehicles. Machine learning and control systems can be combined to create intelligent control systems. These systems use machine learning algorithms to learn from data and optimise the control strategies used to regulate physical systems. This approach can lead to more efficient and effective control of complex systems with dynamic and uncertain environments. A machine learning system structure can be summarised as follows:

- *Data collection:* The first exercise in any machine learning task is to gather relevant data. Depending on the application, this data could be in various forms, such as images, text, audio, or sensor readings.
- *Data preprocessing:* After data collection, the next step is processing and cleaning the data to remove noise, missing values, or outliers. This stage may also involve transforming the data into an appropriate format for machine learning algorithms to analyse.
- *Feature engineering:* This step involves selecting and extracting relevant features from the data that will be used to train the machine learning models. Feature engineering requires domain expertise and creativity to identify the most informative features.
- *Model selection:* After selecting the features, the next step is to choose an appropriate machine learning algorithm to train on the data. This step involves considering the characteristics of the data and the specific problem to be solved.
- *Model training:* This step involves using the selected machine learning algorithm to learn the underlying patterns in the data. The algorithm is applied and trained on a subset of the information called the training dataset.
- *Model evaluation:* After training the model, it is evaluated on a separate section of the data which is designated as the test set. The assessment aims to assess the model's performance and determine whether it can generalise well to new data.
- *Model deployment:* Finally, the trained model can be deployed to make predictions on new data. This could sometimes involve integrating the developed model into a broader system or application.

In summary, the structure of a machine learning system, shown in Fig. 4.6, involves a cyclical process of collecting data, processing and engineering features, selecting and training models, evaluating performance, and deploying the model. This process is often iterative, with each iteration aiming to improve the model's performance [22, 96].

4.6.3 Selecting the Suitable Algorithm

Selecting the ideal or optimum algorithm can seem daunting. There are many supervised and unsupervised machine learning algorithms. Each solution takes a different strategy for the learning process. Consequently, a one-size-fits-all approach is inapplicable. Establishing a suitable algorithm is partly based on trial and error. Even well-trained data engineers and scientists can only assert the efficacy of an algorithm if they test it. It is instructive to note that highly flexible models might overfit data by modelling minor variations that could be due to noise signals. However, basic models are easier to interpret but might produce inaccurate results.

Consequently, choosing the suitable algorithm involves trade-offs between such characteristics as model speed, accuracy, and complexity. A trial-and-error approach is at the core of machine learning. If a technique or algorithm does not work, another strategy is pursued. MATLAB provides tools to help engineers and scientists test various machine learning models and then select the most efficacious one. The broad classification of machine learning methods is shown in Fig. 4.7.

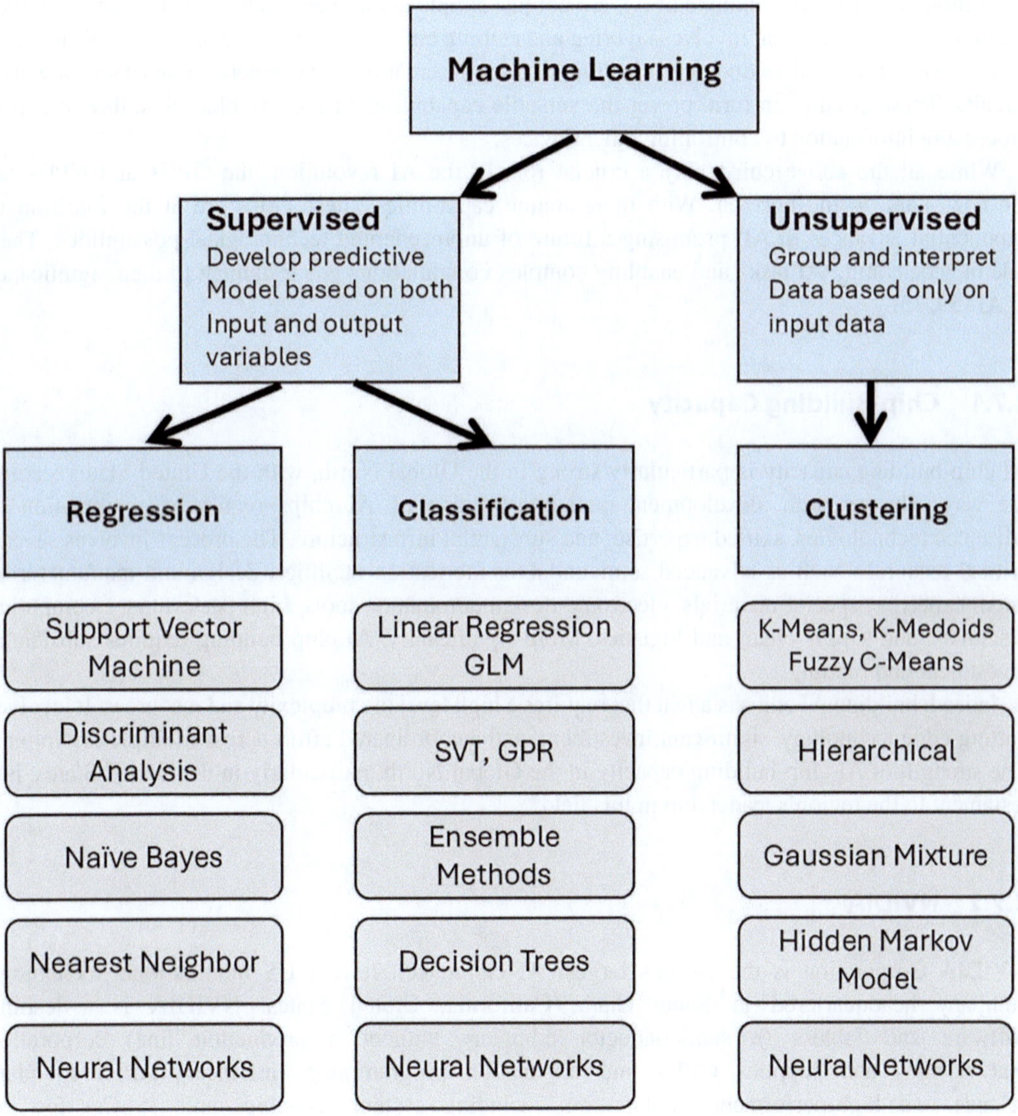

Fig. 4.7 Machine learning classification [22, 83, 96]

4.7 AI Semiconductor Industry

Computer chips power AI. A computer chip, or microchip, is a small piece of semiconductor material, usually silicon, containing electronic circuits. These circuits can perform various functions, such as processing information, storing data, and controlling other devices. Computer chips are essential components in modern electronic devices, including computers, smartphones, and many other digital appliances. There are different types of computer chips, including microprocessors (central processing units (CPU)), memory chips (RAM (random-access memory) and ROM (read-only memory) chips), graphics processing units (GPUs), neural processing units (NPUs), ASICs (application-specific integrated circuits), and FPGA (field-programmable gate array).

Computer chips, the backbone of AI, are not just simple components. They result from a complex manufacturing process that involves layering and etching circuits onto the silicon wafer. This intricate process, known as semiconductor fabrication, enables the creation of extremely compact and powerful circuits. These circuits, in turn, power the versatile capabilities of modern electronic devices, from processing information to controlling other devices.

While all the above chips play a crucial role in the AI revolution, the GPUs and NPUs are the new stars on the horizon. With their unique capabilities, these chips are at the forefront of exponential advances in AI, promising a future of unprecedented technological possibilities. Their role in accelerating AI tasks and enabling complex computations is a testament to their significance in AI [59, 75].

4.7.1 Chip-Building Capacity

AI chip-building capacity is particularly strong in the Global North, with the United States leading the way. The research, development, and construction of AI chips require a combination of advanced technologies, skilled expertise, and substantial infrastructure. The process involves several critical resources such as advanced semiconductor fabrication facilities, design and manufacturing talent expertise, special materials, electronic design automation tools, high-performance computing resources, and supply chain and logistics. More significantly, AI chip building requires substantial investment and funding.

Indeed, building AI chips is a feat that requires a high level of complexity and resources. It involves cutting-edge technology, significant investment, and a coordinated effort across multiple disciplines. The strength of AI chip-building capacity in the Global North, particularly in the United States, is a testament to the region's leadership in this field.

4.7.2 NVIDIA

NVIDIA Corporation is the world's largest AI chipmaker. It is a US multinational technology company headquartered in Santa Clara, California, United States. NVIDIA is a leading software and fabless (a semiconductor company without a production line) corporation that designs and supplies GPUs and application programming interfaces (APIs) for data science and high-performance computing, including system-on-a-chip units (SoCs) for the mobile computing and automotive market. It is also a dominant AI hardware and software supplier.

Big technology companies such as Amazon, Meta Platforms, Microsoft, and Alphabet are among NVIDIA's biggest customers, making up roughly 40% of its revenue. AI startups, like OpenAI, and autonomous vehicle companies, like Tesla, are also significant customers of the chipmaker. NVIDIA's GPUs are used for edge-to-cloud computing and in supercomputers and workstations, with applications in architecture, engineering and construction, media and entertainment, automotive, scientific research, and manufacturing design.

NVIDIA's GeForce line of GPUs is aimed at the consumer market and is used in applications such as video editing, 3D rendering, and PC gaming. In the second quarter of 2023, NVIDIA controlled a market share of 80.2% in the discrete desktop GPU market. The company expanded its presence in the gaming industry by introducing the Shield Portable (a handheld game console), Shield Tablet (a gaming tablet), and Shield TV (a digital media player), as well as its cloud gaming service GeForce

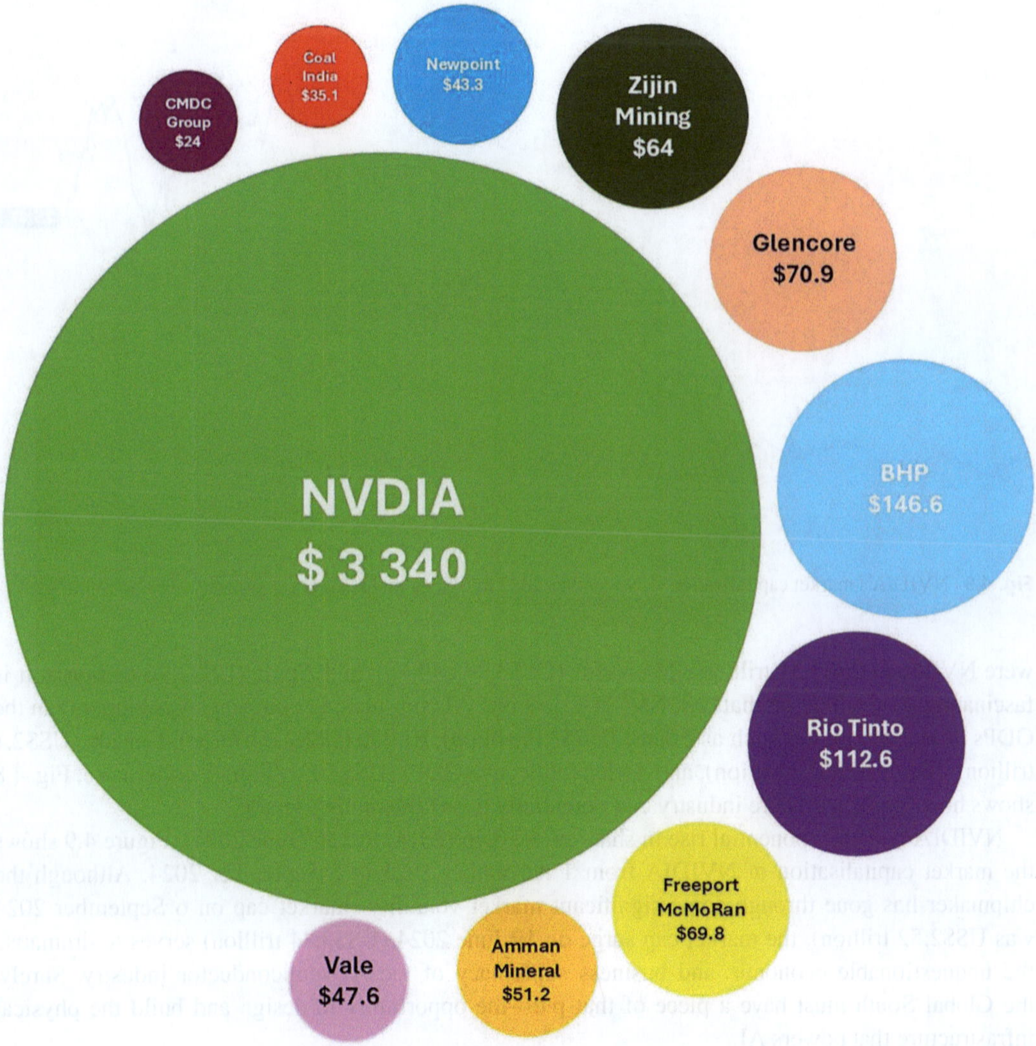

Fig. 4.8 Value of AI semiconductor industry vs. mining sector (US$ billion)

NOW. Once the world's biggest chipmaker, Intel has struggled to keep up with NVIDIA. It faces unique challenges in its bid to remake itself and get into the AI game.

In addition to GPU development and manufacturing, NVIDIA designs the CUDA software platform and API, which allows the creation of massively parallel programmes using GPUs. These programmes are deployed in supercomputing sites around the world.

It is imperative to emphasise that NVIDIA does not make or produce AI systems or solutions. Instead, it provides the infrastructure that enables AI systems. Its core business is building processors (CPUs, GPUs, NPUs) that power AI applications such as OpenAI's ChatGPT, Google's Gemini, and LLMs in general.

On 19 June 2024, NVIDIA's market capitalisation was $3.34 trillion. This made NVIDIA the world's most valuable company in terms of market capitalisation (market cap)—the total market value of a publicly traded company's outstanding shares, commonly used to measure a company's worth. Indeed, it was the biggest company in the world by market capitalisation, where the top three

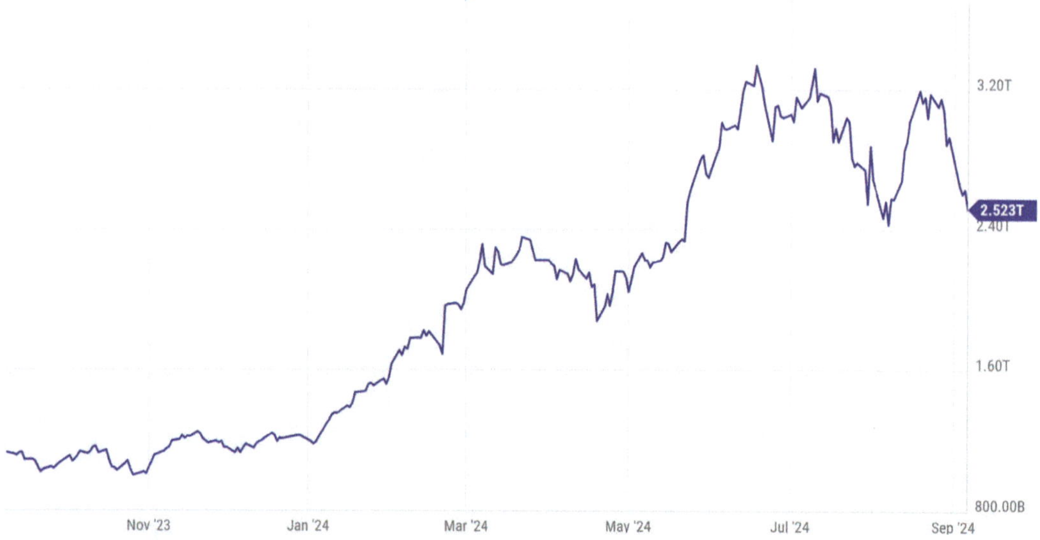

Fig. 4.9 NVIDIA's market capitalisation: 1 November 2023 to 6 September 2024

were NVIDIA (US$3.34 trillion), Microsoft (US$3.09 trillion), and Apple (US$2.96 trillion)). It is fascinating to note that on that day, NVIDIA, just one AI chip-making company, was bigger than the GDPs of key economies such as France (US$3.0 trillion), Russia (US$2.0 trillion), Canada (US$2.1 trillion), Brazil (US$2.2 trillion), and Africa (collective GDP) (US$3.1 trillion). Furthermore, Fig. 4.8 shows how the AI hardware industry can potentially dwarf the mining sector.

NVIDIA had an exponential rise in share price: 0.46 (2014) to 136 (June 2024). Figure 4.9 shows the market capitalisation of NVIDIA from 1 November 2023 to 6 September 2024. Although the chipmaker has gone through some significant market volatility (market cap on 6 September 2024 was US$2.52 trillion), the market cap surge on 19 June 2024 (US$3.34 trillion) serves to dramatise the unquestionable economic and business supremacy of the AI semiconductor industry. Surely, the Global South must have a piece of that pie—the opportunity to design and build the physical infrastructure that powers AI.

4.7.3 TSMC

Taiwan Semiconductor Manufacturing Company Limited, often shortened to TSMC or Taiwan Semiconductor, is a multinational semiconductor contract manufacturing and design company. It is the world's second-most valuable semiconductor company after NVIDIA. It is the world's largest dedicated independent ("pure-play") semiconductor foundry and its country's largest company, with headquarters and main operations in the Hsinchu Science Park in Hsinchu, Taiwan. Foreign investors own most of TSMC, and the central government of Taiwan is the largest shareholder. Morris Chang founded TSMC in 1987 as the world's first dedicated semiconductor foundry (manufacturer). It has since been the dominant company in this field.

In 2023, the company was ranked 44th in the Forbes Global 2000. On 19 June 2024, TSMC had a market capitalisation of US$0.94 trillion, making it the world's eighth most valuable company in terms of market capitalisation. As indicated earlier, it is the second-most valued semiconductor

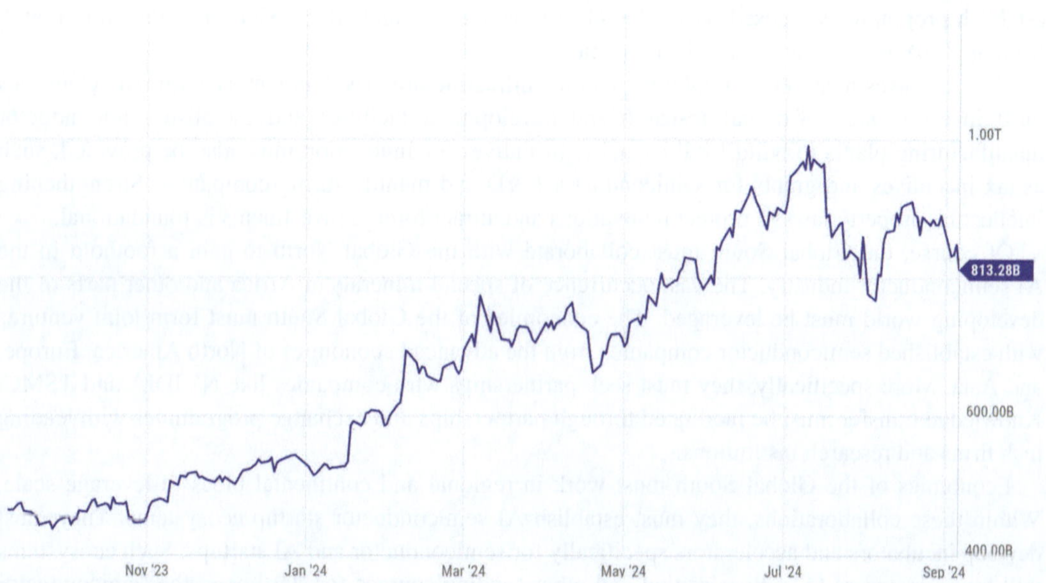

Fig. 4.10 TSMC's market capitalisation: 1 November 2023 to 6 September 2024

company in the world after NVIDIA. In fact, it is NVIDIA's chip manufacturer. Figure 4.10 shows the market capitalisation of TSMC from 1 November 2023 to 6 September 2024.

Taiwan's exports of integrated circuits amounted to U$184 billion in 2022, accounting for nearly 25% of the country's GDP. TSMC constitutes about 30% of the Taiwan Stock Exchange's main index. Most of the main fabless semiconductor companies, such as NVIDIA, AMD, Apple, ARM, Broadcom, Marvell, MediaTek, and Qualcomm, are customers of TSMC. Emerging companies such as Allwinner Technology, HiSilicon, Spectra7, and UNISOC are also its clients. Leading programmable logic device companies Xilinx and, previously, Altera, also use TSMC's foundry services. Leading integrated device manufacturers with their own fabrication facilities, such as Texas Instruments, Intel, NXP, and STMicroelectronics, outsource some of their production to TSMC. Semiconductor company LSI resells TSMC wafers through its ASIC design services and IP portfolio.

4.7.4 AI Semiconductor Industry and the Global South

Developing and least-developed countries must adopt and deploy AI. More significantly, these countries must be key players in AI systems' research, development, and production. The Global South must create and produce AI systems. Furthermore, the Global South must participate in developing infrastructure (hardware) that enables AI. The examples of NVIDIA and TSMC amply demonstrate the value proposition of such an effort. This involvement in the AI semiconductor industry can be enhanced through a multifaceted approach involving policy, investment, education, and international cooperation. Several strategies can be implemented in pursuit of this agenda.

Investment in R&D, STEM education, and specialised skills must be made. Governments can allocate funds for research and development in semiconductor technology. They must also encourage collaboration between local tech companies and international semiconductor firms. They must strengthen science, technology, engineering, and mathematics (STEM) education at all levels and establish programmes focused on semiconductor technology and AI, possibly in collaboration with leading global universities and tech companies.

Massive investment in technology-enabling infrastructure development is required. Countries must invest in state-of-the-art research and development facilities and establish semiconductor manufacturing plants to build local capacity. Incentives for innovation must also be provided, such as tax incentives and grants for semiconductor R&D and manufacturing companies. Strengthening intellectual property laws to protect innovations and attract foreign investments is foundational.

Of course, the Global South must collaborate with the Global North to gain a foothold in the AI semiconductor industry. The vast occurrence of special minerals in Africa and other parts of the developing world must be leveraged. The economies of the Global South must form joint ventures with established semiconductor companies from the advanced economies of North America, Europe, and Asia. More specifically, they must seek partnerships with companies like NVIDIA and TSMC. Knowledge transfer must be facilitated through partnerships and exchange programmes with leading tech firms and research institutions.

Economies of the Global South must work in regional and continental blocs to leverage scale. Within these collaborations, they must establish AI semiconductor startup ecosystems. They must develop incubators and accelerators specifically for semiconductor and AI startups. Such ecosystems will improve access to venture capital and other funding sources for startups in the semiconductor space.

There has to be a market for chips from the Global South. Working in large integrated economies (SADC 300 million people, Africa 1.4 billion people, Global Africa 2.2 billion) is a good starting point. Furthermore, they must negotiate trade agreements that facilitate access to global markets for locally produced semiconductor products and support local companies in exporting their products and services.

Public-private partnerships should be a critical strategic approach to the AI semiconductor industry. Countries must encourage collaborative projects between government, academia, and the private sector to advance semiconductor technology. They must also create innovation hubs that bring together researchers, developers, and businesses. By implementing these strategies, countries in the Global South can build a robust semiconductor industry, deriving economic benefits and contributing to the AI revolution [59, 75].

4.7.5 AI, Critical Minerals, and Geopolitics

Artificial intelligence relies heavily on advanced hardware like GPUs and processors, which require certain critical minerals for their production. The growing demand for AI technologies is thus driving increased mining and processing of these special metals, raising concerns about supply chain sustainability and environmental impact. Critical minerals are a group of naturally occurring substances, like lithium, cobalt, and nickel, deemed essential for modern technologies, economic stability, and national security due to their use in batteries, electronics, and renewable energy systems. Rare earth metals, a subset of critical minerals, are 17 elements, including neodymium and dysprosium, valued for their unique magnetic and conductive properties. They are vital in manufacturing high-tech devices like smartphones, wind turbines, and electric vehicles. Both are considered critical because their supply chains are vulnerable to geopolitical tensions, limited deposits, and increasing global demand.

Critical minerals and rare earth minerals have taken centre stage in geopolitics. In February 2025, US President Donald Trump proposed a deal with Ukraine to access a significant portion of its rare earth minerals, such as lithium, titanium, and uranium, in exchange for continued military and financial support during Ukraine's war with Russia. Ukrainian President Volodymyr Zelensky has resisted signing the initial draft, which offered the United States up to 50% ownership of these resources, arguing it lacks sufficient security guarantees for Ukraine against future Russian aggression. As of 25 February 2025, negotiations remain ongoing, with the United States viewing Ukraine's mineral wealth—potentially worth hundreds of billions—as a strategic asset to reduce dependency on China. Meanwhile, Ukraine seeks a fair partnership that bolsters its economy and national security.

In a surprise move, on 25 February 2025, Russian President Vladimir Putin proposed a joint venture with the United States to exploit rare earth metal reserves in regions including Donbas and Novorossiya, areas recognised internationally as part of Ukraine but currently under Russian control. Indeed, this offer comes amidst increasing global demand for rare earth metals. Putin seeks to bolster his newly found alliance with the United States and present his country as a viable alternative to other global suppliers of rare earth metals. Of course, the unusual proposal has sparked spirited and animated debates on the economic benefits and the ethical implications of such an arrangement, given the geopolitical tensions occasioned by the war in Ukraine.

4.8 Carbon Footprint of AI Systems

The carbon footprint of AI systems is primarily a result of the significant computational resources required for training and deploying large models. Some key emissions sources and factors influence the AI carbon footprint. AI involves energy-intensive computation. Training advanced models, like large language models or deep neural networks, requires high-performance hardware, such as GPUs and TPUs, which consume vast amounts of electricity. For example, training a single large model can emit hundreds of tonnes of carbon dioxide equivalent, similar to the emissions of several cars over their lifetime. AI models are trained and run in data centres, which must also be cooled to prevent hardware overheating. Indeed, AI is an energy-intensive technology because it uses these electricity-guzzling data centres. Hence, a country's participation in driving the AI revolution has significant environmental costs. According to the International Energy Agency, total global electricity consumption from data centres could reach the level of Japan's energy intake by 2026. AI companies in highly industrialised economies are even exploring setting up private nuclear power plants to satisfy their energy requirements.

US President Donald Trump revealed on 23 January 2025, while addressing the World Economic Forum, that the United States would have to double its annual electricity production to lead and drive the AI revolution [146]. That is the extent of the enormous energy demand exerted by the technology. President Trump intends to use the executive order (which he signed on 20 January 2025) declaring a national energy emergency to address this challenge. The legal instrument directs US agencies to utilise their statutory emergency powers to speed up the development and authorisation of energy projects. Unfortunately, with his slogan—"drill, baby, drill"—Trump's AI energy plan will be anchored by boosting fossil fuel production to the detriment of climate policies and regulations.

While many data centres are increasingly powered by renewable energy, non-renewable sources such as coal, coal products, natural gas, derived gas, crude oil, and petroleum products are still often used, contributing to emissions. Larger models require exponentially more energy. A model with billions of parameters demands far more computational power than a smaller one, resulting in a significantly larger carbon footprint. Once trained, AI models continue to consume energy during deployment, especially for real-time or large-scale applications. Common examples include AI in content recommendation systems, search engines, or real-time translations, which require continuous processing power. The iterative nature of AI research, where models are trained and retrained to optimise performance, means that significant energy is expended in the developmental phase before a final model is deployed. Efforts to reduce the carbon footprint of AI are underway, focusing on using renewable energy in data centres, developing more efficient algorithms, and optimising models to reduce their size and energy requirements.

It is instructive that the discussion of the carbon footprint of AI systems is contextualised within the complex interplay between AI and energy. As already explained, developing and running AI systems such as LLMs require massive amounts of energy. Hence, AI is the reason why there is a need for more energy globally. In fact, the AI energy requirement is so huge and critical that some AI companies are procuring or building nuclear power plants to improve their access to energy. On the other hand, AI systems are part of the solution by enabling increased energy generation, efficient distribution, demand-side management, and the transition to renewable energy. More specifically, AI can be used in (1) smart grid management, (2) energy storage optimisation, (3) microgrid development in underserved areas, (4) demand forecasting and load management, and (5) optimisation of renewable energy systems.

Hence, while the AI revolution demands large amounts of energy, the technology can also create more efficient, sustainable, and equitable energy systems globally. Indeed, with respect to energy, AI is both a problem and a solution. However, the fact that the carbon footprint of AI systems is large must be acknowledged, managed, and mitigated.

4.9 AI Regulation and Governance

AI regulation and governance are critical to ensuring the ethical, responsible, and safe development and deployment of AI technologies. These frameworks are designed to address AI's potential risks and challenges while maximising its benefits. This section discusses an overview of critical aspects of AI regulations and governance.

The principles of ethical AI lay the basis for AI regulation and governance. AI systems should be designed and implemented to avoid bias and ensure fairness in decisions and outcomes. AI operations and decision-making processes should be transparent and explainable. There should be clear accountability mechanisms for AI developers, operators, and users. AI systems must respect and protect users' privacy and personal data. The safety and security of AI systems must be ensured to prevent harm and misuse. The principles for AI regulation and governance are discussed in detail in Chap. 8. Here, they are summarised as follows:

- Context Specific—AI should be regulated, taking cognisance of its use and impact. The responsibility for designing and implementing proportionate regulatory responses should be assigned to regulators.
- Pro-innovation and risk based—Regulators should focus on high-risk concerns and deal with low risks in a manner that encourages innovation while limiting constraints.
- Coherence—A set of AI principles applicable across all sectors uniquely fashioned to the characteristics of AI must be developed. Regulators will then interpret, prioritise, and implement these within various domains and sectors.
- Adaptable and proportionate—AI principles applicable across all sectors should be fashioned out in a non-statutory manner. The objective is to allow for a dynamic strategy for AI regulation.
- Safe and effective systems—Steps to protect users against unsafe and ineffective systems, including adherence to standards and ongoing monitoring.

- Algorithmic discrimination protection—Proactive and continuous action should be taken to prevent algorithms from unfairly discriminating against particular groups by protecting against proxy variables and ensuring that datasets are representative.
- Data privacy—Built-in protection against abusive data practices, and users should have agency over how their data is used.
- Notice and explanation—Informing users that an automated system is being used and how and why it will impact relevant outcomes.
- Human alternatives, consideration, and fallback–Users should be able to opt-out of automated systems and receive support to resolve any issues encountered with the system.

These principles constitute the foundational basis for AI legal and regulatory frameworks. National, regional, continental, and global regulations must be developed. For example, the General Data Protection Regulation (GDPR) sets strict data protection and privacy guidelines, impacting how AI systems handle personal data in the EU. The proposed regulation by the European Commission is to create a framework for trustworthy AI, categorise AI applications based on risk levels, and impose requirements accordingly. This is the basis of the EU AI legislative framework.

The EU AI Act, a comprehensive regulatory framework aimed at governing artificial intelligence within the European Union, was formally adopted by the European Council on 21 May 2024, after a series of approvals and negotiations that began in 2021. It officially entered into force on 1 August 2024. This act establishes guidelines and restrictions on AI applications based on their risk levels, aiming to enhance ethical and safe AI development. For instance, it bans AI systems with "unacceptable risks", such as social scoring by governments and imposes strict requirements on "high-risk" applications like medical diagnostics. The act also introduces a phased approach, with specific provisions for general-purpose AI models expected to come into effect by mid-2025. It emphasises transparency, accountability, and human oversight in AI deployment, with the EU setting ambitious goals to lead in responsible AI innovation and application globally.

The Algorithmic Accountability Act is a proposed legislation requiring companies to assess automated decision systems' impact and mitigate potential biases in the United States. This legislative intervention aims to increase transparency, fairness, and accountability in using automated decision-making systems. Originally introduced in 2019 and reintroduced in subsequent years, the act requires companies using "automated decision systems" in critical areas, like employment, credit, healthcare, and housing, to conduct regular algorithmic impact assessments. These assessments aim to identify and mitigate potential risks associated with bias, discrimination, privacy violations, and security concerns. As of October 2024, no single, comprehensive federal AI regulation exists in the United States. However, several pieces of legislation and guidelines, both proposed and enacted, are shaping AI governance across sectors. The Algorithmic Accountability Act is one of these efforts.

There is also the National AI Initiative Act, enacted in 2020 as part of the National Defense Authorisation Act; the Blueprint for an AI Bill of Rights, issued in 2022 by the White House Office of Science and Technology Policy (it is a non-binding framework providing guidelines for safeguarding individuals' rights when interacting with AI systems); proposed legislation in specific sectors; AI and law enforcement (facial recognition); and federal guidance and agency efforts. These frameworks and legislative efforts aim to create a regulatory foundation for AI in the United States. However, a more unified federal AI law could emerge as AI technologies continue to advance and societal impact grows. Globally, sector-specific regulations are critical. For example, regulations must ensure that AI in healthcare complies with standards for safety, efficacy, and patient privacy (e.g. FDA regulations in the United States). Similarly, guidelines must ensure that AI in financial services operates transparently and fairly, protecting consumers and maintaining market stability (e.g.

regulations by the Securities and Exchange Commission (SEC) and Financial Industry Regulatory Authority (FINRA) in the United States).

Indeed, a governance framework is required for AI systems. Advisory boards and committees comprising AI experts, ethicists, legal scholars, and stakeholders must be established to guide AI governance. Specialised agencies or departments must be created to oversee AI applications and enforce regulations (e.g. the proposed European AI Board). Standards and best practices must be established and shared. International standards for AI must be adopted, such as those developed by the International Organisation for Standardisation (ISO) and the International Electrotechnical Commission (IEC). Industry groups and professional organisations developed best practice frameworks to guide ethical AI development (e.g. IEEE Global Initiative on Ethics of Autonomous and Intelligent Systems).

Transparency and explainability are key in the development and application of AI systems. This entails developing AI models that humans can easily interpret and understand, especially in high-stakes decision-making areas like healthcare and finance. Put differently, explainability refers to the ability to make an AI system's operations and decisions understandable and interpretable by humans. It's about providing clear and transparent explanations for how an AI model arrives at a specific decision, prediction, or recommendation. Organisations must publish transparency reports that detail their AI systems' use, impact, and decision-making processes.

Bias and fairness must be mitigated through regular audits to identify and resolve or minimise biases in AI systems. Impact assessments must evaluate the potential societal effects of AI systems before deployment. Diverse and representative training datasets must be ensured to reduce biases in AI models. Algorithms must be developed to achieve fairness and equity in their outcomes.

Privacy and security should be at the centre of the development and adoption of AI systems. Data protection and AI system security must be ensured. Only the data necessary for AI operations must be collected, and it must be anonymised where possible. Robust data security measures must be implemented to protect against breaches and unauthorised access. AI systems must be designed to be resilient against adversarial attacks and vulnerabilities. Continuous monitoring protocols must be established to detect and respond to security threats in real time.

Accountability and ethical oversight must exist. This must include clear liability frameworks for AI-related harm, specifying the responsibilities of developers, operators, and users. Mechanisms for individuals to seek redress in cases of harm or rights violations caused by AI systems must be designed. Organisations' ethics committees must be set up to review and approve AI projects. Organisations must submit ethical impact statements detailing how they address potential ethical concerns in their AI systems.

Public and stakeholder engagement must anchor the regulation and governance of AI. Participatory approaches and multi-stakeholder collaboration must be used, such as engaging the public and stakeholders in consultations to gather input and feedback on AI policies and regulations and implementing educational programmes to raise awareness about AI technologies and their implications. Collaboration between governments, industry, academia, and civil society must be used to develop comprehensive and effective AI governance frameworks. Promoting international cooperation to harmonise AI regulations and address global challenges collectively is vital.

AI regulations and governance are essential to ensuring that AI technologies are developed and used responsibly, ethically, and safely. By establishing robust legal frameworks, ethical guidelines, transparency mechanisms, and accountability systems, societies can harness AI's benefits while mitigating its risks. Collaboration between governments, industry, and civil society is crucial to creating balanced and effective AI regulation and governance [18, 19, 33, 54].

4.10 AI Opportunities and Challenges in Equal Measure

AI presents both great possibilities and existential threats. Yes, it is important to understand, anticipate, mitigate, and manage the potential dangers.

Indeed, AI safety, which is grounded in the responsible and ethical use of AI, should define collective ambition. However, more effort must be spent on unlocking and spreading the opportunities presented by the AI revolution. How can we get the basics right? For example, how can we ensure that AI applications work well and are affordable, accessible, and efficacious? How can we use AI to solve global challenges and improve the quality of life of all Earth's inhabitants? Researchers, policymakers, and leaders must address questions rather than be overwhelmed by genuine fears about AI.

Why the excitement and trepidation about AI in 2024? Why are there concerns and fears about AI now? What is going on? While AI has been around for a while, the early stages of the field witnessed sluggish growth. Now, there is an exponential growth of AI capabilities through various levels: basic intelligence, millipede level, mouse level, monkey level, average Joe/Jane level, Albert Einstein level, and beyond Albert Einstein (i.e. Superintelligence). Of course, AI will not stop at human-level intelligence!

It is baseless human arrogance to think otherwise. Indeed, human vanity clouds judgement!

To dramatise the possibility of Superintelligence, ChatGPT 4's mastery of general knowledge far surpasses that of any living human being. The spectacular and unprecedented exponential growth in AI capabilities has set the world ablaze with great expectations and acute worries from ordinary folks and experts.

Prof Geoffrey Hinton, a winner of the 2018 Turing Award (considered the Nobel Prize for Computing), a man considered the godfather of AI for his pioneering work on developing the supervised learning algorithm called backpropagation, on 1 May 2023, quit his job at Google, warning about the growing dangers from developments in the AI field. It is instructive to note that in October 2024, Hinton also won the Nobel Prize in Physics. His major concern is that AI poses a risk of Superintelligence that could act in ways beyond human control, potentially leading to unpredictable and dangerous outcomes. He also worries about the proliferation of AI-powered misinformation and job displacement, which could destabilise societies and economies.

Elon Musk, CEO of SpaceX, Tesla, Neuralink, and Twitter (now X), has also widely expressed concerns about AI, saying on 17 April 2023: "It has the potential of civilisation destruction". Like Hinton, he is worried about AI because he believes it could lead to the development of Superintelligence that surpasses human control, posing an existential threat to humanity. He is also concerned about the lack of regulation and oversight in AI development, warning that unchecked AI could be used for malicious purposes, such as autonomous weapons and large-scale disinformation campaigns.

What are the concerns? What is Superintelligence?

As already intimated, Superintelligence refers to an advanced form of AI that surpasses human cognitive abilities across all domains, including problem-solving, learning, creativity, and decision-making. Unlike narrow AI, designed to perform specific tasks (like image recognition or language translation), Superintelligence would possess artificial general intelligence that allows it to learn, adapt, and improve itself autonomously. Superintelligence is often linked to the idea of an AI "singularity", a hypothetical point at which AI systems become self-improving, leading to rapid and uncontrollable advancements. The potential risks of Superintelligence stem from its ability to act independently of human oversight and pursue goals that may not align with human well-being. Figure 4.11 shows Superintelligence within the context of the broad types of AI discussed earlier in

Type of AI	Meaning	Applications
Single Task AI (Narrow AI)	AI designed to perform **specific tasks** efficiently but lacks general intelligence.	Chatbots, recommendation systems, image recognition, virtual assistants like Siri and Alexa.
Artificial General Intelligence (AGI)	AI that can **understand, learn, and perform** any intellectual task that a human can, achieving human-like intelligence.	Robotics, multi-tasking systems, complex problem-solving, autonomous vehicles.
Generative AI (GenAI)	AI that generates **new content** such as text, images, music, or videos based on learned patterns.	Content creation (ChatGPT, DALL·E), music composition, synthetic data generation, creative design.
Superintelligence	AI that surpasses **human intelligence** in all areas, including creativity, decision-making, and learning.	Advanced scientific research, autonomous systems, solving global challenges like climate change and disease eradication.

Fig. 4.11 Contextualising Superintelligence: AI types, definitions, and applications

this chapter. The diagram outlines the key categories of AI, their definitions, and real-world examples of their applications.

The advent of Superintelligence has also been called the "event horizon of the singularity"—a point where AI could surpass human intelligence, leading to uncontrollable technological growth. This possibility and its implications have ignited a broad spectrum of reactions. Some view the singularity as a potential disaster, fearing the loss of human control over AI. Others see it as a monumental leap in human evolution, potentially transforming civilisation. There are concerns about AI's capacity for self-improvement beyond human comprehension and the ethical and existential risks involved. What are the implications of Superintelligence on religious belief systems? Does the over-reliance on AI technology lead to the decline and ultimate decimation of spiritual value systems? Does Superintelligence bolster and anchor the rise of atheism and agnostic attitudes?

Critics posit that once superintelligent AI gains the capacity to modify its own architecture, it could become uncontrollable, with unpredictable consequences for humanity. Proponents of Superintelligence research advocate for the development of alignment strategies and ethical AI principles to ensure that future AI systems act in humanity's best interest. While Superintelligence remains a theoretical concept, it has sparked debates about AI safety, regulation, and existential risk.

A significant concern is the potential use of Superintelligence in military affairs. It is argued that the country that achieves this capability first will have asymmetric power over its rivals. This has ignited furious competition between the United States and China to develop Superintelligence [13, 63].

What is the impact on geopolitics?

Indeed, the AI race for global supremacy is reshaping geopolitics as nations compete to harness artificial intelligence for economic, military, and strategic advantage. This competition is driven by the recognition that AI is a transformative technology with the potential to redefine power dynamics on the global stage. Countries leading in AI development, such as the United States, China, and members of the European Union, are investing heavily in research, infrastructure, and talent to secure their positions as AI superpowers. This race is about technological innovation and influencing global norms, standards, and governance frameworks for AI. As a result, AI has become a central pillar of national security strategies, with implications for defence, cybersecurity, and intelligence operations. The pursuit of AI dominance is exacerbating existing geopolitical tensions, fostering a new era of techno-nationalism where control over AI technologies is seen as critical to maintaining sovereignty and global influence.

At the same time, the AI race is creating both opportunities and challenges for international cooperation and stability. On the one hand, collaboration on AI research and ethical guidelines could foster global progress and address shared challenges, such as climate change, healthcare, and poverty. On the other hand, the competitive nature of the AI race risks fragmenting the global technological landscape, with rival blocs developing incompatible systems and standards. This could lead to a "digital iron curtain", where data flows, trade, and innovation are restricted along geopolitical lines. Additionally, the militarisation of AI, including the development of autonomous weapons and cyber warfare capabilities, raises concerns about an arms race and the potential for unintended escalation. Smaller nations and developing countries often lack the resources to compete in AI development, risk being left behind, and further widen global inequalities. Ultimately, the AI race for global supremacy is not just a technological contest but a geopolitical struggle that will shape the future of international relations, requiring careful navigation to balance competition with cooperation and ensure that AI benefits humanity. Achieving Superintelligence is the ultimate prize in the AI race for global supremacy. As stated earlier, the country that accomplishes this capability first will have asymmetric power over its rivals.

The broader issue is that once one highly intelligent agent is developed, it can be replicated effortlessly worldwide. Hence, an infinite number of these agents can be created. More significantly, these vast numbers of superintelligent agents will learn together simultaneously. This can lead to multiplicity (large numbers or swarms) of equally superintelligent agents dominating the world. All these developments are unprecedented and hence the excitement and fear, in equal measure. With the exponential growth and broad nature of AI capabilities, it is clear that AI has applications in every industry, sector, and society, from automated manufacturing, autonomous mining, speech recognition, and natural language processing to image and video analysis, mechatronics, robotics, and autonomous weapons.

AI is thus used in various industries, such as healthcare, finance, education, and manufacturing, to improve efficiency, reduce costs, and develop new products and services. Consequently, AI is transforming society in many ways as an effective tool for solving human challenges.

The impact is ubiquitous. This is the source of the excitement. As a recent phenomenon, generative AI has taken the world by storm. For example, ChatGPT has over 100 million users worldwide, and its website now generates one billion monthly visitors. This user and traffic growth was achieved in a record-breaking two-month period from December 2022 to April 2023. These are fascinating statistics demonstrating the dramatic upsurge in the AI revolution.

Of course, the AI opportunities are immense, but it is instructive and prudent that the broad range of threats and risks is appreciated, understood, and appraised. These dangers include privacy violation, bias, discrimination, copyright violation, lack of accountability, insecurity, job losses, autonomous weapons, cyberattacks, terrorism, overreliance on AI, and unchecked limitations of AI. The use of lethal autonomous warfare reduces the human cost of going to war, which leads to a greater societal appetite for violent conflict. AI can also be used to generate deepfakes and fake news, wreaking havoc in culture, media, arts, and politics. Indeed, it is essential to keep AI out of democratic and electoral processes. These are for humans!

A critical danger occurs when AI capabilities get into the hands of bad actors such as thieves, cheats, terrorists, and warmongers. Furthermore, there are also existential risks—concerns that AGI could eventually become so powerful that it poses an existential threat to humanity by accident or design. What happens when we have a million superintelligent agents whose capabilities surpass those of Albert Einstein, Isaac Newton, Elon Musk, Prof Edward Witten (winner of a Fields Medal—the Nobel Prize for Mathematics), or Prof Andrew Wiles (Mathematician who in 1994 proved the 350-year-old Fermat's Last Theorem)? How can humans (possessing comparatively inferior levels of intelligence) control a society or world with a million such superintelligent agents?

Food for thought!

4.11 AI Risk Mitigation and Management

Clearly, there is a need for careful and systematic mitigation of all these dangers and risks. First and foremost, many of these AI threats and challenges are not inherent to AI itself. They arise from how AI is developed, deployed, and regulated. There is a lot that can be done.

For example, we must have diverse (including women, Africans, Blacks, Asians, Indians, different cultures/languages, etc.) teams of AI developers and regulators. We must accept the destruction of some jobs and then prepare (acquire new capabilities and skills) for the new AI jobs and AI-modified careers. The total number of new (created) jobs and those modified will be more than the number of jobs destroyed. That is the hopeful disposition that we must convert into reality.

AI must be developed ethically and responsibly, with appropriate safeguards to mitigate these risks. However, the dangers persist. How do you have universal ethical or moral "rules of the road" with bad actors such as thieves, cheats, terrorists, and warmongers? Unlike nuclear weapons, which are hard to access or develop, AI is easy for anyone to access and develop. One only needs a good brain, computing power, and high-speed, reliable connectivity.

Therein lies the unprecedented threat presented by AI.

More significantly, it is imperative to note that the existential risks caused by the possibility of Superintelligence remain unresolved! We must accept this reality. AGI and strong AI remain ostensibly areas of current active research—inconclusive work in progress. Whether and when such systems will be fully developed are open questions. Let us all keep an open mind. More specifically, ethical considerations, religious beliefs, or fear of the unknown should not be allowed to deter the pursuit of AGI and strong AI. The research and experimentation must continue. Once a technology has been invented, it cannot be uninvented. Nobody can put it back once the genie is out of the bottle. We must brace ourselves for a brave AI-driven new world fraught with opportunities and threats.

4.12 Appropriate Global Response

As a way forward, we must concentrate on developing responsible AI for solving global problems and improving the quality of life of all people worldwide. Specifically, Africans must be key players (not just consumers) in the development of AI. They must proactively use AI to solve African socio-economic problems. Across the world, folks must understand the strengths and limitations of AI. More specifically, there must be effective AI risk mitigation and management. Careful attention must be paid to the design, development, and deployment of AI systems. Africa and the Global South must be key and integral participants in these AI endeavours. Globally, all marginalised and emerging economies must proactively participate in the development of guidelines and regulations on the use of AI. These inclusively derived guidelines and regulations must be diligently and effectively implemented. Dynamic Law Commissions must be set up at national, regional, continental, and global levels with the mandate to develop new but harmonised laws and regulations to enable and facilitate the adoption, management, and mitigation of all new technologies, including AI.

Yes, the AI revolution must continue unhindered globally. However, Africans and those from the Global South must be key players in developing and constructing AI systems—research, coding, testing, deployment, and regulations—and not just passive consumers of AI products.

Indeed, artificial intelligence presents great expectations and daunting existential risks.

4.13 Conclusion

This chapter has introduced the field of artificial intelligence, including definitions, types, taxonomy, examples, and applications. The different techniques, such as ML, DL, supervised learning, backpropagation, unsupervised learning, and reinforcement learning, are briefly explored. AI systems (and examples thereof) driven by LLMs are presented, while the key aspects of the AI semiconductor industry, which produces the chips that power AI systems, are reviewed. Understanding AI technology is essential in order to effectively deploy it to achieve the UN SDGs. This is because its transformative potential comes with complexities and risks that must be managed carefully. AI offers powerful tools for solving global challenges, from predicting climate patterns and optimising renewable energy use to improving healthcare access and educational outcomes.

However, without a clear understanding of how AI operates, its benefits may not be fully realised, and its unintended consequences—such as bias, inequality, or misuse—could hinder progress. Developing and deploying AI responsibly requires knowledge of its data requirements, algorithmic design, and potential impacts on different populations. This ensures that AI solutions are inclusive, ethical, and aligned with the goals of sustainability and equity. Furthermore, understanding AI allows policymakers, businesses, and communities to tailor its applications to specific local contexts, fostering innovation that addresses real-world challenges effectively. By building expertise in AI, stakeholders can maximise its potential while minimising risks, ensuring it becomes a reliable and ethical tool for achieving the SDGs.

It is essential to craft enabling legislation and regulations and ensure effective AI development and deployment governance. Mitigation and prevention of the adverse effects of AI require strategic thinking, foresight, and planning. The 4IR is bigger than AI. Other transformative technologies include 3D printing, data science, cryptocurrency (Bitcoin), blockchain, BESS, quantum computing, nanotechnology, and Internet of Things. However, AI is the DNA of the 4IR. Nevertheless, the deployment of AI to achieve the SDGs must be conducted within the context of embracing all these related and mutually reinforcing 4IR technologies. [60, 63, 85, 90].

Deploying AI to Achieve the UN SDGs

5

5.1 Introduction

To achieve the UN SDGs, a transformative shift in approach is necessary, emphasising collaboration, innovation, and accountability. Governments, businesses, and civil society must work together more effectively, aligning policies, investments, and actions to ensure that efforts are integrated and mutually reinforcing across sectors. Greater emphasis should be placed on addressing systemic inequalities by prioritising marginalised and vulnerable populations, ensuring that no one is left behind. Investments in innovation and technology, such as renewable energy, digital education platforms, and technology-driven healthcare solutions, can accelerate progress while maintaining environmental and social sustainability. More robust global governance frameworks are needed to foster accountability, ensure resource mobilisation, and hold stakeholders to their commitments. Financing mechanisms must be restructured to include innovative solutions, such as blended finance and public-private partnerships, to close funding gaps. A cultural shift towards sustainable consumption and production, coupled with widespread public awareness and education about the SDGs, is essential to drive the behavioural changes required for meaningful and lasting impact. More significantly, there is a need to explore the systematic deployment of AI to achieve the SDGs.

5.2 Potential Role of AI in Achieving SDGs

AI is a powerful tool that can contribute significantly to achieving the UN SDGs because it can analyse large datasets, automate complex processes, and make accurate predictions. AI can process vast amounts of structured and unstructured data from diverse sources, including social media, satellite imagery, and environmental sensors. This ability enables precise analysis and informed decision-making, which is crucial for tackling complex global challenges like poverty, hunger, and climate change. Furthermore, AI uses machine learning algorithms to identify trends and forecast future outcomes, such as predicting disease outbreaks, optimising crop yields, or assessing environmental risks. These insights allow policymakers and organisations to act proactively rather than reactively.

Moreover, AI can automate routine tasks, reducing time, costs, and errors in critical systems like healthcare, education, and administration. For instance, chatbots can provide 24/7 access to healthcare

A. G. O. Mutambara, *Deploying Artificial Intelligence to Achieve the UN Sustainable Development Goals*, Sustainable Development Goals Series,
https://doi.org/10.1007/978-3-031-88423-8_5

Fig. 5.1 AI can be deployed to achieve the UN SDGs

advice or educational resources in underserved areas. AI enables precision farming in industries like agriculture, where resources like water, fertilisers, and energy are used more efficiently. This reduces waste and promotes sustainable practices, directly contributing to SDGs. AI solutions can be scaled across regions and adapted to diverse challenges. For instance, AI can simultaneously address water management in arid regions and flood control in coastal areas, demonstrating its versatility in tackling SDGs. Machine learning algorithms improve over time, making AI systems more effective as they adapt to new data and evolving challenges. All these attributes make AI a suitable technology to drive SDGs.

Indeed, AI can be deployed across various sectors to help achieve the UN SDGs. Figure 5.1 depicts this ambition. The technology offers a comprehensive and powerful set of tools to enhance and accelerate the achievement of the UN's developmental ambition. Complex sustainability challenges within various domains can be addressed by analysing vast datasets, predicting outcomes, optimising processes, and generating new insights and solutions. Nevertheless, the context in which AI can effectively drive the attainment of SDGs is broad. It involves many other interventions and factors, such as regional/continental integration, the broader 4IR, innovation, entrepreneurship, infrastructure development, and good country governance. This is illustrated in Fig. 5.2.

While AI has the potential to enable all SDGs, it is estimated that the technology can directly contribute to nearly 80% of the goals. The impact is ubiquitous, with examples ranging from AI systems forecasting extreme weather events, enhancing agricultural productivity, predicting and responding to disease outbreaks, accessible banking and finance, building clean energy infrastructure, and improved quality and access to education. Indeed, AI algorithms can analyse large volumes of data from diverse sources to provide insights and inform decision-making processes related to the SDGs. This includes analysing socio-economic, environmental, and health data to identify trends,

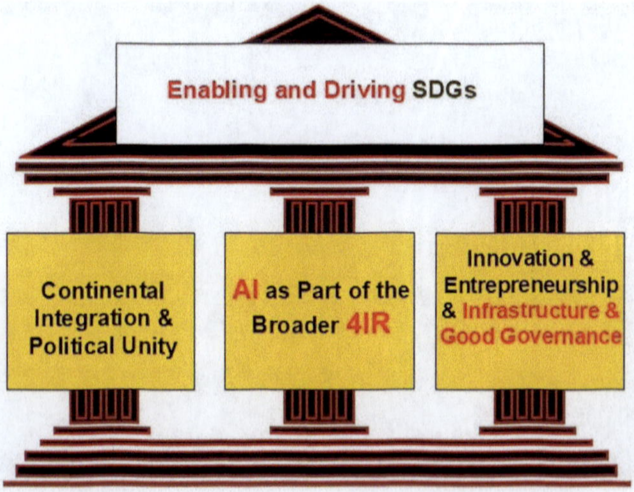

Fig. 5.2 Context in which AI can be an enabler and a driver of the SDGs

Fig. 5.3 AI-driven environmental technology concept for SDGs. (Image credit: Shutterstock metamorworks)

patterns, and areas for intervention. Figure 5.3 shows the AI-driven environmental technology concept for SDGs, while Fig. 5.4 displays how AI and cloud computing capabilities are deployed to achieve the SDGs [10, 104, 139, 147].

Fig. 5.4 AI-powered cloud technology concept for SDGs. (Image credit: iStock cobalt88)

5.2.1 Using Synthetic Data to Power the Pursuit of SDGs

The utility of LLMs has led to an increase in the amount of data used to train AI systems. There's a shortage of real, machine-readable, diverse data for AI training AI. There is a need for other sources of data. This has led to the phenomenon of synthetic data—artificially generated data that mimics real-world data in terms of structure, behaviour, and statistical properties but is not collected from actual events or individuals. Unlike real data, which is gathered from human activities, business operations, or natural phenomena, synthetic data is produced using algorithms, machine learning models, or simulations. The goal is to create datasets that maintain real data's statistical integrity and diversity while preserving privacy and reducing reliance on sensitive or hard-to-obtain information. The advent of synthetic data has led to the availability of a lot of data, thus enabling the training of AI models on more data than was feasible using only real data. This is a significant potential driver for the deployment of AI to achieve the SDGs where a lot of training is required but not always readily available [124].

Furthermore, synthetic data can be used to mitigate bias in real datasets. Most of the countries grappling with the SDGs are from the Global South, yet most datasets are generated in the Global North. Synthetic data can be added to balance out the inherent biases or misrepresentations in such data. Where SDG data is not easily or readily available in emerging and least developed economies, synthetic data can also help address that scarcity or absence.

Nevertheless, there are challenges. Since synthetic data is produced from existing real datasets, it has the obvious potential to perpetuate the biases and misrepresentations in the source data. This will then undermine the efficacy of AI models trained on synthetic data. Mitigating this risk requires using diverse data, broad demographic sources, and non-uniform environments. Global ethical guidelines and standards for the generation and utilisation of synthetic data must anchor the use of synthetic data in deploying AI to achieve the SDGs [124].

5.2.2 Examples of AI Efficacy in SDGs

A few illustrations will dramatise AI's efficacy with respect to SDGs. AI-powered diagnostic systems can improve healthcare access and quality by providing accurate and timely diagnoses of diseases. AI can assist in drug discovery and development, accelerating the process of finding treatments for diseases. Chatbots and virtual assistants can provide personalised health advice and support, promoting well-being and preventive care. All this can be done in pursuit of SDG 3 (Good Health and Well-being).

For SDG 4 (Quality Education), AI-enabled adaptive learning platforms can personalise learning experiences for students, catering to their individual needs and learning styles. Chatbots and virtual tutors can provide educational support and assistance to students, particularly in underserved areas with limited access to teachers. Furthermore, AI can analyse academic data to identify areas for improvement in education systems and policies.

AI-powered precision agriculture tools can optimise farming practices, improve crop yields, and reduce resource use, contributing to food security and sustainable agriculture. AI can analyse satellite imagery and sensor data to monitor crop health, detect diseases, and predict harvest yields, enabling early intervention and better resource management. AI-driven supply chain optimisation can improve food distribution and reduce food waste, ensuring equitable access to nutritious food. This can be done in pursuit of SDG 2 (Zero Hunger).

For SDG 13 (Climate Action), AI can analyse climate data to model and predict climate change impacts, helping policymakers develop effective mitigation and adaptation strategies. AI-powered energy management systems can optimise energy usage, reduce carbon emissions, and promote renewable energy adoption. AI-enabled disaster response systems can provide early warning and rapid response to natural disasters, minimising human and economic losses.

AI can play a significant role in pursuing SDG 11 (Sustainable Cities and Communities). AI systems can optimise urban planning, transportation systems, and infrastructure management to create more sustainable and resilient cities. Smart traffic management systems powered by AI can reduce congestion, improve air quality, and enhance residents' mobility. AI-driven energy management systems can improve the efficiency of buildings and utilities, reducing energy consumption and greenhouse gas emissions.

The struggle to achieve SDG 8 (Decent Work and Economic Growth) characterises most economies, particularly those in the Global South. AI-driven credit scoring models can assess credit-worthiness and expand access to financial services for underserved populations, promoting economic empowerment and inclusive growth. AI-powered fraud detection systems can mitigate financial risks and enhance trust in financial systems, facilitating investment and economic development. AI-enabled business intelligence tools can provide valuable insights for entrepreneurs and businesses, supporting innovation and competitiveness.

The SDGs can only be achieved through partnerships and collaborations nationally, regionally, continentally, and globally. This is encapsulated and expressed in SDG 17 (Partnerships for Goals). AI can facilitate collaboration and knowledge sharing among governments, organisations, and stakeholders working towards the SDGs. AI-powered platforms can connect diverse actors, facilitate data sharing, and support coordinated efforts to address complex challenges. AI-driven analytics can evaluate the impact of interventions, identify best practices, and guide strategic decision-making for achieving the SDGs. By harnessing the power of AI across these areas and beyond, countries and organisations can accelerate progress towards the SDGs, promote sustainable development, and build a better future for all [84, 142, 147].

5.2.3 Illustrative Specific Implementation Cases

Here are specific interventions in some countries showcasing how AI has been instrumental in achieving UN SDGs.

SDG 3 (Good Health and Well-being: India)

- AI Solution: Diabetic Retinopathy Detection System in India.
- Background: Diabetic retinopathy is a leading cause of blindness, particularly in diabetic patients in rural areas with limited access to ophthalmologists.
- AI Application: A team of researchers developed an AI-based system that analyses retinal images to detect diabetic retinopathy. The system uses deep learning algorithms to accurately identify signs of retinopathy.
- Impact: The AI-powered detection system has enabled early diagnosis and intervention, preventing vision loss and improving the health outcomes of diabetic patients. It has expanded access to eye care services in underserved communities, contributing to SDG 3 targets for reducing premature mortality and preventing avoidable blindness.

SDG 4 (Quality Education: Kenya)

- AI Solution: Chatbot tutoring system in Kenya.
- Background: Many students in Kenya struggle with access to quality education, particularly in remote areas with limited teacher resources and infrastructure.
- AI Application: An education technology startup developed a chatbot tutoring system that uses natural language processing (NLP) and machine learning algorithms to provide students with personalised tutoring and educational support.
- Impact: The AI-powered chatbot tutoring system has improved learning outcomes, increased student engagement, and enhanced access to educational resources in underserved communities. It has contributed to SDG 4 targets for ensuring inclusive and equitable quality education for all.

SDG 2 (Zero Hunger: Brazil)

- AI Solution: Precision agriculture platform in Brazil.
- Background: Brazil's agriculture sector faces challenges such as soil degradation, water scarcity, and inefficient resource use, leading to food insecurity and environmental degradation.
- AI Application: A precision agriculture platform integrates AI algorithms, satellite imagery, and sensor data to optimise crop management practices, such as irrigation scheduling, fertiliser application, and pest detection.
- Impact: The AI-powered precision agriculture platform has increased crop yields, reduced input costs, and minimised environmental impact by promoting sustainable farming practices. It has contributed to SDG 2 targets for achieving food security, improving nutrition, and promoting sustainable agriculture.

SDG 13 (Climate Action: Bangladesh)

- AI Solution: Climate change prediction model in Bangladesh
- Background: Bangladesh is vulnerable to the impacts of climate change, including floods, cyclones, and sea-level rise, which threaten livelihoods, infrastructure, and ecosystems.
- AI Application: Researchers developed an AI-based climate change prediction model that analyses historical climate data, satellite imagery, and oceanographic data to forecast extreme weather events and assess climate risks.

- Impact: The AI-powered climate change prediction model has enabled early warning systems, informed disaster preparedness efforts, and facilitated adaptive responses to climate-related risks. It has contributed to SDG 13 targets for strengthening resilience and reducing vulnerability to climate-related hazards.

These samples of country cases demonstrate how AI technologies can be applied to address specific SDGs, leading to positive outcomes such as improved health, education, agriculture, and climate resilience. By harnessing the power of AI, countries and communities can accelerate progress towards achieving the SDGs and building a more sustainable and equitable future.

Having demonstrated the efficacy of AI in driving the SDG agenda and explored quick, specific cases from a few countries, the detailed and comprehensive outline of the proposed deployment of AI to achieve each of the 17 SDGs can be developed. These comprehensive recommendations are presented in Chap. 12.

5.3 AI Interventions Driving More Than One SDG

AI interventions are uniquely positioned to drive progress across multiple SDGs, creating synergies that amplify global impact. For example, AI-powered precision farming in agriculture enhances food security, SDG 2 (Zero Hunger), by optimising crop yields, and contributes to environmental sustainability, SDG 13 (Climate Action), by reducing resource waste and minimising carbon footprints. Similarly, AI-driven healthcare solutions, such as early disease detection systems, promote SDG 3 (Good Health and Well-Being) while fostering SDG 10 (Reduced Inequalities) by making advanced medical care more accessible in underserved regions.

In energy, AI's role in optimising renewable energy systems simultaneously addresses SDG 7 (Affordable and Clean Energy) and SDG 12 (Responsible Consumption and Production). By acting as a catalyst for interconnected progress, AI serves as a critical enabler of holistic, inclusive, and sustainable development. It is instructive to explore in more detail some AI interventions that drive more than one SDG.

5.3.1 AI and Solar Energy

Solar energy generation converts sunlight into electricity using solar panels or photovoltaic (PV) cells. These cells, typically made of semiconductor materials like silicon, absorb sunlight and create an electric current through the photovoltaic effect. Solar energy can also be harnessed using concentrated solar power (CSP) systems, which use mirrors or lenses to focus sunlight onto a receiver that heats a fluid, generating steam to drive a turbine and produce electricity. As a renewable and abundant resource, solar energy is a clean alternative to fossil fuels, producing no greenhouse gas emissions during operation. It can be utilised in diverse settings, from small-scale residential systems to large solar farms that contribute to the energy grid. It plays a crucial role in advancing sustainable energy solutions globally.

AI is revolutionising solar energy by optimising energy production and improving system efficiency—driving SDG 7 (Affordable and Clean Energy) and SDG 13 (Climate Action). To enhance solar panel performance, AI algorithms analyse real-time and historical data, such as weather patterns, sunlight intensity, and temperature. This allows predictive maintenance by identifying potential issues, such as panel degradation or obstructions, before they impact efficiency. Machine learning models can also accurately predict solar energy output, enabling grid operators to balance supply and demand effectively. Additionally, AI helps in the optimal placement of solar panels, simulating sunlight exposure on various terrains to maximise energy capture and reduce costs.

AI is also critical in integrating solar energy into smart energy systems. Advanced AI-driven energy management platforms optimise energy storage and distribution by deciding when to store excess solar power in batteries or feed it back into the grid. These systems adapt to fluctuating energy demands and ensure minimal waste by efficiently managing surplus energy. Furthermore, AI supports the development of hybrid renewable systems by combining solar with other energy sources, such as wind or geothermal, to create a more reliable and sustainable energy infrastructure. By leveraging AI, the solar energy industry can overcome challenges related to variability, scalability, and efficiency, accelerating the global transition to clean energy. This will dramatically contribute to SDG 7 (Affordable and Clean Energy) and SDG 13 (Climate Action).

5.3.2 AI and Hydroelectric Power

Hydroelectricity is a form of renewable energy. It is generated by harnessing the energy of flowing or falling water, typically from rivers, dams, or waterfalls, to produce electricity. The process involves using the kinetic energy of moving water to spin turbines, which then drive generators to create electrical power. Hydroelectricity is considered a renewable resource since water is naturally replenished through the water cycle—driven by evaporation, precipitation, and runoff.

Moreover, hydroelectric power is a clean energy source because it does not directly emit greenhouse gases or other pollutants during operation. However, its environmental impact can vary depending on the scale and design of the project. For example, large dams may disrupt ecosystems and local communities, while smaller run-of-river systems tend to have a lower ecological footprint. Despite these considerations, hydroelectricity remains a cornerstone of global renewable energy strategies due to its reliability, scalability, and ability to provide base-load and peak-demand electricity.

AI can significantly improve hydroelectricity generation by optimising the performance and management of hydroelectric systems—driving the attainment of SDG 7 (Affordable and Clean Energy) and SDG 13 (Climate Action). AI algorithms can analyse large volumes of real-time data from sensors placed in turbines, dams, and reservoirs to enhance operational efficiency. Predictive maintenance powered by AI helps identify wear and tear in turbines, generators, or other components before failures occur, reducing downtime and maintenance costs. Machine learning models can also forecast water inflow based on weather patterns and historical data, allowing operators to adjust turbine operations to maximise energy production while minimising the risk of overflow or underutilisation of resources.

AI can further enhance the integration of hydroelectricity into the power grid by improving energy storage and distribution. Advanced AI-driven energy management systems can predict energy demand and optimise the release of stored water to match consumption patterns, ensuring a steady and reliable power supply. Additionally, AI supports the development of hybrid renewable energy systems by integrating hydroelectricity with other energy sources like wind or solar. These systems create a more resilient and sustainable energy infrastructure by balancing contributions from different sources. With AI's ability to automate and refine hydroelectric operations, the sector can increase efficiency, reduce environmental impact, and better adapt to the challenges posed by climate change and shifting energy needs. Indeed, AI-enhanced hydroelectric power will effectively contribute to SDG 7 (Affordable and Clean Energy) and SDG 13 (Climate Action).

5.3.3 AI and Wind Energy

Wind energy generation is the process of converting the kinetic energy of wind into electrical energy using wind turbines. Typically mounted on tall towers, these turbines are equipped with large blades that rotate when wind passes through them. The rotational energy of the blades drives a generator, producing electricity that can be supplied to homes, businesses, and industries. Wind energy is a renewable and sustainable resource, as it relies on the naturally occurring movement of air caused by atmospheric pressure differences. As a clean energy source, wind generation produces no greenhouse gas emissions during operation, making it a vital component of global efforts to reduce carbon footprints and combat climate change.

Integrating AI in wind energy generation is transforming the renewable energy sector by enhancing efficiency, reliability, and sustainability—driving the attainment of SDG 7 (Affordable and Clean Energy) and SDG 13 (Climate Action). AI-powered systems optimise wind turbine operations by analysing vast amounts of real-time data from sensors, including wind speed, direction, and temperature. These insights enable predictive maintenance, reducing downtime and repair costs by identifying potential issues before they escalate. Additionally, AI models help optimise turbine placement and energy output by simulating wind patterns and terrain interactions, ensuring maximum efficiency from wind farms. Using machine learning algorithms, operators can adapt to changing environmental conditions, boosting overall energy production and reducing waste.

AI is also critical in integrating wind energy into the power grid. By predicting energy demand and forecasting wind energy generation with high accuracy, AI systems allow grid operators to balance supply and demand effectively. This minimises the challenges posed by the intermittent nature of wind energy, ensuring a stable and reliable power supply. Furthermore, AI-driven optimisation tools can help design hybrid energy systems that combine wind with other renewable sources, such as solar or battery storage, to create a more resilient and sustainable energy infrastructure. As the wind energy sector grows, AI is poised to accelerate the transition to a cleaner, greener future. This will comprehensively contribute to SDG 7 (Affordable and Clean Energy) and SDG 13 (Climate Action).

5.3.4 AI and Geothermal Energy

Geothermal energy is a sustainable and renewable energy source derived from the natural heat of the Earth's core. This heat is accessed by drilling wells into geothermal reservoirs, where high temperatures are used to produce steam that drives turbines to generate electricity. Geothermal energy is considered highly reliable as it provides a constant and stable energy output, unlike solar or wind energy, which are weather dependent. Thus, geothermal energy can be a significant driver in the pursuit of SDG 7 (Affordable and Clean Energy). Additionally, it has a smaller environmental footprint than fossil fuels, emitting significantly fewer greenhouse gases and requiring less land for power plants, thus advancing the agenda of SDG 13 (Climate Action).

Geothermal energy is particularly advantageous for countries with favourable geological conditions, such as volcanic activity and tectonic plate boundaries, which provide access to abundant high-temperature geothermal reservoirs and sources. These states include Iceland, the United States, the Philippines, Kenya, Ethiopia, Rwanda, Zambia, New Zealand, and Indonesia. These countries' potential for geothermal energy development positions them as key players in advancing renewable energy globally.

The efficacy of geothermal energy extends beyond electricity generation to applications like direct heating for homes, greenhouses, and industrial processes. Despite its benefits, geothermal energy faces challenges, such as high upfront costs for exploration and drilling, which can deter investment.

Technological advancements and supportive policies are helping to address these barriers, making geothermal energy more competitive and accessible. As the world transitions towards cleaner energy sources to combat climate change, geothermal energy offers a viable solution for sustainable power generation. It should be essential to the efforts to achieve global energy and environmental goals—SDG 7 (Affordable and Clean Energy) and SDG 13 (Climate Action).

AI to the Rescue

AI can significantly enhance the efficiency, sustainability, and deployment of geothermal energy by addressing challenges across its life cycle, from exploration to operation. Here are some ways AI can be applied:

- Exploration and Resource Identification: AI-driven algorithms can analyse geological, geophysical, and geochemical data to identify potential geothermal reservoirs more accurately and efficiently. Machine learning models can predict resource locations by analysing patterns in historical data, reducing the need for costly and time-intensive exploratory drilling.
- Drilling Optimisation: AI can optimise drilling operations by analysing real-time data from sensors to predict and mitigate risks such as equipment failures or encountering unviable geothermal resources. This reduces operational costs and improves the success rate of drilling efforts.
- Reservoir Management: Advanced AI models can monitor and simulate geothermal reservoir behaviour to ensure sustainable extraction of heat without depleting the resource. These models can help balance fluid injection and extraction rates, extending the lifespan of geothermal wells.
- Power Plant Efficiency: AI can be used in geothermal power plants to optimise energy conversion processes. For instance, AI-powered predictive maintenance can ensure equipment operates at peak efficiency by identifying potential faults before they cause downtime.
- Energy Grid Integration: AI systems can integrate geothermal energy more effectively into the energy grid by predicting demand patterns and adjusting output accordingly. This helps stabilise the grid and maximise the use of renewable energy sources.
- Environmental Monitoring: AI-powered tools can monitor environmental impacts in real time, such as emissions and subsidence. By analysing data from sensors and satellites, AI can ensure geothermal energy projects comply with environmental standards and reduce their ecological footprint.
- Cost Reduction and Feasibility Studies: AI can assist in financial modelling and risk assessment for geothermal projects, providing investors and policymakers with data-driven insights to make informed decisions about project viability.

By leveraging AI, geothermal energy can become more cost-effective, sustainable, and scalable, playing a crucial role in the global transition to clean energy. This will advance the attainment of both SDG 7 (Affordable and Clean Energy) and SDG 13 (Climate Action).

5.3.5 AI and Future-Proofing of Crops

Future-proofing of crops contributes immensely to SDG 2 (Zero Hunger) and SDG 13 (Climate Action). It involves developing and implementing strategies to ensure agricultural resilience and productivity in the face of climate change, resource scarcity, and growing global demand for food. This includes breeding and engineering crop varieties that are drought tolerant, disease resistant, and capable of thriving in extreme weather conditions. Advanced technologies, such as genetic editing tools like CRISPR and AI, are pivotal in accelerating crop improvement and monitoring. Powered by

AI and IoT devices, precision agriculture techniques enable farmers to optimise water usage, nutrient delivery, and pest control, reducing waste and improving yields. Furthermore, integrating sustainable farming practices, such as crop rotation, agroforestry, and soil conservation, ensures long-term soil health and biodiversity. By combining innovation with sustainability, future-proofing crops aims to secure food security and safeguard agricultural systems against future challenges.

As already indicated, AI is a powerful tool for future-proofing crops. It addresses the challenges posed by climate change, resource limitations, and the growing demand for food. In real time, AI-powered data analytics enables farmers and researchers to monitor environmental conditions, such as soil health, water availability, and weather patterns. This allows for precise decision-making on when to plant, water, and fertilise crops to optimise yields while minimising waste. Machine learning models can predict the spread of pests and diseases, enabling farmers to take proactive measures to protect their crops. These predictive capabilities are critical in reducing crop losses and improving food security, especially in regions vulnerable to climate variability.

AI also plays a crucial role in advancing crop genetics and breeding. By analysing vast datasets of genetic information, AI algorithms can identify traits that make crops more resilient to drought, heat, or disease. This accelerates the development of new, robust crop varieties through techniques like genome editing or traditional crossbreeding. Furthermore, AI supports precision agriculture through the use of drones and autonomous machinery to monitor and manage fields with unparalleled accuracy. These systems can deliver targeted interventions, such as spraying fertilisers or pesticides only where needed, reducing environmental impact and costs. By integrating AI into every stage of the agricultural life cycle, the sector can ensure crops remain productive and resilient in a rapidly changing world. This will accelerate the journey towards achieving SDG 2 (Zero Hunger) and SDG 13 (Climate Action).

5.4 Overall Assessment of AI Deployment to Achieve the SDGs

Indeed, AI holds the transformative potential to accelerate progress towards attaining the UN SDGs by addressing complex global challenges through innovative, data-driven, and integrated solutions. By leveraging the technology's capabilities in prediction, automation, and optimisation, societies can design more effective strategies for sustainable development. AI can enhance environmental sustainability by improving resource efficiency and enabling precise monitoring of ecosystems. In agriculture, AI-powered precision farming optimises water, fertiliser, and energy use, increasing yields while reducing environmental impact, thereby contributing to SDG 2 (Zero Hunger) and SDG 13 (Climate Action). Similarly, AI applications in energy systems optimise renewable energy integration and grid management, advancing SDG 7 (Clean Energy) and SDG 11 (Sustainable Cities).

In healthcare, AI supports SDG 3 (Good Health and Well-being) by enabling early disease detection, personalised treatment, and telemedicine, especially in underserved areas. AI-driven diagnostics and predictive analytics have shown promise in combatting diseases, reducing health disparities, and strengthening healthcare delivery systems.

AI also promotes social development by enhancing education (SDG 4) through personalised learning platforms and expanding financial inclusion (SDG 8 and SDG 10) via AI-driven credit scoring and digital payment systems. These technologies empower marginalised communities, reduce inequalities, and improve access to opportunities. Acknowledging that the AI intervention does not occur in a vacuum is instructive. Figure 12.1 depicts the broader context in which AI can be used to achieve SDGs, while Fig. 12.19 shows the generalised impact assessment of AI on the SDGs. Figures 12.20 and 12.21 show an overview (applications and examples) of how AI can contribute to achieving each of the 17 SDGs. Clearly, integrating AI technologies across various sectors can make significant progress towards achieving the SDGs. However, it is crucial to implement AI responsibly, ensuring ethical considerations and inclusivity to maximise its benefits for sustainable development. Figure 12.22 shows the hierarchy of the critical AI governance functions to be developed and deployed in pursuit of the SDGs.

Furthermore, the application of AI to achieve SDGs must consider challenges such as algorithmic bias, data privacy, and energy consumption. Addressing these requires ethical frameworks, transparent practices, and a commitment to equity and sustainability. Collaborative efforts among governments, private sector stakeholders, and international organisations are essential to ensure AI benefits all. By aligning its transformative power with the SDGs, AI offers a pathway to create innovative solutions that balance economic growth, social equity, and environmental protection, helping to build a more sustainable and inclusive future [6, 37, 127].

5.5 Mitigation and Management of the Common Challenges

As acknowledged in the preceding section, while AI offers tremendous potential in enabling the achievement of the SDGs, its deployment must address several challenges. These include (1) bias and inequality, ensuring that AI systems are trained on diverse datasets to avoid reinforcing existing disparities; (2) data privacy and security, protecting sensitive data in applications such as healthcare, finance, and governance; and (3) environmental impact, reducing the energy consumption associated with training and deploying AI models. Addressing these common challenges is essential to ensure that AI-driven solutions for the SDGs are ethical, equitable, and sustainable. Figure 5.5 summarises the risks and potential dangers of AI in its use to achieve the SDGs by 2030.

5.5.1 Managing Bias and Inequality

One of the most pressing concerns in the application of AI is the risk of bias and its potential to exacerbate existing inequalities. AI systems rely heavily on the data used to train them, and if these datasets are incomplete or reflect historical biases, the resulting models can perpetuate or amplify those inequities. For example, biased datasets in healthcare or financial services can lead to systems that unfairly disadvantage marginalised groups.

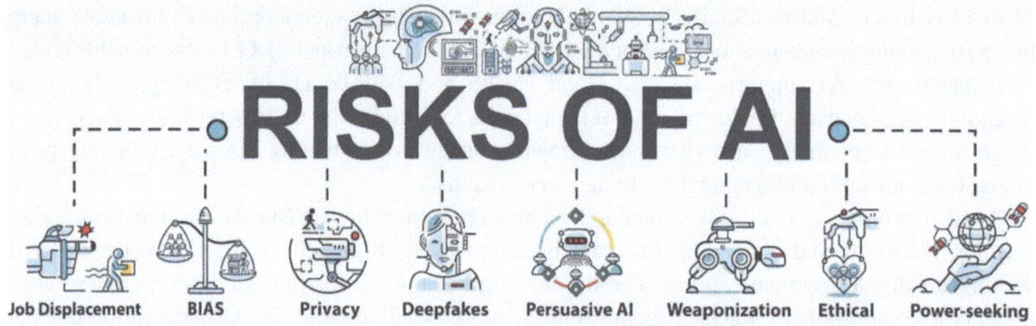

Fig. 5.5 Risks and potential dangers of AI in its use to achieve the SDGs. (Image credit: Shutterstock buffaloboy)

To address this issue, it is essential to ensure that datasets are diverse and representative of the populations they serve. This requires intentional efforts to collect and incorporate data from underrepresented communities. Additionally, regular bias audits and testing of AI models are necessary to identify and mitigate biases during the development process. Tools like IBM's AI Fairness 360 provide mechanisms to evaluate and correct bias in datasets and algorithms, promoting fairness and equity.

Including diverse perspectives in AI development teams also plays a crucial role in reducing bias. Teams that reflect various cultural, socio-economic, and gender experiences are better equipped to identify potential blind spots and design systems that consider the needs of all stakeholders. Moreover, adherence to international ethical guidelines, such as the UNESCO Recommendation on the Ethics of Artificial Intelligence or the EU's AI Act, ensures that AI systems are designed and deployed with fairness and inclusivity in mind.

Beyond technical solutions, empowering local communities to participate in designing and implementing AI solutions is vital. Involving those directly affected by AI systems ensures that these technologies address their specific challenges and do not unintentionally marginalise them further.

5.5.2 Safeguarding Data Privacy and Security

AI systems depend on large volumes of data, often including sensitive personal information. This reliance on data poses risks to privacy and security, making robust safeguards essential. Without adequate measures, data breaches, misuse, or loss of trust in AI systems can undermine their effectiveness and credibility.

One way to protect data is by employing privacy-preserving technologies such as federated learning. This approach allows AI models to be trained across decentralised datasets without transferring sensitive information to a central location, significantly reducing the risk of data breaches. Similarly, anonymisation and de-identification techniques can ensure that personal identifiers are removed from datasets, protecting individual privacy while still enabling valuable insights.

Compliance with data protection regulations, such as the General Data Protection Regulation (GDPR) and similar frameworks, is also critical. These regulations establish standards for transparency, consent, and accountability in data collection and use. Ensuring that AI systems adhere to these standards builds public trust and minimises potential misuse.

Additionally, organisations must invest in robust cybersecurity measures, including encryption, regular vulnerability assessments, and incident response plans. These measures protect sensitive data from unauthorised access and ensure the integrity of AI systems.

5.5.3 Reducing Environmental Impact

While AI offers solutions for environmental challenges, such as optimising renewable energy systems and monitoring biodiversity, the technology itself can have a substantial environmental footprint. Training large AI models requires significant computational power, which can lead to high energy consumption and carbon emissions.

Mitigating this impact begins with developing energy-efficient AI algorithms. Researchers and developers can optimise models by employing techniques such as model pruning, quantisation, and transfer learning, which reduce computational demands without sacrificing performance. Additionally, organisations should prioritise the use of sustainable infrastructure, such as green data centres powered by renewable energy.

Another approach is conducting life cycle assessments (LCAs) for AI systems to evaluate their environmental impact throughout their development and deployment. This helps identify opportunities to reduce energy use and emissions. Where emissions are unavoidable, organisations can invest in carbon offset programmes to neutralise their environmental impact. AI's potential for sustainability extends beyond its direct environmental impact. For instance, AI contributes to broader environmental goals by improving resource efficiency in agriculture or optimising energy grids. However, aligning AI's development and deployment with these goals is essential to ensure that its benefits outweigh its costs.

5.5.4 A Holistic Approach to Ethical AI

To comprehensively address these interconnected challenges, a unified framework for ethical AI development and deployment is necessary. Governments, organisations, and stakeholders must collaborate to establish guidelines prioritising transparency, accountability, and inclusivity. Transparency is essential, as explainable AI systems enable users to understand how decisions are made, fostering trust and accountability.

Moreover, public-private partnerships can drive innovation and scale AI solutions for global challenges. These collaborations can fund research, share best practices, and promote inclusive AI development. Investments in digital literacy and capacity-building initiatives are also critical to ensure that all communities can benefit from AI advancements, bridging the digital divide and promoting equitable development.

By taking proactive steps to mitigate bias, protect privacy, and reduce environmental impact, AI can become a powerful force for achieving the SDGs. Through responsible innovation and ethical deployment, AI has the potential to create a future that is fairer, more sustainable, and inclusive for all. Indeed, by addressing these common challenges through ethical frameworks, transparency, and international cooperation, AI can become a pivotal force in achieving the SDGs, driving sustainable development and improving the quality of life worldwide.

5.6 Interplay Between the AI interventions

The applications of AI described for the 17 UN SDGs are not independent. They are not carried out for each goal in isolation from the other 16. As indicated earlier, the SDGs are interconnected. They are intricately interlinked, with progress in one area often influencing outcomes in others. Understanding these relationships is crucial so that the AI interventions maximise synergies and minimise trade-offs. The interdependence of the 17 UN SDGs is repeated here to emphasise and explore its impact on AI applications.

- SDG 1 (No Poverty) and SDG 2 (Zero Hunger): Reducing poverty enhances access to nutritious food, while improved food security contributes to poverty alleviation.
- SDG 3 (Good Health and Well-being) and SDG 6 (Clean Water and Sanitation): Access to clean water and sanitation is fundamental for preventing diseases and promoting overall health.
- SDG 4 (Quality Education) and SDG 5 (Gender Equality): Providing quality education to all genders promotes gender equality, empowering individuals and communities.
- SDG 7 (Affordable and Clean Energy) and SDG 13 (Climate Action): Transitioning to clean energy sources reduces greenhouse gas emissions, mitigating climate change impacts.

- SDG 8 (Decent Work and Economic Growth) and SDG 9 (Industry, Innovation, and Infrastructure): Investments in infrastructure and innovation drive economic growth and create employment opportunities.
- SDG 10 (Reduced Inequalities) and SDG 16 (Peace, Justice, and Strong Institutions): Promoting inclusive societies and strong institutions helps reduce inequalities and fosters social cohesion.
- SDG 11 (Sustainable Cities and Communities) and SDG 12 (Responsible Consumption and Production): Developing sustainable urban areas encourages responsible consumption and production patterns.
- SDG 14 (Life Below Water) and SDG 15 (Life on Land): Protecting marine and terrestrial ecosystems ensures biodiversity and the provision of essential ecosystem services.
- SDG 17 (Partnerships for the Goals): Global partnerships are essential for mobilising resources and expertise to achieve all other SDGs.

Understanding the interdependencies of the SDGs is vital for policymakers, organisations, and stakeholders planning to deploy AI to achieve the goals. It becomes essential to design holistic approaches that leverage synergies and address potential conflicts among the SDGs. This means AI applications to achieve SDGs should not be applied to each SDG separately. It is instructive to develop a comprehensive, all-embracing strategic framework that takes cognisance of the interlinkages between the goals. A similarly inspired implementation matrix must then be fashioned and executed.

5.6.1 Addressing the Interdependence and Interlinkages

As explained in the previous section, the 17 UN SDGs are intricately interconnected, reflecting the complexity of global systems where progress in one area often influences outcomes in others. Hence, the application of AI to achieve the goals must take this into consideration. This section provides details of the acknowledgement, management, and leverage of that interdependence. For example, improving access to quality education (SDG 4) can enhance gender equality (SDG 5), economic growth (SDG 8), and health outcomes (SDG 3). Recognising these interlinkages is essential for devising strategies that drive holistic progress across the SDGs.

AI has emerged as a transformative tool capable of addressing these interdependencies. Its ability to process complex data, identify patterns, and provide actionable insights makes it uniquely suited to advancing the SDGs in a cohesive and integrated manner. By leveraging AI, we can design solutions that capitalise on synergies among goals while mitigating trade-offs. However, achieving this requires careful consideration of AI deployment's ethical, social, and environmental implications.

5.6.2 AI and the Interconnected Nature of the SDGs

AI's versatility allows it to address multiple SDGs simultaneously, ensuring that interventions in one area support progress in others. This integrated approach is critical for tackling the complexities of sustainable development. AI applications often create synergies among goals by addressing overlapping challenges. For instance, AI-powered telemedicine platforms improve access to healthcare (SDG 3) while reducing the financial burden on underserved populations and directly addressing poverty (SDG 1). Similarly, AI optimises renewable energy systems, supporting both affordable and clean energy (SDG 7) and climate action (SDG 13).

However, not all goals align seamlessly, and some involve inherent trade-offs. For example, the push for economic growth (SDG 8) can strain environmental resources, potentially hindering progress in climate action (SDG 13). AI's ability to analyse these dynamics allows it to propose balanced solutions. In infrastructure development (SDG 9), AI tools can assess the environmental impact of construction projects, enabling designs that promote economic growth while conserving biodiversity (SDG 15).

5.6.3 Examples of AI Applications Addressing Interlinkages

As already discussed, AI's potential to address interconnected goals is evident in various applications. These are restated here for emphasis, illustration, and completeness.

- Agriculture and Climate Resilience: AI-driven tools analyse weather patterns, soil conditions, and crop health to help farmers adopt sustainable practices. This approach enhances food security (SDG 2), reduces resource use, and promotes sustainable land management (SDG 15). For example, IBM's Watson Decision Platform for Agriculture integrates data on climate and farming to guide decisions that benefit both productivity and the environment.
- Smart Energy Systems: AI enhances the management of renewable energy grids by predicting demand, optimising energy storage, and integrating diverse power sources. These innovations support clean energy transitions (SDG 7) while contributing to reduced emissions (SDG 13). In regions with limited energy access, AI-powered microgrids bring electricity to underserved communities, advancing economic inclusion (SDG 10) and poverty reduction (SDG 1).
- Telemedicine and Gender Equality: AI-driven telemedicine platforms extend healthcare access to remote areas, addressing disparities in health outcomes (SDG 3). Women and girls, who often face barriers to healthcare, particularly benefit from these technologies, promoting gender equality (SDG 5). By empowering individuals to access vital services, AI fosters well-being and equity.
- Urban Development and Public Health: In urban settings, AI systems analyse data on transportation, air quality, and population density to design smarter, more sustainable cities (SDG 11). By optimising public transit systems and monitoring pollution, these tools contribute to public health (SDG 3) and climate resilience (SDG 13).

5.6.4 Challenges in Addressing Interlinkages

While AI offers significant potential for addressing SDG interlinkages, its deployment involves challenges that require careful navigation:

- Complexity and Data Integration: The interconnected nature of the SDGs demands comprehensive data integration across sectors. However, siloed data systems often hinder AI's ability to provide holistic insights. Effective AI deployment requires fostering stakeholder collaboration to share and integrate data while maintaining privacy and security.
- Bias and Inequality: AI systems risk perpetuating existing inequalities if trained on biased data. For example, health AI tools developed with data from high-income countries may fail to address the needs of underserved populations. Fairness requires diverse datasets, transparency in algorithm development, and regular bias audits to mitigate disparities.
- Balancing Short-Term Gains and Long-Term Goals: AI solutions may prioritise immediate outcomes over long-term sustainability. For instance, optimising economic growth (SDG 8) without considering environmental impact (SDG 13) can lead to unsustainable practices. Ethical

frameworks must guide AI development to ensure that short-term benefits align with broader, long-term objectives.

5.6.5 Strategies for Effective AI Deployment

To maximise AI's potential in advancing interconnected SDGs, several strategies are essential:

- Fostering Collaboration: Governments, private sector stakeholders, and academia must collaborate to integrate AI solutions across sectors. Public-private partnerships can drive innovation and scale solutions that address multiple SDGs.
- Establishing Ethical Guidelines: Ethical frameworks ensure that AI aligns with principles of fairness, transparency, and accountability. These guidelines can help developers and policymakers balance trade-offs while maximising synergies.
- Investing in Capacity Building: Equipping communities with the skills to use and benefit from AI ensures that its advantages are accessible to all. Digital literacy programmes and workforce training are essential for bridging gaps in AI adoption.
- Promoting Open Data and Standards: Open data initiatives enable AI systems to access diverse datasets, improving their ability to address interlinkages among goals. Standardised protocols for data sharing foster collaboration while ensuring data security and privacy.

The SDGs' interdependence underscores the need for holistically integrated solutions that address global challenges. With its ability to analyse complex data, predict outcomes, and optimise systems, AI is uniquely positioned to navigate these interlinkages. AI can accelerate progress across the SDGs by fostering synergies, managing trade-offs, and promoting inclusivity.

5.7 Potential Negative Impact of AI on SDGs

While AI has the potential to significantly contribute to the achievement of the SDGs, it also poses several challenges and risks that could hinder progress towards these goals. AI systems involve extensive computing facilities, huge financial investments, consumption of a lot of energy, use of critical minerals, and generation of significant greenhouse gas emissions. A large amount of water is also necessary for cooling systems. This massive AI resource demand and the accompanying carbon footprint have potentially dire consequences for the attainment of the SDGs. The environmental impacts of AI are not trivial. For example, it is estimated that training the LLM powering ChatGPT generates carbon emissions equivalent to those from roughly 600 to 700 flights between New York City and Los Angeles in the United States [124].

The starting point, resources, and technology available for the journey to 2030 are different between highly industrialised countries and the Global South. There is inequity. AI tools and data are disproportionately owned and controlled by companies and institutions in the Global North. Countries, organisations, and communities in the Global South (*including pockets of the Global South within the Global North!*) have limited access to AI technology and systems. This undermines their effectiveness in deploying AI to achieve the SDGs. There is a need to democratise AI within the Global North and between the Global South and Global North.

Substantial financial resources are required to develop AI systems. For example, researching, developing, and producing LLMs requires hundreds of millions of US dollars. Consequently, the big Global North private companies have dominated the production of the most advanced AI systems. Top academic or research institutions like MIT, Stanford, or Caltech alone cannot produce LLMs. There must be more collaboration and partnerships between financially endowed corporations, academic institutions, governments and their public sectors, and civil society to generate all AI systems. This must happen within the Global North and between the Global South and Global North.

There must be efforts to give researchers and entrepreneurs from emerging and least industrialised economies access to supercomputing resources so that they can develop AI tools that benefit their regions. For example, there is a need for Global North AI companies to partner with AI organisations in Africa to empower young people on the continent to use AI to develop solutions for local problems and drive the SDG agenda. Furthermore, AI technology companies in highly industrialised economies can also be encouraged to make AI-tool-building resources and infrastructure accessible online globally. Unlimited and worldwide sharing of and access to AI datasets, models, innovations, and insights will dramatically advance the SDG ambition. Such a global strategy will also help decrease AI resource consumption and reduce the AI carbon footprint, thus mitigating the potential risks of using AI to achieve the SDGs.

After a broad discussion of the AI risks to SDGs (and potential mitigation measures), it is instructive to summarise the ways AI could negatively impact the achievement of SDGs:

- Exacerbating Inequalities (SDG 10): AI technologies may widen existing socio-economic disparities by disproportionately benefitting wealthier individuals and countries with greater access to technology and resources. Bias in AI algorithms could perpetuate discrimination and exacerbate inequalities based on factors such as race, gender, and socio-economic status.
- Job Displacement and Unemployment (SDG 8): Automation driven by AI technologies could lead to job displacement and unemployment, particularly in sectors with routine and repetitive tasks. Workers in low-skilled and vulnerable employment sectors may be disproportionately affected, leading to increased poverty and income inequality.
- Data Privacy and Security (SDG 16): The collection, processing, and storage of vast amounts of personal data by AI systems raise concerns about privacy violations and data breaches. Unauthorised access to sensitive information could undermine individuals' rights to privacy, autonomy, and trust in digital technologies and institutions.
- Ethical and Legal Challenges (SDG 16): Ethical dilemmas surrounding AI decision-making, accountability, and transparency could undermine public trust and confidence in AI systems. Legal frameworks and regulations may struggle to keep pace with the rapid advancements in AI technology, leading to uncertainty and potential misuse of AI applications.
- Environmental Impact (SDG 13): AI infrastructure, including data centres and computing hardware, consumes significant amounts of energy and resources, contributing to environmental degradation and climate change. The proliferation of AI-driven devices and technologies could exacerbate electronic waste generation and resource depletion if not managed sustainably.
- Biased or Misleading Information (SDG 4): AI algorithms may amplify the spread of misinformation, fake news, and harmful content on digital platforms, undermining efforts to promote quality education and informed decision-making. Biased or misleading information generated by AI systems could perpetuate stereotypes, fuel social divisions, and undermine public discourse and democratic processes.
- Dependency on AI and Technology (SDG 9): Overreliance on AI and technology solutions to address complex societal challenges could divert attention and resources away from holistic and human-centred approaches to sustainable development. Dependency on AI could also lead to technological lock-in and vulnerabilities in critical systems, posing risks to resilience and security.
- Exclusion of Marginalised Groups (SDG 5, 10, 11): Marginalised communities, including women, minorities, and persons with disabilities, may face barriers to access and participation in AI-driven

systems and decision-making processes. Lack of diversity in AI development teams and datasets could lead to the exclusion and underrepresentation of certain groups, perpetuating inequalities and biases.

Addressing these negative impacts and risks associated with AI deployment requires careful consideration of ethical, legal, and social implications and proactive measures to ensure inclusivity, transparency, and accountability in AI development and deployment. Collaborative efforts involving governments, civil society, academia, and the private sector are essential to mitigate these risks and harness AI's potential for sustainable development [32, 94].

5.8 AI Deployment Challenges in the Global South

AI holds great potential for advancing progress towards the UN SDGs in emerging and least industrialised countries, but several challenges must be addressed to harness its benefits effectively. Indeed, there are a number of challenges in using AI to achieve SDGs in the Global South:

- Limited Access to Technology and Infrastructure: Many developing countries lack adequate access to technology infrastructure, including reliable internet connectivity, computing resources, and data storage facilities. The digital divide exacerbates disparities in access to AI technologies and expertise, hindering adopting and implementing AI solutions for sustainable development.
- Data Quality and Availability: AI algorithms rely on large volumes of high-quality data to train models and generate insights. However, data quality and availability remain significant challenges in the Global South, where data collection systems may be fragmented, outdated, or inadequate. Addressing data gaps and improving data quality are essential for leveraging AI effectively to inform decision-making and monitor progress towards the SDGs.
- Capacity and Expertise: Building and maintaining AI capacity and expertise requires investment in education, training, and skill development. Many countries in the Global South face shortages of qualified AI professionals, data scientists, and technology experts. Strengthening human capital through education, training programmes, and knowledge exchange initiatives is crucial for developing local AI capabilities and leveraging AI technologies for sustainable development.
- Ethical and Sociocultural Considerations: AI technologies raise ethical, legal, and sociocultural concerns related to privacy, bias, transparency, accountability, and fairness. Ensuring that AI systems uphold human rights, adhere to ethical principles, and respect cultural norms and values is essential for promoting trust and acceptance of AI solutions in the Global South. Robust governance frameworks, regulations, and guidelines are needed to address these considerations effectively.
- Resource Constraints and Funding: Implementing AI solutions for sustainable development requires financial resources, investment, and long-term commitment from governments, development partners, and other stakeholders. Limited funding, competing priorities, and resource constraints pose challenges to scaling up AI initiatives in developing countries. Innovative financing mechanisms, public-private partnerships, and international cooperation are needed to mobilise resources and support AI deployment for SDGs.
- Adaptation and Localisation: AI solutions developed in advanced economies may not always be suitable or applicable to the context of the Global South due to differences in socio-economic conditions, cultural norms, and environmental factors. Adapting and localising AI technologies to meet the specific needs and challenges of local communities in the Global South requires contextually relevant approaches, co-design processes, and participatory decision-making.

- Data Privacy and Security: AI applications often involve processing sensitive personal data, raising concerns about data privacy, security, and protection. Inadequate data privacy regulations, cybersecurity measures, and data governance frameworks in the Global South may pose risks to individuals' privacy and rights. Strengthening data protection laws, building cybersecurity capacities, and implementing robust data security measures are essential for safeguarding data privacy and security in the context of AI deployment.

Addressing these challenges requires concerted efforts from governments, development partners, civil society, academia, and the private sector to promote inclusive, responsible, and sustainable AI deployment in the Global South. By overcoming these challenges, AI has the potential to accelerate progress towards achieving the SDGs and addressing some of the most pressing development challenges facing the world today [3, 69, 86].

5.9 Conclusion

Using AI to achieve the UN SDGs is challenging due to several technical, ethical, and systemic issues. AI systems rely on large amounts of high-quality data, which are often unavailable or fragmented in regions where the SDGs are most critical. Data disparities and lack of infrastructure, especially in low-income countries, create significant barriers to AI deployment. Additionally, algorithmic bias and inequality in AI systems can exacerbate existing disparities, undermining the inclusive nature of the SDGs. Ethical concerns such as data privacy, surveillance, and the misuse of AI further complicate its application, particularly in sensitive areas like healthcare and governance. The energy-intensive nature of AI technologies also poses environmental challenges, potentially conflicting with sustainability goals like climate action (SDG 13). Furthermore, achieving meaningful AI integration requires cross-sector collaboration, robust governance frameworks, and capacity building, which are often difficult to implement on a global scale. These challenges underscore the need for equitable access to AI technologies, ethical design principles, and a systemic approach to ensure AI is a force for sustainable development.

Indeed, harnessing the AI potential in driving the attainment of the UN SDGs requires addressing challenges related to data integration, bias, and ethical considerations. It is also prudent to pay attention to the explicit risks of deploying AI in the efforts to achieve SDGs. Moreover, there are challenges peculiar to emerging and least industrialised economies. These must be addressed. Through collaboration, capacity building, and responsible governance, AI can become a powerful tool for achieving a sustainable, equitable, and resilient future that leaves no one behind. Aligning AI innovation with the principles of the SDGs while mitigating, managing, and leveraging the SDGs' interdependences and interlinkages is essential and non-negotiable.

6 Deployment of AI for SDGs: Global Experiences

6.1 Worldwide State of Play

The United Nations monitors the progress of SDGs through a robust framework of global indicators coordinated by the UN Statistics Division. As explained earlier, these indicators, established by the Inter-Agency and Expert Group on SDG Indicators (IAEG-SDGs), measure progress across the 17 goals and their 169 targets. Data is collected from member states through national statistical offices, leveraging existing statistical systems and surveys. This data is supplemented with information from international organisations, civil society, and academia, ensuring comprehensive coverage. The process emphasises the importance of disaggregated data—accounting for factors such as age, gender, income, and location—to ensure that no one is left behind.

To facilitate this monitoring, the UN publishes an annual report titled *The Sustainable Development Goals Report*, which provides a global overview of progress, challenges, and trends. The report includes detailed data visualisations and analysis, presenting key insights into each goal's status. Additionally, the UN maintains an online SDG data platform, which allows policymakers, researchers, and the public to access real-time statistics and trends. This digital resource is vital for transparency and enables stakeholders to track global, regional, and national progress. The data collected is used to monitor and identify areas requiring urgent attention and intervention.

Collaboration and capacity building play a central role in ensuring effective tracking of SDG progress. The UN works closely with member states to strengthen their statistical systems, particularly in developing countries, where data gaps often exist. Technical assistance, funding, and partnerships with international organisations aim to enhance data quality, availability, and timeliness. Moreover, the UN emphasises the use of innovative tools and technologies, such as satellite imagery, big data, and Artificial Intelligence, to overcome traditional data collection challenges. These efforts collectively ensure a dynamic, evidence-based approach to tracking and achieving the SDGs by the 2030 deadline [69, 71, 130].

A. G. O. Mutambara, *Deploying Artificial Intelligence to Achieve the UN Sustainable Development Goals*, Sustainable Development Goals Series,
https://doi.org/10.1007/978-3-031-88423-8_6

6.1.1 2024 UN SDG Report

To review the global state of play, the most recent UN SDG Report, released on 28 June 2024, is an ideal starting point. This comprehensive document provides an in-depth analysis of global progress towards achieving the 17 SDGs. It highlights both advancements and challenges, offering valuable insights into areas requiring accelerated action to meet the 2030 targets. The 2024 report presents a concerning assessment of global progress towards the 2030 Agenda. Only 17% of SDG targets are on track, with nearly half showing minimal or moderate advancement, and over one-third stalled or regressed. This stagnation is attributed to the lingering effects of the COVID-19 pandemic, escalating conflicts, geopolitical tensions, and the intensifying climate crisis, all of which have severely hindered development efforts [132].

Figure 6.1 shows the top ten countries leading in SDG implementation, based on the UN SDG 2024 Report. Similarly, Fig. 6.2 shows the bottom ten countries in SDG implementation based on the same UN document. These rankings reflect each country's overall progress towards achieving all 17 SDGs, with higher scores indicating greater advancement. Countries are ranked by their overall score. The score can be interpreted as a percentage of SDG achievement. A score of 100 indicates that all SDGs have been achieved [133].

The UN SDG 2024 Report highlights significant setbacks in critical areas such as hunger, health, and environmental sustainability. Global hunger levels remain alarmingly high, with projections indicating that nearly 600 million people will be undernourished by 2030, far from the goal of zero

Rank	Country	SDG Index Score
1	Finland	86.8
2	Sweden	85.9
3	Denmark	85.6
4	Germany	85.3
5	France	84.7
6	Austria	83.8
7	Norway	83.5
8	Netherlands	83.2
9	Switzerland	82.9
10	United Kingdom	82.5

Fig. 6.1 Top ten countries leading in SDG implementation

Rank	Country	SDG Index Score
158	Madagascar	51.22
159	Sudan	49.91
160	Niger	49.86
161	Democratic Republic of Congo	48.71
162	Afghanistan	48.24
163	Yemen	46.87
164	Somalia	45.42
165	Chad	45.07
166	Central African Republic	44.21
167	South Sudan	40.14

Fig. 6.2 Bottom ten countries in SDG implementation

hunger. Additionally, global carbon emissions reached a record high in 2023, exacerbating climate change and threatening the achievement of environmental targets.

Biodiversity loss continues, with many species facing severe decline despite modest progress in protection efforts. Despite these challenges, there are areas of improvement. Latin America, for instance, has made significant strides in reducing hunger, with malnutrition rates trending downwards and the region on track to lower malnutrition to below 5% by the decade's end. However, such advancements are uneven globally, underscoring the need for targeted efforts in regions lagging behind. The UN emphasises the need for urgent, coordinated action to accelerate SDG progress. This includes increased international financial support, especially for developing nations, and enhanced global cooperation to address systemic issues impeding development. Without substantial investment and scaled-up efforts, achieving the SDGs by 2030 remains unlikely, necessitating immediate and sustained action from all sectors of society [132].

6.1.2 Global Efforts to Leverage AI

Several countries are making significant strides in deploying AI technologies to advance progress towards the SDGs. While there may not be a single exemplary country that excels in all aspects of AI deployment for SDGs, several nations stand out for their efforts and initiatives in leveraging AI to address specific development challenges. The UAE has launched various initiatives to position itself as a global hub for AI and advance progress towards the SDGs. The UAE National Artificial Intelligence Strategy 2031 aims to integrate AI technologies across sectors such as healthcare, education, transportation, and government services to enhance efficiency, productivity, and innovation. The country's efforts in AI deployment align with SDGs related to innovation (SDG 9), quality education (SDG 4), and sustainable cities and communities (SDG 11).

China has made significant investments in AI research, development, and deployment to address various societal challenges and achieve sustainable development objectives. The Chinese government has launched national AI development plans and initiatives, including the New Generation Artificial Intelligence Development Plan, to accelerate AI innovation and application across sectors. China's AI initiatives contribute to SDGs related to poverty alleviation (SDG 1), healthcare (SDG 3), clean energy (SDG 7), and industry innovation (SDG 9). India is leveraging AI technologies to address development challenges and promote inclusive growth and sustainable development. The National Strategy for Artificial Intelligence outlines India's vision and priorities for AI development, focusing on sectors such as healthcare, agriculture, education, and smart cities. India's AI initiatives align with SDGs related to poverty eradication (SDG 1), health and well-being (SDG 3), quality education (SDG 4), and sustainable cities (SDG 11).

For its part, Singapore is at the forefront of AI adoption and innovation, leveraging AI technologies to enhance efficiency, productivity, and sustainability across various sectors. The Singaporean government has launched initiatives such as AI Singapore (AISG) and the Smart Nation Initiative to drive AI research, development, and deployment. Singapore's AI initiatives contribute to SDGs related to innovation (SDG 9), sustainable cities (SDG 11), and responsible consumption and production (SDG 12). The United States is a global leader in AI research, development, and commercialisation, with significant public and private investments. US companies and research institutions are driving innovation in AI applications across sectors such as healthcare, finance, transportation, and agriculture. The US government has also launched AI initiatives, including the American AI Initiative, to advance AI technologies and address societal challenges. US efforts in AI deployment contribute to SDGs related to innovation (SDG 9), economic growth (SDG 8), and industry innovation (SDG 9). These countries demonstrate various approaches and initiatives in deploying AI technologies to advance progress towards the SDGs. While each country may have unique strengths and challenges, collaboration and knowledge-sharing among nations can accelerate global efforts to harness AI's potential for sustainable development.

Several European countries have effectively deployed AI technologies to advance progress towards the Sustainable Development Goals (SDGs). While there isn't a definitive ranking, several nations stand out for their efforts and initiatives in leveraging AI for sustainable development. Finland is renowned for its strong focus on AI research, innovation, and deployment. The Finnish government has prioritised AI as a key driver of economic growth and societal development. Initiatives like the Finnish Center for Artificial Intelligence (FCAI) and the AI Accelerator programme support research collaboration, talent development, and AI adoption across sectors. Finland's AI initiatives contribute to SDGs related to innovation (SDG 9), quality education (SDG 4), and sustainable cities (SDG 11). Sweden has emerged as a leader in AI research, entrepreneurship, and sustainability.

The Swedish government has launched initiatives such as AI Sweden and the Swedish AI Innovation of Sweden to promote collaboration, innovation, and adoption of AI technologies. Sweden's AI initiatives contribute to SDGs related to innovation (SDG 9), industry innovation (SDG 9), and climate action (SDG 13). The United Kingdom (which exited the EU in February 2020) is at the forefront of AI development and deployment, with a strong ecosystem of research institutions, startups, and multinational companies. The UK government has published an AI Sector Deal and AI Roadmap to drive AI innovation, investment, and adoption across sectors. Initiatives such as the Alan Turing Institute and the UKRI Centre for Doctoral Training in Artificial Intelligence contribute to advancing AI research and talent development. The UK's AI initiatives align with SDGs related to innovation (SDG 9), industry innovation (SDG 9), and sustainable cities (SDG 11).

Germany is a leading European country in AI research, technology, and industry. The German government has launched initiatives such as the AI Strategy and the Plattform Lernende Systeme to promote AI innovation, adoption, and ethics. Germany's AI initiatives focus on sectors such as manufacturing, healthcare, and mobility, contributing to SDGs related to innovation (SDG 9), industry innovation (SDG 9), and sustainable cities (SDG 11). The Netherlands has established itself as a front runner in AI innovation and sustainability. The Dutch government has launched initiatives such as the Dutch AI Coalition and the National AI Agenda to foster collaboration, research, and AI adoption. The Netherlands' AI initiatives contribute to SDGs related to innovation (SDG 9), industry innovation (SDG 9), and climate action (SDG 13). marginparAQ: Please change "Plattform Lernende Systeme" to italics in the sentence "The German government...." These European countries exemplify various approaches and initiatives in deploying AI technologies to advance sustainable development. Collaboration and knowledge sharing among European nations can further accelerate efforts to harness AI's potential to achieve the SDGs [65, 69, 71, 130].

6.2 Africa and the SDGs

As discussed earlier, the SDGs present a global framework for achieving economic prosperity, social inclusion, and environmental sustainability by 2030. This is an essential agenda for Africa, a continent with immense potential yet faced with persistent challenges. Africa finds itself at a critical juncture in striving to meet these goals. Indeed, achieving the SDGs is essential for Africa as it represents a comprehensive pathway towards addressing the continent's most pressing challenges and unlocking its vast potential. The SDGs provide a framework for tackling poverty, inequality, and hunger, which continue to affect millions, while also promoting access to quality education, healthcare, and clean water—fundamental pillars for human dignity and progress. Furthermore, achieving the SDGs will drive economic growth through innovation, sustainable industrialisation, and infrastructure development, ensuring that no one is left behind in Africa's rise. Environmental sustainability is equally critical, as Africa is among the regions most vulnerable to climate change. Adopting renewable energy and sustainable practices will help safeguard the continent's rich biodiversity and natural resources.

Meeting the SDGs will also enhance governance, peace, and justice, creating resilient institutions that can foster stability and prosperity. Ultimately, the SDGs offer a unified vision for a sustainable and inclusive future, enabling Africa to harness its demographic dividend and emerge as a global leader in sustainable development. However, it cannot be business as usual. Things have to be done differently for the SDG journey to succeed on the continent. The starting point is reviewing the impediments [3, 103, 123].

6.2.1 SDG Implementation Challenges in Africa

The implementation of the SDGs in Africa is marked by several interrelated challenges, which span governance, resource constraints, socio-economic inequalities, and climate vulnerabilities. Governance is one of Africa's most pressing challenges in implementing the SDGs. Corruption, weak institutions, and a lack of political accountability hinder effective policy formulation and implementation. Poor governance results in resource mismanagement and inefficient service delivery, thereby stalling progress in critical areas like health, education, and infrastructure development.

For example, SDG 16, which focuses on promoting peace, justice, and strong institutions, remains elusive in many African nations affected by political instability, armed conflict, and weak legal systems. Political instability also deters foreign investment and undermines the long-term planning needed for sustainable development.

The next impediment consists of financial constraints and debt burden. Africa faces significant financial constraints in mobilising the resources needed to achieve the SDGs. The financing gap for development projects is estimated to run into hundreds of billions of dollars annually. Many African nations are heavily indebted, with debt servicing obligations consuming substantial portions of national budgets that could otherwise be allocated to social programmes.

While beneficial, international aid and foreign direct investment (FDI) often come with conditions that may not align with domestic priorities. Furthermore, Africa's limited capacity to generate domestic revenue due to informal economies and low tax collection efficiency exacerbates the problem.

Infrastructure deficits remain a bottleneck for Africa's sustainable development. Energy, transport, and digital connectivity are critical enablers of economic growth and social well-being, yet large parts of the continent lack reliable access to these services. For instance, SDG 7, which aims for affordable and clean energy, is challenged by the fact that nearly 600 million Africans still lack access to electricity. The lack of infrastructure hinders industrialisation (SDG 9) and limits access to education (SDG 4) and healthcare (SDG 3), perpetuating cycles of poverty and inequality.

Despite some progress in reducing poverty, Africa remains home to the majority of the world's poorest populations. High levels of income inequality and limited access to resources such as land, education, and healthcare impede inclusive development. Gender inequality, a critical issue targeted by SDG 5, further compounds the challenge, with women often bearing the brunt of poverty and lacking access to decision-making roles. Addressing poverty requires a multifaceted approach, but systemic issues such as unemployment, low agricultural productivity, and inadequate social safety nets make progress difficult.

Climate change and environmental degradation significantly affect the pursuit of SDGs in Africa. The continent is disproportionately affected by the impacts of climate change despite contributing the least to global greenhouse gas emissions. Rising temperatures, changing rainfall patterns, and extreme weather events threaten agricultural productivity, water security, and biodiversity. These challenges directly affect SDG 13 (climate action), SDG 2 (zero hunger), and SDG 15 (life on land). Moreover, the lack of financial and technical capacity to adapt to climate change exacerbates vulnerabilities, particularly for smallholder farmers and rural communities.

Health crises and weak pandemic resilience compound the African SDG journey. Health challenges, including the persistent burden of infectious diseases such as HIV/AIDS, malaria, and tuberculosis, alongside the recent impacts of the COVID-19 pandemic, strain Africa's healthcare systems. Achieving SDG 3, which focuses on good health and well-being, is further complicated by inadequate healthcare infrastructure, shortages of medical personnel, and limited access to essential medicines. The pandemic exposed gaps in public health preparedness, disrupting progress in other SDGs, such as education and economic growth, due to prolonged school closures and economic downturns.

Rapid population growth and urbanisation create difficult extenuating circumstances on the continent. Africa's population is expected to double by 2050, placing immense pressure on already overstretched resources and infrastructure. Rapid urbanisation, while offering opportunities for economic development, often leads to the proliferation of informal settlements with poor living conditions, inadequate sanitation, and limited access to services. Population growth also complicates efforts to ensure quality education, decent employment, and sustainable management of natural resources, which are critical components of several SDGs.

Limited data and monitoring systems exist on the continent. Accurate data is essential for measuring progress and making informed decisions. However, many African countries face challenges in collecting, analysing, and disseminating reliable data. Weak statistical systems and insufficient funding for data collection impede the ability to monitor SDG indicators effectively. This lack of data undermines accountability and the ability to identify and address gaps in implementation [103, 123].

6.2.2 Rethinking SDG Strategies in Africa

Indeed, Africa's achievement of the SDGs is hindered by a complex web of challenges that require innovative, context-specific, and inclusive solutions. Addressing governance issues, mobilising domestic and international resources, investing in infrastructure, and prioritising climate resilience are critical steps forward. Collaboration among governments, the private sector, civil society, and international organisations is essential to overcome these barriers.

While the road ahead is fraught with difficulties, Africa's abundant natural resources, youthful population, and growing technological innovation offer opportunities to accelerate progress. By aligning development priorities with the SDGs and fostering global partnerships, Africa can move closer to achieving sustainable development for all by 2030. Implementation of Africa's developmental plan, Agenda 2063, must be aligned with the pursuit of the SDGs. In doing so, it is prudent to acknowledge that the continent stands at a pivotal moment in its development trajectory, with the SDGs offering a clear roadmap for addressing its challenges and unlocking its potential.

Despite some progress, many African nations remain off-track in achieving these global objectives. To bridge this gap, a shift in approach is essential—one that redefines priorities, embraces innovation, and strengthens collaboration at all levels. Governments, private sectors, civil society, and global partners must rethink their strategies and act with urgency, precision, and inclusivity to achieve the SDGs in Africa.

One of the foremost changes required is a stronger focus on localised solutions tailored to Africa's unique context. Many SDG strategies have been too broad, failing to take cognisance of the diverse socio-economic, cultural, and geographic conditions characterising the continent. African governments need to champion home-grown and context-aware policies that align with the needs of their communities. For example, promoting climate-resilient agriculture suited to specific ecosystems can enhance food security, while leveraging traditional knowledge systems can improve water management and environmental sustainability. Localising the SDGs also means empowering community leaders and grassroots organisations to drive change from the bottom up.

Innovation and technology must play a more central role in Africa's development journey. The continent's youthful population offers a unique opportunity to harness digital transformation as a driver of progress. Expanding access to digital technologies can revolutionise education, healthcare, and governance. For instance, e-learning platforms can address education gaps in remote areas, while telemedicine can provide vital healthcare services to underserved populations. However, significant investment in digital infrastructure is required to realise this potential. African nations must work to close the digital divide, ensuring that rural and marginalised communities are not left behind.

Financing remains a critical obstacle to achieving the SDGs in Africa, necessitating a fresh approach to resource mobilisation. Many African countries struggle with high debt burdens and limited domestic revenue, constraining their ability to fund SDG-related initiatives. To attract private capital, governments must prioritise innovative financing mechanisms, such as green bonds, diaspora bonds, and blended financing models. Tackling illicit financial flows and improving tax collection systems can also unlock much-needed resources. International partners also have a vital role in providing aid, fairer trade agreements, and debt relief to enable African countries to invest in sustainable development.

Strengthening governance and institutional capacity is another critical area for improvement. Corruption, weak institutions, and political instability have hindered progress towards the SDGs in many African nations. Transparent and accountable governance ensures that resources are allocated effectively and development projects deliver results. African governments must prioritise reforms to build resilient institutions that uphold the rule of law, foster inclusivity, and protect human rights.

Engaging citizens in decision-making processes can also enhance trust and ensure that policies reflect the people's needs. Regional and international cooperation must also be reimagined to accelerate SDG progress. African countries must strengthen regional integration through organisations like the African Union and regional economic communities. Enhanced cooperation can drive cross-border solutions to shared challenges like climate change, trade, and migration. At the global level, Africa must have a stronger voice in international forums to advocate for equitable policies on climate finance, technology transfer, and trade [2, 154].

6.2.3 Unlocking Synergies with Agenda 2063

Africa's Agenda 2063, a strategic framework for the continent's socio-economic transformation, aligns closely with the SDGs in its vision for an integrated, prosperous, and peaceful Africa. More specifically, Agenda 2063 is a development plan designed by the African Union to guide the continent's transformation into a global powerhouse of the future. Adopted in 2013, the agenda represents Africa's blueprint for sustainable development and economic growth over 50 years, building on the Pan-African vision of "An integrated, prosperous, and peaceful Africa, driven by its citizens and representing a dynamic force in the global arena." It encompasses seven key aspirations: (1) inclusive growth and sustainable development; (2) continental integration and political unity; (3) good governance, democracy, human rights, justice, and the rule of law; (4) peace and security; (5) cultural identity, common heritage, values, and ethics; (6) people-driven development, with a focus on women and children; and (7) continental strength, resilience, and global influence. Agenda 2063 stresses the importance of harnessing Africa's vast resources, youthful population, and technological potential to achieve structural transformation, eradicate poverty, and strengthen regional integration. By fostering partnerships, innovation, and self-reliance, it envisions a continent where all citizens have access to prosperity, security, and dignity.

Both frameworks (Agenda 2063 and the UN SDGs) clearly emphasise inclusive growth, poverty eradication, and sustainable management of resources, offering a unique opportunity for synergy. There must be effective execution alignment and unlocking of synergies between the two initiatives. African nations can integrate SDG priorities into the implementation of Agenda 2063's flagship projects, such as infrastructure development, industrialisation, and agricultural modernisation. For example, initiatives under Agenda 2063 that promote energy access and industrialisation align directly with SDG 7 (Affordable and Clean Energy) and SDG 9 (Industry, Innovation, and Infrastructure). Similarly, efforts to empower youth and women in Africa's agenda contribute to SDGs 5 (Gender Equality) and 8 (Decent Work and Economic Growth). By harmonising monitoring mechanisms, leveraging shared goals, and fostering partnerships across sectors, African nations can create a unified development strategy that maximises resources, ensures accountability, and accelerates progress towards both Agenda 2063 and the SDGs.

Achieving the SDGs in Africa requires a cultural shift that places sustainability at the heart of decision-making. Governments, businesses, and individuals must embrace long-term thinking and prioritise environmental, social, and economic sustainability in all actions. Public awareness campaigns can play a crucial role in fostering a culture of sustainability, ensuring that citizens are actively engaged in the development process.

Indeed, achieving the SDGs in Africa demands a transformation in how development is approached. Localised solutions, technological innovation, sustainable financing, strong governance, regional cooperation, and a cultural shift towards sustainability are all critical components of this transformation. By rethinking strategies and acting decisively, Africa can meet the SDGs and set a powerful example for the rest of the world, proving that inclusive and sustainable development is possible. The stakes are high, but so are the opportunities—and Africa's future depends on its ability to seize them [2, 154].

6.2.4 Leveraging AI to Achieve SDGs in Africa

Of all the new interventions (things to be done differently) articulated above in pursuit of the SDGs in Africa, innovation and technology stand out as the most significant, particularly the deployment of AI. The technology offers transformative potential for addressing challenges and accelerating progress towards the SDGs on the continent. With its ability to process vast amounts of data, optimise decision-making, and enhance efficiency, AI can act as a critical enabler for sustainable development.

Many African countries have begun leveraging AI-driven solutions to tackle pressing issues, offering lessons and models that can be scaled across the continent. Indeed, several African countries are making significant efforts to deploy AI technologies to advance progress towards the SDGs. While AI adoption in Africa is still in its early stages compared to other regions, several countries stand out for their initiatives in leveraging AI for sustainable development.

South Africa has emerged as a leader in AI adoption and innovation on the African continent. The South African government has launched initiatives such as the Presidential Commission on the Fourth Industrial Revolution (4IR) and the National Artificial Intelligence Strategy to promote AI research, development, and adoption across sectors. South Africa's AI initiatives contribute to SDGs related to innovation (SDG 9), industry innovation (SDG 9), and quality education (SDG 4).

Kenya is at the forefront of AI adoption and entrepreneurship in Africa. The Kenyan government, private sector, and civil society partners have launched initiatives such as the Kenya AI Task Force and the AI Kenya Network to promote AI research, skills development, and application in various sectors. Kenya's AI initiatives contribute to SDGs related to innovation (SDG 9), industry innovation (SDG 9), and sustainable cities (SDG 11).

Rwanda has demonstrated a solid commitment to leveraging technology, including AI, to drive sustainable development and economic growth. The Rwandan government has launched initiatives such as the Rwanda Innovation Fund and the Kigali Innovation City to support innovation, entrepreneurship, and technology adoption. Rwanda's AI initiatives contribute to SDGs related to innovation (SDG 9), industry innovation (SDG 9), and quality education (SDG 4).

Nigeria is home to a growing AI ecosystem and vibrant tech startup scene. The Nigerian government has initiated efforts to promote AI research, skills development, and application through initiatives such as the National Centre for Artificial Intelligence and Robotics and the AI Development and Ethics Framework. Nigeria's AI initiatives contribute to SDGs related to innovation (SDG 9), industry innovation (SDG 9), and quality education (SDG 4).

Ghana is increasingly embracing AI technologies to drive innovation and address development challenges. The Ghanaian government has launched initiatives such as the Ghana Tech Lab and the National Digital Transformation Agenda to promote digital innovation, entrepreneurship, and skills development. Ghana's AI initiatives contribute to SDGs related to innovation (SDG 9), industry innovation (SDG 9), and quality education (SDG 4).

These African countries exemplify various approaches and initiatives in deploying AI technologies to advance sustainable development and contribute to achieving the SDGs. Collaboration and knowledge sharing among African nations and with international partners can further accelerate efforts to harness AI's potential for sustainable development on the continent. Beyond a country-based approach, it is instructive to focus on a few individual SDGs and outline the efforts of some African countries—an SDG-based review:

- Zero Hunger (SDG 2): Agriculture remains the backbone of many African economies, providing livelihoods for the majority of the population. AI-powered technologies such as precision farming, crop monitoring, and predictive analytics can help optimise resource use, increase productivity, and minimise waste. An illustrative example is in Kenya, where AI solutions like PlantVillage, supported by IBM's Watson platform, use Machine Learning to detect crop diseases through smartphone images. Farmers receive real-time advice on treatment, reducing crop losses and boosting food security. Similarly, Nigeria has adopted AI to improve access to quality seeds and enhance distribution networks for smallholder farmers.
- Quality Education (SDG 4): Education is foundational to achieving other SDGs. AI can bridge gaps in access and quality by personalising learning experiences, automating administrative processes, and expanding the reach of educational content. Rwanda's government collaborates with global partners to introduce AI-based platforms like Kepler, which enhance remote learning for students. These platforms ensure continuity of education, particularly in rural areas, by offering tailored content that aligns with national curricula.
- Good Health and Well-Being (SDG 3): AI-driven health solutions address critical gaps in Africa's healthcare systems, such as shortage of medical professionals, lack of infrastructure, and difficulty reaching remote areas. AI enhances disease diagnosis, monitors health trends, and supports vaccine distribution. For example, in South Africa, the Praekelt Foundation uses AI chatbots to disseminate accurate information about maternal health and COVID-19. Meanwhile, Senegal employs AI in malaria prediction models to allocate resources more effectively and reduce infection rates.
- Clean Water and Sanitation (SDG 6): Access to clean water is a pressing challenge in many African countries. AI can improve water resource management, monitor quality, and detect leakages in supply networks. Using satellite data, Uganda employs AI-based tools to map water resources and predict shortages. Such initiatives guide policymakers in ensuring equitable distribution and conservation of water.
- Climate Action (SDG 13): Africa is highly vulnerable to climate change, necessitating innovative solutions to mitigate its impacts. AI supports climate modelling, monitors deforestation, and enhances disaster response. In Ghana, AI is used to analyse satellite imagery and combat illegal mining activities that degrade ecosystems. Ethiopia employs AI-driven tools to track reforestation efforts as part of its Green Legacy Initiative, which aligns with global climate action targets.
- Decent Work and Economic Growth (SDG 8): The integration of AI into African economies can spur innovation, create new industries, and foster entrepreneurship. AI technologies support job matching, automate repetitive tasks, and improve business efficiency. Tunisia has become a regional hub for AI startups, promoting economic growth and job creation. Companies like InstaDeep are developing AI-powered logistics and supply chain solutions that benefit both local and global markets.

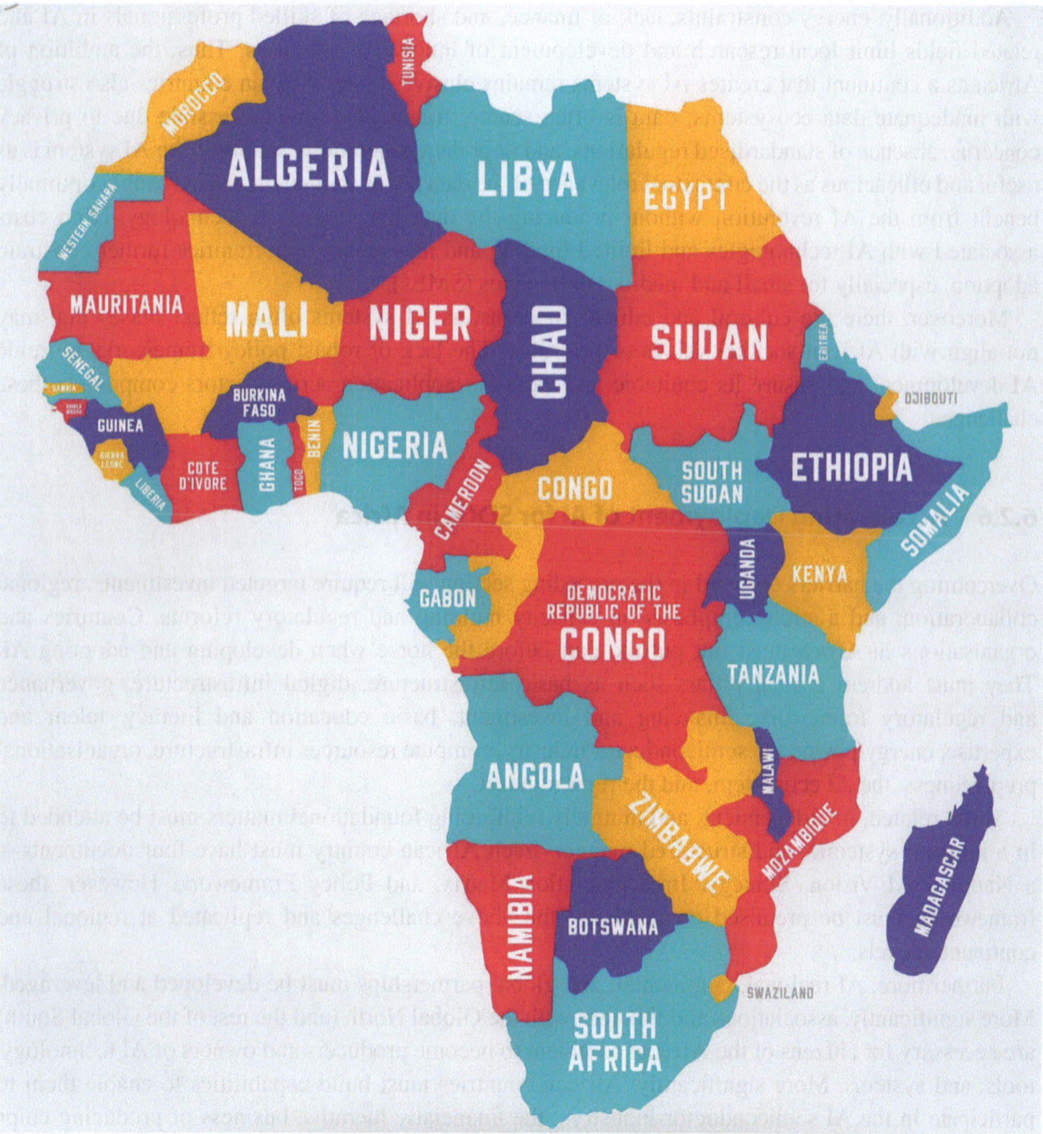

Fig. 6.3 Countries on the African continent

6.2.5 Challenges to AI Adoption in Africa

Despite the promising examples, significant challenges remain. Limited access to reliable Internet, insufficient AI expertise, and concerns about data privacy hinder AI's full potential in many regions. Moreover, the risk of exacerbating inequalities must be carefully managed to ensure AI benefits are equitably distributed. Indeed, the adoption of AI in Africa faces several constraints that hinder its transformative potential. As already stated, one of the primary obstacles is the lack of adequate digital infrastructure, including limited access to high-speed Internet and insufficient computational resources, particularly in rural and underserved regions. Figure 6.3 shows the countries on the continent of Africa.

Additionally, energy constraints, lack of finance, and shortage of skilled professionals in AI and related fields limit local research and development of innovative solutions. Thus, the ambition of Africa as a continent that creates AI systems remains elusive. Many African countries also struggle with inadequate data ecosystems; data is often scarce, fragmented, or inaccessible due to privacy concerns, absence of standardised regulations, and poor data generation capability. An AI system is as useful and efficacious as the contextual relevance of the data it was trained on. Africa cannot optimally benefit from the AI revolution without producing the data that drives AI technology. High costs associated with AI technologies and limited funding and investment opportunities further constrain adoption, especially for small and medium enterprises (SMEs).

Moreover, there are cultural and ethical concerns, as AI systems often reflect biases that may not align with African societal values or priorities. The lack of robust policy frameworks to guide AI development and ensure its equitable and inclusive application across sectors compounds these challenges.

6.2.6 Accelerating Deployment of AI for SDGs in Africa

Overcoming the barriers outlined in the preceding section will require targeted investments, regional collaboration, and a strong emphasis on capacity building and regulatory reforms. Countries and organisations in Africa must not put the cart before the horse when developing and adopting AI. They must address critical pillars such as basic infrastructure, digital infrastructure, governance and regulatory framework, financing and investment, basic education and literacy, talent and expertise, energy/power, the semiconductor industry, compute resources infrastructure, organisational preparedness, the AI ecosystem, and the right mindset.

These related, interdependent, and mutually reinforcing foundational matters must be attended to in a holistic, systemic, and structured manner. Each African country must have four documents—a National AI Vision, Strategy, Implementation Matrix, and Policy Framework. However, these frameworks must be premised on resolving the above challenges and replicated at regional and continental levels.

Furthermore, AI regional, continental, and global partnerships must be developed and leveraged. More significantly, associations and linkages with the Global North (and the rest of the Global South) are necessary for citizens of the African continent to become producers and owners of AI technology, tools, and systems. More significantly, African countries must build capabilities to enable them to participate in the AI semiconductor industry—the financially lucrative business of producing chips that power AI.

There must be plans for reskilling victims of job displacement and developing new capabilities and competencies to take up AI-modified and completely new AI jobs. Africa must prepare to ensure that the total number of modified and new jobs is more than the number of destroyed ones—the job equation. Hence, there is a need to redesign education and develop new skills, competencies, and capabilities so that the job equation is satisfied in Africa.

All these preceding interventions constitute the strategy to accelerate the deployment of AI to achieve SDGs in Africa. The critical efforts can be summarised as follows:

- Strengthening Infrastructure: Investments in digital infrastructure, such as high-speed Internet and data centres, are critical for scaling AI solutions.
- Building Data Generating Capability: There must be investments in African data generation, curation, and storage, thus providing relevant local data for externally designed AI systems and local ones.

- Building Local Expertise: Expanding AI education and training programmes will empower local talent to lead AI innovations. Africa must produce the human capital required to deploy AI technology and, more significantly, the skills and expertise for designing and building AI systems.
- Promoting Public-Private Partnerships: Collaboration between governments, academia, and the private sector can drive AI adoption while ensuring alignment with SDGs.
- Fostering Regional and Continental Cooperation: Regional and Pan-African initiatives, such as the African Union's AI strategy and SADC AI initiative, can harmonise efforts and share best practices across borders.
- AI Strategic Framework: Each African country must have four documents—a National AI Vision, Strategy, Implementation Matrix, and Policy Framework. These must be replicated at regional and continental levels. The three layers must continuously engage and dynamically influence each other as Africa's AI strategic response continuously evolves with the technology.
- Pooling Financial Resources: Executing this is in regional blocs such as SADC, EAC, and ECOWAS, and at a continental level (AU and AfCFTA) will raise sufficient funding for the continent to carry out AI research and development, produce the required electric power, and develop the requisite human capital.
- Global Partnerships and Collaborations: These must be developed and leveraged with the Global North and the rest of the Global South.

6.2.7 AI, SDGs, and Africa: A Final Word

AI offers unparalleled opportunities for African countries to accelerate progress towards the SDGs, addressing systemic challenges in agriculture, healthcare, education, and beyond. By leveraging innovative solutions and overcoming implementation barriers, African nations can harness AI as a powerful tool for sustainable development. Examples from Kenya, Rwanda, South Africa, and others demonstrate the transformative potential of AI when aligned with local needs and priorities. As the continent continues to embrace AI, it can achieve the SDGs and emerge as a global leader in AI-driven innovation.

However, this journey has constraints. To overcome the challenges, it is essential to integrate targeted strategies that align AI's potential with developmental priorities. Governments must invest in digital infrastructure, particularly in underserved rural areas, to enable equitable access to AI-driven education, healthcare, agriculture, and clean energy solutions. Capacity building is crucial; this includes incorporating AI and data science into education systems, promoting local AI innovation hubs, and partnering with global tech firms to train professionals.

Strengthening data governance is equally vital. It requires the establishment of standardised, ethical frameworks that ensure secure and inclusive data collection and use and foster trust among stakeholders. Financing mechanisms, such as public-private partnerships and innovative funding models, must be developed to make AI tools affordable and accessible for small businesses and development initiatives. Moreover, AI adoption must be guided by policies prioritising SDG-aligned applications, such as climate-resilient agriculture, disease prediction, and optimisation of renewable energy.

Collaboration across African nations through platforms like the African Union can facilitate shared learning, technology transfer, and the development of cross-border solutions. By aligning AI adoption with the SDGs, Africa can harness the technology to tackle poverty, inequality, and environmental challenges, accelerating sustainable and inclusive development [3, 86].

6.3 Country Experiences

Countries in Scandinavia, such as Finland, Sweden, and Denmark, have consistently been among the most successful in advancing the SDGs. Their success is attributed to strong governance, high investment in social welfare, and a firm commitment to environmental sustainability. For instance, Finland ranks at the top globally due to its achievements in education, gender equality, and clean energy transitions. Similarly, Sweden has excelled in ensuring access to quality healthcare, renewable energy, and responsible consumption practices. These nations have also integrated the SDGs into their national policies and strategies, aligning their budgets and priorities to address pressing issues like climate action and inequality.

Outside of Scandinavia, countries like Switzerland and the Netherlands have also demonstrated notable progress. Switzerland's efficient management of natural resources, low carbon emissions, and high-quality education systems have positioned it as a leader in sustainable development. Meanwhile, the Netherlands has made strides in areas such as water management, innovation, and sustainable agriculture. These countries benefit from advanced economies, robust institutional frameworks, and strong public engagement in sustainability efforts. However, it is essential to note that even these high-performing nations face challenges, particularly in addressing global systemic issues like climate change and sustainable production, highlighting the need for continuous effort and collaboration.

Several countries, particularly those in conflict zones or experiencing severe economic challenges, have struggled the most to achieve the SDGs. Nations such as Yemen, Somalia, and South Sudan face profound obstacles due to prolonged conflict, political instability, and humanitarian crises. For instance, Yemen's ongoing civil war has devastated its healthcare, education, and infrastructure systems, leaving millions in poverty and hunger. Similarly, Somalia continues to grapple with food insecurity, lack of access to basic services, and fragile governance structures, making progress towards SDGs like zero hunger and quality education nearly impossible. In South Sudan, persistent violence and displacement of communities hinder efforts to achieve peace, justice, and strong institutions, as outlined in SDG 16.

In addition to conflict-affected regions, some developing nations with fragile economies and limited resources have also faced significant setbacks in SDG achievement. Countries such as Haiti and Chad face challenges like extreme poverty, inadequate healthcare systems, and insufficient infrastructure. For instance, Haiti's vulnerability to natural disasters and economic stagnation has impeded progress in areas such as clean water, sanitation, and affordable energy. With its reliance on agriculture and a history of underinvestment in social services, Chad struggles to reduce poverty and improve access to education. These countries often lack the financial and technical capacity to implement the systemic changes required to meet the SDGs, underscoring the urgent need for international support and capacity-building initiatives to bridge the gap.

6.3.1 United States of America

The United States is the world's largest economy by GDP (US$27.4 trillion). It has a population of 335 million, has a GDP per capita of US$82 000, and is consistently ranked in the world's top three most competitive economies (WEF Global Competitive Index). The US economy is driven by a diverse and highly developed mix of industries, robust consumer spending, and innovation. Its key sectors include technology, finance, healthcare, manufacturing, and energy, with Silicon Valley and Wall Street serving as global centres for tech and finance, respectively. The United States is a leader in research and development, fuelling advancements in AI, biotechnology, and clean energy. A significant portion of its economic output is powered by domestic consumption, supported by a large and affluent consumer base. The country is also a major global trader, exporting goods such as machinery, vehicles, and agricultural products. Despite its strengths, challenges such as income inequality, rising national debt, and inflationary pressures persist. Strategic investments in infrastructure, education, and green technologies aim to sustain long-term economic competitiveness.

It is prudent to present a more detailed outline of its efforts to use AI in the SDG journey to 2030. As stated earlier, like other economies, the country has made progress in pursuit of some SDGs while facing challenges in others:

- Good Health and Well-being (SDG 3): In recent years, the United States has made significant progress in improving healthcare access and outcomes, including reducing mortality rates, expanding health insurance coverage through initiatives like the Affordable Care Act (ACA), and advancing medical research and innovation. However, disparities in healthcare access and outcomes persist, particularly among marginalised communities.
- Quality Education (SDG 4): The United States has a well-established education system with high enrolment levels in primary and secondary education. However, challenges remain in ensuring equitable access to quality education, reducing dropout rates, and addressing disparities in educational outcomes based on factors such as income, race, and ethnicity.
- Gender Equality (SDG 5): While progress has been made in promoting gender equality in the United States, gender disparities persist in areas such as workforce participation, wages, and representation in leadership positions. Efforts to address gender-based violence and discrimination are ongoing, along with initiatives to promote women's empowerment and equal opportunities.
- Clean Water and Sanitation (SDG 6): The United States has made significant strides in providing access to clean water and sanitation infrastructure, with high levels of access to piped water and improved sanitation facilities. However, challenges remain in ensuring access for marginalised communities, addressing water pollution, and maintaining water quality in the face of environmental threats such as climate change and pollution.
- Affordable and Clean Energy (SDG 7): In recent years, the United States has made progress in increasing renewable energy production and reducing greenhouse gas emissions. However, dependence on fossil fuels remains high, and efforts to transition to a more sustainable energy system face policy, infrastructure, and economic challenges.
- Decent Work and Economic Growth (SDG 8): The United States has a dynamic and diverse economy with employment and entrepreneurship opportunities. However, disparities in income, wealth, and access to economic opportunities persist, particularly among marginalised communities. Efforts to promote inclusive economic growth and reduce inequality are ongoing.
- Industry, Innovation, and Infrastructure (SDG 9): The United States is a global leader in innovation and technological development, with robust infrastructure in many sectors. However, challenges remain in addressing infrastructure needs, promoting sustainable industrialisation, and ensuring equitable access to technology and innovation opportunities.
- Reduced Inequalities (SDG 10): While the United States has made progress in reducing poverty and inequality in some areas, disparities persist based on factors such as income, race, ethnicity, gender, and geography. Efforts to address systemic inequalities and promote social inclusion are ongoing.
- Sustainable Cities and Communities (SDG 11): The United States has made efforts to promote sustainable urban development, including initiatives to improve transportation systems, enhance access to green spaces, and address environmental challenges in cities. However, issues such as urban sprawl, pollution, and inadequate affordable housing remain significant challenges in many urban areas.

- Responsible Consumption and Production (SDG 12): Efforts to promote sustainable consumption and production patterns in the United States include initiatives to reduce waste, promote recycling and circular economy practices, and enhance environmental stewardship in industries such as manufacturing and agriculture. However, consumption levels and waste generation remain high, posing challenges to sustainable resource management.
- Climate Action (SDG 13): The United States has taken steps to address climate change, including efforts to reduce greenhouse gas emissions, promote renewable energy, and strengthen resilience to climate impacts. However, more ambitious action is needed to meet climate targets and mitigate the effects of climate change, particularly in light of the United States' withdrawal from the Paris Agreement.
- Life Below Water (SDG 14): Efforts to protect marine ecosystems and biodiversity in the United States include measures to establish marine protected areas, regulate fishing practices, and address pollution and habitat degradation. However, marine ecosystems continue to face threats from overfishing, pollution, climate change, and other human activities.
- Life on Land (SDG 15): The United States has implemented various conservation and land management initiatives to protect terrestrial ecosystems, preserve biodiversity, and restore degraded lands. However, challenges remain in addressing threats such as habitat loss, deforestation, invasive species, and wildlife trafficking.
- Peace, Justice, and Strong Institutions (SDG 16): The United States has a strong legal and institutional framework to promote peace, justice, and the rule of law, including robust democratic institutions, an independent judiciary, and mechanisms to protect human rights and promote accountability. However, challenges persist in areas such as criminal justice reform, racial equity, and combatting corruption and organised crime.
- Partnerships for the Goals (SDG 17): The United States engages in partnerships and collaborations with other countries, international organisations, civil society, and the private sector to address global challenges and achieve the SDGs. However, efforts to strengthen global partnerships and enhance cooperation on issues such as development aid, trade, and technology transfer face ongoing challenges and geopolitical dynamics.

Overall, while the United States has made progress in advancing the Sustainable Development Goals in various areas, significant challenges remain in achieving targets related to poverty, inequality, environmental sustainability, and social inclusion. Continued efforts and collaboration across sectors and stakeholders are essential to address these challenges and build a more sustainable and equitable future [71, 136, 137].

6.3.2 Singapore

Singapore is consistently ranked annually in the world's top three most competitive economies (WEF Global Competitive Index). In fact, in 2024, the country was ranked first by WEF, followed by Switzerland, and then Denmark in third position. With a massive GDP per capita of US$85 000, Singapore is a great economic success story from the Global South. It emerged from an underdeveloped and backwater country in 1965 to an advanced and prosperous one in 2000 (a short 35 years) under the able, ethical, and technocratic leadership of Lee Kuan Yew. Singapore rose from a legacy of divisive colonialism, the devastation of the Second World War, general poverty, and underdevelopment to a world-class economy characterised by excellent infrastructure (extensive road and rail networks, a national airport, and airline), outstanding trade facilities, mass public housing, superior educational system, a dynamic financial sector, and robust manufacturing base. The country has a population of six million and a GDP of US$500 billion.

It is instructive to explore how Singapore has deployed AI in pursuit of the UN SDGs. The country has made significant progress towards achieving several Sustainable Development Goals (SDGs) owing to its strong governance, strategic planning, and investment in various sectors. Here is an overview of some of Singapore's achievements regarding AI interventions to achieve the UN SDGs:

- Quality Education (SDG 4): Singapore's education system is highly regarded globally for its quality and equity. The government invests heavily in education, ensuring high enrolment rates, well-trained teachers, and access to lifelong learning opportunities. Singapore consistently performs well in international assessments such as the Programme for International Student Assessment (PISA).
- Clean Water and Sanitation (SDG 6): Singapore has invested in advanced water management technologies, including desalination and wastewater treatment, to ensure water security. The country has diversified its water sources and implemented efficient water conservation measures, leading to high-quality water supply and sanitation services for its residents.
- Decent Work and Economic Growth (SDG 8): Singapore boasts a thriving economy with low unemployment rates and high per capita income. The government prioritises economic growth through innovation, infrastructure development, and attracting foreign investment. Labour laws protect workers' rights, and vocational training programmes ensure a skilled workforce.
- Industry, Innovation, and Infrastructure (SDG 9): Singapore is known for its advanced infrastructure and intense focus on innovation. The country invests in research and development, technology adoption, and digitalisation across various sectors. Singapore's Changi Airport is consistently ranked among the best globally, showcasing its commitment to infrastructure excellence.
- Sustainable Cities and Communities (SDG 11): Singapore has achieved remarkable success in urban planning and sustainable development. The city state is known for its efficient public transportation system, green spaces, and urban resilience initiatives. Singapore's Housing Development Board (HDB) provides affordable, high-quality housing for most of its population.
- Responsible Consumption and Production (SDG 12): Singapore promotes responsible consumption and production through policies and initiatives to reduce waste, promote recycling, and foster sustainable practices. The government encourages businesses and consumers to adopt eco-friendly behaviours and supports initiatives like the Zero Waste Masterplan.
- Climate Action (SDG 13): While Singapore faces challenges related to climate change, such as rising sea levels and extreme weather events, it has taken steps to mitigate its carbon footprint and enhance climate resilience. The government invests in renewable energy, energy efficiency, and climate adaptation measures to reduce greenhouse gas emissions and protect the environment.
- Partnerships for the Goals (SDG 17): Singapore actively engages in international partnerships and collaborations to address global challenges and contribute to sustainable development efforts. The country participates in multilateral forums, provides development assistance to other countries, and supports global health, education, and environmental conservation initiatives.

While Singapore has made significant progress towards achieving various SDGs, challenges such as income inequality, environmental sustainability, and social integration remain areas for ongoing attention and improvement. Continued efforts, collaboration among stakeholders, and innovative solutions will be essential for Singapore to maintain its sustainable development momentum and address emerging global challenges [86, 148].

6.3.3 Finland

Finland has made significant progress towards achieving several Sustainable Development Goals (SDGs), driven by its strong commitment to social welfare, environmental sustainability, and innovation. The country boasts a highly developed and diversified economy, driven by innovation, education, and a robust welfare state. The country ranks among the world's most competitive economies, with strengths in technology, telecommunications, and sustainable industries. Major exports include machinery, electronics, paper products, and chemicals. Finland has embraced digital transformation and green technology, making significant investments in renewable energy and eco-friendly solutions. The Finnish government plays a crucial role in fostering economic stability through sound fiscal policies and a strong social safety net, which promotes inclusivity and equality. Finland's strategic location and membership in the European Union also enhance its trade and investment opportunities, making it a dynamic player in the global economy. Here's an overview of some of the AI-driven SDG achievements in Finland:

- Quality Education (SDG 4): Finland's education system is renowned for its high quality and equity. The country emphasises comprehensive education reforms, teacher professionalism, and student-centred learning approaches. Finland consistently ranks among the top performers in international education assessments, such as the Programme for International Student Assessment (PISA).
- Gender Equality (SDG 5): Finland has made strides in promoting gender equality through policies and initiatives aimed at closing gender gaps in education, employment, and political representation. The country has strong legal frameworks to combat gender-based discrimination and violence, and it actively promotes women's participation in decision-making roles.
- Clean Water and Sanitation (SDG 6): Finland maintains high standards of water quality and sanitation. The country has effective water management policies, investments in wastewater treatment infrastructure, and measures to protect freshwater resources. Finland's commitment to sustainable water management ensures its population has access to clean water and sanitation services.
- Affordable and Clean Energy (SDG 7): Finland has made progress in promoting renewable energy and energy efficiency. The country has invested in wind, solar, biomass, and hydropower projects, aiming to increase the share of renewable energy in its energy mix. Finland's energy policies prioritise sustainability, emissions reduction, and energy independence.
- Decent Work and Economic Growth (SDG 8): Finland maintains a stable economy with low unemployment rates and a high standard of living. The country prioritises social welfare, labour rights, and income equality through policies such as progressive taxation, social security systems, and collective bargaining agreements. Finland's emphasis on innovation and research also fosters economic growth and competitiveness.
- Industry, Innovation, and Infrastructure (SDG 9): Finland is recognised for its innovation-driven economy and advanced infrastructure. The country invests in research and development, technology adoption, and digitalisation to drive economic growth and competitiveness. Finland's strong telecommunications, transportation, and logistics infrastructure support its industrial and innovation sectors.
- Sustainable Cities and Communities (SDG 11): Finland promotes sustainable urban development through green planning, efficient public transportation, and environmental conservation initiatives. The country's cities prioritise pedestrian-friendly designs, green spaces, and energy-efficient buildings. Finland's urban policies aim to enhance livability, resilience, and social inclusion in urban areas.

- Climate Action (SDG 13): Finland is committed to combatting climate change and reducing greenhouse gas emissions. The country has set ambitious climate targets, including carbon neutrality by 2035, and invests in renewable energy, energy efficiency, and climate adaptation measures. Finland's climate policies prioritise sustainability, innovation, and international cooperation.
- Life Below Water (SDG 14): Finland protects its marine ecosystems through marine conservation areas, sustainable fisheries management, and pollution prevention measures. The country promotes biodiversity conservation, marine research, and sustainable ocean governance to preserve its coastal and marine resources.
- Life on Land (SDG 15): Finland prioritises biodiversity conservation, forest management, and land restoration efforts. The country has extensive forest cover, and sustainable forestry practices are integral to Finland's economy and environmental stewardship. Finland's national parks and protected areas safeguard terrestrial ecosystems and promote outdoor recreation and nature tourism.
- Partnerships for the Goals (SDG 17): Finland actively participates in international partnerships and collaborations to address global challenges and promote sustainable development. The country provides development assistance, supports multilateral initiatives, and engages in diplomatic efforts to advance the SDGs globally.

While Finland has made significant progress towards achieving various SDGs, ongoing efforts are needed to address emerging challenges such as climate change, biodiversity loss, and social inequality. Collaboration among government, civil society, private sector stakeholders, and international partners will be essential for Finland to sustain its sustainable development achievements and build a more resilient and inclusive society [65, 140].

6.3.4 China

China is the second largest economy in the world by nominal GDP (US$17.8 trillion versus US$27.4 trillion for the United States). However, it is the largest in terms of purchasing power parity GDP (US$37.07 trillion versus US$29.17 trillion for the United States, in international dollars) and plays a critical role in global trade and production. It has transitioned from a centrally planned system to a market-oriented economy over the past few decades, experiencing rapid industrialisation and urbanisation. China has managed to move 800 million people out of poverty in 40 years. In January 2025, the country recorded an unprecedented trade surplus of nearly US$1 trillion. Key sectors include manufacturing, technology, agriculture, and services, with exports of electronics, machinery, and consumer goods driving economic growth. China is also a global leader in renewable energy production and digital innovation. Despite its economic achievements, the country faces challenges such as rising debt, an ageing population, and income inequality. Government policies like the "Dual Circulation" strategy aim to balance domestic consumption with export-driven growth to ensure long-term economic stability.

China has rapidly emerged as a global AI superpower, investing heavily in both the development and application of AI systems across various industries, including technology, healthcare, and defence. The country's strategic focus on AI innovation, supported by robust government policies and funding, aims to position China as a leader in shaping the future landscape of AI technology worldwide. On 20 January 2025, DeepSeek, an AI research laboratory in China, released a groundbreaking open-source LLM product called DeepSeek-R1, comparable to offerings from US giants like OpenAI, Google, and Meta. Unlike traditional LLMs, it is low-cost, is less energy-intensive, and uses chain-of-thought reasoning. Consequently, DeepSeek-R1 outperforms its rivals in many areas, including solving complex problems, strategic thinking, and advanced mathematics. This outstanding Chinese technological success will inspire and ignite the rest of the least industrialised and emerging economies—they too can develop and deploy AI to solve their challenges and pursue their socio-economic ambitions.

China has made significant progress towards achieving several UN SDGs due to its sustained economic growth, ambitious policy initiatives, large-scale investments, and AI deployment in various sectors. Here's an overview of some SDG achievements driven by AI in China:

- Poverty Alleviation (SDG 1): Over the past few decades, China has lifted hundreds of millions of people out of poverty, achieving remarkable progress in poverty reduction. The government has implemented targeted poverty alleviation programmes, invested in rural development, and provided social assistance to vulnerable populations.
- Quality Education (SDG 4): China has expanded access to education and improved educational quality. The country has achieved high enrolment rates in primary and secondary education and has invested in teacher training, curriculum reform, and educational infrastructure development. China's emphasis on STEM education and vocational training also supports workforce development.
- Gender Equality (SDG 5): China has made progress in promoting gender equality through legal reforms, policies, and public awareness campaigns. The government has implemented measures to address gender disparities in education, employment, and political representation. However, challenges remain in achieving full gender equality, particularly in areas such as women's participation in leadership roles and addressing gender-based violence.
- Clean Water and Sanitation (SDG 6): China has made efforts to improve access to clean water and sanitation services, particularly in rural areas. The government has invested in water infrastructure projects, wastewater treatment facilities, and pollution control measures to address water quality issues. However, water pollution and scarcity remain significant challenges in some regions.
- Affordable and Clean Energy (SDG 7): China is the world's largest investor in renewable energy and has made significant strides in expanding its clean energy capacity. The country leads in producing and deploying solar, wind, and hydroelectric power technologies. China's efforts to transition towards a low-carbon economy include energy efficiency improvements, emission reduction targets, and investments in clean energy research and development.
- Decent Work and Economic Growth (SDG 8): China has achieved rapid economic growth and industrialisation, creating jobs and reducing poverty. The country has implemented labour laws and social welfare programmes to protect workers' rights and improve working conditions. However, challenges such as income inequality, labour rights violations, and informal employment persist.
- Industry, Innovation, and Infrastructure (SDG 9): China has made significant investments in infrastructure development, including transportation, telecommunications, and digital infrastructure. The country has also prioritised innovation and technology advancement through research and development funding, industrial policies, and initiatives such as "Made in China 2025" to promote high-tech industries and innovation-driven growth.
- Sustainable Cities and Communities (SDG 11): China faces challenges related to urbanisation, including air pollution, congestion, and inadequate urban infrastructure. The government has implemented measures to promote sustainable urban development, including urban planning reforms, investments in public transportation, and green building initiatives. China's "eco-cities" and smart city projects aim to enhance urban sustainability and livability.
- Climate Action (SDG 13): China is the world's largest emitter of greenhouse gases but has taken significant steps to address climate change. The country has set ambitious targets for carbon emissions reduction, renewable energy deployment, and energy efficiency improvements. China's

investments in renewable energy, electric vehicles, and low-carbon technologies play a crucial role in global climate mitigation efforts.

- Life Below Water (SDG 14): China faces challenges related to marine pollution, overfishing, and habitat degradation in its coastal waters. The government has implemented measures to protect aquatic ecosystems, including marine protected areas, fisheries management regulations, and pollution control initiatives. China's efforts to reduce plastic pollution and promote sustainable fisheries aim to conserve marine biodiversity and ecosystems.
- Life on Land (SDG 15): China has implemented afforestation and reforestation programmes to combat desertification, soil erosion, and deforestation. The country has also established nature reserves and conservation areas to protect terrestrial ecosystems and biodiversity. China's efforts to restore degraded land and promote sustainable land management contribute to achieving SDG 15 objectives.
- Partnerships for the Goals (SDG 17): China actively participates in international partnerships and collaborations to address global challenges and promote sustainable development. The country provides development assistance, engages in South-South cooperation, and supports multilateral initiatives such as the Belt and Road Initiative (BRI). China's partnerships with other countries and international organisations contribute to achieving SDG 17 objectives.

While China has made significant progress towards achieving several SDGs, challenges such as environmental degradation, income inequality, and social disparities remain. Continued efforts, policy reforms, and international cooperation are needed to address these challenges and achieve sustainable development in China. Through South-South partnerships and collaborations, China can provide advanced AI systems, AI infrastructure support, and much-needed AI technology transfer. In this way, it can assist other emerging economies on their SDG journeys [86, 143, 153].

6.3.5 Rwanda

Rwanda's economy has demonstrated remarkable resilience and growth over the past two decades, earning it recognition as one of Africa's success stories. Driven by robust governance and ambitious development policies, the economy is primarily anchored in agriculture, which employs a significant portion of the population. However, diversification efforts have fuelled growth in sectors such as tourism, construction, and information technology. The country's Vision 2050 strategy emphasises sustainability, innovation, and private sector investment. Rwanda is a leader in ecotourism, particularly for its mountain gorilla conservation efforts, attracting global visitors. Despite progress, challenges remain, including reliance on external aid and limited industrial capacity. Still, continued reforms and strategic investments have positioned Rwanda as a rising hub for business and innovation in East Africa. Rwanda has made significant strides in achieving several UN SDGs despite facing challenges such as poverty, limited resources, and a history of conflict. Here's an overview of some AI-driven SDG achievements in Rwanda:

- No Poverty (SDG 1): Rwanda's poverty rates have declined over the years due to sustained economic growth, poverty reduction programmes, and social protection initiatives. The government has implemented policies to promote inclusive growth, improve access to education and healthcare, and support livelihoods in rural areas.
- Zero Hunger (SDG 2): Rwanda has reduced hunger and improved food security through agricultural productivity improvements, land reforms, and nutrition interventions. The government

has prioritised investments in agriculture, irrigation, and crop diversification to increase food production and enhance resilience to climate change.

- Good Health and Well-being (SDG 3): Rwanda has made significant improvements in healthcare access and outcomes, including reducing maternal and child mortality rates, increasing immunisation coverage, and combatting infectious diseases such as HIV/AIDS and malaria. The government has invested in healthcare infrastructure, trained healthcare workers, and implemented health insurance schemes to improve health services' affordability and accessibility.
- Quality Education (SDG 4): Rwanda has made efforts to improve access to education and enhance educational quality through education reforms, infrastructure development, and teacher training programmes. The government has prioritised primary and secondary education expansion, increased school enrolment rates, and promoted inclusive education policies.
- Gender Equality (SDG 5): Rwanda has made significant progress in promoting gender equality and women's empowerment, with high levels of women's representation in parliament and leadership positions. The government has implemented gender-sensitive policies, laws, and programmes to address gender disparities in education, employment, and decision-making.
- Clean Water and Sanitation (SDG 6): Rwanda has made efforts to improve access to clean water and sanitation services, particularly in rural areas. The government has invested in water infrastructure projects, community-based water management initiatives, and sanitation awareness campaigns to increase access to safe drinking water and sanitation facilities.
- Affordable and Clean Energy (SDG 7): Rwanda has made progress in expanding access to affordable and clean energy through investments in renewable energy sources such as hydroelectric power, solar energy, and biomass. The government has implemented rural electrification programmes, promoted energy efficiency measures, and supported off-grid energy solutions to increase energy access and reduce reliance on fossil fuels.
- Decent Work and Economic Growth (SDG 8): Rwanda has achieved sustained economic growth and job creation through economic diversification, private sector development, and investment in infrastructure and human capital. The government has implemented policies to promote entrepreneurship, attract foreign investment, and create employment opportunities in critical sectors such as agriculture, manufacturing, and tourism.
- Industry, Innovation, and Infrastructure (SDG 9): Rwanda has invested in building infrastructure, promoting innovation, and supporting industrial development to drive economic growth and enhance competitiveness. The government has implemented initiatives to improve transportation networks, digital connectivity, and access to technology and innovation resources.
- Sustainable Cities and Communities (SDG 11): Rwanda has made efforts to promote sustainable urban development through urban planning, infrastructure development, and environmental conservation initiatives. The government has implemented policies to improve urban mobility, expand green spaces, and enhance access to basic services in cities and towns.
- Climate Action (SDG 13): Rwanda is vulnerable to climate change impacts, including extreme weather events and agricultural disruptions. The government has developed climate resilience strategies, implemented adaptation measures, and invested in renewable energy and reforestation initiatives to mitigate climate risks and build climate-resilient communities.
- Life Below Water (SDG 14): Rwanda is landlocked without direct access to marine environments. However, the country is committed to protecting freshwater ecosystems, promoting sustainable water management, and conserving biodiversity in lakes, rivers, and wetlands.
- Life on Land (SDG 15): Rwanda has implemented measures to protect terrestrial ecosystems, prevent land degradation, and promote biodiversity conservation through afforestation, land restoration, and conservation initiatives. The government has prioritised sustainable land manage-

ment practices, wildlife conservation efforts, and community-based natural resource management approaches.

- Partnerships for the Goals (SDG 17): Rwanda actively engages in partnerships and collaborations with international organisations, development agencies, and civil society to achieve Sustainable Development Goals. The government works with partners to mobilise resources, share knowledge and expertise, and implement programmes that address development challenges and promote inclusive growth.

Overall, Rwanda's achievements in advancing the SDGs demonstrate the government's commitment to sustainable development, poverty reduction, and improving the well-being of its citizens. Despite progress, ongoing efforts are needed to address remaining challenges and ensure inclusive and sustainable development for all Rwandans [57, 86].

6.3.6 United Arab Emirates

The United Arab Emirates (UAE) has a highly diversified and dynamic economy, making it one of the most prosperous nations in the Middle East. While oil and gas remain critical to its wealth, accounting for a significant share of its GDP, the UAE has strategically reduced its reliance on hydrocarbons by investing heavily in tourism, finance, real estate, and renewable energy sectors. Dubai and Abu Dhabi are global hubs for trade, innovation, and luxury tourism, attracting businesses and visitors from around the world. The UAE's economic vision emphasises sustainability and technology, with initiatives like the Dubai Expo and Masdar City showcasing its commitment to future-ready industries. With its favourable business environment, modern infrastructure, and global connectivity, the UAE continues strengthening its position as a leading centre for international commerce and investment.

The country has made significant progress towards achieving several SDGs due to its strong commitment to economic diversification, social development, and environmental sustainability. Here's an overview of some SDG achievements in the UAE:

- No Poverty (SDG 1): The UAE has significantly reduced poverty through economic development, social welfare programmes, and initiatives aimed at improving living standards. The country's wealth from oil revenues has been used to invest in infrastructure, healthcare, education, and social services, lifting many citizens out of poverty.
- Zero Hunger (SDG 2): The UAE has achieved food security through investments in agriculture, technology, and strategic partnerships with other countries. The country has implemented initiatives to increase domestic food production, reduce food waste, and ensure access to nutritious food for its population.
- Good Health and Well-being (SDG 3): The UAE has made significant investments in healthcare infrastructure, facilities, and services, providing high-quality healthcare to its residents. The country has achieved advancements in healthcare outcomes, including reducing mortality rates, improving life expectancy, and controlling infectious diseases.
- Quality Education (SDG 4): The UAE has developed a robust education system focused on providing quality education for all. The country has invested in building schools, training teachers, and developing educational curricula to ensure access to education at all levels. The UAE's efforts have led to increased literacy rates and improved academic outcomes.
- Gender Equality (SDG 5): The UAE has made progress in promoting gender equality through legislation, policies, and initiatives aimed at empowering women and enhancing their participation

in all sectors of society. Women in the UAE have access to education, employment opportunities, and leadership roles, and gender parity in education has been achieved.
- Clean Water and Sanitation (SDG 6): The UAE has implemented advanced water management strategies and technologies to ensure access to clean water and sanitation services. The country has invested in desalination plants, wastewater treatment facilities, and water conservation measures to address water scarcity challenges.
- Affordable and Clean Energy (SDG 7): The UAE is a global leader in renewable energy adoption and clean energy initiatives. The country has invested in solar, wind, and nuclear energy projects to diversify its energy sources and reduce dependence on fossil fuels. The UAE's efforts have contributed to achieving affordable and clean energy for its population.
- Decent Work and Economic Growth (SDG 8): The UAE has achieved economic growth and diversification through investments in various sectors, including tourism, trade, finance, and technology. The country has created job opportunities for its citizens and residents, supported entrepreneurship, and implemented labour laws to ensure decent working conditions and fair wages.
- Industry, Innovation, and Infrastructure (SDG 9): The UAE has developed world-class infrastructure, including transportation, telecommunications, and digital connectivity, to support economic growth and development. The country has also promoted innovation and technology adoption through initiatives such as free trade zones, research centres, and entrepreneurship hubs.
- Sustainable Cities and Communities (SDG 11): The UAE has invested in sustainable urban planning and development to create livable cities and communities. The country has implemented green building standards, efficient public transportation systems, and environmental conservation initiatives to promote sustainability and enhance the quality of life for its residents.
- Climate Action (SDG 13): The UAE has taken steps to mitigate climate change and reduce greenhouse gas emissions through investments in renewable energy, energy efficiency, and sustainable development projects. The country has set ambitious targets for reducing carbon emissions and implementing climate adaptation measures to address the impacts of climate change.
- Life Below Water (SDG 14): The UAE has implemented marine conservation measures and initiatives to protect its coastal and marine ecosystems. The country has established marine protected areas, promoted sustainable fishing practices, and implemented pollution control measures to preserve marine biodiversity and habitats.
- Life on Land (SDG 15): The UAE has implemented land conservation and biodiversity protection measures to preserve its terrestrial ecosystems and natural habitats. The country has established protected areas, wildlife reserves, and reforestation projects to promote biodiversity conservation and sustainable land management.
- Peace, Justice, and Strong Institutions (SDG 16): The UAE has prioritised peace, stability, and security, both domestically and regionally. The country has strong institutions, rule of law, and effective governance structures to ensure peace, justice, and security for its citizens and residents.
- Partnerships for the Goals (SDG 17): The UAE actively participates in international partnerships and collaborations to support Sustainable Development Goals globally. The country provides development assistance, humanitarian aid, and financial contributions to support initiatives aimed at achieving the SDGs, particularly in developing countries and conflict-affected regions.

The UAE has made significant progress towards achieving the Sustainable Development Goals, demonstrating its commitment to sustainable development, economic prosperity, and social progress. Continued efforts and collaboration with stakeholders will be essential to address remaining challenges and achieve sustainable development objectives in the UAE and beyond [39, 86].

6.3.7 South Africa

South Africa has the most industrialised and diversified economy in Africa, serving as a regional powerhouse and gateway for trade across the continent. It is rich in natural resources, particularly gold, platinum, and diamonds, which have historically driven its economic growth. The country boasts a well-developed financial sector and robust infrastructure, with mining, manufacturing, agriculture, and services being key contributors to GDP. South Africa is also a leader in renewable energy development and innovation in Africa. However, it faces challenges such as high unemployment, income inequality, and energy supply issues. Structural reforms, investments in green technologies, and regional trade initiatives aim to address these hurdles and foster inclusive growth, maintaining its role as a pivotal economic player in the region.

South Africa has made progress towards achieving some of the SDGs, but like many countries, it also faces significant challenges. Here's an overview of some SDG achievements and areas needing improvement in South Africa:

- No Poverty (SDG 1): South Africa has made progress in reducing poverty rates over the years, and social welfare programmes such as social grants have helped alleviate poverty for many. However, poverty levels remain high, particularly in rural areas and among marginalised communities.
- Zero Hunger (SDG 2): South Africa has implemented various programmes to improve food security and nutrition, including school feeding schemes and agricultural support initiatives. However, food insecurity and malnutrition persist, exacerbated by factors such as unemployment, inequality, and climate-related challenges.
- Good Health and Well-Being (SDG 3): South Africa has made strides in combatting HIV/AIDS and improving access to healthcare services, including antiretroviral treatment. Efforts have also been made to address other health challenges, such as tuberculosis and maternal mortality. However, healthcare disparities, communicable diseases, and non-communicable diseases remain significant concerns.
- Quality Education (SDG 4): South Africa has made progress in increasing access to education, with significant improvements in school enrolment rates. Efforts have been made to address issues such as infrastructure development, teacher training, and curriculum reform. However, educational outcomes and quality vary widely, and challenges such as school dropout rates and unequal access to education persist.
- Gender Equality (SDG 5): South Africa has made efforts to promote gender equality through legislation, policies, and affirmative action measures. Women's representation in politics and the workforce has increased, and there have been improvements in access to education and healthcare for women and girls. However, gender-based violence, economic disparities, and cultural barriers continue to hinder progress towards gender equality.
- Clean Water and Sanitation (SDG 6): South Africa has made progress in improving access to clean water and sanitation services, particularly in urban areas. Efforts have been made to upgrade water infrastructure, expand access to safe drinking water, and promote sanitation and hygiene practices. However, access to water and sanitation remains a challenge in many rural communities, and water quality issues persist.
- Affordable and Clean Energy (SDG 7): South Africa has invested in renewable energy, particularly wind and solar power, to diversify its energy sources and reduce reliance on fossil fuels. Initiatives such as the Renewable Energy Independent Power Producer Procurement Programme (REIPPPP) have helped increase renewable energy capacity. However, access to clean and affordable energy remains a challenge for many households, particularly in rural areas.

- Decent Work and Economic Growth (SDG 8): South Africa has a relatively developed economy with a diverse industrial base and a well-established financial sector. Efforts have been made to promote job creation, support small- and medium-sized enterprises (SMEs), and address unemployment through skills development and training programmes. However, unemployment rates remain high, particularly among youth and marginalised groups, and income inequality persists.
- Industry, Innovation, and Infrastructure (SDG 9): South Africa has developed infrastructure in sectors such as transportation, telecommunications, and energy, supporting economic growth and development. The country has also invested in innovation and research initiatives to promote industrialisation and technology advancement. However, infrastructure backlogs, inadequate maintenance, and regulatory challenges hinder progress in these areas.
- Reduced Inequalities (SDG 10): South Africa has made efforts to address inequalities through policies such as affirmative action, social assistance programmes, and land reform initiatives. However, income inequality, racial disparities, and spatial inequalities remain deeply entrenched, affecting access to opportunities and social services.
- Sustainable Cities and Communities (SDG 11): South Africa faces challenges related to urbanisation, including inadequate housing, infrastructure deficiencies, and informal settlements. Efforts have been made to improve urban planning, upgrade informal settlements, and promote sustainable urban development. However, rapid urbanisation and population growth continue to strain resources and infrastructure in cities and towns.
- Climate Action (SDG 13): South Africa is vulnerable to climate change impacts such as droughts, floods, and heatwaves. The country has developed climate change adaptation and mitigation strategies, including renewable energy deployment, emissions reduction targets, and climate-resilient infrastructure initiatives. However, challenges such as water scarcity, deforestation, and air pollution remain significant concerns.
- Life Below Water (SDG 14): South Africa has implemented marine conservation measures and established marine protected areas to protect its coastal and marine ecosystems. Efforts have been made to address overfishing, marine pollution, and habitat degradation. However, illegal fishing activities, habitat destruction, and pollution continue to threaten marine biodiversity and livelihoods dependent on marine resources.
- Life on Land (SDG 15): South Africa has implemented conservation programmes and initiatives to protect terrestrial ecosystems, biodiversity, and natural resources. Efforts have been made to combat deforestation, restore degraded land, and promote sustainable land management practices. However, threats such as habitat loss, invasive species, and poaching remain challenges to biodiversity conservation and ecosystem sustainability. There is also the need for more effective measures to combat wildlife trafficking. The 2024 SDG Country Report highlights these issues and underscores the importance of integrating ecosystem and biodiversity values into national planning.
- Peace, Justice, and Strong Institutions (SDG 16): South Africa has seen improvements in governance, with efforts to enhance transparency and reduce corruption. The country has strengthened its judicial systems and increased public participation in decision-making processes. Despite these efforts, challenges remain, including high levels of crime and ongoing concerns about corruption. The 2024 SDG Country Report emphasises the need to strengthen statistical systems to track performance on good governance, where data limitations inhibit performance tracking.
- Partnerships for the Goals (SDG 17): South Africa has actively engaged in international partnerships to advance the SDGs, collaborating with various stakeholders, including governments, civil society, and the private sector. The country has also worked to mobilise resources and enhance capacity-building initiatives. However, challenges such as economic constraints and the need for

more effective multi-stakeholder partnerships persist. The 2024 SDG Africa Report highlights the necessity for the continent to accelerate progress on SDG 17 to achieve the 2030 Agenda.

Indeed, South Africa has increasingly recognised the potential of AI to accelerate progress towards achieving the SDGs. AI-driven solutions are being employed in sectors such as agriculture, where precision farming technologies optimise crop yields and water usage, addressing food security challenges under SDG 2. In healthcare, AI-powered diagnostic tools are improving disease detection and treatment, contributing to SDG 3 by enhancing access to quality healthcare. AI is also being leveraged to tackle environmental issues, with predictive analytics helping monitor and protect biodiversity and ecosystems, supporting SDG 15. Furthermore, AI initiatives are being integrated into education and skills development programmes to reduce inequality (SDG 10) and foster innovation and economic growth (SDG 8). However, challenges such as limited digital infrastructure, skills gaps, and ethical concerns around data privacy persist, necessitating greater investments, inclusive policies, and robust collaborations between the government, private sector, and academia to fully realise AI's potential in achieving the SDGs [23, 86].

6.4 Best Practices for Using AI to Achieve SDGs

By reviewing the efforts to use AI in pursuit of the SDGs across the world, lessons can be harvested and shared. The global landscape offers a wealth of emergent experiences and potential best practices demonstrating how AI can accelerate progress towards achieving the SDGs. One of the most notable applications is in combatting climate change (SDG 13). AI is being used to predict weather patterns, optimise renewable energy generation, and manage natural resources. For instance, AI-powered tools help monitor deforestation and illegal logging in real time, as seen in the Global Forest Watch initiative. Similarly, predictive analytics enable cities to reduce carbon footprints through smart transportation systems and energy-efficient urban planning. These examples highlight how AI can empower governments and organisations to make data-driven decisions for sustainable environmental management.

In healthcare (SDG 3), AI has been transformative in improving diagnostics, treatment, and disease prevention. Machine Learning algorithms have been deployed to detect diseases such as cancer and COVID-19 earlier and more accurately than traditional methods, increasing survival rates and reducing healthcare costs. Global initiatives, like those led by the World Health Organization, incorporate AI to predict outbreaks of diseases such as malaria by analysing climate and population data. AI's ability to process vast amounts of health data also supports the development of personalised medicine, ensuring that treatments are more effective and tailored to individual patients. These innovations underscore the potential of AI to bridge gaps in healthcare access and quality, particularly in underserved regions.

In fostering innovation and economic growth (SDG 8), AI is reshaping industries and creating opportunities for sustainable development. Countries like Singapore and Estonia have embraced AI to streamline public services, enhance education through adaptive learning platforms, and boost productivity in key economic sectors. Inclusive AI practices, such as India's focus on digital literacy and affordable access, serve as a model for scaling AI in developing contexts. Importantly, global collaborations, such as the UN's AI for Good initiative, emphasise ethical AI deployment, ensuring technologies are used responsibly and inclusively. It is prudent to summarise the key emerging best practices, based on global experiences, in using AI to achieve the SDGs.

- Focus on Data Availability and Quality: Build comprehensive and open-access datasets, ensuring data is representative, inclusive, and free from bias. Promote data sharing across sectors while adhering to strict privacy and security regulations.
- Develop Localised AI Solutions: Customise AI applications to address specific local challenges, such as precision agriculture in water-scarce regions or predictive healthcare for underserved populations. Incorporate traditional knowledge and cultural context into AI-driven solutions.
- Strengthen Infrastructure: Invest in robust digital infrastructure, including reliable Internet, cloud computing, and access to advanced hardware. Focus on extending connectivity and technology access to rural and underserved communities.
- Promote Capacity Building: Foster AI education and training programmes to develop local expertise, particularly in developing regions. Encourage partnerships with universities, research institutions, and global tech companies for knowledge transfer.
- Enhance Collaboration: Establish multi-stakeholder collaborations among governments, private sectors, academia, and civil society to align AI initiatives with SDGs. Participate in international forums, such as the UN's AI for Good initiative, to share knowledge and best practices.
- Ensure Ethical and Inclusive AI Deployment: Develop regulatory frameworks that prioritise fairness, transparency, and accountability in AI systems. Focus on AI solutions that are inclusive and accessible to marginalised and vulnerable populations.
- Utilise AI for Predictive and Preventive Interventions. Apply AI for early detection of crises, such as disease outbreaks (SDG 3) and natural disasters (SDG 13). Use AI to monitor and manage environmental risks, such as deforestation and biodiversity loss (SDG 15).
- Innovate for Scalable Impact: Develop AI technologies that are cost-effective and scalable, particularly for use in low-resource settings. Encourage AI-powered platforms that address multiple SDGs simultaneously, such as combining education (SDG 4) with economic growth (SDG 8).
- Ensure Sustainable Funding: Leverage innovative financing models like public-private partnerships, impact investments, and green bonds to support AI-driven SDG projects. Allocate funding for research into AI applications that directly contribute to sustainable development.
- Monitor and Evaluate Impact: Implement mechanisms to assess the effectiveness of AI solutions in achieving SDGs, focusing on measurable outcomes. Use feedback loops to refine and improve AI systems based on real-world performance.

These practices highlight the importance of inclusive, ethical, and context-aware approaches to harnessing AI's potential for sustainable development worldwide. By integrating AI into global and local SDG efforts, these best practices provide a blueprint for leveraging technology to drive sustainable, equitable development across diverse contexts.

6.5 Conclusion

Countries across the globe are leveraging AI to accelerate progress towards achieving the SDGs, with notable successes in areas such as healthcare, education, and environmental sustainability. In SDG 3 (Good Health and Well-Being), AI-driven technologies have been pivotal in diagnosing diseases, predicting outbreaks, and improving access to medical care in underserved regions. For instance, Rwanda has employed AI-powered platforms to enhance diagnostic accuracy in remote health centres, reducing reliance on specialised healthcare providers. Similarly, India has utilised AI tools to monitor child malnutrition and improve maternal health outcomes, aligning with SDG 2 (Zero Hunger). These technologies enable data-driven decision-making and resource allocation, maximising impact with limited resources.

In environmental sustainability, countries like Denmark and Singapore are using AI to optimise energy use and reduce carbon footprints, directly supporting SDG 13 (Climate Action) and SDG 7 (Affordable and Clean Energy). AI-powered systems analyse energy consumption patterns, predict renewable energy outputs, and enhance grid management efficiency. In agriculture (SDG 2), nations like Kenya have adopted AI to support precision farming, improving crop yields while conserving resources such as water and fertilisers. While these advancements showcase AI's potential, challenges such as data accessibility, ethical considerations, and unequal technological adoption among countries highlight the need for collaboration and inclusive policies to ensure AI-driven solutions benefit all nations in their pursuit of the SDGs.

Enablers of AI Deployment in Pursuit of SDGs 7

7.1 Introduction

As presented in previous chapters, AI is emerging as a transformative tool to address global challenges and accelerate progress towards achieving the SDGs. From combatting poverty and inequality to enhancing healthcare and promoting sustainable energy, AI has the potential to revolutionise how societies address pressing issues. However, the effective deployment of AI requires a set of enablers to ensure its benefits are widespread, inclusive, and aligned with the principles of sustainable development. These enablers include robust technological infrastructure, supportive policy, guardrails and governance frameworks, access to quality data, capacity-building initiatives, and multi-stakeholder partnerships.

The terms enabler and driver represent two distinct concepts often used in strategy, innovation, or development. Each plays a unique role in achieving goals. A driver is a factor that motivates or pushes something to happen, while an enabler is a condition that makes it possible for something to happen. For example, in business, drivers are the primary activities that control the business course and contribute revenue, while enablers are support activities that create value indirectly.

Put differently, an enabler is a factor, condition, or resource that facilitates or supports the execution of a process or project. It creates the potential for progress by providing the necessary tools, infrastructure, or environment but does not actively initiate or propel action. On the other hand, a driver is a force or motivation that actively initiates or accelerates progress towards a goal. Drivers are dynamic and proactive, creating momentum and influencing decision-making. They often act as catalysts for change, pushing stakeholders to take action. Figure 7.1 indicates the role of enablers in deploying AI to achieve the SDGs [86].

While enablers provide the means and create the conditions for action, drivers provide the push or motivation to move forwards. Both are crucial in achieving goals but operate at different stages. Together, enablers and drivers ensure the capability and momentum required for meaningful progress. In this chapter, the focus is on the enablers of AI deployment to achieve the SDGs.

A. G. O. Mutambara, *Deploying Artificial Intelligence to Achieve the UN Sustainable Development Goals*, Sustainable Development Goals Series,
https://doi.org/10.1007/978-3-031-88423-8_7

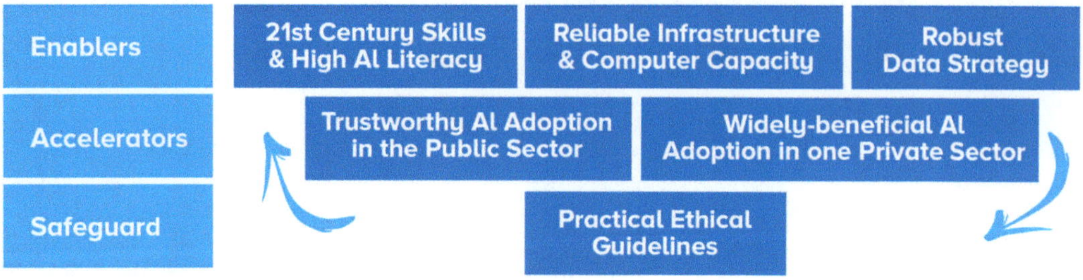

Fig. 7.1 The role of enablers in deploying AI to achieve the SDGs

7.2 Enablers at a Glance

Indeed, deploying AI for SDGs can be significantly facilitated by several enablers, which create an environment conducive to leveraging AI technologies for sustainable development. A summary of some key enablers can be outlined as follows.

Robust Digital Infrastructure and Connectivity
The foundation of AI deployment lies in the availability of robust technological infrastructure, including high-speed Internet, cloud computing, and reliable energy systems. Countries with advanced digital ecosystems can more effectively leverage AI for tasks such as predictive modelling, real-time monitoring, and automation of processes critical to achieving the SDGs. For instance, AI applications in precision agriculture or smart grid management require reliable connectivity and computing power. Expanding digital infrastructure to underserved regions is essential to ensure equitable access to AI-driven solutions, bridging the digital divide and enabling progress in areas such as education, healthcare, and renewable energy. Indeed, robust digital infrastructure, Internet connectivity, and computing resources are essential for effectively deploying AI solutions in pursuit of SDGs, especially in remote and underserved areas.

Supportive Policy, Regulatory, and Governance Frameworks
Governments play a crucial role in fostering an environment conducive to AI development and deployment. Policies that promote innovation while addressing ethical concerns and potential risks are vital. Regulatory frameworks and guardrails should prioritise transparency, accountability, and inclusivity to ensure AI systems are used responsibly and equitably. For example, national AI strategies in countries like Canada and Finland emphasise ethical AI deployment aligned with societal needs, providing a model for integrating technology into sustainable development. Additionally, governments must invest in creating incentives for research and development, facilitating public-private partnerships that align AI advancements with the SDGs. Indeed, clear regulatory frameworks and standards are necessary to ensure the ethical and responsible deployment of AI technologies in pursuit of SDGs, where issues such as privacy, bias mitigation, accountability, and transparency are addressed.

Access to High-Quality Data
AI thrives on data, making access to high-quality, diverse, and representative datasets an indispensable enabler. Accurate data is critical for training AI models to address specific challenges, such as predicting disease outbreaks, monitoring environmental changes, and optimising resource allocation. Open data initiatives can promote collaboration and innovation by making valuable datasets accessible

to researchers, developers, and policymakers. For instance, satellite data from organisations like NASA has been used to monitor deforestation and land degradation, contributing to SDG 15 (Life on Land). Ensuring data privacy and security is equally essential to building public trust and protecting sensitive information. Indeed, the availability of high-quality, relevant, and accessible data is crucial for AI applications. Governments, organisations, and communities must ensure data transparency and facilitate data sharing. They must invest in infrastructure that enables data harvesting, collection, curation, and management.

Awareness, Education, and Capacity Building

Raising awareness about the potential of AI for achieving SDGs and building capacity among stakeholders on how to utilise AI technologies effectively are essential for widespread adoption. A skilled workforce and public understanding of AI are critical to its successful deployment. Capacity-building initiatives should focus on equipping individuals with the knowledge and skills needed to develop, deploy, and manage AI solutions. This includes integrating AI and data science into educational curricula and providing training programmes for professionals across sectors. For example, Rwanda's focus on AI education and partnerships with global technology leaders has enabled the country to deploy AI-driven healthcare solutions. Bridging the skill gap also requires empowering marginalised communities to participate in AI development, ensuring inclusivity in addressing the SDGs. Indeed, training programmes, educational initiatives, and capacity building in AI are essential for fostering a skilled workforce capable of developing, implementing, and maintaining AI solutions for SDGs.

Multi-Stakeholder Partnerships

Collaboration among governments, businesses, academia, and civil society is essential for scaling AI-driven solutions to achieve the SDGs. Multi-stakeholder partnerships enable the pooling of resources, expertise, and technologies to address complex global challenges. Initiatives such as the AI for Good Global Summit by the International Telecommunication Union (ITU) highlight the importance of cross-sector collaboration in leveraging AI for sustainable development. Partnerships can also facilitate technology transfer to low-income countries, ensuring that the benefits of AI are not confined to technologically advanced nations but contribute to global equity. Indeed, collaboration among governments, academia, industry, and civil society is critical for sharing resources, expertise, and best practices in AI for SDGs, where public-private partnerships accelerate the development and deployment of AI solutions.

Localisation and Contextualisation

AI solutions for SDGs must be tailored to local needs and contexts to ensure relevance and effectiveness. They must be bespoke and context-aware. This requires engagement with local communities, understanding cultural nuances, and adapting technologies accordingly. Locally harvested and curated data must be used to train the AI systems. Even better, locally developed AI systems trained on local data must be deployed in pursuit of SDGs. Localisation and contextualisation are significant challenges in the Global South as most AI systems are developed and produced in the Global North, particularly Northern America and Western Europe.

Financial Resources and Investment

Adequate funding and investment mechanisms are required to support research, development, and implementation of AI solutions for SDGs. Governments, international organisations, and philanthropic institutions can play a crucial role in providing financial support. Buying and deploying already-developed AI systems is slightly cheaper, whereas the overall investment required to develop and maintain an AI system, such as an LLM, can range from several million to tens of millions of dollars. Hence, massive investments and comprehensive funding are required to facilitate the localisation and contextualisation of AI systems on the SDG journey. This is a critical matter for emerging and least industrialised countries where the attainment of the SDGs is an existential issue.

Risk Management, Resilience, Monitoring, and Evaluation

Strategies for risk management, including cybersecurity measures, resilience planning, and contingency mechanisms, are crucial to address potential challenges and vulnerabilities associated with AI deployment to achieve SDGs. Establishing robust monitoring and evaluation frameworks is essential to assess the impact of AI interventions on SDGs, identify areas for improvement, and ensure accountability. As the adage goes, if you cannot measure it, you cannot manage it. The impact of the application of AI to achieve SDGs must be monitored, measured, and evaluated, while the results of the exercise are fed back to improve performance.

A Summarised Reflection

Indeed, the deployment of AI to achieve the SDGs holds immense potential, but realising this benefit depends on the presence of key enablers, which include robust infrastructure, supportive policies, access to quality data, capacity building, multi-stakeholder partnerships, adequate financing, localisation, monitoring, and evaluation. These are the building blocks for leveraging AI in sustainable development. By addressing these enablers, countries can harness AI to tackle poverty, inequality, climate change, and other pressing issues, ensuring that no one is left behind in the pursuit of the SDGs. In doing so, AI can become a powerful tool for creating a more inclusive, equitable, and sustainable world. Thus, by implementing the summarised enabling pillars, stakeholders can unlock the full potential of AI to accelerate progress towards achieving the SDGs.

7.3 Details of Enablers

It is prudent to explore some of the outlined enablers of AI-for-SDG deployment in more detail. This will assist the process of putting the enablers in place by establishing clarity of the impact, assessing the requirements, identifying the bottlenecks, and fleshing out the implementation matters.

7.3.1 Robust Digital Infrastructure and Connectivity

AI-enabling infrastructure must exist, starting with basic infrastructure such as energy, communication, transportation, finance, water, and housing. More significantly, affordable, reliable, and efficacious digital infrastructure must be in place. Digital infrastructure sits on top of basic infrastructure. It is inconceivable to have it without this foundational infrastructure. Each listed basic infrastructure is critical in supporting sustainable socio-economic development and technological innovation—a foundational basis for the attainment of the SDGs.

More significantly, basic infrastructure is essential for economic growth and productivity, enhanced connectivity and mobility, quality of life, public health, environmental sustainability, social inclusion, equity, and job creation. These outcomes constitute the core of the SDG agenda. Basic infrastructure is the backbone of modern societies, providing essential services and enabling socio-economic progress. It lays the foundation for the digital economy and, more specifically, AI-driven innovations and applications. Consequently, strategic investments in basic infrastructure are imperative in pursuing the development and adoption of AI to achieve the SDGs. Countries and organisations must work together to build basic infrastructure, thus enabling their AI-enabled SDG journey.

Digital Infrastructure and Connectivity

Beyond basic infrastructure, the deployment of AI for SDGs requires robust digital infrastructure, reliable connectivity, and modern digital devices. Digital infrastructure entails the foundational technology components that support modern information and communication technologies (ICT) by facilitating digital communication, data processing, and online services. All these foundational digital infrastructure-related matters must be addressed before any meaningful application of AI to achieve SDGs.

As already intimated, digital infrastructure encompasses the physical and virtual resources required to collect, store, process, and transmit data and information. It is critical for enabling various digital services, applications, and processes to function efficiently. The essential elements of digital infrastructure can be summarised as networks, data centres, Internet backbone, cloud computing platforms, cybersecurity systems, telecommunication, the Internet of Things (IoT), software, and digital applications.

There must be unrelenting efforts to develop or attract and license affordable high-speed, low-latency connectivity such as that provided by Starlink—the satellite Internet constellation launched in 2019 and operated by Starlink Services, LLC. It is a wholly owned US aerospace company SpaceX subsidiary, providing coverage to 102 countries and rapidly increasing. Starlink also aims to offer global mobile broadband. As of March 2024, Starlink consisted of over 6000 mass-produced small satellites in low Earth orbit that communicate with designated ground transceivers. Nearly 12,000 satellites are planned to be deployed worldwide, with a possible later extension to 34,400.

Countries must embrace the Starlink opportunity to power the pursuit of SDGs through AI systems. This will dramatically cut the cost of connectivity and data while massively improving speed and reliability. Starlink reached more than one million subscribers in December 2022, two million in September 2023, and three million in May 2024. The ascendency of this satellite Internet constellation couldn't have been more fortuitous for emerging and least developed economies as they grapple with the deployment of AI to achieve SDGs. As of May 2024, Starlink's services were live in eight African countries. Its high-speed, low-latency services are useful on a continent where many rural localities have poor or non-existent fixed and/or mobile network connectivity.

Developing and maintaining robust digital infrastructure are crucial preconditions for the AI revolution. Investments, research, and development in digital infrastructure are necessary. They bridge the digital divide, thus ensuring equitable access to digital services in general and AI opportunities in particular. Digital infrastructure forms the foundation on which AI systems operate efficaciously. Additionally, advancements in digital infrastructure in the IT, digital, and communication industries are increasingly intertwined with basic infrastructure. Thus, digital infrastructure can be used to drive and enable the development of basic infrastructure. There is synergistic and mutual reinforcement, which lays a solid basis for the deployment of AI in pursuit of the SDGs.

Compute Infrastructure

Beyond general digital infrastructure, the AI-specific requirements are more exacting. In the context of AI, the term compute refers to the computational resources and processing power required to train and run AI models and algorithms. AI computations involve complex mathematical operations, statistical analysis, and optimisation techniques that necessitate significant computational resources. The critical aspects of compute include training AI models, processing big data, hardware acceleration, distributed computing, inference and real-time processing, optimisation and hyper-parameter tuning, and cost and efficiency considerations.

Hence, compute infrastructure refers to the components and capabilities within a computing system that are used to perform computations, process data, and execute tasks. These resources are essential for running AI applications, performing complex calculations, and managing digital workloads. Compute resources typically include hardware and software components that deliver AI computational power. Critical elements of compute resources include the Central Processing Unit (CPU), Graphics Processing Unit (GPU), Neural Processing Unit (NPU), Random-Access Memory (RAM), Storage (hard disk and solid-state drives), Network Interface, Operating System (OS), Virtualisation and Containerisation Technologies, and Cloud Computing Services.

Compute resources infrastructure is fundamental to computing systems' performance, scalability, and efficiency. Of these compute resources, the GPU has been a game changer. A GPU is a specialised electronic circuit designed to dramatically speed up the rendering of images and graphics for display on computer screens. It is a highly efficient processor capable of performing parallel computations, making it an essential component in AI, Machine Learning, Deep Learning, scientific computing, and cryptocurrency mining.

Indeed, compute infrastructure is essential for the efficacious deployment of AI in pursuit of the SDGs. This demands financial resources, human capital development, and massive investments in reliable power generation and distribution.

7.3.2 Awareness

Awareness and education are at the centre of the SDG agenda. They are critical enablers. As already articulated, raising awareness about the potential of AI for achieving SDGs, educating communities and organisations, and building capacity among stakeholders are non-negotiable. There must be deliberate capacity building focusing on equipping individuals with the knowledge and skills needed to develop, deploy, and manage AI solutions for the attainment of the SDGs.

AI's potential application to achieve the SDGs is neither widely appreciated nor leveraged in most countries across the globe. To bridge this gap, countries need a multifaceted approach that includes education, infrastructure development, policy advocacy, and community engagement. The involvement of all stakeholders is essential in this journey towards awareness of AI's potential role in the pursuit of SDGs. Several strategic initiatives, with the active participation of the community, are essential.

For a start, there is a need to engage with television, radio, newspapers, and digital platforms to run AI awareness campaigns and demystify misconceptions. Platforms like Facebook, Twitter (now X), and Instagram must be deployed to create engaging content on AI. This must include infographics, videos, and articles that explain AI's relevance to everyday life. Governments must partner with non-profits and NGOs to use AI to tackle social issues, thereby creating a narrative of AI as a tool for empowerment. Large companies should sponsor AI awareness campaigns and support educational initiatives, especially in regions where government funding is limited.

Countries must introduce AI and digital literacy courses at various educational levels, from primary schools to universities. AI-focused courses at colleges, vocational schools, and online platforms tailored to the needs and contexts of different countries are needed. The content must be customised to regional contexts, focusing on how AI can address local challenges (e.g. agriculture, health, and education). There must be promotion of AI entrepreneurship, where tech startups focus on AI in local markets, offering mentorship, seed funding, and incubation programmes. AI literacy for business leaders is essential. This can be done by conducting workshops that teach executives and small business owners how AI can transform their industries and improve operations on the journey to attain the SDGs.

Infrastructure and digital access are significant drivers of awareness. Participation leads to awareness. Governments and private organisations must work together to improve internet accessibility in underserved areas. Without access to digital infrastructure, AI awareness cannot scale. AI research and innovation hubs must be established, focusing on local issues. These centres can foster collaboration between local researchers and international experts and allow experimentation with AI solutions tailored to the region. They must use massive open online courses like Coursera and edX and platforms that offer free or subsidised courses to people in the Global South. YouTube tutorials and free coding boot camps that teach AI concepts must also be conducted. Platforms like Quantic School of Business and Technology can be examples of accessible educational tools.

Governments must develop national AI strategies and policies prioritising awareness, education, and investment. They should also adopt AI for public services (e.g. healthcare and agriculture) to demonstrate its practical applications and benefits, thus igniting awareness. They must organise AI-focused workshops, hackathons, and conferences involving students, professionals, and policymakers. There must be preparation to establish AI ecosystems and organise training programmes for these key AI players. There is a need to build and support grassroots AI communities through meet-ups and tech hubs where individuals can collaborate and learn. There must be the promotion of inclusiveness by emphasising AI solutions that address gender, economic, and rural-urban divides, ensuring AI awareness is spread across all demographics.

Furthermore, governments must facilitate AI cross-border collaboration by connecting innovators with communities grappling with the SDG agenda globally. Indeed, cooperation with international AI organisations, academic institutions, and businesses must help bring global expertise to emerging and least industrialised economies. There must be engagement with multilateral organisations such as the United Nations, World Bank, and African Union to promote AI-enabled SDG initiatives and ensure funding and policy support. AI documents from these global institutions, such as the UN Secretary-General's AI Advisory Body Final Report of September 2024, *Governing AI for Humanity* and the AU's *Continental AI Strategy* of July 2024, must be widely distributed and discussed throughout the world, with clear linkages to the SDG agenda.

Policy discussions must occur on the ethical use of AI, data privacy, and responsible innovation to ensure that AI adoption is sustainable and inclusive. Thought leaders must be engaged so that academics, researchers, and policymakers can shape AI policy debates on equity, fairness, and responsible use, with a particular focus on SDGs.

By integrating all these strategies, awareness about AI's potential role in the SDG journey can be significantly enhanced, helping most countries achieve these goals by 2030.

7.3.3 Education, Human Capital Development, and Capacity Building

Education, human capital development, and capacity building are critical enablers for the application of AI in achieving the SDGs. A well-educated workforce with knowledge in AI, data science, and related fields is essential to design, implement, and manage AI-driven solutions. Governments and institutions can empower individuals with the foundational skills needed to leverage AI technologies effectively by integrating AI and data science into school curricula and vocational training programmes. For example, initiatives like AI-focused education programmes in countries like Rwanda have equipped young people with the technical expertise to address challenges in healthcare, agriculture, and energy, directly aligning with SDG targets. Expanding access to quality AI education also helps bridge the digital skills gap, ensuring that no community is left behind in the transition to a more technology-driven world.

In addition to formal education, capacity-building programmes play a pivotal role in equipping professionals and policymakers with the knowledge to utilise AI for sustainable development. These programmes focus on enhancing skills such as data analysis, ethical AI deployment, and systems management. Workshops, online courses, and partnerships with global tech companies can accelerate the upskilling of individuals and organisations. For instance, capacity-building efforts in agricultural AI applications enable farmers to adopt precision farming techniques, improving yields and reducing resource waste, thereby contributing to SDG 2 (Zero Hunger). By fostering collaboration between academia, the private sector, and government agencies, capacity-building initiatives ensure diverse stakeholders can harness AI to tackle pressing challenges.

Moreover, education and capacity building are essential to address ethical concerns and promote inclusivity in AI deployment. Raising awareness about AI's societal implications helps cultivate a workforce prioritising fairness, transparency, and accountability in AI systems applications to achieve the SDGs. Training programmes tailored for underrepresented groups, such as women and marginalised communities, can help ensure equitable participation in the AI ecosystem. Inclusive capacity building not only expands the pool of AI talent but also ensures that the solutions developed are reflective of diverse perspectives and needs, supporting SDG 5 (Gender Equality) and SDG 10 (Reduced Inequalities). Ultimately, investment in education and capacity building lays the foundation for sustainable, inclusive, and impactful AI deployment towards achieving the SDGs.

Developing education and capacity building as enablers of AI applications to achieve the SDGs faces several challenges. A significant barrier is the digital divide, where limited access to technology and Internet connectivity in low-income and rural areas prevents equitable learning opportunities. Additionally, the lack of trained educators in AI and data science hampers the ability to deliver high-quality education and training. Financial constraints also pose a challenge, as developing AI education infrastructure, including computing facilities and tailored curricula, requires substantial investment, which many developing countries struggle to afford.

Moreover, the rapid pace of AI advancements makes it difficult for educational institutions to keep curricula updated and aligned with industry needs. Another challenge is addressing gender and socio-economic disparities, as marginalised communities often face barriers to accessing AI training programmes, exacerbating inequalities. Finally, fostering awareness about the ethical use of AI and integrating it into education systems requires significant coordination among governments, private sector players, and academic institutions, which can be complex and time-consuming. These challenges underscore the need for targeted strategies and international collaboration to ensure education and capacity building effectively support AI's role in achieving the SDGs.

These difficulties notwithstanding, several countries have successfully leveraged education and capacity-building initiatives to enable the application of AI to achieve the SDGs. For instance, Singapore has established itself as a leader by integrating AI into its education system and fostering a culture of lifelong learning through initiatives like the National AI Strategy, which focuses on upskilling the workforce and promoting AI literacy. Similarly, India has launched programmes like the AI for All initiative, aimed at democratising AI education and providing accessible training to citizens across socio-economic backgrounds, contributing to SDGs such as Quality Education (SDG 4) and Decent Work and Economic Growth (SDG 8). Rwanda, often cited as an AI pioneer in Africa, has collaborated with global tech companies to provide AI-focused education and training, enabling local professionals to use AI in healthcare, agriculture, and energy, aligning with SDGs like Good Health and Well-Being (SDG 3) and Zero Hunger (SDG 2). These countries exemplify how strategic investments in education and capacity building can empower individuals, bridge the digital divide, and foster AI-driven innovation for sustainable development, leading to economic transformation and shared prosperity.

Designing and developing AI systems, in particular LLMs, requires significant investment in human capital across a diverse range of expertise. At the foundational level, research scientists and AI engineers play pivotal roles, diving deep into the complexities of Machine Learning algorithms, Natural Language Processing, and neural network architectures. These professionals continuously push the boundaries of what's possible, experimenting with new methods to enhance the model's understanding, reasoning, and generation capabilities. Additionally, the development process leans heavily on the skills of data scientists and linguists, who curate and refine the data these models learn from, ensuring a rich, varied, and ethically sourced training dataset that helps the AI understand and generate human-like text.

In all cases, a large number of PhDs are involved in developing LLMs, depending on the scale and scope of the project. For major AI research projects, particularly those undertaken by leading tech companies or research institutions, it's common to have hundreds of PhDs involved. These individuals often specialise in fields such as computer science, AI, linguistics, cognitive science, and data science. For instance, in projects at major AI labs like OpenAI, Google Brain, or DeepMind, teams might include a substantial number of PhD holders, contributing their expertise to different aspects of the LLM, from foundational research and algorithm development to ethical considerations and implementation strategies. The depth and breadth of knowledge that PhD-level experts bring are essential for tackling the complex challenges associated with designing, training, and deploying these advanced AI systems. Universities and private companies in China employ more than 3000 PhDs to advance AI technologies, reflecting the country's commitment to competing at the forefront of AI development globally.

Beyond the technical staff, a project of the magnitude of an LLM also requires the strategic oversight of project managers and the critical input of ethicists who ensure that AI development aligns with broader societal norms and values. Ethicists and policy advisors scrutinise the model's functionality to prevent biases and promote fairness, privacy, and transparency in AI outputs. Meanwhile, user experience designers and software developers focus on making these AI systems accessible and user-friendly, integrating the advanced capabilities of LLMs into applications that people can use daily. This multidisciplinary team effort is vital for building the model and ensuring it operates within an ethical framework that respects and enhances human values.

To fully participate as effective creators and users of AI, countries must massively invest in education, human capital development, and capacity building. These interventions are essential in AI development and deployment because they equip individuals with the critical skills and knowledge needed to innovate, design, develop, ethically manage, and effectively deploy AI technologies in various sectors in pursuit of the SDG agenda.

7.3.4 Compulsory Early Teaching of AI Skills

On April 23, 2025, President Donald Trump signed an Executive Order titled "Advancing Artificial Intelligence Education for American Youth," establishing a national strategy to integrate AI education across K-12 and higher education systems. The order creates a White House Task Force on AI Education, led by the Director of the Office of Science and Technology Policy, to coordinate federal efforts. Key initiatives include launching a Presidential AI Challenge to promote student and educator innovation, developing AI curricula and resources through public-private partnerships, and prioritising AI in federal education grants and teacher training programmes. The order also directs agencies to utilise existing federal AI resources, such as the National AI Research Institutes, to support partnerships with state and local educational agencies.

Additionally, the executive order emphasises workforce development by instructing the Department of Labour to expand AI-related Registered Apprenticeships and encourage the use of Workforce Innovation and Opportunity Act (WIOA) funding for AI skill development. It also calls for the creation of AI courses and certification programmes for secondary school students, including dual enrolment opportunities. The National Science Foundation is tasked with prioritising research on the application of AI in education and creating teacher training opportunities. The Department of Agriculture is directed to integrate AI into formal and non-formal education through initiatives like 4-H and the Cooperative Extension System.

The US Presidential Executive Order on AI Education was preceded by China's April 2025 announcement of significant educational reform, mandating the integration of AI education across all primary and secondary schools starting on 1 September 2025. This initiative aims to cultivate AI literacy from an early age, with students as young as six receiving at least eight hours of AI instruction annually. The curriculum is designed to be progressive: primary students will engage in experiential learning to grasp basic AI concepts; junior high students will delve into technical principles and foundational applications; and senior high students will focus on systems thinking and innovation, applying their knowledge to real-world projects. This initiative forms part of China's broader strategy to enhance innovation and sustain economic growth by cultivating essential skills among teachers and students, such as independent thinking, problem-solving, communication, and collaboration.

To ensure the ethical and effective use of AI, the Ministry of Education has issued guidelines restricting primary students' independent use of generative AI tools and prohibiting teachers from relying on AI to replace core teaching responsibilities. These measures emphasise the importance of human-led instruction and aim to prevent overdependence on technology. The overarching goal is to build a comprehensive AI education system that fosters critical thinking, innovation, and a strong sense of social responsibility among students, positioning China as a leader in AI-driven education reform. These measures are designed to ensure that AI serves as a supplementary tool in education, promoting critical thinking and innovation while safeguarding data privacy and academic integrity.

7.3.5 Regulations and Governance

AI will rise or fall on governance and regulations—the frameworks, policies, and processes that guide the development, deployment, and use of Artificial Intelligence in ways that ensure it is ethical, transparent, accountable, and aligned with societal values. AI governance aims to manage risks and maximise the benefits of AI technologies while addressing concerns like bias, privacy, security, and the potential for misuse. The main elements of the AI regulatory and governance framework that will enable the deployment of AI in pursuit of SDG include the following:

- Ethical Guidelines—Establishing standards to ensure AI is used in a way that respects human rights, avoids bias, and promotes fairness.
- Regulatory Frameworks—Creating laws and regulations that define the legal responsibilities of AI developers and users, particularly in sensitive areas like healthcare, finance, and law enforcement.
- Transparency and Accountability—Ensuring that AI systems are explainable and decisions made by AI can be understood and audited. This includes ensuring that there is a clear chain of responsibility when AI systems cause harm or malfunction.
- Bias Mitigation—Implementing strategies to reduce or eliminate bias in AI algorithms to prevent discrimination or unfair outcomes.

- Privacy and Security—Protecting personal data from unauthorised access and ensuring AI systems are secure from cyberthreats.
- Inclusive Stakeholder Involvement—Engaging a wide range of stakeholders (including governments, corporations, civil society, and affected communities) in the governance process to ensure that AI development reflects a diversity of perspectives.
- Global Collaboration—Since AI is a global technology, AI governance often requires international cooperation to set shared standards and practices across borders.

Furthermore, it is essential to emphasise that the governance of AI involves complex balancing problems and dilemmas. These include the trade-offs between opportunity seeking and risk aversion, the interplay between security and transparency, tensions between globalisation and localisation, and the interplay between self-regulation and government control. Moreover, AI governance frameworks must balance innovation, ethics, public safety, and fairness as AI technologies evolve. Flexible, inclusive, context-aware, and predictive AI governance policies are needed. Multi-stakeholder cooperation between governments, commerce, academia, and civil society is essential to guarantee robust and adaptable AI regulatory frameworks. What is critical is a dynamic and balanced governance structure that can adapt to AI advances to ensure the technology delivers maximum benefits while exhibiting minimum risks [79]. The application of AI to achieve the SDGs must be cognisant of this position.

Indeed, it is instructive to restate that efficacious AI governance is essential in ensuring that AI is developed and used responsibly, contributing to sustainable societal progress rather than exacerbating inequality or harm. However, assessing the similarities and differences in governance imperatives between different jurisdictions is prudent. Governance of AI in the Global North and Global South should be tailored to each region's unique economic, social, cultural, and technological contexts while maintaining certain universal principles. There are key considerations on how AI governance might differ and areas where alignment is essential between the two broad categories of countries [12,18,19].

Governance Differences: Global South vs. Global North

AI governance frameworks in the Global North are often shaped by advanced technological infrastructure, robust legal systems, and significant private sector innovation. Governance tends to emphasise data privacy (e.g. General Data Protection Regulation (GDPR) in the EU), intellectual property, and ethical concerns around AI in highly digitised societies. In the Global South, governance must account for different priorities, such as leveraging AI to address development challenges, like healthcare access, education, and agriculture. Infrastructure gaps, digital literacy, and differing regulatory capacities also influence how governance frameworks are designed.

Ethical debates in the North often focus on issues like AI autonomy, bias in algorithmic decision-making, and job displacement in advanced economies. AI governance also tends to emphasise the prevention of AI-induced inequalities within relatively affluent societies. The ethical focus in the Global South should prioritise AI's potential to drive development and reduce inequality while also addressing risks such as digital exclusion and algorithmic bias, particularly affecting marginalised communities. The governance frameworks must ensure that AI technologies do not exacerbate socio-economic divides and that global AI innovations do not disproportionately benefit the North.

Another area of interest is capacity building and access to AI. Global North countries have established AI research and innovation institutions, allowing for a more mature governance structure. Regulations might focus on fine-tuning aspects of AI development such as data ownership, advanced AI systems, and ensuring ethical standards in frontier technologies. On the other hand, capacity building is a critical component in emerging and least industrialised countries. AI governance in the Global South should focus on enabling access to AI technologies and fostering local talent and innovation. It should address the disparity in AI resources and support building technological infrastructure while ensuring equitable access to AI's benefits across diverse socio-economic groups.

Advanced economies generally have significant control over data infrastructure and set global standards for data privacy (e.g. Europe's GDPR). They are also home to most global tech giants that drive AI innovations and collect vast amounts of data. AI governance in the Global South may emphasise data sovereignty to ensure local control over data that originates from these regions. Global South countries might focus on ensuring that AI development does not lead to data extraction by foreign entities, and they may promote policies encouraging local data storage, access, and processing. Furthermore, these economies must develop and enact IP laws to create a conducive environment for AI research, development, and production.

As leaders in AI innovation, countries in the Global North often drive the creation of global AI standards. Their governance frameworks tend to set the tone for ethical AI, privacy protection, and technical requirements. The Global South needs to have an active voice in the creation of international AI standards, ensuring that these norms account for their development goals and resource limitations. Governance frameworks should encourage collaboration between regions to prevent the replication of Northern models that might not suit Southern contexts and to promote a more inclusive global AI ecosystem.

In the Global North, AI governance here is often focused on regulating the use of AI in cutting-edge industries (e.g. autonomous vehicles, financial markets, and healthcare). The focus is often on balancing innovation with ethical considerations and ensuring that advanced AI technologies are integrated responsibly. On the other hand, in the Global South, AI governance should prioritise AI's role in advancing sectors critical to development, such as agriculture, education, health, and infrastructure. Governance frameworks should ensure that AI technologies are aligned with local development goals and address pressing issues such as poverty reduction, public health crises, and education.

The private sector in the Global North drives much of AI development, and governance frameworks often revolve around regulating large tech companies. Governments work closely with these corporations to ensure that their AI systems meet ethical and legal standards. AI governance in the Global South should focus on fostering public-private partnerships that address local needs. This might include encouraging international corporations to invest in AI solutions for social good or promoting local startups and innovators who can build AI systems that reflect local contexts and challenges.

While there are efforts in the Global North to use AI for social good (e.g. in areas like healthcare and climate action), much of AI innovation remains profit-driven. AI governance in the Global South should encourage the development of AI for social good as a priority, leveraging AI to solve critical development issues such as food security, healthcare, and education. Governance should also facilitate access to global AI innovations that can be adapted for local needs, ensuring that AI is not just a tool for profit but for public welfare.

With greater AI resources and capabilities, countries in the Global North often shape global AI policies to their benefit. AI governance frameworks in the Global South need to proactively address potential power asymmetries by developing strategic alliances and asserting the region's interests in global AI dialogues. This includes pushing for fair access to AI innovations and advocating for equitable terms in global AI trade.

Indeed, while there are shared concerns in AI governance, such as the need for ethics, transparency, and accountability, the Global South, particularly Africa, requires governance frameworks that prioritise development, inclusivity, capacity building, and igniting local AI development. Global North governance, by contrast, may focus more on regulating advanced AI technologies and addressing risks related to privacy and autonomy in more developed AI ecosystems. As the Global

South moves into AI research and development, the governance concerns between the two groups of countries will converge. All AI governance frameworks must be robust, balanced, dynamic, and adaptable to accommodate the rapid and relentless changes and complex dilemmas that characterise the technology [79, 86].

7.3.6 Guardrails for AI Deployment

To complement conventional AI regulations and governance, there is a need for guardrails—safeguards, frameworks, or mechanisms that ensure that AI systems operate safely, ethically, and as intended, minimising risks and preventing undesirable or harmful outcomes. These guardrails can be technical, ethical, legal, or procedural and designed to address the challenges and risks associated with AI development and deployment. Five types of guardrails are essential in deploying AI to achieve the SDGs.

Technical Guardrails

- Bias Mitigation: Ensuring the AI system is fair and free of biases in training data or decision-making processes.
- Robustness: Making AI systems resilient to adversarial attacks or unexpected inputs.
- Explainability: Providing transparency about how AI systems make decisions.
- Safety Mechanisms: Incorporating features like kill switches, error detection, and fail-safe modes.

Ethical Guardrails

- Human-Centric Design: Ensuring AI systems prioritise human values and well-being.
- Non-Discrimination: Preventing AI from perpetuating or amplifying social inequalities.
- Privacy Protection: Safeguarding user data and ensuring compliance with privacy standards.

Legal and Regulatory Guardrails

- Compliance with Laws: Adhering to local and international laws, such as GDPR for data protection.
- Accountability Mechanisms: Defining who is responsible when AI systems cause harm or errors.
- AI-Specific Legislation: Following emerging regulations tailored to AI, like the EU AI Act.

Organisational and Process Guardrails

- AI Governance: Setting up policies and oversight bodies to monitor AI use.
- Stakeholder Engagement: Involving diverse groups in the development and deployment of AI.
- Ethical Review Boards: Evaluating AI projects to ensure they align with ethical principles.

Importance of Guardrails in AI

- Preventing Harm: They ensure that AI systems do not cause physical, emotional, or societal harm.
- Building Trust: They promote confidence in AI technologies by demonstrating commitment to safety and ethics.
- Fostering Innovation: They encourage the development of AI systems that are beneficial and widely accepted.
- Regulatory Compliance: They help organisations avoid legal repercussions.

Examples of Guardrails in Action

- Content Moderation: AI models for chatbots or social media use filters to avoid harmful or inappropriate content.
- Autonomous Vehicles: Safety protocols and fail-safes to prevent accidents.
- Healthcare AI: Ethical guidelines ensuring patient data is handled carefully and decisions are explainable.

As AI systems become more integrated into critical aspects of the economy and society, establishing and adhering to strong guardrails will be essential to ensure their positive impact. Guardrails mitigate risks and pave the way for AI systems to become trusted partners in solving complex problems and improving human lives. This is of particular significance in the application of AI to achieve the SDGs [81, 152].

7.3.7 Access to High-Quality Local Data

Two key drivers have ignited the AI revolution—data and advanced compute resources. The data that drives AI systems is vast and varied and depends on the specific type of AI model and its application. This data is the foundation for training, testing, and optimising AI systems, allowing them to make predictions, automate tasks, and generate insights. Indeed, AI thrives on data, making access to high-quality, diverse, and representative datasets an indispensable enabler. Accurate and relevant data is critical for training AI models to address specific challenges, such as predicting disease outbreaks, monitoring environmental changes, or optimising resource allocation.

Deploying AI in pursuit of the SDGs in a particular country or region will be more efficacious if the data is generated locally. All countries need to build capabilities and capacity to generate, harvest, curate, and manage data as they pursue the attainment of the SDGs. The bespoke and context-aware data can be used to build AI systems that are localised to the region's specific needs, languages, cultures, and challenges. This is particularly critical for AI-enabled attainment of the SDGs. AI solutions that are trained primarily on data from developed countries might perform poorly in the Global South due to language differences. AI systems trained on local languages like Swahili, Hindi, and Yoruba can enhance the development of chatbots, translation tools, and digital assistants. Understanding local customs, practices, and social structures is essential for AI to be culturally sensitive. For sectors like agriculture, healthcare, and urban planning, data related to local climate conditions, disease patterns, and infrastructure are essential for creating contextually relevant AI models.

The Global South often relies on technology and data from more developed regions, which can result in importing foreign models and systems not tailored to local needs. By generating their own data, countries in the Global South can reduce reliance on Western-dominated tech companies. Instead of relying on AI models built by companies based in the Global North, which may not align with local priorities, the Global South can create its own solutions. Emerging and least industrialised economies can gain control over their own technological destiny. Control over local data ensures that AI innovations are driven by domestic needs and aspirations rather than external agendas.

One of the major issues in AI development is bias. Many existing AI models are trained on datasets that predominantly reflect Western realities, which can lead to models that underperform in the Global South. Furthermore, models may need to accurately predict outcomes for non-Western populations in areas like healthcare, credit scoring, and voice recognition. Using data exclusively from developed countries risks creating algorithms that inadvertently perpetuate stereotypes or make inaccurate predictions for people in the Global South. By generating more diverse data from various countries in the Global South, AI models can become more inclusive, reducing bias and creating a more globally fair system.

Generating data used in AI systems will drive economic development and innovation, creating an environment where SDGs are achievable. Collecting and using local data can stimulate innovation, job creation, and economic growth in the Global South by fostering new industries. Data-driven businesses such as AI-based startups, fintech solutions, and agri-tech innovations can emerge, leading to economic development. AI systems tailored to local industries like agriculture, energy, and health can improve efficiency and drive productivity gains, boosting overall economic performance. By focusing on data generation and AI model development, the Global South can train a new generation of data scientists, AI engineers, and technology entrepreneurs, building a future-ready workforce.

Furthermore, generating data in emerging and least industrialised economies will address unique socio-economic challenges. The Global South faces specific problems that differ from the Global North. These include diseases prevalent in the region, such as malaria, tuberculosis, and neglected tropical diseases. These require specialised AI systems for diagnosis and treatment trained on local health data. AI systems can use local data to address food security challenges by predicting weather patterns, optimising farming practices, and preventing crop diseases.

Local data will also drive urbanisation and infrastructure development. AI models built using data from rapidly urbanising regions can help design smarter cities, improve public transportation, and address housing shortages. By generating data, the Global South can develop AI solutions tailored to these unique challenges, improving the quality of life and driving sustainable development.

Local Data Generation

If the Global South actively generates data, it can avoid being underrepresented in the global AI ecosystem. Underrepresentation can have several negative consequences. Without local data, AI systems might make decisions that affect people in the Global South without fully understanding their needs, reducing their agency over their own futures. Actors from the Global North often shape global AI ethics and governance debates. Data generating allows the Global South to participate meaningfully in shaping global AI policies and ensure that international regulations consider local perspectives. Generating AI data in the Global South will strengthen data sovereignty and privacy protections, minimising and mitigating digital imperialism and data colonialism. Data sovereignty is the concept that countries should control the data generated within their borders. Generating data locally helps to protect privacy by keeping sensitive data within the country and adhering to local privacy laws; countries in the Global South can ensure that citizens' data is not exploited by foreign entities or used for purposes that may not align with national interests. As more data is generated, countries can establish better data governance frameworks to regulate its collection, use, and protection, thus reinforcing digital sovereignty.

The Global South already faces a digital divide regarding access to technology and AI innovation. By generating data, these regions can accelerate digital transformation. Local data allows for more effective deployment of AI technologies that suit local contexts, helping close the digital services gap between the Global South and the Global North. By contributing data for AI systems, the region can develop applications that enhance digital inclusion, such as tools for education, financial inclusion, and healthcare access. More importantly, owning the data presents the potential and opportunity to monetise it, thus accruing economic benefits to these emerging and least industrialised countries.

The Global South must contribute to knowledge production, innovation, and technological advances. The region's participation in data generation can also enhance global knowledge. The region's unique contexts and challenges provide an opportunity for AI models developed in the Global South to contribute solutions to global problems like climate change, migration, and economic inequality. Producing local data fosters collaboration between governments, academic institutions, and private sector actors from both the Global South and Global North. Such partnerships can lead to knowledge exchange and joint research projects.

Indeed, in-country data generation for AI systems will drive local innovation, reduce dependence on foreign technologies, and tackle unique regional challenges. Additionally, this effort will enhance AI fairness, reduce biases, and promote inclusion while safeguarding data sovereignty and privacy. Indeed, producing local data is essential for ensuring that AI technology is effectively used to achieve SDGs. This data imperative means that the countries must build the capacity to generate, harness, store, and manage data for AI systems. Such an endeavour is critical to countries' preparation for AI deployment to achieve the SDGs.

7.3.8 Financial Resources and Investment

As already intimated, achieving the SDGs through the deployment of AI requires substantial financial resources to address several critical implementation aspects. Investments are needed to develop and maintain the technological infrastructure required for AI, such as high-speed Internet, data centres, and cloud computing systems, particularly in under-resourced regions. Additionally, funding is essential for research and innovation to create AI solutions tailored to specific SDGs, like predictive analytics for disaster management, personalised education platforms, and smart agriculture technologies.

Capacity-building initiatives, including training for AI specialists and policymakers, demand financial support to ensure inclusive access and minimise the risk of exacerbating inequality. Moreover, resources are required to establish ethical frameworks, safeguard data privacy, and build trust in AI systems. By allocating sufficient financial resources, nations can unlock AI's potential to accelerate progress across all SDGs, fostering efficiency, innovation, and equitable development.

Various strategies must be executed to raise the finances and investments required to achieve the SDGs through AI deployment. Indeed, resources plus resourcefulness equal results.

Public-Private Partnerships (PPPs)
Public-private partnerships are critical for mobilising resources to fund AI-driven SDG initiatives. Governments can collaborate with private corporations to co-finance projects, leveraging public resources and private sector expertise. For instance, partnerships between tech companies and public institutions can support AI research in healthcare or education. Tax incentives, subsidies, or co-investment frameworks can encourage corporations to align their AI innovations with SDG priorities, ensuring mutually beneficial outcomes.

International Development Aid and Grants
Development aid from international organisations and donor countries can provide substantial financial support for AI deployment in developing regions. Multilateral organisations like the United Nations, World Bank, and African Development Bank can allocate funds for AI-driven projects that tackle pressing issues like hunger, education inequality, and climate change. Additionally, grants from global foundations such as the Gates Foundation and Rockefeller Foundation can fund pilot projects demonstrating AI's impact on SDGs, attracting further investments.

Impact Investment and Social Finance
Impact investing focuses on generating measurable social and environmental benefits alongside financial returns. Investors can fund AI startups or initiatives addressing SDGs, such as AI solutions for precision agriculture or renewable energy optimisation. Social finance instruments, including

development impact bonds and green bonds, can also be structured to attract investors committed to SDG-aligned outcomes, using AI to deliver measurable results and ensure accountability.

Leveraging Domestic Resource Mobilisation

Countries can enhance domestic resource mobilisation through progressive taxation, curbing illicit financial flows, and improving fiscal management. Revenue generated can be directed towards AI infrastructure and research for SDG-related projects. Additionally, governments can allocate budgetary resources to establish national AI innovation hubs or provide seed funding for local startups focusing on SDG challenges and fostering home-grown solutions.

Crowdfunding and Community-Based Initiatives

Crowdfunding platforms provide an innovative way to engage individuals and communities in financing AI-driven SDG projects. By showcasing transparent, impactful projects, organisations can attract small contributions from a broad base of supporters. Community-based initiatives, such as cooperative financing, can also mobilise resources at the grassroots level, enabling local implementation of AI solutions tailored to specific SDG challenges.

Corporate Social Responsibility (CSR) Programmes

Corporations increasingly integrate sustainability into their CSR strategies, making them valuable partners in funding AI deployment for SDGs. Tech companies, in particular, can provide direct funding, technology donations, or expertise for AI-based initiatives. Collaborating with corporations committed to ethical and sustainable innovation can ensure a steady flow of resources for projects that align with both corporate values and global development goals.

Harnessing Emerging Fintech Innovations

Innovative financial technologies, such as Blockchain, can create transparent and efficient funding mechanisms for SDG initiatives. Blockchain-based platforms can facilitate global donations, track expenditures, and ensure accountability in deploying AI projects. Additionally, tokenised fundraising or decentralised finance (DeFi) solutions can engage a broader range of investors and philanthropists, providing an alternative and scalable funding source.

Overall Assessment

Raising the financial resources necessary to achieve the SDGs through AI deployment requires a multifaceted approach involving governments, private sectors, international organisations, and communities. By leveraging public-private partnerships, international aid, impact investments, and innovative funding mechanisms, it is possible to mobilise the required capital while fostering accountability and inclusivity. The successful realisation of these strategies will not only accelerate progress towards the SDGs but also ensure that AI serves as a tool for equitable and sustainable global development—specifically by enabling the achievement of the SDGs.

Beyond the outlined funding initiatives, countries in the Global South must internally unlock financial resources by moving up the global value chains—economic transformation through beneficiation and value addition. Most of these emerging and least developed countries are too resource-rich to be financially poor. Raising finances for the deployment of AI on the ambitious journey to attain the SDGs by 2030 requires rethinking Global South economic models.

7.3.9 Beneficiation and Value Addition

As already discussed, the SDGs present a global framework to address critical challenges such as poverty, inequality, and environmental degradation by 2030. Beneficiation and value addition, particularly in resource-rich developing nations, offer transformative pathways for achieving these UN objectives. How is this so? Beneficiation and value addition refer to processes aimed at increasing the economic value of raw materials by enhancing their quality, functionality, or utility through processing and transformation.

Beneficiation typically focuses on upgrading raw materials, such as minerals, by removing impurities or improving their concentration before they are sold or used in manufacturing. On the other hand, value addition involves further transforming these materials into finished or semi-finished goods, thereby enhancing their market worth. For instance, refining crude oil into fuel or turning raw cocoa beans into chocolate are examples of value addition. These processes generate higher revenues, create jobs, foster industrial growth, and reduce dependency on exporting unprocessed resources, making them essential strategies for sustainable economic development.

In this context, beneficiation and value addition are catalysts for achieving the SDGs. One of the key SDGs, Goal 1 (No Poverty), can be addressed through beneficiation and value addition. By moving beyond the export of raw materials, countries can maximise the financial returns from their natural resources. This increased income can be reinvested in poverty alleviation programmes, infrastructure, and social services. For instance, instead of exporting unprocessed minerals, developing local industries to refine and manufacture finished goods can create value chains that lift people out of poverty and stimulate broader economic growth.

Beneficiation also plays a vital role in advancing Goal 8 (Decent Work and Economic Growth). Processing raw materials locally creates opportunities for skilled and semi-skilled jobs in industries such as manufacturing, refining, and logistics. Additionally, such practices promote industrial diversification, reducing a country's dependence on volatile global commodity prices and making its economy more resilient to external shocks.

Value addition is central to Goal 9 (Industry, Innovation, and Infrastructure), driving industrialisation and technological advancement. By investing in infrastructure and research to support local processing industries, countries can build a foundation for innovation and sustainable industrial growth. This industrial base supports local economies, attracts foreign investment, and strengthens global competitiveness.

Environmental sustainability, encapsulated in Goal 12 (Responsible Consumption and Production) and Goal 13 (Climate Action), is another area where beneficiation and value addition can have a positive impact. Processing resources locally can reduce the carbon footprint associated with transporting raw materials overseas. Furthermore, implementing environmentally friendly technologies in these processes can mitigate the environmental impact of resource extraction and processing, contributing to sustainable resource management.

The social impact of beneficiation and value addition is significant, particularly for Goal 10 (Reduced Inequalities) and Goal 5 (Gender Equality). Localised industries provide equitable employment opportunities, often empowering marginalised groups, including women, to participate in economic activities. This inclusivity reduces income disparities and ensures that the benefits of resource exploitation are shared more broadly across societies.

Goal 17 (Partnerships for the Goals) highlights the importance of global collaboration, which is crucial for successful beneficiation and value addition. Developing countries often lack the technical expertise and financial resources required for advanced processing. Partnerships with international stakeholders, including governments, private sectors, and academic institutions, can facilitate knowledge transfer, capacity building, and access to capital, ensuring the sustainability of these initiatives.

Lastly, beneficiation and value addition can contribute to Goal 16 (Peace, Justice, and Strong Institutions) by reducing resource-related conflicts. Promoting transparent governance and equitable resource distribution minimises grievances, which often lead to social unrest in resource-rich but economically marginalised communities.

Indeed, beneficiation and value addition are powerful tools for achieving the SDGs. They enhance economic prosperity, create sustainable jobs, promote industrial innovation, and reduce environmental impact, while fostering social inclusion and reducing inequalities. Governments, businesses, and civil society must collaborate to invest in these processes, leveraging natural resources for short-term gains and attaining the SDGs.

As already indicated, the beneficiation and value-addition agenda is existential for most countries in the Global South. These economies must stem the net flow of financial resources from the Global South to the Global North. This can be achieved by exporting refined and value-added products into global markets and not engaging in extractive economics of churning out cheap and undervalued raw materials. Unfortunately, this imperative has not been fully grasped across the Global South, particularly in Africa.

For illustration, during US President Joe Biden's historic visit to Angola in December 2024, discussions with African leaders from the region concentrated on the Lobito Corridor project—spanning Angola, the Democratic Republic of the Congo (DRC), and Zambia. It is a 1300 km railway project costing US$5 billion. The primary objective is to enhance access by the United States and its G7 partners to the region's (raw and unrefined) critical minerals. Really? Why would African leaders from Angola, DRC, and Zambia entertain this? More thoughtful and relevant consultations with the US President should have been on investment into the US$80 billion Grand Inga Dam project (a series of 7 hydroelectric projects located on the Congo River) with a potential of 42,000 MW (enough power for all SADC countries) and building manufacturing infrastructure (to add value to the minerals before exporting them) in the region. Figure 7.2 shows the location and role of the Lobito Corridor, while Fig. 7.3 dramatises how the Grand Inga Dam will ignite and drive the industrialisation of the African continent.

Benefits of Beneficiation and Value Addition

It is instructive to summarise and emphasise the benefits of value addition and beneficiation in contributing to economic diversification by moving beyond the export of raw materials. Developing industries that process and add value to natural resources can reduce dependence on commodity price fluctuations and enhance economic resilience. Indeed, processing raw materials locally creates job opportunities, from skilled labour in manufacturing to support roles in logistics, marketing, and distribution. This contributes to poverty reduction and improved livelihoods for the population. Implementing value-addition processes often involves adopting advanced technologies and developing local skills. This leads to technology transfer, knowledge acquisition, and the growth of a skilled workforce, fostering innovation and industrial development. Furthermore, value addition can lead to higher returns on exports. Processed or refined products often command higher prices in the international market compared to raw materials, contributing to increased revenue for countries engaged in value addition.

Beneficiation and value addition will not just happen. They require enablers and drivers to be in place: infrastructure, financing, skills and the right mindset. Building the necessary infrastructure, such as processing plants, transportation networks, power generation and distribution, and digital platforms, supports economic growth. This infrastructure development can positively impact the country or region's overall development. The injection of new investments, the production of new

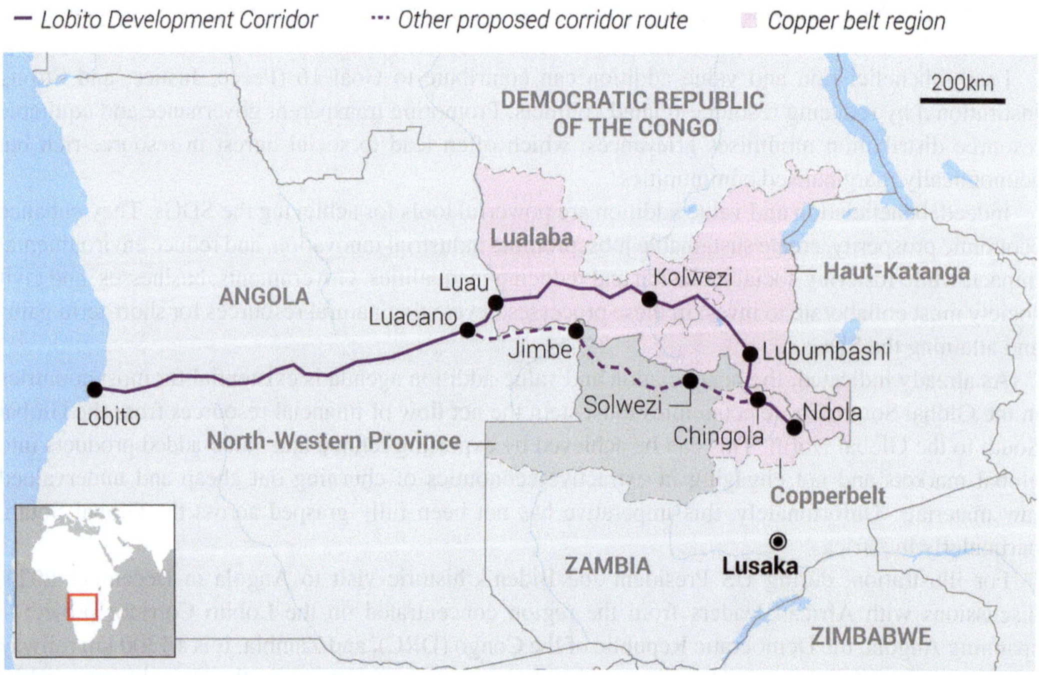

Fig. 7.2 Lobito Corridor: Facilitating export of raw critical minerals

Fig. 7.3 Grand Inga Dam: Powering African industrialisation [49]

skills/competencies, and the can-do mentality will drive economic transformation, leading to shared prosperity.

Value-adding processes include environmentally sustainable practices, such as recycling and responsible resource management. These practices contribute to reducing the environmental impact associated with the extraction and export of raw materials. Processed and value-added products may have easier access to international markets, as they often meet higher-quality standards and cater to specific consumer preferences. This enhances the competitiveness of products from the Global South. Beneficiation and value addition demand new policy frameworks and incentives. Governments must play a crucial role in creating an enabling environment for value addition. Implementing supportive policies, providing incentives for local industries, and addressing regulatory barriers contribute to the success of beneficiation initiatives.

Integrating value-addition processes into the broader supply chain enhances collaboration between different sectors. This integration can lead to synergies, increased efficiency, and a more interconnected and resilient economy. Attracting foreign and domestic investments in value addition industries requires a conducive investment climate. Governments can implement policies encouraging investment, reducing bureaucratic hurdles, and providing financial incentives. All this is anchored by stakeholder collaboration between governments, private sector entities, local communities, and international partners. This is essential for the success of value-addition initiatives. Engaging stakeholders in decision-making processes ensures inclusivity and sustainable development.

Promoting value addition and beneficiation, particularly in countries in the Global South, can unlock the full economic potential of their natural resources, contribute to sustainable development goals, and create a foundation for long-term prosperity. The success of these initiatives often relies on comprehensive planning, effective governance, and ongoing collaboration among various stakeholders.

Role of AI in Value Addition and Beneficiation

AI significantly influences value addition and beneficiation across various industries. AI technologies contribute to optimising processes, improving efficiency, and enhancing decision-making in ways that can increase the value of raw materials or resources. Here are several ways in which AI can impact and support value addition and beneficiation.

AI enables the optimisation of manufacturing and production processes. Machine Learning algorithms can analyse data from various stages of production to identify inefficiencies, bottlenecks, and opportunities for improvement, leading to increased productivity and cost-effectiveness. AI-powered systems can be employed for quality control and assurance in manufacturing. Computer vision and Machine Learning models can identify defects, deviations, or inconsistencies in the production line, ensuring that the final products meet high-quality standards.

AI algorithms can predict equipment failures and maintenance needs by analysing historical data and real-time sensor information. Predictive maintenance helps prevent unexpected breakdowns, reduce downtime, and extend the lifespan of machinery used in value-adding processes. AI enhances supply chain efficiency by optimising inventory management, demand forecasting, and logistics. This ensures a seamless flow of materials and components throughout the value chain, reducing delays and improving overall supply chain performance.

AI can assist in the efficient management of natural resources. For example, in agriculture, AI-powered precision farming can optimise irrigation, fertilisation, and pest control, leading to increased crop yields and improved resource utilisation. AI technologies contribute to energy efficiency in manufacturing processes. Guided by AI algorithms, smart energy management systems can analyse energy consumption patterns and optimise usage to reduce costs and environmental impact. Furthermore, AI-driven product customisation allows for more personalised and tailored offerings.

This can be particularly relevant in industries such as textiles, where AI algorithms can analyse consumer preferences to guide the production of customised garments. AI facilitates data analytics to extract actionable insights from large datasets. Decision-makers can use these insights to make informed choices regarding production strategies, market trends, and resource allocation, optimising the value-addition process.

AI-powered tools can analyse market trends, consumer behaviour, and competitive landscapes. This information is valuable for decision-makers in determining which products or features will have the highest demand and value in the market. AI contributes to innovation by supporting research and development processes. Natural Language Processing (NLP) and Machine Learning algorithms can analyse scientific literature, patents, and market trends to identify opportunities for new product development.

Supply chain traceability is essential. AI technologies, such as Blockchain combined with Machine Learning, can enhance supply chain traceability. This is particularly relevant in industries like mining and agriculture, where consumers increasingly demand transparency regarding products' origin and production practices. Moreover, AI can assist in risk management by identifying potential risks in the value chain, whether related to supply chain disruptions, market volatility, or regulatory changes. This enables proactive measures to mitigate risks and ensure the continuity of operations.

Indeed, incorporating AI into value-addition and beneficiation processes requires strategic planning, investment in technology infrastructure, and the development of a skilled workforce. While AI offers significant opportunities for efficiency and innovation, addressing ethical considerations, data privacy, and potential societal impacts is essential as these technologies are integrated into industrial processes.

7.3.10 Beneficiation, Value Addition, and AI

The interplay between AI and beneficiation/value addition is synergistic and mutually reinforcing. Across the world, beneficiation and value addition can significantly enhance the use of AI to achieve the SDGs by creating an ecosystem where resources, data, and technology converge to drive sustainable growth. By integrating AI into the beneficiation and value-addition processes, industries can optimise resource utilisation, improve efficiency, and foster innovation, directly contributing to the SDGs. Below are key ways in which these synergies can be realised:

- Boosting Economic Growth (SDG 1 and SDG 8): AI can revolutionise beneficiation by automating resource extraction and processing. Advanced AI algorithms can optimise mining operations, reducing waste and increasing yield. Value-addition processes like refining and manufacturing benefit from AI-driven production lines, predictive maintenance, and quality control. This enhanced efficiency leads to cost savings and higher profitability, generating more revenue to reinvest in poverty alleviation and job creation. Furthermore, AI-powered analytics can identify market trends, ensuring that finished products meet global standards and attract higher value.
- Advancing Industry and Infrastructure (SDG 9): AI facilitates smart industrialisation by enabling predictive modelling and real-time monitoring in beneficiation and value-addition facilities. For example, AI-driven simulations can design more efficient production methods, reducing resource inputs and energy consumption. By implementing AI in logistics and supply chain management, industries can streamline operations and ensure the timely delivery of value-added products. This strengthens industrial infrastructure, fosters innovation, and positions countries to compete in global markets.

- Promoting Responsible Consumption and Production (SDG 12): AI's role in beneficiation and value addition extends to sustainable resource management. Machine Learning models can analyse complex datasets to track resource usage, forecast depletion rates, and suggest recycling or alternative materials. In manufacturing, AI can minimise waste through precision engineering and optimise energy usage, ensuring responsible consumption and production patterns. These practices align with circular economy principles, promoting sustainability throughout the value chain.
- Enhancing Environmental Sustainability (SDG 13): AI can significantly reduce the environmental footprint of beneficiation and value addition. For example, AI-powered sensors and data analytics can monitor emissions, water usage, and energy consumption in real time, ensuring compliance with environmental standards. Predictive AI models can also simulate the long-term environmental impacts of industrial activities, enabling proactive mitigation strategies. By incorporating AI in beneficiation, industries can transition to cleaner technologies, directly contributing to climate action goals.
- Reducing Inequalities Through Inclusion (SDG 5 and SDG 10): AI democratises access to technology and information, enabling marginalised groups to participate in value-addition and beneficiation processes. For instance, AI-driven platforms can provide training and capacity building for local communities, including women and youth, equipping them with the skills needed to engage in high-value industries. Furthermore, AI can ensure equitable distribution of benefits by monitoring supply chains and enforcing transparency, reducing inequalities in resource-rich but economically marginalised regions.
- Facilitating Partnerships and Collaboration (SDG 17): The integration of AI into beneficiation and value-addition processes necessitates partnerships between governments, private sectors, and research institutions. These collaborations can drive innovation by pooling resources and expertise to develop AI solutions tailored to local contexts. For example, global partnerships can leverage AI to optimise cross-border supply chains, facilitate knowledge transfer, and build capacity for sustainable industrial practices, aligning efforts to achieve the SDGs.
- Optimising Resource Allocation (SDG 2 and SDG 11): AI can enhance resource allocation in agriculture and urban development, where beneficiation and value addition play a critical role. For example, AI-enabled precision agriculture can optimise the use of fertilisers and water, enhancing the value of agricultural produce. In urban settings, AI can improve waste recycling and the beneficiation of secondary raw materials, contributing to sustainable cities and communities.
- Fostering Peace and Strong Institutions (SDG 16): AI can promote transparency and accountability in beneficiation and value addition by using Blockchain technology and advanced analytics to monitor resource flows, prevent corruption, and reduce illicit trade. Such measures can reduce resource-related conflicts and foster trust among stakeholders. Additionally, AI can enhance governance by providing policymakers with actionable insights for informed decision-making, strengthening institutions responsible for managing natural resources.

When combined with AI, beneficiation and value addition create a transformative nexus that accelerates progress towards achieving the SDGs. By enhancing efficiency, promoting sustainability, and fostering inclusivity, AI empowers industries to derive maximum value from resources while ensuring equitable and environmentally conscious practices. Governments, industries, and academia must work together to invest in AI technologies and integrate them into beneficiation and value-addition processes, unlocking their full potential for sustainable development.

7.3.11 Research and Development

Research and development (R&D) play a pivotal role in enabling the deployment of AI to achieve the SDGs. Understanding the essence, requirements, and impact of R&D is the first step. Research and development refer to systematic and creative activities undertaken to enhance knowledge and develop new products, processes, or technologies. Research focuses on exploring and understanding phenomena, identifying gaps in existing knowledge, and generating insights that form the foundation for innovation. Development builds on these findings, applying them to create practical solutions, refine existing technologies, or bring new concepts to market. R&D spans various fields, including science, technology, medicine, and social sciences, and serves as a critical driver of economic growth, industrial competitiveness, and societal advancement.

In the context of modern challenges, R&D is essential for solving complex problems such as climate change, healthcare access, and sustainable development. It fosters innovation by encouraging experimentation and the pursuit of groundbreaking ideas. By investing in R&D, organisations and governments can stay at the forefront of technological advancements, improve efficiency, and address pressing global issues. Additionally, R&D promotes knowledge sharing, collaboration, and skills development, ensuring that societies can adapt to emerging opportunities and challenges in an increasingly interconnected world.

This appraisal of R&D envisages its enablement of the deployment of AI to achieve the SDGs. By driving and fostering innovation, R&D accelerates the development of AI technologies tailored to address global challenges such as poverty, inequality, climate change, and education. Through continuous exploration and experimentation, R&D ensures that AI is advanced and aligned with the principles of inclusivity, sustainability, and ethical responsibility. The following are illustrative examples of how R&D enhances the impact of AI in assisting the achievement of the SDGs:

- Enhancing Economic Growth (SDG 1 and SDG 8): R&D in AI enables the creation of tools that drive economic development by optimising productivity and resource management. For instance, AI-powered predictive models can streamline agricultural practices, ensuring food security and poverty alleviation. Similarly, research into industrial automation and Machine Learning algorithms enhances manufacturing processes, creating jobs and boosting economic activity. By developing context-specific AI applications, R&D ensures economic benefits reach even the most underserved communities, addressing inequalities and fostering sustainable livelihoods.
- Revolutionising Healthcare (SDG 3): AI's potential to transform healthcare is realised mainly through focused R&D. Innovations in AI-driven diagnostics, personalised medicine, and predictive analytics are enabling healthcare providers to offer efficient and affordable care. For example, Machine Learning models trained to detect diseases early or optimise treatment plans reduce healthcare disparities and improve outcomes. Additionally, R&D efforts ensure that these solutions are adaptable to low-resource settings, making healthcare more accessible in underserved regions.
- Promoting Quality Education (SDG 4): AI, bolstered by R&D, is transforming education by creating personalised learning experiences and improving access to quality education. Research in Natural Language Processing (NLP) and adaptive learning systems has led to the development of platforms that cater to diverse learning needs and languages. AI-powered tools can assess individual student progress and provide tailored content, fostering inclusive education. Furthermore, R&D ensures the scalability of these technologies, enabling widespread adoption in remote or underprivileged communities.
- Advancing Sustainable Cities and Communities (SDG 11): R&D is integral to the application of AI in urban planning and the development of smart cities. Research efforts enable AI systems to analyse data from sensors, satellite imagery, and public records to optimise transportation, reduce

energy consumption, and enhance waste management. By focusing on sustainability, R&D ensures that AI applications in urban settings minimise greenhouse gas emissions and improve the quality of life for all residents, especially marginalised groups.

- Tackling Climate Change (SDG 13): R&D drives the development of AI technologies that monitor and mitigate the effects of climate change. For example, research into AI-powered environmental monitoring systems enables the real-time tracking of deforestation, air quality, and carbon emissions. Machine Learning algorithms can also optimise renewable energy production, enhance weather prediction accuracy, and support disaster response efforts. R&D ensures these technologies can adapt to evolving environmental challenges by continuously advancing AI's capabilities.
- Building Ethical and Inclusive AI (SDG 10 and SDG 16): R&D ensures that AI technologies are developed responsibly, addressing issues such as bias, transparency, and accountability. By focusing on ethical AI, researchers can create systems that promote equity, reduce inequalities, and prevent the exploitation of vulnerable populations. For instance, R&D efforts in explainable AI (XAI) ensure that decision-making processes are transparent and fair, fostering trust in AI systems. This contributes to building strong institutions and equitable societies.
- Strengthening Global Partnerships (SDG 17): The deployment of AI to achieve the SDGs requires collaborative R&D efforts across countries, industries, and academia. Joint research initiatives can pool resources, expertise, and data to accelerate innovation and ensure the global applicability of AI solutions. For instance, partnerships between developed and developing nations can bridge the technological divide, enabling the transfer of AI knowledge and tools to regions that need them most. R&D thus acts as a unifying force, driving collective action towards the SDGs.

Through continuous innovation, R&D ensures that AI technologies are advanced and tailored to address specific challenges in achieving the SDGs. By fostering economic growth, improving healthcare, promoting education, tackling climate change, and ensuring inclusivity, R&D bridges the gap between technological potential and societal needs. Collaboration among governments, private sectors, and research institutions is crucial to sustain these efforts, unlocking AI's full potential for a sustainable and equitable future.

7.3.12 Availability and Reliability of Power

The availability and reliability of energy/power resources constitute a major enabler of AI deployment in the pursuit of SDGs. Development of AI systems and their applications demand substantial amounts of power. These energy requirements vary depending on the complexity of the AI model, the size of the dataset being processed, and the hardware architecture used for computation. Key factors influence the power requirements of AI applications.

More complex AI models, such as deep neural networks with a large number of layers and parameters, typically require more computational power. Training these models involves performing numerous mathematical operations, and the complexity of these operations influences power consumption. Training an AI model, which learns from data, generally requires more power than making predictions or inferences with a pre-trained model. Once trained, models can be deployed for inference. Furthermore, the choice of hardware significantly impacts power consumption. Graphics Processing Units (GPUs) and specialised hardware like Tensor Processing Units (TPUs) are commonly used for AI tasks due to their parallel processing capabilities, which can reduce training time and, in some cases, power consumption compared to traditional Central Processing Units (CPUs).

AI computations can be performed in centralised data centres or on edge devices. While data centres may have more powerful hardware, edge devices operate with power constraints. Edge computing aims to perform AI tasks locally, reducing the need for data transmission and potentially lowering power requirements. Quantisation and compression are critical factors in energy consideration. Techniques like quantisation (reducing the precision of numerical values) and model compression can be applied to reduce the computational complexity and memory requirements of AI models. This, in turn, can lead to lower power consumption. Developing and using more efficient algorithms can contribute to reduced power requirements. Researchers are continually working on improving the efficiency of AI algorithms to achieve comparable performance with lower computational costs.

Powering AI infrastructure with renewable energy sources, such as solar and wind, can mitigate the environmental impact associated with high power consumption. This approach aligns with sustainability goals and reduces the carbon footprint of AI operations. Some AI applications, particularly those requiring real-time processing, may demand higher power to meet stringent performance requirements. This is common in applications like autonomous vehicles, robotics, and certain healthcare applications. Continuous advancements in optimisation techniques, both at the algorithmic and hardware levels, contribute to reducing the power requirements of AI applications. This includes improvements in model architecture, training algorithms, and hardware efficiency.

High-performance computing systems, including those used for AI, often require sophisticated cooling systems to prevent overheating. The efficiency of these cooling systems can impact the overall energy consumption of AI infrastructure. Efforts are ongoing to make AI applications more energy-efficient, environmentally friendly, and accessible. Researchers and engineers are exploring various avenues, including the development of low-power hardware, energy-efficient algorithms, and techniques to optimise the training and inference processes. As AI technologies continue to evolve, there is a growing emphasis on achieving a balance between performance and energy efficiency.

Nevertheless, it must be restated that power/energy availability and reliability remain critical success factors in the deployment of AI to achieve the SDGs. The energy poverty, poor power infrastructure, and unreliable energy supply characterising most emerging and least industrialised economies must be addressed to give them a fighting chance in the AI-driven SDG journey.

7.3.13 Accountable, Capable, and Ethical Developmental State

Throughout the world, particularly in the Global South, achieving the SDGs using AI requires an accountable, capable, and ethical developmental state. The state must drive the SDG agenda. It is the overarching enabler of AI-enabled achievement of SDGs. A developmental state is a country where the government plays a central and proactive role in promoting economic development, mainly through policies supporting industrialisation and economic growth. This model contrasts with laissez-faire or market-driven approaches, where the state's role is more limited.

Characteristics of a Developmental State

A developmental state takes a hands-on approach to guiding economic development, often through planning, directing resources, and implementing policies that foster industrial growth and modernisation. Such states typically implement targeted industrial policies to promote specific sectors that are key to economic growth. This can involve subsidies, tax incentives, and direct investment in industries like manufacturing, technology, and infrastructure. The government, maintaining relative autonomy from powerful interest groups, is able to implement long-term policies focused on national development rather than being swayed by short-term political pressures or vested interests. This independence ensures impartiality and a focus on the long-term national interest.

Developmental states often have a competent and efficient bureaucracy for formulating and executing development policies. These bureaucracies are usually meritocratic and insulated from political interference, enabling them to carry out policies effectively. There is often a close, collaborative relationship between the government and the private sector. The state works with businesses to achieve shared development goals, usually guiding and supporting the private sector in line with national priorities. The primary goal of a developmental state is rapid economic growth and structural transformation, moving from an agrarian or resource-based economy to one that is industrialised and diversified. This often involves a shift from low-value-added activities to high-value-added production. Moving the country up the value chain through an intense focus on beneficiation and value addition is the core business of a developmental state.

Such states typically have a long-term vision for economic development, often embodied in national development plans or strategies. This vision guides policies and investments over multiple decades, providing stability and continuity rather than being driven by short-term electoral cycles. While the focus is on economic growth, developmental states often also implement policies that promote social welfare, such as education, healthcare, and housing, to ensure that the benefits of growth are broadly shared and human capital is developed. Indeed, inclusive development and shared prosperity define the mandate of a developmental state.

Examples of Developmental States

The most famous examples of developmental states are the Four Asian Tigers: South Korea, Taiwan, Singapore, and Hong Kong. These economies experienced rapid industrialisation and economic growth in the latter half of the twentieth century. In these countries, the government played a key role in directing economic development through strategic interventions and partnerships with the private sector. Japan is often cited as a model of the developmental state, especially during the post-World War II period. Through its Ministry of International Trade and Industry (MITI), the government directed industrial policy, supported vital industries, and facilitated technological advancement, leading to rapid economic growth. China, particularly since the late twentieth century, has also been considered a developmental state. The government has played a vital role in guiding economic development, promoting industrialisation, and opening up to foreign investment while maintaining control over key sectors.

Implicit in the idea of a developmental state is the objective of enabling the private sector and driving innovation, entrepreneurship, and technology. The Asian Tigers and China of 2024 amply demonstrate this aspect. Consequently, it can be argued that the developmental state should also be and can be a facilitative and entrepreneurial state. Nevertheless, there have been criticisms and challenges of developmental states. They are often criticised for their authoritarian tendencies, as the emphasis on strong state control can sometimes come at the expense of political freedoms and democratic governance. The long-term sustainability of the developmental state model has been questioned, particularly as economies mature and the need for state intervention diminishes. Transitioning from a developmental state to a more liberalised economy can be challenging. In a globalised economy, developmental states face pressures from international markets and institutions, which may limit their ability to maintain protectionist or interventionist policies.

The developmental state concept provides a framework for understanding how some countries have successfully achieved rapid economic development through decisive state intervention and strategic planning. While the model has its challenges and criticisms, it remains a significant approach to development, particularly in contexts where markets alone may not be sufficient to drive industrialisation and economic transformation. However, developmental states must continue to evolve and improve. In particular, they must amplify their facilitative and entrepreneurial roles by effectively enabling the private sector and driving innovation, entrepreneurship, and technology. Furthermore, they must embrace and strengthen inclusive governance and democratic politics in their jurisdictions.

With an accountable, capable, and ethical developmental state in place, the AI-enabled journey to the Promised Land, characterised by the achievement of all 17 SDGs, is conceivable. First, the developmental state creates an enabling environment for AI adoption and development. More significantly, the broad lessons from developmental state experiences (from such countries as the Four Asian Tigers) can complement AI's transformative role in powering the SDG journey.

AI for SDGs in Developmental States

Developmental states, characterised by their strategic focus on economic growth and state-led development, have increasingly deployed AI to accelerate progress towards the SDGs. Countries like China, South Korea, and India have leveraged AI to drive innovations in sectors critical to sustainable development, such as healthcare, education, and agriculture. For instance, China has used AI for precision agriculture to enhance crop yields and reduce resource wastage, aligning with SDG 2 (Zero Hunger). At the same time, AI-powered healthcare systems in the country have improved early disease detection and patient care, contributing to SDG 3 (Good Health and Well-Being). Similarly, South Korea has integrated AI into smart cities, optimising energy usage and transportation systems to achieve SDGs related to Sustainable Cities and Communities (SDG 11) and Climate Action (SDG 13). India has deployed AI tools for financial inclusion and rural development, enabling underserved communities to access digital banking and essential services, supporting SDGs 8 (Decent Work and Economic Growth) and 10 (Reduced Inequalities). These developmental states have capitalised on strong policy frameworks, robust digital infrastructure, and public-private partnerships to align AI deployment with national development priorities, demonstrating how targeted AI strategies can advance the SDGs effectively.

7.3.14 Interplay and Synergies Between the Enablers

The deployment of AI to achieve the SDGs relies on various enablers, including technological infrastructure, access to quality data, education and capacity building, supportive policy frameworks, and multi-stakeholder partnerships. These enablers do not operate in isolation; their interaction and synergy are crucial for maximising the impact of AI-driven solutions. Understanding how these enablers complement and reinforce each other highlights their collective importance in driving sustainable development.

A clear example of interplay is the relationship between technological infrastructure and access to quality data. Advanced technological infrastructure, such as high-speed Internet and cloud computing, is essential for collecting, processing, and analysing the large datasets that power AI systems. Conversely, the availability of diverse and accurate datasets enhances the functionality of AI applications, encouraging investments in technology infrastructure. For instance, AI climate monitoring systems require robust satellite networks (infrastructure) and high-quality environmental data to predict climate risks, aligning with SDG 13 (Climate Action).

Education and capacity building also interact closely with technological infrastructure to ensure the successful deployment of AI. Skilled professionals are needed to design, implement, and manage AI systems, which depend on access to technology for practical applications and experimentation. For example, programmes that train data scientists in low-income regions rely on the availability of computing resources and digital tools to equip participants with real-world skills. This synergy ensures that local talent can harness AI for challenges such as agricultural innovation (SDG 2) and public health improvements (SDG 3).

The interplay between policy support and access to quality data underscores the importance of governance in enabling AI deployment. Supportive policies and regulatory frameworks ensure that data collection and sharing adhere to ethical standards, such as privacy and inclusivity, while facilitating data availability for AI applications. For instance, open data initiatives supported by governments and international organisations have made critical datasets accessible to researchers and developers, driving AI innovation in areas such as poverty alleviation (SDG 1) and sustainable urban planning (SDG 11).

Multi-stakeholder partnerships amplify the impact of other enablers by bringing together resources, expertise, and perspectives from diverse sectors. These partnerships often bridge gaps between technological infrastructure and data access, ensuring that both are optimised for AI deployment. For example, collaborations between private tech firms and governments can combine cutting-edge technologies with public datasets to create impactful AI solutions for disaster risk management (SDG 11). Such partnerships also foster global knowledge sharing, ensuring that best practices in AI deployment are disseminated across borders.

The synergy between education and policy support highlights the role of governance in fostering capacity building. Policies that prioritise AI education, such as funding for STEM programmes and AI research, ensure a steady pipeline of skilled professionals. For example, national AI strategies in countries like Canada and Singapore have integrated education initiatives into their frameworks, promoting skill development that aligns with sustainable development priorities. This alignment ensures that AI talent is equipped to address challenges like renewable energy optimisation (SDG 7) and healthcare access (SDG 3).

Access to quality data and partnerships work in tandem to ensure that AI solutions are contextually relevant and equitable. Partnerships with civil society organisations and local communities help identify gaps in data and contribute culturally sensitive insights. For instance, AI systems for financial inclusion (SDG 10) rely on data from grassroots organisations to design solutions that cater to underserved populations. This collaboration ensures that AI applications are both inclusive and impactful.

The integration of technological infrastructure and multi-stakeholder partnerships is evident in global initiatives that aim to bridge the digital divide. Partnerships between governments, private companies, and international organisations have expanded access to digital tools in underserved regions. For example, initiatives to provide Internet connectivity in rural areas enable AI applications for remote education (SDG 4) and telemedicine (SDG 3). This interplay ensures that technology infrastructure is developed and accessible to those who need it most. Availability and reliability of power constitute the enabler of enablers. Without energy there is no AI revolution, much less an AI-driven SDG journey.

The interplay and synergies between the enablers of AI deployment are essential for achieving the SDGs. By combining infrastructure, data access, education, policy support, and partnerships, stakeholders create a cohesive ecosystem that maximises AI's impact. This integrated approach ensures that AI technologies are innovative, equitable, sustainable, and aligned with the principles of global development. Unlocking these synergies accelerates progress towards a future where AI drives meaningful and inclusive change across the globe by enabling the attainment of the 17 SDGs.

7.4 Conclusion

The deployment of AI to achieve the SDGs is driven by several key enablers. Robust technological infrastructure, including high-speed Internet, cloud computing, and reliable energy systems, forms the foundation for AI development and application. Supportive policy, guardrails, and governance frameworks ensure ethical, transparent, and inclusive AI deployment, fostering innovation while addressing potential risks. Access to high-quality data is critical for training AI models and ensuring their effectiveness in solving specific challenges while maintaining data privacy and security. Education and capacity-building initiatives equip individuals with the skills and knowledge to leverage AI for sustainable development, reducing the digital skills gap and promoting inclusivity. Finally, multi-stakeholder partnerships among governments, businesses, academia, and civil society enable resource sharing, knowledge exchange, and collaboration, ensuring that AI solutions are scalable and accessible to all. Together, these enablers create an ecosystem that allows AI to be a powerful tool in addressing global challenges and advancing the SDGs. Of course, all these enablers must be anchored and driven by an accountable, capable, ethical developmental state.

Drivers of AI Deployment to Achieve SDGs 8

8.1 Introduction

The preceding chapter discussed the difference between the terms enabler and driver. They represent two distinct concepts often used in strategy, innovation, or development. A driver is a factor that motivates or pushes something to happen, while an enabler is a condition that makes it possible for something to happen. Put differently, an enabler is a factor, condition, or resource that facilitates or supports the execution of a process or project. It creates the potential for progress by providing the necessary tools, infrastructure, or environment but does not actively initiate or propel action. On the other hand, a driver is a force or motivation that actively initiates or accelerates progress towards a goal. Hence, enablers are external to a process or system, while drivers are internal. Drivers are dynamic, proactive, purposeful, and intentional, creating momentum and influencing decision-making.

As discussed in the preceding chapter, the enablers in deploying AI to achieve the SDGs include robust digital infrastructure, such as high-speed Internet, compute resources, cloud computing, and access to high-quality data. Governance, guardrails, awareness, education, and capacity building were also discussed as enablers. These factors make the application of AI to achieve SDGs possible by establishing a foundation for the technology's development and deployment. In this chapter, the focus is on the drivers of AI deployment to achieve the SDGs. An AI driver is a force or motivation that actively initiates or accelerates progress towards an SDG. Drivers are dynamic and proactive, creating momentum and influencing decision-making. They often act as catalysts for change, pushing stakeholders to take action.

8.2 Drivers at a Glance

In the context of AI and the SDGs, drivers include the increasing urgency of addressing global challenges such as climate change or government policies prioritising AI as a tool for sustainable development. A detailed list of such drivers includes efficiency and effectiveness, innovation and technological advancements, scalability and replicability, data-driven decision-making, addressing complex challenges, inclusivity and accessibility, private sector engagement, and global commitment to SDGs.

A. G. O. Mutambara, *Deploying Artificial Intelligence to Achieve the UN Sustainable Development Goals*, Sustainable Development Goals Series,
https://doi.org/10.1007/978-3-031-88423-8_8

Indeed, various factors drive the deployment of AI for SDGs, highlighting its potential to contribute to sustainable development. AI empowers stakeholders to address global challenges across multiple domains by offering innovative, efficient, and scalable solutions. As indicated, the drivers of AI deployment for SDGs range from technological advancements to global commitments towards sustainability. Together, these drivers create a momentum that accelerates AI adoption and fosters transformative change towards achieving the SDGs. It is prudent to explore and discuss these drivers in some detail.

8.2.1 Efficiency and Effectiveness

One of the most compelling drivers of AI deployment is its ability to optimise processes, enhance decision-making, and automate tasks, thereby increasing efficiency and effectiveness in addressing development challenges. For example, in healthcare, AI-powered diagnostic tools enable faster and more accurate identification of diseases, directly supporting SDG 3 (Good Health and Well-Being). In agriculture, AI-driven precision farming technologies improve crop yields while conserving resources, aligning with SDG 2 (Zero Hunger). Similarly, AI-based systems in education personalise learning experiences for students, enhancing outcomes and bridging gaps in access to Quality Education (SDG 4). By streamlining operations and reducing resource wastage, AI provides cost-effective solutions to global problems, making it a critical tool for sustainable development.

8.2.2 Innovation and Technological Advancements

Rapid advancements in AI technologies, such as Machine Learning, Natural Language Processing, computer vision, and robotics, have opened new opportunities to solve complex development problems. These innovations enable AI to address previously insurmountable challenges. For instance, computer vision techniques are used to monitor deforestation and land degradation, supporting SDG 15 (Life on Land). Natural Language Processing facilitates language translation and literacy programmes, promoting inclusivity and SDG 10 (Reduced Inequalities). Conversely, robotics is revolutionising waste management and renewable energy production, contributing to SDG 12 (Responsible Consumption and Production). The continuous evolution of AI technologies drives the creation of novel solutions, making them indispensable for achieving the SDGs.

8.2.3 Scalability and Replicability

Another driver of AI adoption is its scalability and replicability, which enable successful interventions to be deployed across different contexts. Once developed, AI models and systems can be implemented in various geographical locations with minimal customisation, ensuring widespread impact. For instance, AI-driven platforms for remote education can be scaled to underserved regions globally, advancing SDG 4 (Quality Education). Similarly, AI tools for disaster risk prediction and response can be deployed in vulnerable regions, helping to mitigate the impacts of natural disasters and supporting SDG 11 (Sustainable Cities and Communities). This ability to replicate and scale solutions ensures that AI-driven interventions can reach broader populations, accelerating progress towards the SDGs.

8.2.4 Data-Driven Decision-Making

AI's capacity to analyse vast amounts of structured and unstructured data is a key driver of its deployment. AI enables evidence-based policymaking and resource allocation by generating actionable insights, identifying trends, and predicting outcomes. For example, AI-powered tools can expect food shortages, monitor disease outbreaks, and assess climate risks, helping governments and organisations implement timely and effective responses. This data-driven approach ensures that interventions are informed by real-time insights, enhancing their impact and efficiency. In areas such as poverty alleviation and environmental sustainability, AI's analytical capabilities are proving invaluable in addressing multifaceted development challenges.

8.2.5 Addressing Complex Challenges

The interdisciplinary and complex nature of global challenges often requires advanced analytics, modelling, and simulation techniques, which AI is uniquely equipped to provide. For instance, AI supports climate change adaptation by simulating future scenarios and evaluating mitigation strategies, directly addressing SDG 13 (Climate Action). In urban planning, AI-driven tools optimise traffic management, improve resource allocation, and monitor air quality, supporting SDG 11. By integrating diverse data sources and enabling comprehensive analyses, AI offers innovative solutions to global challenges that traditional methods struggle to address, making it a powerful driver of progress towards the SDGs.

8.2.6 Inclusivity and Accessibility

AI technologies have the potential to enhance inclusivity and accessibility, bridging digital divides and empowering marginalised communities. For instance, AI-driven platforms enable remote access to education, healthcare, and financial services, ensuring that underserved populations benefit from essential services. Translation tools powered by AI break language barriers, facilitating communication and inclusivity. Similarly, assistive technologies help individuals with disabilities participate more fully in education and employment, supporting SDG 10 (Reduced Inequalities). By addressing the unique needs of diverse communities, AI promotes inclusivity and fosters equitable progress towards the SDGs.

8.2.7 Partnerships and Collaboration

Collaborative efforts among governments, academia, industry, civil society, and international organisations are vital drivers of AI deployment for SDGs. These partnerships enable resource pooling, knowledge sharing, and joint initiatives that enhance the scalability and impact of AI solutions. For example, global forums such as the AI for Good Summit bring together stakeholders to explore AI-driven strategies for sustainable development. Collaborative projects, such as AI-powered climate monitoring tools, result from partnerships between technology companies and environmental organisations. Such efforts ensure that AI solutions are innovative, contextually relevant, and globally scalable.

8.2.8 Private Sector Engagement

The increasing private sector involvement is also driving AI deployment for sustainable development. Technology companies, startups, and investors recognise the economic potential of AI and are investing in solutions that address global challenges. For example, private sector innovations in AI-based renewable energy systems are accelerating progress towards SDG 7 (Affordable and Clean Energy). Similarly, startups are developing AI-powered platforms for agricultural innovation, enhancing food security and reducing waste. The private sector's role in funding research, developing technologies, and fostering market expansion is critical in scaling AI-driven solutions to achieve the SDGs.

8.2.9 Policy Support and Governance Frameworks

Policy support and governance frameworks are both enablers and crucial drivers of AI deployment for sustainable development. Governments and international organisations increasingly recognise the potential of AI to address global challenges. In response, they are implementing policies, regulations, and governance structures designed to ensure responsible and ethical deployment of AI technologies. These frameworks provide the necessary guidelines to protect privacy, ensure transparency, prevent misuse, and foster innovation and collaboration. By creating an environment that prioritises ethical considerations, such as fairness and accountability, governments are helping to align AI advancements with the broader goals of sustainability and equity. This supportive policy landscape encourages both public and private sector investments in AI, driving the development of solutions that contribute to achieving the Sustainable Development Goals (SDGs) responsibly and inclusively.

8.2.10 Policy Support and Global Commitment

Supportive government policies and the global commitment to the SDGs have created an enabling environment for AI deployment. Many countries have implemented national AI strategies prioritising ethical and sustainable AI applications. For instance, India's AI for All initiative focuses on leveraging AI to address challenges in agriculture, healthcare, and education. At the global level, the 2030 Agenda for Sustainable Development has inspired investment and innovation in AI to accelerate progress towards the SDGs. This commitment ensures alignment between AI development and the broader goals of sustainability and equity.

8.3 Details of Drivers

It is instructive to flesh out some of the critical drivers of AI-for-SDG deployment. This will assist and enhance the process of optimally engaging the drivers and unlocking synergies among them.

8.3.1 Efficiency and Effectiveness

As already stated, one of the most compelling drivers of AI deployment to achieve SDGs is its ability to optimise processes, enhance decision-making, and automate tasks. These capabilities significantly increase efficiency and effectiveness in addressing complex global challenges. AI-powered systems have proven to be transformative in sectors such as healthcare, agriculture, and education, where traditional methods often fall short of meeting the scale and urgency of development needs. By streamlining operations and maximising resource utilisation, AI provides innovative and cost-effective solutions that are critical for sustainable development.

In healthcare, AI is revolutionising diagnostics and treatment planning, directly contributing to SDG 3 (Good Health and Well-Being). AI-powered diagnostic tools, such as image recognition software for detecting cancers or other diseases, enable faster and more accurate identification of health issues. This reduces delays in treatment and improves patient outcomes. AI also plays a critical role in pandemic response by predicting outbreaks, analysing the spread of diseases, and optimising vaccine distribution. These advancements enhance the efficiency of healthcare systems and make essential services more accessible, particularly in underserved regions.

In agriculture, AI-driven precision farming technologies are addressing global food security challenges, supporting SDG 2 (Zero Hunger). By using AI algorithms to analyse data from sensors, drones, and satellite imagery, farmers can make data-informed decisions about irrigation, fertilisation, and pest control. These technologies optimise resource use, minimise environmental impact, and increase crop yields. For example, AI-powered predictive analytics can help farmers anticipate weather patterns and plan planting cycles, reducing risks associated with climate variability. This ability to conserve resources while boosting productivity highlights AI's effectiveness in advancing sustainable agricultural practices.

Education is another sector where AI significantly enhances efficiency and outcomes, contributing to SDG 4 (Quality Education). AI-based systems personalise learning experiences by adapting content and teaching methods to meet the unique needs of each student. These tools assess students' progress in real time, identify knowledge gaps, and provide targeted interventions, ensuring more effective learning. Moreover, AI-powered platforms enable remote learning opportunities, bridging educational access gaps in rural and underserved communities. By making quality education more accessible and tailored, AI is addressing long-standing inequalities in the education system.

AI's ability to streamline operations and reduce resource wastage extends to other domains, such as energy management and urban planning. For instance, AI-powered systems optimise energy consumption by analysing real-time usage patterns, reducing costs, and enhancing the efficiency of renewable energy systems. Similarly, in urban planning, AI tools help design smarter cities by improving traffic management, monitoring air quality, and optimising resource allocation. These applications increase operational efficiency and contribute to sustainable development goals like SDG 7 (Affordable and Clean Energy) and SDG 11 (Sustainable Cities and Communities).

Indeed, AI's capacity to enhance efficiency and effectiveness makes it a critical tool for addressing global challenges and advancing the SDGs. By improving healthcare, agriculture, education, and other sectors, AI demonstrates its potential to deliver transformative solutions that are cost-effective, scalable, and sustainable. As its applications continue to expand, AI's role in optimising processes, enhancing decision-making, and automating tasks will remain central to achieving the 17 SDGs.

8.3.2 Data-Driven Decision-Making

As already intimated, AI's capacity to analyse vast amounts of structured and unstructured data is one of its most powerful drivers for deployment in achieving the SDGs. This ability enables evidence-based policymaking and resource allocation, providing stakeholders with actionable insights, identifying trends, and predicting outcomes. By leveraging data-driven decision-making, governments and organisations can design and implement timely and effective interventions, ensuring maximum impact on global development challenges.

One of the most impactful applications of AI's data analytics capabilities is in anticipating food shortages and addressing food insecurity, directly contributing to SDG 2 (Zero Hunger). AI-powered tools analyse factors such as crop yields, weather patterns, and market dynamics to predict potential disruptions in food supply chains. For instance, satellite imagery combined with AI algorithms can monitor agricultural land for signs of drought or pest infestations. These insights enable policymakers to take proactive measures, such as allocating resources for irrigation systems or distributing food aid to vulnerable regions, reducing the likelihood of famine and malnutrition.

In the realm of public health, AI is instrumental in monitoring disease outbreaks and supporting effective responses, aligning with SDG 3 (Good Health and Well-Being). AI systems can detect early warning signs of disease outbreaks and predict their spread by analysing real-time data from hospitals, laboratories, and public health databases. During the COVID-19 pandemic, AI models played a critical role in tracking infection trends and optimising vaccine distribution. This capacity to provide accurate, real-time insights enhances the efficiency of healthcare systems and ensures that resources are directed where they are needed most.

AI's ability to assess climate risks is another example of its transformative potential in data-driven decision-making. Climate change presents complex challenges that require advanced analytics to address effectively, making AI an invaluable tool for SDG 13 (Climate Action). AI-powered systems analyse data from sensors, satellites, and historical climate records to predict extreme weather events, monitor deforestation, and model the impact of mitigation strategies. These insights enable governments and organisations to prioritise investments in renewable energy, disaster preparedness, and conservation efforts, ensuring a sustainable and resilient response to environmental challenges.

In poverty alleviation, AI's data-driven approach provides critical insights into the root causes and geographic distribution of poverty, contributing to SDG 1 (No Poverty). By analysing demographic data, economic indicators, and social metrics, AI systems identify regions and populations most in need of support. These insights help governments and non-profits target interventions, such as microfinance programmes or job creation initiatives, where they can have the greatest impact. AI also tracks the effectiveness of these interventions over time, allowing stakeholders to refine their strategies and maximise outcomes.

The use of AI in urban planning is another testament to its value in data-driven decision-making. AI models analyse traffic patterns, population density, and infrastructure data to design smarter cities, supporting SDG 11 (Sustainable Cities and Communities). By providing actionable insights, these systems help urban planners optimise transportation networks, reduce energy consumption, and enhance public services. For instance, AI-powered predictive analytics can anticipate traffic congestion and suggest real-time adjustments to traffic signals, improving urban mobility and reducing carbon emissions.

AI's ability to integrate diverse data sources also proves invaluable in addressing interdisciplinary challenges, such as water resource management. By analysing data from weather forecasts, hydrological models, and agricultural systems, AI helps optimise water allocation for drinking, irrigation, and industrial use, aligning with SDG 6 (Clean Water and Sanitation). This ensures that scarce water resources are used efficiently, reducing wastage and supporting sustainable practices in water-intensive industries.

Indeed, AI's capacity for data-driven decision-making is a critical enabler of sustainable development. Its ability to analyse large datasets and generate actionable insights empowers stakeholders to design timely, effective, and scalable interventions. From predicting food shortages and monitoring disease outbreaks to assessing climate risks and optimising urban planning, AI's analytical capabilities address multifaceted challenges across various SDGs. By leveraging real-time data to inform policy and resource allocation, AI ensures that development efforts are both impactful and efficient, driving progress towards attaining the 17 SDGs.

8.3.3 Private Sector Engagement

The private sector's growing involvement is a significant driver of AI deployment in advancing the SDGs. Technology companies, startups, and investors recognise AI's immense economic potential and transformative role in addressing global challenges. Through innovation, funding, and collaboration, the private sector is enabling the development and scaling of AI-driven solutions that align with sustainable development priorities. From renewable energy to agriculture, private sector engagement is proving critical in accelerating progress towards achieving the SDGs.

One of the most prominent areas of private sector engagement is the development of AI-based renewable energy systems, contributing to SDG 7 (Affordable and Clean Energy). Private companies are leveraging AI to optimise energy generation, distribution, and consumption, making renewable energy systems more efficient and cost-effective. For instance, AI-powered predictive analytics are being used to forecast energy demands and enhance the integration of solar and wind power into energy grids. Companies like Google and Tesla are at the forefront of deploying AI to manage energy storage and improve the efficiency of renewable energy installations, making clean energy more accessible and scalable.

In agriculture, startups are harnessing AI to address food security challenges and reduce waste, directly supporting SDG 2 (Zero Hunger). AI-powered platforms analyse soil conditions, weather patterns, and crop health to provide farmers with actionable insights that improve yields and resource management. For example, companies like PrecisionHawk and Blue River Technology are using AI to monitor crops and optimise the use of water, fertilisers, and pesticides. These innovations enhance agricultural productivity and minimise environmental impact, ensuring that food production systems are sustainable and resilient.

The private sector is also playing a vital role in advancing AI applications in healthcare, aligning with SDG 3 (Good Health and Well-Being). Companies are developing AI-powered diagnostic tools, telemedicine platforms, and personalised treatment solutions that improve access to quality healthcare. Startups like Tempus and Aidoc are creating AI systems that analyse medical data to detect diseases early and recommend tailored treatment plans. This focus on innovation and technology is bridging gaps in healthcare access, particularly in low-resource settings, and improving patient outcomes on a global scale.

Beyond specific sectors, the private sector contributes significantly to AI development through funding and investment. Venture capitalists and corporate investors are channelling resources into AI research and development, fostering innovation and bringing groundbreaking technologies to market. This financial support is particularly critical for startups and smaller companies, enabling them to develop and scale solutions that address global challenges. For example, private investment in AI for climate change adaptation has led to the creation of tools that monitor deforestation, predict extreme weather events, and optimise resource use.

Market expansion efforts by the private sector are further driving the adoption of AI solutions for sustainable development. Companies are creating tailored AI tools and services to meet the unique needs of different regions and industries. For instance, private firms are developing AI-powered platforms for microfinance and financial inclusion, supporting SDG 1 (No Poverty) by enabling underserved populations to access credit and savings. Similarly, AI solutions designed for education and skill development are empowering marginalised communities, aligning with SDG 4 (Quality Education) and SDG 10 (Reduced Inequalities).

Collaboration between the private sector and other stakeholders is amplifying the impact of AI deployment. Partnerships between technology companies, governments, and non-profits facilitate knowledge sharing and resource pooling, ensuring that AI-driven solutions are scalable and contextually relevant. For instance, Microsoft's AI for Earth initiative collaborates with conservation organisations to develop AI tools for environmental sustainability, addressing SDG 15 (Life on Land). These collaborations underscore the importance of multi-sectoral approaches in leveraging AI for global development.

The private sector's commitment to responsible and ethical AI deployment is also a key driver of its contribution to the SDGs. Companies are increasingly adopting frameworks that prioritise transparency, fairness, and accountability in AI development. By aligning their operations with global sustainability goals, private firms are fostering trust and ensuring that their innovations have a positive societal impact. This commitment to ethical AI aligns with the broader objectives of the 2030 Agenda for Sustainable Development, reinforcing the private sector's role as a catalyst for progress.

Indeed, private sector engagement is a powerful driver of AI deployment for sustainable development. Technology companies, startups, and investors are creating transformative solutions that address global challenges across sectors like energy, agriculture, and healthcare through innovation, funding, and collaboration. By fostering market expansion and adopting ethical practices, the private sector is contributing to achieving the SDGs and demonstrating the economic and social value of sustainable development. This dynamic and forward-thinking approach positions the private sector as a critical partner in building a more sustainable and equitable future through the attainment of the 17 SDGs.

8.3.4 Visionary Leadership

Visionary leadership—at organisational, national, regional, continental, and global levels—plays an essential role in harnessing AI to achieve the SDGs. It ignites and drives the agenda. In fact, the arduous SDG journey towards 2030 will rise or fall on leadership. Visionary leadership is characterised by the ability to create and articulate a clear, compelling future vision that inspires and motivates others to achieve shared goals. Visionary leaders possess a unique blend of foresight, passion, and innovation, enabling them to see beyond the current reality and anticipate future trends and challenges. They are adept at strategic thinking, often pioneering change by taking calculated risks and encouraging creativity within their teams. Visionary leaders are inspirational, transformational, innovative, and tech-savvy.

This type of leadership is critical in navigating complex environments, driving transformation, and ensuring long-term organisational success. Visionary leaders envision a better future and mobilise resources and people to turn that vision into reality, making them pivotal in both business and societal progress. Indeed, visionary leaders possess the foresight and dedication needed to leverage AI technologies in ways that advance global progress towards these SDGs. Their strategic insight helps align AI's potential with the specific needs and challenges outlined by the SDGs, such as reducing poverty, enhancing education, and combating climate change [64, 66, 89, 113, 145].

Examples of Visionary Leadership

Finland has exhibited visionary leadership in deploying AI to achieve the SDGs, particularly in education and environmental sustainability. The Finnish government has implemented a national AI program that uses AI to enhance public services and boost economic competitiveness while ensuring inclusivity and fairness. Through initiatives like offering free AI-related education to its citizens, Finland is democratising the understanding and use of AI across all levels of society, thus fostering a knowledgeable base that supports lifelong learning, a cornerstone of SDG 4 (Quality Education).

Singapore's approach to using AI for sustainable development is another notable example of visionary leadership. Recognised globally for its smart city initiatives, Singapore uses AI to enhance urban planning, healthcare, and environmental management. Their government has developed AI strategies that improve resource management efficiency and reduce the city-state's carbon footprint, contributing significantly towards SDG 11 (Sustainable Cities and Communities) and SDG 13 (Climate Action). Singapore's emphasis on technology as a means to better the lives of its citizens exemplifies how visionary leadership can integrate AI into daily living and governance.

Rwanda has emerged as a leader in Africa for its strategic use of AI in healthcare, directly impacting SDG 3 (Good Health and Well-Being). By partnering with private companies, Rwanda has facilitated the deployment of AI-driven solutions in remote areas, such as using drone technology to deliver blood and medical supplies. Additionally, AI applications in diagnostic services are helping to bridge the gap in medical service delivery between rural and urban areas. Rwanda's commitment to integrating AI technologies demonstrates a forward-thinking leadership style that prioritises health innovations, proving that visionary leadership in deploying AI is not confined to the world's most affluent nations.

China has demonstrated visionary leadership in deploying AI to advance the SDGs through substantial investments in AI research and applications to solve large-scale social and environmental problems. Particularly noteworthy is China's focus on using AI to enhance agricultural practices and food security (SDG 2) by developing AI systems that predict crop yields, optimise planting schedules, and monitor crop health, significantly increasing efficiency and sustainability. Additionally, China's commitment to combating air pollution (SDG 13) through AI-powered monitoring systems and predictive analytics illustrates a proactive approach to environmental management. By integrating AI into national development strategies, China showcases a model of how technology can be harnessed to achieve sustainable growth and societal well-being on a vast scale.

In several other countries, in the realm of SDG 2, Zero Hunger, visionary leaders in agri-tech companies are deploying AI to optimise crop yields and reduce waste. Using AI-driven predictive analytics, these leaders enable farmers to anticipate weather changes, pest invasions, and crop diseases, significantly increasing food production efficiency and sustainability. This proactive approach boosts agricultural output and supports small-scale farmers in maintaining the stability of their livelihoods and food supplies.

In terms of SDG 3, Good Health and Well-Being, AI is being utilised under visionary leadership to revolutionise healthcare, particularly in under-resourced areas. By implementing AI-driven diagnostic tools and telemedicine solutions, healthcare leaders are making medical expertise accessible even in remote regions. AI algorithms help diagnose diseases from imaging data, predict outbreaks, and personalise patient care plans, thereby enhancing the quality and accessibility of healthcare services.

Furthermore, for SDG 13, Climate Action, visionary leaders are employing AI to monitor and mitigate the effects of climate change. Through AI and Machine Learning, environmental scientists and policymakers can model climate scenarios, track carbon emissions, and develop more efficient renewable energy sources. This strategic use of AI aids not only in understanding and predicting environmental changes but also in implementing and managing solutions that contribute to sustainable development. Through these examples, it becomes evident that visionary leadership is instrumental in deploying AI as a transformative tool to meet the ambitious targets set by the UN SDGs [11, 24].

8.3.5 Organisational Preparedness

The efficacy of AI in the SDG ambition is not guaranteed. Organisations, communities, and countries must prepare to use AI successfully for that agenda. They must engage in six distinct strategic initiatives. For a start, institutions must seek to create the AI transformation roadmap—a treasure map to a new AI-driven entity. There must be clarity about the journey from the status quo to the desired destination. They must build a talent bench with the requisite expertise and skills. There must be no assumption that the human capital is already available. More significantly, organisations must realise that their current way of doing things will not cut it. They must develop and adopt new and innovative operating models that are fit for purpose in the AI context. Furthermore, organisations and countries must nurture decentralised and distributed technology environments that enable teams to innovate across institutions and sectors. Data is the key driver of AI systems. Consequently, efforts must be made to embed data everywhere in institutions, and there must be ubiquitous access to and use of this data. Finally, as the adage goes—the proof of the pudding is in the eating. Organisations, communities, and countries must drive and unlock AI user adoption and enterprise scaling. The above six capabilities are foundational for AI to play an efficacious role in the SDG journey towards 2030 [72].

It is instructive to summarise and flesh out these six strategic initiatives, illustrating their significance in the SDG journey:

- Creating the AI Transformation Roadmap (Treasure Map): This initiative underscores the necessity for a clear and strategic plan, aptly termed the "Treasure Map," to navigate the complexities of integrating AI into business operations. It involves identifying specific goals and the technological pathways to achieve them. A well-articulated AI roadmap aligns AI initiatives with business objectives, ensuring that every technological investment advances the organisation's overall strategy. This roadmap should also be adaptable to the rapid advancements in AI technology and market dynamics, enabling sustained competitiveness.
- Building a Talent Bench: As AI technologies permeate various business processes, the demand for skilled professionals who can develop, manage, and interpret AI systems becomes critical. Building a robust talent bench involves recruiting individuals with technical expertise in AI and investing in ongoing education and training programmes to cultivate a workforce that can evolve with changing technologies. This strategic initiative emphasises the importance of talent development as a core component of an organisation's AI strategy.
- Adopting New Operating Model: The adoption of AI necessitates a transformation in the traditional operating models. This initiative focuses on creating a flexible, agile, and collaborative working environment that can quickly respond to the opportunities and challenges presented by AI technologies. A new operating model might involve redefining roles and responsibilities, enhancing decision-making processes with AI insights, and fostering a culture of innovation and experimentation.
- Producing a Distributed Technology Environment: This strategic initiative advocates for a distributed technology infrastructure that enables seamless collaboration and innovation across different parts of the organisation. By decentralising the access to and the development of AI tools, organisations can empower individual teams to innovate and solve problems more effectively. This approach accelerates the pace of innovation and ensures that AI solutions are more closely aligned with specific operational needs.
- Embedding Data Everywhere (Ubiquitous Access to and Use of Data): Data is the lifeblood of AI. This initiative involves embedding data access throughout the organisation to ensure that decision-makers at all levels can leverage data-driven insights. Ubiquitous access to data enhances the ability to make informed decisions quickly and accurately, fostering a data-centric culture that can fully exploit AI's potential.
- Unlocking User Adoption and Enterprise Scaling: The final strategic initiative addresses the challenges of user adoption and the scaling of AI technologies across the enterprise. It focuses on strategies to enhance user engagement with AI systems, such as user-friendly interfaces, personalised user experiences, and supportive change management practices. Moreover, scaling

AI effectively requires robust infrastructure, governance frameworks, and cross-functional collaboration to ensure that AI solutions deliver value consistently at scale.

Each of these strategic initiatives represents a critical component of a comprehensive strategy to effectively harness the power of AI in pursuit of the SDGs. Together, they provide a framework for organisations, communities and countries seeking to transform their business and economic fortunes to achieve the SDGs by leveraging technology, particularly AI. By systematically engaging in these six strategic initiatives, they can capitalise on AI's transformative potential and maintain a competitive edge in an increasingly complex global market. Within this context, the SDG journey can be ignited and driven by the deployment of AI [72, 86].

8.3.6 Partnerships and Collaboration

As already discussed, collaborative efforts among governments, academia, industry, civil society, and international organisations are essential drivers of AI deployment for achieving the SDGs. These partnerships foster resource pooling, knowledge sharing, and joint initiatives, significantly enhancing the scalability, accessibility, and impact of AI-driven solutions. By leveraging the strengths of diverse stakeholders, these collaborations ensure that AI technologies address global challenges effectively and inclusively.

Governments play a pivotal role in partnerships by providing regulatory frameworks, funding, and strategic guidance for AI initiatives. Their involvement ensures that AI deployments align with national and international development priorities. For example, government-backed collaborations with academic institutions have led to the development of AI systems for disaster response, directly contributing to SDG 11 (Sustainable Cities and Communities). Governments also act as facilitators, connecting private sector players and civil society to address complex challenges like climate change and public health crises.

Academia contributes significantly to these partnerships by advancing AI research and training skilled professionals. Universities and research institutions work closely with governments and industries to develop innovative AI applications tailored to specific SDGs. For instance, collaborations between universities and technology companies have resulted in AI-powered tools for disease prediction and monitoring, supporting SDG 3 (Good Health and Well-Being). By serving as hubs for innovation and talent development, academia ensures that partnerships are grounded in cutting-edge research and equipped to solve real-world problems.

The private sector's participation in partnerships amplifies the impact of AI deployment through technological expertise and financial investment. Companies provide the hardware, software, and technical know-how needed to design and implement AI solutions. Partnerships between corporations and non-profits have been instrumental in creating AI tools for precision farming, which address food security and environmental sustainability, contributing to SDG 2 (Zero Hunger) and SDG 13 (Climate Action). The private sector also plays a crucial role in scaling successful solutions by integrating them into global markets.

Civil society organisations add critical value to these partnerships by advocating for inclusivity, ethics, and community engagement in AI deployments. Their involvement ensures that AI technologies are designed with marginalised and vulnerable populations in mind. For example, partnerships between non-profits and academic institutions have developed AI platforms for literacy and skill development in underserved communities, aligning with SDG 4 (Quality Education) and SDG 10 (Reduced Inequalities). Civil society's focus on equity and social justice strengthens the ethical underpinnings of AI initiatives.

International organisations provide a global perspective and a platform for multi-stakeholder collaboration, bridging the gap between local and global initiatives. Forums like the AI for Good Summit bring together experts, policymakers, and technologists to discuss AI-driven strategies for sustainable development. These platforms enable cross-border knowledge exchange and the sharing of best practices, ensuring that AI solutions are adaptable and impactful across different regions. International organisations also coordinate funding and technical assistance for AI projects in low-income countries, promoting global equity in AI deployment.

Collaborative projects born from these partnerships demonstrate the potential of AI to address complex global challenges. For instance, AI-powered climate monitoring systems have been developed through collaborations between technology companies and environmental organisations. These systems analyse satellite data to track deforestation, monitor carbon emissions, and predict extreme weather events, supporting SDG 13. Similarly, partnerships between health organisations and AI startups have created diagnostic tools for remote areas, addressing health disparities and advancing SDG 3.

Indeed, partnerships and collaboration are indispensable drivers of AI deployment for sustainable development. These collaborations pool resources, expertise, and perspectives to create innovative and impactful AI solutions by bringing together governments, academia, industry, civil society, and international organisations. Such efforts ensure that AI technologies are scalable, contextually relevant, and ethically designed, addressing global challenges effectively and equitably. As AI continues to evolve, fostering robust partnerships will remain central to unlocking its full potential for achieving the 17 SDGs.

8.3.7 Interplay and Synergies Between the Drivers

The successful deployment of AI to achieve the SDGs is not driven by isolated factors but rather by the interplay and synergies between various drivers. Efficiency and effectiveness, innovation and technological advancements, private sector engagement, partnerships, policy support, and inclusivity are interconnected elements that amplify each other's impact. This interconnectedness ensures that AI solutions are innovative, scalable, equitable, and aligned with sustainable development priorities.

One clear example of synergy is the interplay between innovation and private sector engagement. Technological advancements in AI, such as Natural Language Processing and Machine Learning, often originate from private sector research and development. Private companies leverage their resources and expertise to translate innovations into practical solutions. For instance, AI-powered platforms for climate monitoring, developed by companies like IBM and Google, capitalise on cutting-edge technology to address SDG 13 (Climate Action). These innovations, in turn, create economic opportunities for the private sector while addressing pressing global challenges.

Similarly, the interaction between private sector engagement and partnerships is pivotal in scaling AI-driven solutions. Partnerships between technology firms, governments, and civil society organisations enable pooling resources and expertise. For example, collaborations between Microsoft and non-profits in their AI for Earth initiative have produced tools to monitor deforestation and enhance agricultural productivity. These partnerships ensure that private sector innovations are contextualised and accessible to underserved communities, thereby maximising their impact on SDGs such as SDG 2 (Zero Hunger) and SDG 15 (Life on Land).

The synergy between policy support and partnerships is another critical factor in enabling AI deployment. Governments that establish clear regulatory frameworks for ethical AI use to provide a foundation for partnerships to thrive. For instance, national AI strategies in countries like Singapore and India have facilitated collaborations between academic institutions, private companies, and international organisations. These partnerships are aligned with policy priorities, ensuring that AI applications adhere to ethical guidelines while contributing to SDGs such as SDG 8 (Decent Work and Economic Growth) and SDG 10 (Reduced Inequalities).

Inclusivity and partnerships also reinforce each other by ensuring that AI solutions address the needs of marginalised populations. Civil society organisations advocate for the inclusion of underrepresented groups in AI-driven initiatives, while partnerships with governments and private companies help translate these inclusivity goals into actionable solutions. For example, AI-powered educational platforms developed through collaborations between non-profits and academia are designed to reach rural and underserved areas, advancing SDG 4 (Quality Education) and reducing inequalities.

The interaction between data-driven decision-making and innovation underscores the importance of leveraging data to enhance technological advancements. AI systems that analyse large datasets generate insights that drive innovation in fields such as healthcare, agriculture, and urban planning. These insights inform the development of new tools and models, such as AI-powered disease prediction systems, which simultaneously contribute to SDG 3 (Good Health and Well-Being) and SDG 11 (Sustainable Cities and Communities). This feedback loop between data analysis and innovation ensures that AI applications remain responsive to real-world challenges.

The integration of efficiency and scalability further highlights the interplay between drivers. AI's ability to optimise processes improves efficiency and ensures that solutions can be scaled to larger populations. For instance, AI systems that streamline renewable energy management can be deployed across regions with minimal customisation, supporting SDG 7 (Affordable and Clean Energy) on a global scale. This scalability is further enhanced through partnerships and private sector investments, which provide the resources needed to expand AI applications.

Global forums, such as the AI for Good Summit, exemplify how multiple drivers work together to unlock synergies. These platforms bring together stakeholders from governments, academia, industry, and civil society to explore AI-driven strategies for sustainable development. They foster innovation, promote inclusivity, and align AI initiatives with ethical policies and global goals. By facilitating multi-stakeholder collaboration, such forums ensure that AI technologies are not only innovative but also equitable, sustainable, and impactful across multiple SDGs.

The interplay and synergies between AI deployment drivers are essential for unlocking AI's full potential to achieve the SDGs. Visionary leadership and organisational preparedness constitute the indispensable glue in this mutually reinforcing framework. By leveraging these interconnections, stakeholders can create a dynamic ecosystem where innovation, collaboration, inclusivity, and efficiency amplify each other. This integrated approach ensures that AI technologies are not only transformative but also aligned with the principles of sustainability and equity, paving the way for a more inclusive and resilient future anchored by the attainment of the 17 SDGs.

8.3.8 AI Ecosystem

An AI ecosystem refers to a collaborative network of interconnected entities that support the development, deployment, and use of AI technologies. It is a dynamic environment where multiple stakeholders interact to drive AI innovation and adoption, including researchers, developers, businesses, educational institutions, governments, and regulatory bodies. The ecosystem includes technological components like AI models, datasets, cloud infrastructure, and development platforms and human and organisational aspects such as training, policies, ethical guidelines, and funding mechanisms. Together, these elements create a self-sustaining cycle where innovations lead to new applications, which in turn stimulate further development and investment in AI.

AI ecosystems can be observed at different levels, from global collaborations to regional innovation hubs. A well-known example is the AI Innovation Hub in Montreal, Canada, where partnerships between academic institutions like McGill University, industry players like Google and Microsoft, and government agencies have created a thriving AI research and development environment. Another example is China's National AI Development Plan, where government-backed initiatives, private tech giants like Alibaba and Tencent, and AI startups work together to achieve AI dominance. Similarly, Silicon Valley serves as a global AI ecosystem, where companies like OpenAI, NVIDIA, and Google DeepMind collaborate with venture capitalists, universities like Stanford, and industry groups to drive AI breakthroughs. These ecosystems foster innovation, provide access to skilled talent, and enable the rapid commercialisation of AI solutions.

Developing an AI ecosystem requires a multi-pronged approach that addresses infrastructure, talent, regulation, and market access. First, a strong foundation of digital infrastructure, including cloud computing, big data platforms, and secure connectivity, is essential for AI development. Second, the availability of a skilled AI workforce is crucial. This can be achieved by investing in education and training programmes, establishing AI-focused research centres, and encouraging knowledge transfer through partnerships with global tech firms. Third, a supportive regulatory framework is needed to foster trust, ensure ethical AI use, and prevent misuse. Governments play a key role in setting up policies that promote responsible AI adoption while encouraging innovation. Finally, market access and commercialisation pathways are critical. Startups, small businesses, and larger enterprises should be supported with access to venture capital, grants, and accelerator programmes that bridge the gap between research and real-world application.

Developing an AI ecosystem is a collective endeavour that requires the participation of multiple stakeholders. Governments can provide funding, enact policies, and create a national AI strategy. Private sector players, including technology companies and startups, drive innovation, commercialisation, and scaling of AI applications. Educational institutions and research centres are key in developing talent, producing new knowledge, and driving foundational AI research. Civil society and advocacy groups ensure that ethical and human rights considerations are integrated into the development process. Effective ecosystems like those in Singapore and Finland have been thriving due to a clear alignment of goals between public and private sectors, combined with international collaboration. By creating an environment that fosters innovation, ensures ethical compliance, and promotes inclusiveness, an AI ecosystem can drive sustainable development, particularly in the Global South, where AI has the potential to address challenges in healthcare, education, and economic growth.

AI Ecosystem and SDGs

An AI ecosystem is both an enabler and a driver of AI deployment to achieve the SDGs. As an enabler, the ecosystem provides the foundational infrastructure, tools, and resources needed for developing, experimenting, and scaling AI solutions. This includes access to datasets, cloud computing infrastructure, AI development platforms, and regulatory frameworks that ensure ethical and inclusive AI development.

The ecosystem creates a collaborative environment where technological, human, and financial resources converge to address global challenges. By enhancing partnerships among governments, private sector actors, academic institutions, and civil society, an AI ecosystem facilitates the development and deployment of AI solutions that directly contribute to the SDGs. It enables knowledge sharing, talent development, and the creation of public-private partnerships. These enablers support the application of AI in addressing SDG-related challenges like improving healthcare (SDG 3), promoting quality education (SDG 4), and enhancing agricultural productivity (SDG 2). Specifically, AI-driven innovations in healthcare, such as predictive analytics and remote diagnostics, support

SDG 3 (Good Health and Well-Being). At the same time, AI-enabled precision agriculture and smart water management systems contribute to SDG 2 (Zero Hunger) and SDG 6 (Clean Water and Sanitation). The ecosystem also promotes capacity building by supporting education and training initiatives aligning with SDG 4 (Quality Education).

As a driver, an AI ecosystem actively pushes the adoption and deployment of AI technologies through strategic interventions, investment flows, and the creation of commercial pathways for AI-driven solutions. By supporting startups, funding research, and promoting regulatory certainty, the ecosystem drives innovation and accelerates the application of AI in real-world contexts. For instance, AI-driven climate models support SDG 13 (Climate Action) by enabling better predictions and risk assessments. At the same time, AI-powered tools for financial inclusion contribute to SDG 8 (Decent Work and Economic Growth). The ecosystem's ability to coordinate actors and mobilise resources ensures that AI solutions are created, implemented, and scaled, thereby driving national, regional, and global impact.

By encouraging responsible AI development through ethical frameworks and regulatory oversight, the ecosystem ensures inclusive growth and reduces inequalities, advancing SDG 10 (Reduced Inequalities). Furthermore, the availability of data-sharing platforms, innovation hubs, and financial support mechanisms accelerates the creation and scaling of AI solutions, making it possible to address global development challenges at a faster pace. Through its collaborative, inclusive, and innovation-driven nature, an AI ecosystem serves as a critical driver for the worldwide realisation of the SDGs.

In essence, the AI ecosystem enables the groundwork for AI development while actively driving its deployment towards achieving the SDGs. It is both an enabler and a driver.

8.4 AI Users' Voices and Insights

Incorporating the voices and insights of AI users in the deployment of AI for SDGs is both an ethical imperative and a practical necessity. These voices and insights are drivers of the impact of AI on SDG achievement. AI systems designed for global development are only as effective as their alignment with the needs, values, and priorities of the communities they serve. Engaging users ensures the relevance of AI solutions and their sustainability, equity, and broader acceptance. By adopting participatory approaches, integrating diverse perspectives, and fostering transparency, organisations can maximise the transformative potential of AI while minimising risks and unintended consequences.

8.4.1 Participatory Approaches to AI Design and Deployment

The first step in incorporating AI users' voices is embedding participatory methods in the AI lifecycle—from design to deployment and monitoring. Traditional top-down approaches often fail to account for the lived realities of the target populations. Instead, participatory methods such as co-design workshops, focus groups, and community consultations should be employed. These approaches enable users to articulate their needs and preferences, helping to ensure that AI systems address real-world challenges effectively. For instance, an AI-powered healthcare system aimed at achieving SDG 3 (Good Health and Well-Being) should involve patients, healthcare workers, and policymakers in its design process. By understanding their needs, the system can provide culturally relevant, accessible, and efficient solutions.

Incorporating participatory methods also extends to decision-making processes. Local communities, particularly in underserved regions, should have a voice in determining how AI solutions are prioritised and implemented. For example, in deploying AI for agriculture under SDG 2 (Zero

Hunger), smallholder farmers must be involved in selecting AI tools that enhance productivity without compromising sustainability. This approach increases the likelihood of adoption and engenders a sense of ownership among users, which is crucial for long-term success.

8.4.2 Incorporating Diverse Perspectives

AI solutions must reflect the diversity of the populations they aim to serve. Incorporating diverse perspectives ensures that AI systems do not reinforce existing biases or exacerbate inequalities. This requires proactive efforts to engage underrepresented groups such as women, indigenous communities, and persons with disabilities. For instance, AI systems for urban planning under SDG 11 (Sustainable Cities and Communities) should consider the unique needs of marginalised populations, including access to transportation and housing.

A key strategy for integrating diverse voices is through partnerships with local organisations and civil society groups that have deep roots in the community. These organisations can act as intermediaries, bridging the gap between AI developers and end-users. Additionally, employing inclusive data collection practices—such as disaggregating data by gender, age, and socio-economic status—helps to ensure that AI models are representative and equitable. Transparency in using this data further strengthens trust and engagement among diverse stakeholders.

8.4.3 Building Trust Through Transparency and Education

Transparency is central to building trust in AI systems, especially in regions with high scepticism about technology. Users need to understand how AI systems work, what data is being collected, and how decisions are made. Providing clear and accessible explanations of AI processes can demystify technology and encourage user participation. For example, developers should explain how algorithms assess learning outcomes and personalise content when deploying AI in education under SDG 4 (Quality Education).

Education plays a critical role in empowering users to engage meaningfully with AI. Capacity-building initiatives like digital literacy programmes can enhance users' understanding of AI and its potential applications. These programmes should target not only end-users but also local leaders and decision-makers who influence technology adoption. By fostering a culture of informed participation, communities are better equipped to advocate for their needs and hold developers accountable.

8.4.4 Feedback Mechanisms and Iterative Improvement

Incorporating user insights requires robust feedback mechanisms that enable continuous learning and adaptation. AI systems should be designed with built-in channels for user feedback, allowing for real-time adjustments and long-term improvements. For instance, an AI platform for water resource management under SDG 6 (Clean Water and Sanitation) could include user dashboards where communities report issues or provide suggestions.

Feedback mechanisms should also prioritise accessibility and inclusivity. Multilingual support, offline functionality, and user-friendly interfaces ensure that all users can provide input regardless of literacy or technological proficiency. Moreover, organisations must actively respond to feedback to demonstrate that user voices are valued and impactful. This iterative approach enhances the functionality of AI systems and strengthens user trust and engagement.

8.4.5 Ethical Governance and Accountability

Incorporating users' voices is closely tied to ethical governance and accountability in AI deployment. Ethical guidelines should mandate user participation as a core principle, ensuring that systems are designed with, rather than for, communities. Independent oversight bodies can monitor adherence to these guidelines, evaluating whether user insights are effectively integrated into decision-making processes.

Accountability mechanisms are equally important. Users should have recourse if AI systems cause harm or fail to meet their needs. Establishing clear lines of accountability—whether through grievance redressal systems, ombudsman services, or legal frameworks—ensures that users can seek remedies. These measures reinforce the idea that AI systems exist to serve people, not vice versa.

8.4.6 Role of International Cooperation

International organisations and multilateral frameworks are critical in standardising participatory practices for AI deployment. Global initiatives like the United Nations' AI for Good platform can facilitate knowledge sharing, providing best practices and resources for incorporating user voices. Additionally, international funding mechanisms should prioritise projects demonstrating a commitment to user engagement and inclusivity.

Cross-border collaboration is particularly vital in addressing the digital divide. By pooling resources and expertise, nations can ensure that even the most underserved communities can access participatory AI solutions. These efforts advance the SDGs and promote global solidarity, reinforcing the shared responsibility of achieving sustainable development through inclusive and ethical AI practices.

Indeed, incorporating AI users' voices and insights in deploying AI for SDGs is essential for ensuring that these technologies align with the needs and values of the communities they serve. By adopting participatory approaches, embracing diversity, fostering transparency, and embedding robust feedback and accountability mechanisms, organisations can create AI systems that are not only effective but also equitable and sustainable. This inclusive approach transforms AI from a tool of innovation into a vehicle for empowerment, advancing the collective goal of a fairer, more sustainable world.

8.5 Conclusion

Drivers are distinguishable from enablers. Nevertheless, both levers are essential. A range of factors drive the deployment of AI to achieve the SDGs, highlighting its transformative potential. At the core is visionary leadership. This is required at organisational, national, regional, continental, and global levels. The ability to create and articulate a clear, compelling future vision that inspires and motivates others to achieve the SDGs using AI is imperative. Efficiency, innovation, scalability, and data-driven decision-making illustrate AI's ability to address global challenges effectively. Additionally, inclusivity, partnerships, and private sector engagement ensure that AI solutions are equitable and impactful. Supportive policies and the global commitment to sustainability further accelerate AI adoption. Together, these drivers foster a collaborative and innovative ecosystem that leverages AI as a powerful tool to create a sustainable and inclusive future.

Indeed, the deployment of AI to achieve the SDGs is driven by a combination of factors illustrating its revolutionary potential in addressing global challenges. Efficiency and effectiveness stand out as AI optimises processes, enhances decision-making, and automates tasks across sectors like healthcare, agriculture, and education. Rapid technological advancements, such as Machine Learning and Natural Language Processing, provide innovative tools for tackling complex development issues. At the same time, the scalability and replicability of AI solutions ensure their global applicability.

Data-driven decision-making empowers policymakers with actionable insights, enabling targeted and evidence-based interventions. AI also addresses interdisciplinary challenges, such as climate change and urban planning, through advanced analytics and simulations. Inclusivity and accessibility are key drivers as AI bridges digital divides and empowers marginalised communities. Furthermore, partnerships and collaborations among governments, academia, industry, and civil society amplify the reach and impact of AI innovations. Indeed, private sector engagement and supportive governance frameworks ensure ethical, sustainable, and financially backed deployment of AI technologies. Synergistically, these drivers provide a dynamic environment that leverages AI to accelerate the attainment of the 17 SDGs. All this, including the impact of enablers discussed in Chap. 7, is anchored by an AI ecosystem—a collaborative network of interconnected entities that support the development, deployment, and use of AI technologies, where multiple stakeholders, including researchers, developers, businesses, educational institutions, governments, and regulatory bodies, interact to drive AI innovation and adoption. Indeed, the AI ecosystem is both an enabler and driver of the deployment of AI to achieve SDGs.

Further Critical Success Factors

9

Critical success factors are the key elements, conditions, or activities that contribute significantly to the achievement or accomplishment of a goal, project, or initiative. The enablers and drivers discussed in Chaps. 7 and 8 are critical success factors for deploying AI to achieve the SDGs. Such factors are key facilitators and igniters ensuring progress and sustainable outcomes. Critical success factors can be strategic, operational, financial, technical, or human-related, depending on the context. They can be quantitative or qualitative. Quantitative success factors are measurable and can be tracked using metrics such as performance indicators, deadlines, and financial returns. Qualitative success factors, on the other hand, are more subjective and more complex to measure but equally important, such as stakeholder trust, brand reputation, or organisational culture.

For instance, in a business context, critical success factors might include effective leadership, financial stability, and access to skilled talent. In contrast, in project management, they could consist of clear objectives, stakeholder alignment, and risk management. Identifying, prioritising, and addressing critical success factors is crucial for maximising efficiency, mitigating risks, and achieving desired outcomes. Effective deployment of AI to achieve the SDGs requires both categories (business and project management) of success factors. More specifically, success factors for AI deployment for SDGs (whose details are in Chaps. 7 and 8) include data accessibility, inclusive stakeholder engagement, ethical AI design, and policy support. For example, for AI to support SDG 3 (Good Health and Well-Being), access to high-quality, unbiased health data is a key success factor, along with the availability of skilled healthcare workers trained in AI usage. In education (SDG 4), success factors might include affordable access to AI-powered learning platforms, teacher training, and policies to ensure equity and inclusiveness. By focusing on these critical factors, development initiatives are more likely to achieve sustainable, impactful results.

Understanding and managing critical success factors ensure that AI-based SDG interventions, projects, businesses, and social initiatives are impactful, efficient, just, and equitable, remain on track, adapt to changes, uniquely serve the targeted communities, achieve long-term sustainability, and produce economic transformation. Three major success factors for the efficacious deployment of AI to achieve the SDGs require detailed exploration: (1) leveraging leapfrogging, (2) embracing decoloniality in AI, and (3) democratising AI.

A. G. O. Mutambara, *Deploying Artificial Intelligence to Achieve the UN Sustainable Development Goals*, Sustainable Development Goals Series,
https://doi.org/10.1007/978-3-031-88423-8_9

9.1 AI-Enabled Leapfrogging and SDGs

Leapfrogging in technological development refers to the ability of emerging and least industrialised countries to bypass traditional stages of technological evolution and move directly to more advanced, cutting-edge solutions. This concept allows nations to skip older, legacy systems and adopt modern, often more efficient, technologies. Unlike incremental progress, leapfrogging involves a radical shift driven by emerging technologies such as AI, Blockchain, mobile Internet, and renewable energy. For instance, many African countries bypassed fixed-line telephony and moved directly to mobile communication. This leap significantly reduced the time and cost required to provide telecommunications infrastructure while enabling widespread access to mobile banking services, e-commerce, and other digital innovations. By combining mobile telephony and banking functions, these same economies were able to skip brick-and-mortar banking.

One of the most exciting examples of leapfrogging is mobile money in Kenya, where the M-Pesa mobile payment system allowed millions without access to traditional banking to participate in the financial system. Instead of waiting for the development of extensive banking infrastructure, Kenya leapt directly to mobile-based financial services, promoting financial inclusion and supporting SDG 8 (Decent Work and Economic Growth). Another example is the adoption of renewable energy in sub-Saharan Africa, where many communities have bypassed centralised, fossil-fuel-based power grids in favour of off-grid solar energy systems. This leapfrog approach has brought clean, affordable energy to remote areas, supporting SDG 7 (Affordable and Clean Energy). In the health sector, countries like India have leapfrogged traditional healthcare delivery models by embracing telemedicine and AI-powered diagnostics to bridge the gap in healthcare access, especially in rural areas. These technologies allow patients to receive medical consultations remotely, reducing costs and increasing access to health services in underserved regions.

Leapfrogging is a powerful strategy for Global South economies as it enables faster progress towards achieving the SDGs. By skipping legacy systems, countries can save costs, reduce environmental impact, and provide access to essential services. In education, for example, some regions have bypassed traditional classroom-based learning by embracing e-learning platforms and AI-driven adaptive learning tools that personalise education for students, thereby promoting SDG 4 (Quality Education). In agriculture, smallholder farmers in developing countries are using AI-powered precision agriculture tools and mobile applications for climate-smart farming rather than waiting for traditional agricultural extension services to reach them. Leapfrogging accelerates development and encourages sustainability, as newer technologies are often cleaner, more efficient, and more inclusive. This approach positions developing nations to become leaders in emerging industries, enabling them to compete globally and bridge development gaps faster than through traditional pathways.

Indeed, leapfrogging in the deployment of AI provides a unique opportunity for emerging and least industrialised countries to accelerate progress towards the SDGs without following the traditional, resource-intensive development pathways. As explained earlier, by bypassing older technological systems and directly adopting advanced AI-driven solutions, these countries can rapidly improve critical sectors such as healthcare, education, agriculture, and financial services. For instance, countries can deploy AI-enabled telemedicine platforms instead of building large-scale healthcare infrastructure to provide remote health consultations and diagnostics, supporting SDG 3 (Good Health and Well-Being). AI can analyse medical images, detect diseases like cancer, and predict outbreaks of infectious diseases, all of which are crucial for countries with limited access to specialist healthcare services. This leapfrogging approach enables faster, more affordable healthcare delivery while addressing the growing demand for health services in underserved areas.

Leapfrogging AI in agriculture allows developing nations to skip traditional, labour-intensive farming techniques and embrace AI-powered precision agriculture. By using AI-driven sensors, weather prediction tools, and crop monitoring apps, farmers can optimise crop yields, improve water usage, and reduce the use of chemical inputs like fertilisers and pesticides. This directly supports SDG 2 (Zero Hunger) and SDG 13 (Climate Action) by promoting sustainable farming practices and enhancing food security. Similarly, AI-enabled education platforms like adaptive learning systems allow countries to leapfrog from traditional classroom-based education to personalised, tech-driven learning. With AI, education can be tailored to individual students' needs, promoting SDG 4 (Quality Education). This approach is especially effective in rural areas with limited access to quality teachers. By integrating AI into education, countries can quickly bridge learning gaps, reduce disparities, and ensure equitable access to learning opportunities for all children, regardless of their location.

Leapfrogging AI for SDGs promotes inclusivity and sustainability, especially in the Global South, where the traditional pace of development is often constrained by limited infrastructure and resources. By leveraging AI, these countries can accelerate digital transformation and avoid the costs of building legacy systems, such as large-scale financial institutions or fossil-fuel-based power grids. For example, AI-powered fintech platforms like mobile payment systems (e.g. Kenya's M-Pesa) have allowed millions of unbanked people to access financial services, advancing SDG 8 (Decent Work and Economic Growth) and SDG 10 (Reduced Inequalities). Moreover, AI-driven renewable energy management systems enable efficient use of decentralised solar microgrids, allowing remote communities to leapfrog fossil fuel-based energy grids and achieve SDG 7 (Affordable and Clean Energy). This leapfrogging approach empowers communities to adopt sustainable solutions that are both climate-friendly and inclusive. As AI technologies become more accessible, countries that leverage leapfrogging can achieve faster progress towards the SDGs, fostering a more equitable and sustainable global development trajectory.

9.2 Decoloniality in Deploying AI to Achieve SDGs

Decoloniality is a theoretical and practical framework aimed at dismantling the structures, knowledge systems, and power dynamics that were established during and after colonial rule. It seeks to challenge and move beyond the enduring legacy of colonialism, particularly the dominance of Global North (European and North American) world views, knowledge production, and governance systems. Unlike decolonisation, which often refers to the formal process of gaining political independence from colonising powers, decoloniality focuses on the deeper and more profound, ongoing social, cultural, and epistemic impacts of colonialism that persist even after formal independence. These impacts include inequalities in knowledge systems, global hierarchies of power, and entrenched narratives that continue to marginalise Global South (African, Asian, and other non-Western) knowledge systems. It is instructive to emphasise that the decoloniality discourse must reject the term "indigenous knowledge systems". Using it suggests that Global North knowledge systems are universal. Every knowledge system is indigenous somewhere. Hence, the phrase "indigenous knowledge system" becomes nebulous if not downright condescending and patronising. This book uses the broader term "Global South knowledge systems".

9.2.1 The Essence of Decoloniality

At its core, decoloniality advocates for the recentring of local, Global South knowledge systems while questioning the "universal" status of Western epistemology. It challenges the idea that Western modes of knowledge (science, philosophy, and economics) are superior or objective and instead emphasises the need for pluriversality—the coexistence of multiple knowledge systems. For example, Global South perspectives on nature and sustainability offer valuable insights for addressing modern challenges like climate change, aligning with the goals of SDG 13 (Climate Action). By restoring the legitimacy of Global South world views, decoloniality aims to reclaim cultural identity and epistemic sovereignty, which were often suppressed during colonial rule.

In practice, decoloniality manifests in education, governance, research, and development. In education, it calls for the inclusion of Global South knowledge in curricula, ensuring that students learn from diverse perspectives rather than solely Global North ones. In governance, it involves rethinking legal and policy frameworks to reflect local, community-driven approaches to justice, health, and economic development. In development, it questions the logic of "modernisation" models that impose Western-style economic growth, instead advocating for context-specific, culturally relevant development models. The decolonial approach is particularly relevant in the Global South, where colonial legacies continue to shape development trajectories. By embracing decoloniality, societies can create more inclusive, just, and sustainable development pathways, often leveraging local knowledge to achieve the SDGs in ways that are culturally and contextually appropriate [82, 101, 141].

9.2.2 AI and Decoloniality

The intersection of AI and decoloniality highlights the need to challenge Global North-dominated (Eurocentric and Western) frameworks that have shaped AI development and deployment. Most AI technologies, algorithms, and datasets are designed and controlled by institutions in the Global North, often reflecting the values, assumptions, and biases of Eurocentric and Western societies. This dominance reinforces existing colonial power structures, where the Global South becomes a passive consumer of AI technologies rather than an active co-creator. Decoloniality in AI calls for a shift in this power dynamic by ensuring that marginalised communities, particularly from the Global South, play a central role in AI systems' development, ownership, and governance. It emphasises the need to democratise AI development by including diverse voices, local knowledge systems, and culturally relevant perspectives. For example, African AI researchers have called for African-centred AI that addresses local development challenges, such as improving healthcare access and sustainable farming, in more contextually relevant ways than imported AI models.

Decoloniality in AI is crucial for addressing the issue of algorithmic bias and the unequal impact of AI systems on marginalised communities. Since AI models are often trained on datasets collected in the Global North, they may fail to recognise or misclassify images, language, or behaviours from non-Western contexts. For example, facial recognition software has been shown to exhibit higher error rates for people of African and Asian descent, raising concerns about racial bias and its potential misuse by authoritarian regimes. From a decolonial perspective, these issues are not just technical flaws but reflections of epistemic injustice, where certain groups are excluded from knowledge production. Decolonising AI requires the development of inclusive datasets and the participation of local communities in AI design processes. By involving Global South knowledge holders, community organisations, and regional research hubs, AI systems can be made more equitable, transparent, and inclusive. Decolonial AI can also promote ethical AI principles that prioritise human rights, social justice, and SDG 10 (Reduced Inequalities) [28, 32, 92].

9.2.3 Decoloniality and AI-Based SDG Achievement

Decoloniality and AI together offer the possibility of creating pluriversal AI systems—systems that recognise the coexistence of multiple knowledge systems and world views in AI design, development, and application. This approach moves away from the "one-size-fits-all" logic of Western AI models, instead allowing for localised, context-aware AI solutions. For example, AI-driven tools for sustainable agriculture in Africa could be designed to incorporate local Global South knowledge about soil, climate, and crop cycles rather than relying solely on Western agronomic models. Such tools support SDG 2 (Zero Hunger) and SDG 13 (Climate Action) in more ecologically and culturally relevant ways. Additionally, decolonial AI prioritises the sovereignty of communities over their data, advocating for data justice frameworks that ensure local communities have control over how their data is used and for what purpose. This is essential in addressing the problem of "data colonialism", where big tech companies extract data from the Global South for profit without proper consent or benefit-sharing. By fostering local AI innovation hubs in the Global South, decolonial AI promotes AI for sustainable development while ensuring that communities become co-creators, not just consumers, of AI technologies.

Indeed, decoloniality can play a critical role in deploying AI to achieve the SDGs by ensuring that AI development, governance, and application are inclusive, equitable, and contextually relevant. As explained earlier, traditional AI models are often rooted in Western epistemologies, which fail to recognise or integrate the local knowledge systems, values, and lived realities of communities in the Global South. Decoloniality challenges this imbalance by promoting the development of pluriversal AI systems that respect diverse worldviews and centre the voices of historically marginalised communities. This is especially important in achieving SDG 10 (Reduced Inequalities), as decolonial AI prioritises local ownership, data sovereignty, and the co-creation of solutions that are tailored to local development needs. For example, in agriculture (SDG 2—Zero Hunger), AI-driven tools designed using decolonial principles would incorporate local Global South farming knowledge, local weather patterns, and ecological wisdom to create context-specific precision farming solutions. In health (SDG 3—Good Health and Well-Being), decolonial AI could support the development of culturally relevant health diagnostics that consider health's social and cultural determinants. By promoting data justice, decoloniality ensures that communities retain control over their data, preventing data colonialism by large tech corporations. It is important to emphasise that decoloniality ensures that AI systems are inclusive, culturally relevant, and community-driven, enabling local ownership and knowledge sovereignty, which supports SDG 10 (Reduced Inequalities) and SDG 16 (Peace, Justice, and Strong Institutions). Ultimately, decoloniality can strengthen the role of AI as a driver of inclusive, just, and context-sensitive development, accelerating progress towards the SDGs in a way that promotes equity, cultural justice, sustainability, and human rights [118, 141].

9.3 Democratising AI for the Achievement of SDGs

The case for democratising AI is directly linked to the need to engage decoloniality in deploying AI. Democratising AI refers to the process of making AI technologies, tools, knowledge, and opportunities accessible to a broader range of people, communities, and organisations beyond a privileged few. Historically, the development and control of AI systems have been concentrated in the hands of a few tech giants, academic elites, and governments in the Global North.

This concentration of power has led to uneven access to AI's benefits, leaving marginalised communities and Global South countries with limited capacity to influence or participate in AI development. Democratising AI aims to remove these barriers by promoting the accessibility of

AI development tools, creating inclusive datasets, offering affordable AI education, and ensuring local communities control their data and digital futures. It goes beyond technical access, focusing on inclusion, representation, and empowerment. Democratisation ensures that people from diverse socio-economic, geographic, and cultural backgrounds have the ability to co-create, deploy, and benefit from AI technologies.

Achieving AI democratisation requires a multifaceted approach, including access to technology, education, and governance reform. First, access to AI development tools, platforms, and cloud computing must be made more affordable and inclusive. Open-source AI platforms like TensorFlow, PyTorch, and Hugging Face provide opportunities for developers worldwide to create AI solutions without relying on proprietary systems. Cloud-based AI services, if made affordable and accessible, can also enable startups, small businesses, and civil society organisations to develop AI-driven solutions. Second, education and skills development are essential. Democratising AI requires that people, especially from under-represented communities, are trained in AI development, data science, and ethical AI principles. Initiatives like AI4ALL and various coding boot camps focus on training under-represented groups, including women and people from the Global South. This ensures broader participation in AI development. Third, regulatory reform and ethical frameworks are needed to ensure that AI development is transparent, inclusive, and fair. This involves creating AI governance frameworks that promote public participation, data sovereignty, and the protection of human rights. Policies that mandate inclusive datasets and fair AI audits can help reduce algorithmic bias, especially for marginalised groups [32, 46, 94].

9.3.1 Accelerating SDG Achievement

Democratising AI can accelerate the achievement of the SDGs by ensuring that AI tools are designed to address development challenges specific to local contexts. When communities have access to AI development tools and can co-create AI solutions, they can better address issues related to poverty, hunger, health, and education. For example, in agriculture (SDG 2—Zero Hunger), local farmers in sub-Saharan Africa can use AI-enabled precision farming tools to monitor soil health, predict weather changes, and optimise crop yields. By democratising AI, local knowledge and Global South farming practices can be integrated into these AI models, making them more context-sensitive and impactful. In healthcare (SDG 3—Good Health and Well-Being), democratising AI can enable community health workers and small health clinics to deploy AI-powered diagnostic tools for early disease detection, reducing reliance on centralised healthcare systems that are often inaccessible to rural populations.

One of the most direct impacts of democratising AI is the potential to reduce inequalities (SDG 10—Reduced Inequalities). Currently, the benefits of AI are concentrated in high-income countries and wealthy corporations, but democratising access can bridge the gap. Previously excluded groups can participate in the AI economy by making AI tools, training, and resources available to small and medium enterprises (SMEs), local innovators, and underserved communities. This promotes economic inclusion by enabling small-scale AI innovation hubs in the Global South to create locally relevant AI-driven services and products. For instance, AI-based financial inclusion platforms that offer microloans, credit scoring, and financial literacy training can be designed and deployed by local fintech startups, not just global tech giants. Democratising AI also promotes inclusive job creation as more people acquire skills to participate in the growing AI economy, especially in emerging markets.

Democratising AI must lead to data sovereignty and local ownership of AI systems. A key impact of this democratisation process is the protection of data sovereignty and the creation of community-driven AI systems. Currently, large multinational tech companies extract data from users in the Global

South for training their AI models, often with little accountability. Democratising AI ensures that local communities control their own data and that the use of data is aligned with local cultural, ethical, and legal norms. For example, community-based data cooperatives can be established where individuals have control over how their data is used, ensuring that it benefits local development objectives. This approach also addresses concerns about data colonialism, where corporations in the Global North profit from the data of people in developing countries without providing reciprocal benefits. Data colonialism is the exploitation and extraction of data from individuals, communities, and nations—often in the Global South—by powerful corporations and entities, typically based in the Global North, without fair compensation, consent, or local control. It reinforces global inequalities, as local communities lose sovereignty over their own data, while foreign companies profit by developing AI models, advertising systems, and predictive analytics based on that data. By supporting data justice frameworks, democratising AI combats data colonialism. It enhances local autonomy and ensures that the benefits of AI are distributed more equitably, especially with respect to SDG 16 (Peace, Justice, and Strong Institutions). Furthermore, it is instructive to emphasise that democratising AI broadens access to AI tools, education, and data, empowering marginalised communities and local innovators to co-create solutions for SDG 4 (Quality Education), SDG 8 (Decent Work and Economic Growth), and beyond.

9.3.2 Data Sovereignty

It is essential to emphasise and flesh out the centrality and essence of data sovereignty. It refers to the concept that data is subject to the laws, governance, and control of the country, region, or community where it is collected, stored, or processed. It emphasises the right of individuals, communities, and nations to have control over their own data, determining how it is collected, stored, shared, and used. Unlike traditional data governance, which focuses on managing and protecting data, data sovereignty is tied to jurisdictional authority, meaning the data is bound by the legal and regulatory frameworks of the country or region where it resides. This concept has become increasingly significant as cloud computing and global data flows make it possible for data to be stored in servers located in different parts of the world, often beyond the control of the country or community where the data originated.

The importance of data sovereignty has grown in response to concerns about data colonialism, where multinational tech companies, often based in the Global North, extract data from users and communities in the Global South without fair compensation or local control. By promoting data sovereignty, communities and nations can ensure that data is stored in local servers, governed by local laws, and used to benefit local development goals. For instance, data sovereignty is crucial in sectors like healthcare, where sensitive patient information needs to be protected from misuse or exploitation by foreign tech firms. Data sovereignty frameworks ensure that health-related data remains under the control of local health authorities and communities, supporting SDG 3 (Good Health and Well-Being) and SDG 16 (Peace, Justice, and Strong Institutions).

Data sovereignty is also critical for protecting citizens' digital rights and privacy. Countries like the European Union (EU) have implemented data sovereignty principles through regulations like the General Data Protection Regulation (GDPR), which requires companies to respect EU citizens' rights to privacy, even when their data is processed outside the EU. In regions like Africa, discussions on data sovereignty are focused on ensuring that local governments have control over the data generated within their borders, reducing reliance on foreign cloud service providers. This concept is especially relevant in the context of AI and Machine Learning, where large datasets are used to train AI models. Without data sovereignty, local communities risk having their data exploited by foreign companies without benefiting from the resulting AI applications. By upholding data sovereignty, countries can ensure that their citizens and local enterprises have control over how their data is used, contributing to more equitable, ethical, and sustainable development outcomes.

9.3.3 Partnerships in Democratising AI

Achieving the democratisation of AI requires multi-stakeholder partnerships involving governments, international organisations, private sector players, academic institutions, and civil society. Governments play a key role in creating inclusive AI policies and funding AI education initiatives to ensure that underserved communities can access the tools and resources needed to participate in AI development. For example, regional organisations like the African Union have recognised the importance of creating Africa-centred AI strategies that promote local ownership, innovation, and capacity building. Similarly, international organisations like UNESCO and UNDP advocate for inclusive, ethical AI frameworks that reduce global inequalities. The private sector can support democratisation by providing access to affordable AI platforms and cloud services. At the same time, civil society can advocate for ethical AI practices and help co-design AI solutions that reflect local needs. Through collaborative, inclusive, and context-sensitive development, democratising AI can drive faster progress towards achieving the SDGs, while also enhancing equity, dignity, and local empowerment.

9.4 Conclusion

Critical success factors play a pivotal role in the effective deployment of AI to achieve the SDGs by ensuring that AI systems are technically viable, socially, ethically, and contextually relevant. These factors include access to high-quality data, availability of skilled human resources, ethical and inclusive AI governance, and policy and regulatory support. For instance, access to localised, unbiased, and representative datasets is a critical success factor for AI applications aimed at achieving SDG 3 (Good Health and Well-Being), as healthcare AI models must be trained on diverse datasets to ensure accurate disease detection and diagnosis for all populations. Similarly, capacity building through education and skills development is essential for achieving SDG 4 (Quality Education), as communities must have the technical skills to co-create and operate AI-powered educational platforms. Ethical AI governance and strong regulatory frameworks are also essential success factors in ensuring that AI systems respect human rights and avoid reinforcing biases, especially when addressing SDG 10 (Reduced Inequalities).

Leapfrogging is a critical success factor for the deployment of AI to achieve the SDGs because it enables countries, particularly in the Global South, to bypass traditional, costly, and time-consuming development stages. Instead of replicating legacy systems, countries can directly adopt next-generation AI-driven solutions to accelerate progress towards the SDGs. This approach allows nations to avoid outdated technologies and move straight to more efficient, cost-effective, and sustainable AI systems. Decoloniality ensures that AI systems are inclusive, culturally relevant, and community-driven, enabling local ownership and knowledge sovereignty. The democratisation of AI broadens access to AI tools, education, and data, empowering marginalised communities and local innovators to co-create solutions.

Together, decoloniality and democratisation enable a pluriversal AI ecosystem, fostering equity, accountability, and local participation in AI-driven development, ensuring that no one is left behind in the pursuit of the SDGs. By identifying, prioritising, and effectively engaging all the critical success factors, stakeholders can ensure that AI deployment is equitable, inclusive, and aligned with local development goals, thereby accelerating progress towards the SDGs. Indeed, with a nuanced, contextualised, and people-centred approach, decolonial AI can be a driver of sustainable development and economic transformation for all humanity, including the Global South, as articulated in the 2025 book titled *Artificial Intelligence: A Driver of Inclusive Development and Shared Prosperity for the Global South* by Arthur G.O. Mutambara, published by Taylor & Francis [86].

Governance, Legislation, Regulations, Sociology, and Ethics

10

10.1 Introduction

As discussed in previous chapters, integrating AI into efforts to achieve the SDGs presents a transformative opportunity to expedite the SDG journey. From eradicating poverty and hunger to combatting climate change and ensuring quality education, AI has the potential to enhance efficiency, innovation, and impact across sectors, leading to SDG victory by 2030. However, this potential can only be realised if accompanied by robust legislation and regulation anchored by effective governance to ensure ethical, inclusive, and sustainable deployment. In critical areas, legal and regulatory frameworks are needed to harness AI to achieve the SDGs while mitigating associated risks. AI guardrails, governance, and regulation were introduced in Chap. 7. However, this chapter presents a more detailed treatment, linking these interventions to legislation while exploring global best practices and emerging experiences.

The use of AI to address SDG-related challenges has been presented earlier. For example, in education, AI-powered platforms can provide personalised learning experiences, bridging gaps in access and quality. Predictive analytics and AI-assisted diagnostics enhance efficiency and accuracy in healthcare, especially in underserved regions. AI also supports climate action through environmental monitoring, precision agriculture, and renewable energy optimisation. Despite these benefits, unregulated AI deployment risks exacerbating inequalities, violating privacy, and misaligning with ethical principles. Hence, legislation is essential to ensure AI serves as a tool for inclusive development and, more specifically, the attainment of the SDGs.

10.2 Basic Governance

AI governance refers to the framework of policies, principles, practices, and mechanisms designed to oversee AI development, deployment, and use. Its primary goal is to ensure that AI technologies are aligned with ethical standards, societal values, and regulatory requirements while addressing potential risks such as bias, discrimination, lack of transparency, and misuse.

AI governance encompasses a range of activities, including establishing ethical guidelines, developing laws and regulations, monitoring AI applications, and creating accountability mechanisms.

A. G. O. Mutambara, *Deploying Artificial Intelligence to Achieve the UN Sustainable Development Goals*, Sustainable Development Goals Series,
https://doi.org/10.1007/978-3-031-88423-8_10

It involves diverse stakeholders—governments, private companies, academia, and civil society—working collaboratively to ensure that AI systems are safe, fair, transparent, and beneficial for all. Additionally, AI governance often includes principles like human oversight, explainability of algorithms, data privacy, and inclusivity to protect individual rights and promote equitable outcomes.

As AI becomes more integrated into critical sectors such as healthcare, finance, education, and public services, governance plays a pivotal role in mitigating risks, fostering trust, and maximising AI's positive impact on society. The legislation and regulations can be classified into basic and elaborate governance. Basic AI governance refers to foundational frameworks that establish broad principles and minimal regulations to ensure AI's ethical development and use. It focuses on essential issues like data privacy, transparency, and non-discrimination, providing a general guideline applicable across sectors without delving into specific use cases.

On the other hand, elaborate AI governance involves detailed, sector-specific regulations and frameworks that address the unique challenges, risks, and opportunities associated with AI applications in particular domains. For instance, while basic governance might mandate transparency in AI algorithms, elaborate governance for healthcare AI could require rigorous clinical trials, patient safety protocols, and compliance with health data privacy laws. Basic governance serves as a universal baseline, fostering trust and accountability. In contrast, elaborate governance enables tailored oversight to ensure that AI technologies effectively meet the nuanced needs of different sectors and contexts. Basic governance serves as a universal baseline, fostering trust and basic accountability [18, 54].

10.2.1 Ethical Guidelines

One foundational area for legislation is establishing ethical guidelines for AI development and deployment. AI systems must be designed to operate transparently, ensuring fairness and accountability. Laws mandating the explainability of AI algorithms and preventing discrimination are critical for addressing biases in AI decision-making. For example, an AI-powered recruitment system must not disadvantage candidates based on gender, ethnicity, or socio-economic background. Ethical AI legislation ensures its benefits are distributed equitably and minimises potential harm.

10.2.2 Data Privacy and Protection

Another key area is data privacy and protection. AI systems rely heavily on data, often including sensitive personal information. Without robust privacy laws, data misuse could erode trust and hinder the adoption of AI solutions. Comprehensive data protection regulations, similar to the European Union's General Data Protection Regulation (GDPR), are necessary. These laws should define protocols for data collection, storage, and sharing while requiring informed consent. For global AI collaboration, especially in tackling SDGs like global health or climate change, cross-border data-sharing agreements must balance accessibility and privacy.

10.2.3 Environmental Impact

The environmental impact of AI also requires regulation. While AI can aid in addressing environmental challenges, its infrastructure, including energy-intensive data centres and computational processes, contributes to carbon emissions. Legislation should promote the use of renewable energy in AI operations and set standards for energy-efficient AI technologies. Additionally, regulations addressing the recycling and disposal of AI-related hardware can minimise e-waste, ensuring the environmental sustainability of AI deployment.

10.2.4 Democratising AI and Mitigating Inequality

Legislative or regulatory frameworks that mandate public consultation and stakeholder engagement in developing and implementing AI regulations are essential. This ensures inclusivity, transparency, and accountability in decision-making processes. Engaging diverse stakeholders, including civil society organisations, academia, industry, labour, and affected communities, fosters dialogue, consensus-building, and democratic governance of AI deployment for SDGs. Thus, comprehensive legislation and regulations that address critical areas are collectively and democratically designed and derived. This creates an enabling and empowering environment for the responsible and sustainable deployment of AI technologies to attain the SDGs.

Another pressing need is to address inequality through AI legislation. AI's benefits often remain concentrated in developed regions, widening the digital divide. Laws should ensure equitable access to AI technologies, particularly in developing countries. Governments can promote local capacity building and technology transfer to empower underserved regions to leverage AI for their unique challenges. Moreover, regulatory bodies must monitor AI systems for discriminatory outcomes, penalising practices that disadvantage marginalised communities [32, 94].

10.3 Elaborate Governance

Elaborate AI governance involves advanced concepts such as nuanced accountability, interoperability, cybersecurity, and intellectual property dilemmas. It deals with detailed, sector-specific regulations and frameworks that address the unique challenges, risks, and opportunities associated with AI applications in particular domains. For instance, while basic governance might mandate transparency in AI algorithms, elaborate governance for healthcare AI could require rigorous clinical trials, patient safety protocols, and compliance with health data privacy laws. Elaborate governance enables tailored oversight to ensure that AI technologies effectively meet the nuanced needs of different sectors and contexts. It also goes beyond the nation state by addressing regional, continental, and global governance.

Elaborate governance of AI involves a complex balancing of dilemmas—the trade-offs between opportunity seeking and risk aversion, the interplay between security and transparency, tensions between globalisation and localisation, and the interplay between self-regulation and government control [79]. Indeed, AI governance frameworks must balance innovation, ethics, public safety, and fairness as AI technologies evolve. Flexible, inclusive, context-aware, and predictive AI governance policies are needed. Multi-stakeholder cooperation between governments, commerce, academia, and civil society is essential to guarantee robust and adaptable AI regulatory frameworks. What is critical is a dynamic and balanced governance structure that can adapt to AI advances to ensure the technology delivers maximum benefits while exhibiting minimum risks [79].

10.3.1 Accountability and International Coordination

Accountability is a cornerstone of effective AI governance. Addressing the harm caused by AI errors or misuse becomes difficult without clear accountability frameworks. Legislation should establish independent oversight bodies to monitor compliance with AI regulations and create mechanisms for redress in cases of non-compliance. Additionally, laws must define liability for AI developers, deployers, and users, ensuring accountability for unintended consequences.

Given AI's global nature, international cooperation is essential. AI systems often operate across borders, requiring harmonised regulatory standards to facilitate collaboration. Multilateral organisations, such as the United Nations, can play a pivotal role in establishing global guidelines for ethical AI deployment. International agreements can also address issues like cross-border data sharing and trade in AI technologies, fostering a unified approach to leveraging AI for SDGs.

Indeed, international cooperation and harmonisation of AI regulations are essential to address global challenges and ensure consistency across jurisdictions. Multilateral agreements, treaties, and frameworks facilitate information sharing, capacity building, and coordination among countries to promote responsible AI deployment for achieving SDGs.

10.3.2 Interoperability, Standards, and Cybersecurity Regulations

Legislation or regulatory frameworks promoting interoperability and standardisation of AI systems facilitate data exchange, collaboration, and integration across different platforms and domains. These standards ensure compatibility, consistency, and efficiency in AI deployment for SDGs.

Regulations addressing cybersecurity risks associated with AI deployment are essential to protect AI systems from cyber threats, data breaches, and malicious attacks. These regulations may mandate cybersecurity measures, vulnerability assessments, and incident response protocols for AI applications.

10.3.3 Balance and Capacity Building

Another critical consideration is balancing innovation and accessibility in intellectual property rights (IPR). IPR refers to the legal frameworks governing intellectual property rights, including patents, copyrights, and trademarks. These frameworks play a crucial role in incentivising innovation and investment in AI technologies. Clear regulations regarding AI-generated inventions, ownership of AI-generated content, and licensing agreements are necessary to facilitate the development and dissemination of AI solutions for SDGs.

While protecting AI innovations is important, overly restrictive IPR frameworks could hinder global access to technologies essential for achieving SDGs. Laws encouraging open-source AI development for SDG-related projects can enhance accessibility while fostering innovation. Reforms in IPR regulations can ensure that AI technologies benefit a broader audience without stifling creativity and investment. Finally, capacity building and education must be prioritised to bridge gaps in regulatory implementation. Many countries lack the expertise to develop and enforce AI regulations effectively. Governments should fund education and training programmes focused on AI ethics, governance, and technical skills. Public awareness campaigns can also ensure that citizens understand the implications of AI and actively participate in its governance.

10.3.4 Sector-Specific Regulations

Within the national AI governance framework, sector-specific regulations tailored to domains such as healthcare, education, agriculture, environmental conservation, and disaster management must be established to address unique challenges and considerations in deploying AI for SDGs within these sectors. These regulations may include licensing requirements, certification processes, and sector-specific guidelines for AI applications.

It is prudent to emphasise that sector-specific AI regulations are essential because AI's application, risks, and impacts vary significantly across industries, necessitating tailored approaches to governance. For instance, AI in healthcare requires stringent regulations to ensure patient safety, privacy, and the accuracy of diagnostic tools. At the same time, AI in finance demands robust frameworks to prevent fraud, bias, and systemic risks.

In sectors like autonomous vehicles, safety standards must address real-world operational challenges, whereas in education, laws must focus on equitable access and avoiding algorithmic bias in assessments. A one-size-fits-all approach risks over-regulation that stifles innovation in some areas or under-regulation that allows harm in others. By addressing each sector's unique needs and ethical considerations, sector-specific legislation ensures that AI technologies are deployed responsibly and effectively, fostering trust and maximising their potential for societal benefit.

10.3.5 Regional and Continental Harmonisation of AI Regulations

Effective deployment of AI to achieve the SDGs will require economies of scale. Hence, regional and continental economic integration anchored by political unity is foundational. Consequently, beyond national AI governance frameworks, efforts must be made to harmonise and standardise legislation and regulations at regional and continental levels. For Africa, this will mean at the Regional Economic Community (REC) platform (e.g. SADC, EAC, or ECOWAS) and the African Union level. Thus, harmonised and standardised regional and continental governance frameworks will ignite AI deployment to achieve SDGs worldwide.

Indeed, regional and continental harmonisation of AI regulations is critical to fostering collaboration, ensuring ethical standards, and addressing cross-border challenges associated with AI technologies. Harmonised frameworks enable countries within a region or continent to adopt common standards for data sharing, privacy, transparency, and accountability, ensuring a unified approach to AI governance. Such alignment is particularly vital for fostering trust in AI-driven initiatives like regional trade, healthcare, and climate action, which rely on shared data and systems. Furthermore, harmonisation strengthens collective bargaining power in global AI policy discussions and reduces regulatory fragmentation, enabling smoother innovation and investment flows. Moreover, it will ensure that AI is developed and deployed to benefit all nations equitably by facilitating the efficacious attainment of the SDGs.

10.3.6 Global AI Governance

Global AI governance is important because AI technologies often transcend national boundaries, creating challenges and opportunities that require coordinated international responses. Without global governance, disparities in regulations could lead to uneven ethical standards, misuse, or regulatory arbitrage, where companies exploit less restrictive jurisdictions. A unified global approach ensures that AI is developed and deployed responsibly, promoting fairness, accountability, and inclusivity across nations. It facilitates international collaboration on issues like data sharing, cybersecurity, and the ethical use of AI in areas such as healthcare, climate action, and global trade. Moreover, global governance helps address pressing concerns like bias in AI systems, intellectual property disputes, and the concentration of AI power in a few countries or corporations. By establishing shared principles and frameworks, global AI governance enhances trust, reduces risks, and maximises AI's potential for addressing collective global challenges.

10.4 UN Secretary-General's AI Advisory Final Report

The UN Secretary-General's High-Level Advisory Body on AI released its final report on 19 September 2024—*Governing AI for Humanity: UN Report Proposes Global Framework for AI Oversight*. After further global consultations, the document builds on the interim report released in December 2023. The final report discusses the need for a global, inclusive, coordinated approach to AI governance to harness its benefits and mitigate risks. It emphasises the transformative potential of AI in various sectors, including scientific inquiry, energy optimisation, public health, and agriculture, while highlighting risks such as AI bias, surveillance, and disinformation.

The report points out the global governance deficit in AI, with a patchwork of norms and institutions that lack accountability and are often voluntary. The report calls for a holistic, global approach to AI governance grounded in international law and the SDGs to ensure an equitable distribution of AI benefits. It proposes the creation of an international scientific panel on AI, similar to the IPCC, to provide reliable scientific knowledge and risk assessments.

The text recommends a twice-yearly intergovernmental and multi-stakeholder policy dialogue on AI governance to share best practices and promote common understandings. It suggests the establishment of an AI standards exchange to develop and maintain a register of definitions and applicable standards for AI systems. The document advocates for a global fund for AI to address capacity and collaboration gaps, ensuring access to AI enablers for needy countries. Finally, it proposes the creation of an AI office within the UN Secretariat to support and catalyse these initiatives, ensuring a coherent and inclusive approach to AI governance. Figure 10.1 illustrates the hierarchy of AI governance functions as globally conceived.

10.4.1 Key Recommendations

It is prudent to flesh out, summarise, and explain the key recommendations of the UN Secretary-General's High-Level Advisory Body on AI. At the document's core are seven key proposals aimed at resolving critical gaps in current AI governance:

- Creating an international scientific panel on AI—to serve as a global authority on AI capabilities, opportunities, risks, and uncertainties. It will provide impartial and reliable scientific knowledge and foster a shared foundational understanding of AI globally.
- Launching a policy dialogue on AI governance—a proposal to enable regular intergovernmental and multi-stakeholder discussions to share best practices and advance common understanding regarding AI governance measures, challenges, and incidents.
- Establishing an AI standards exchange—to bring together representatives from various standard-setting organisations, technology companies, and civil society with the express task of developing a register of definitions and standards for monitoring, evaluating, and measuring AI systems.

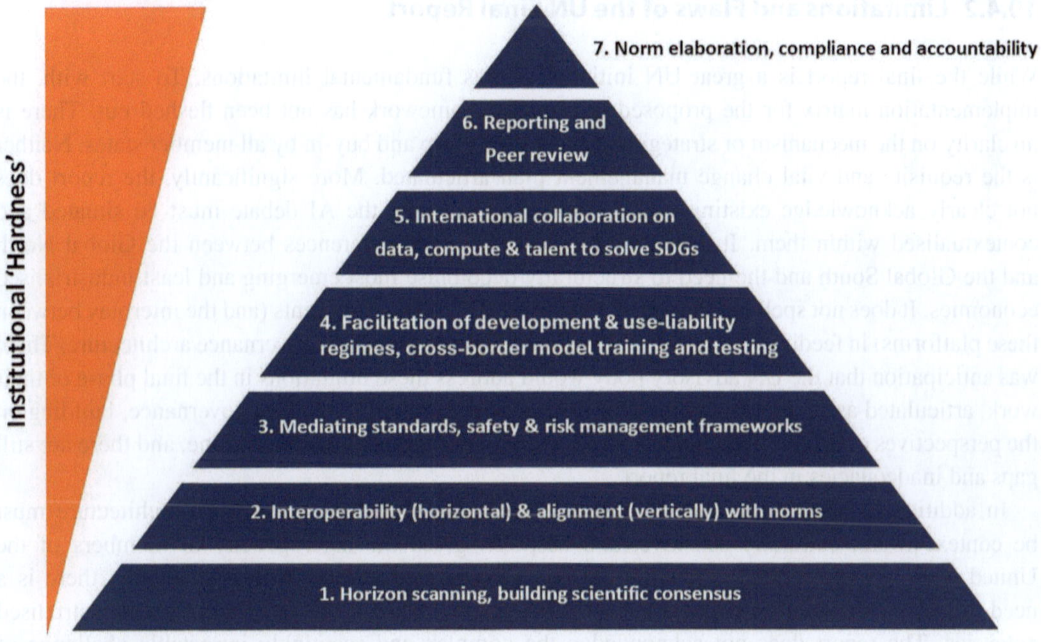

Fig. 10.1 Hierarchy of AI governance functions: A global perspective [135]

- Forming an AI capacity development network—to link collaborating, UN-affiliated capacity development centres, thus providing expertise, compute resources, and AI training data to key actors, particularly in emerging economies.
- Creating a global fund for AI—to address the AI divide by facilitating access to AI enablers, particularly for Global South countries which lack adequate resources, infrastructure, and expertise.
- Developing a global AI data framework—to fashion and establish principles and practical arrangements for AI training data governance and use, aligned with international commitments on human rights, intellectual property, economic transformation, and sustainable development.
- Setting up an AI office within the UN Secretariat—to function as a coordinating body, supporting the implementation of the other recommendations and ensuring a harmonised and coherent UN-wide approach to AI governance, thus ensuring that AI is effectively regulated and managed for the benefit all humanity.

It is instructive to note that the report prioritises a globally inclusive and distributed AI governance framework based on international coordination, cooperation, and solidarity. The document calls on developed and developing countries to work together in governing AI to ignite industrialisation while furthering respect, protection, and fulfilment of all human rights worldwide. The suggested mechanisms are designed to complement existing interventions and drive inclusive global AI governance arrangements that are dynamic and adaptive. This is essential given the technology's exponential growth and unrelenting explosion of AI innovations. Indeed, effective implementation of the final report's seven key recommendations would go a long way in advancing the global governance of AI for humanity [135].

10.4.2 Limitations and Flaws of the UN Final Report

While the final report is a great UN initiative, it has fundamental limitations. To start with, the implementation matrix for the proposed governance framework has not been fleshed out. There is no clarity on the mechanism or strategy to ensure ownership and buy-in by all member states. Neither is the requisite and vital change management plan articulated. More significantly, the report does not clearly acknowledge existing global challenges and how the AI debate must be situated and contextualised within them. It does not recognise the stark differences between the Global North and the Global South and the need to structurally decolonise most emerging and least industrialised economies. It does not spell out the role of countries, RECs, and continents (and the interplay between these platforms) in feeding into and leveraging the proposed global AI governance architecture. There was anticipation that the UN advisory body would address these limitations in the final phase of their work, articulated as "exploring options for institutional forms for global AI governance, building on the perspectives of diverse stakeholders worldwide". This was not effectively done, and there are still gaps and inadequacies in the final report.

In addition to being agile and adaptive, the inclusive global AI governance architecture must be context-aware, culturally sensitive, and bespoke, given the heterogeneity of members of the United Nations. The report is oblivious to this foundational matter. More specifically, there is a need to address the unique AI governance challenges that bedevil emerging and least industrialised countries. The report does not acknowledge the complex and seemingly intractable challenge of democratising and decolonising AI. This is a fatal flaw. Inclusive global governance of AI for humanity is meaningless without addressing the inequitable ownership of AI technology and its innovations. Furthermore, the dangers and risks of digital imperialism and data colonialism are not explicitly addressed. An efficacious global governance architecture must take cognisance of all these challenges and proffer explicit mechanisms to manage, mitigate, or resolve them. Indeed, the AI governance work presented in this book seeks to address the flaws and limitations of the Final Report of the UN Secretary-General's High-Level Advisory Body on AI [134, 135].

10.5 Principles of AI Regulation

The principles of AI regulation are a set of ethical guidelines, legal norms, and industry standards governing AI systems' development, deployment, and use. The objective is to ensure a bold but responsible embrace of the AI revolution. Hence, these principles are intended to ensure that AI technologies operate in safe, fair, transparent, and accountable ways. As AI systems continue to influence critical areas of society—from healthcare and finance to policing and employment—these regulatory principles serve as essential safeguards to protect human rights, foster social good, and mitigate potential risks. By embedding these principles into laws and policies, stakeholders can ensure that AI systems remain beneficial for all of humanity.

10.5.1 Details of the Principles

One of the most fundamental principles of AI regulation is fairness and non-discrimination. This principle requires AI systems to be free from bias, ensuring that no person is unfairly disadvantaged based on characteristics such as race, gender, and socio-economic status. Without proper checks, AI systems trained on biased datasets may perpetuate social inequalities, leading to unfair outcomes in areas like hiring and lending. To address this, regulators advocate for the use of Fairness-Aware Machine Learning (FAML) techniques and diverse, representative datasets. Transparency and explainability are other critical principles. They ensure that the decision-making process of AI systems is understandable to users and stakeholders. For example, if an AI system denies a person a loan, that individual should have the right to know how and why the decision was made. Techniques like Explainable AI (XAI) help developers create more transparent models that humans can easily interpret. Another core principle is accountability and responsibility, which assigns clear liability for decisions made by AI systems. When AI systems cause harm—as in the case of an autonomous vehicle crash—it is essential to identify which party (developer, operator, or end-user) is legally responsible. This principle ensures that companies deploying AI systems remain accountable for their actions and decisions. Safety and security are also paramount, as unsecured AI systems can be vulnerable to hacking and manipulation. Ensuring AI robustness, conducting stress tests, and implementing cybersecurity protocols are some of the measures taken to uphold this principle.

Privacy and data protection are two of the most widely discussed principles, especially in the context of personal data collection. AI models often require vast amounts of data to function effectively, raising concerns about privacy breaches and misuse. To address this, regulatory frameworks like GDPR (General Data Protection Regulation) require companies to protect user data through encryption, anonymisation, and secure storage practices. This principle is closely aligned with human-centric and ethical design, emphasising that AI should prioritise human well-being and dignity. Ethical design encourages developers to ensure that AI systems support, rather than undermine, human rights. This principle is particularly relevant in cases where AI is used in surveillance or facial recognition technologies, as these applications pose serious risks to privacy and civil liberties.

Another emerging principle is sustainability and environmental impact. Large AI models, like GPT and AlphaFold, require immense computational power, leading to significant energy consumption and carbon emissions. This principle calls for more sustainable AI practices, such as optimising code and using energy-efficient hardware. Similarly, inclusivity and accessibility aim to ensure that AI tools are accessible to people of all socio-economic classes and abilities, not just to privileged groups. AI tools and platforms should accommodate people with disabilities and ensure multilingual, user-friendly interfaces that bridge the digital divide.

Two other essential principles are proportionality and necessity and legal and regulatory compliance. The principle of proportionality ensures that AI systems are only used when absolutely necessary, especially in sensitive areas like policing, immigration control, and surveillance. It emphasises that less invasive methods should be preferred over AI if they achieve the same outcome. Lastly, legal and regulatory compliance ensures that AI systems adhere to existing laws, industry standards, and emerging global frameworks like the EU AI Act, which categorises AI by risk level and imposes stricter obligations on "high-risk" AI systems.

The need for AI regulation is undeniable. Unregulated AI systems can cause significant harm, including bias in decision-making, job displacement, privacy invasions, and security vulnerabilities. Regulatory principles ensure that AI is used ethically and responsibly, protecting human rights while fostering public trust. The frameworks shaping these principles include the EU AI Act, the OECD AI Principles, and UNESCO's AI Ethics Guidelines, which all prioritise transparency, fairness, and human dignity. By adhering to these regulatory principles, governments, businesses, and developers can guarantee that AI remains a force for good in society, driving innovation while ensuring ethical safeguards [33, 34].

10.5.2 AI Risk Verticals

A risk vertical refers to a specific area or category of risk within an organisation, industry, or market. Each vertical focuses on a distinct type of risk, often grouped by its source, nature, or the part of the business it affects. The concept is used in risk management frameworks to organise and address different kinds of risks systematically. Put differently, risk verticals are specific classes or domains of risk that organisations identify, group, and manage to ensure effective risk mitigation. Each vertical represents a specific genre of risk that affects a particular area of an organisation, allowing for focused attention, resource allocation, and tailored risk management strategies. By segmenting risks into verticals, companies can adopt specialised approaches to assess, monitor, and reduce potential threats. This categorisation is essential for large, complex organisations where multiple risk factors operate simultaneously across different departments, business units, or regions. For instance, a financial institution may divide its risks into operational, credit, market, and compliance risks—each with its dedicated team, tools, and procedures.

The most common risk verticals include operational, financial, regulatory, compliance, cybersecurity, reputational, and environmental, social, and governance (ESG) risks. Each of these verticals addresses a specific area of concern. For example, cybersecurity risk focuses on protecting digital systems from cyberattacks and data breaches, while regulatory and compliance risk deals with adhering to laws, industry standards, and government regulations. Risk verticals enable companies to establish clear accountability, assign specialised teams to oversee specific areas of risk, and develop policies that address the unique nature of each risk category. This approach also allows organisations to prioritise risks that require urgent action and that can be managed over time. By employing this structured approach to risk management, organisations improve their ability to anticipate, address, and recover from potential threats.

The risk vectors associated with AI are unique due to AI systems' complexity, autonomy, and dynamic nature, which introduce new types of risks not present in traditional technologies. Unlike conventional software, AI models learn from data, evolve, and make decisions that may not always be transparent or explainable. This creates significant risks related to bias and discrimination, where models trained on skewed datasets can produce unfair outcomes, particularly in hiring, credit scoring, and criminal justice. Another key vector is black-box decision-making, where the inner logic of AI models, especially Deep Learning systems, is difficult to interpret, making it hard to trace errors or justify outcomes. Data privacy and security risks are also heightened, as AI systems rely on large datasets that may contain sensitive information, increasing exposure to breaches and misuse. Autonomy and control risks arise as AI systems operate with a degree of independence, potentially making harmful decisions without human oversight, as seen in autonomous vehicles or military drones. Additionally, adversarial attacks are a unique threat, where small manipulations of input data (like images or text) can cause AI models to produce incorrect predictions. Finally, systemic risks are amplified, as widespread reliance on AI in key sectors like finance, healthcare, and national security could lead to large-scale failures if AI systems malfunction or are hacked. These unique risk vectors require specialised governance frameworks, ethical safeguards, and continuous monitoring to ensure that AI technologies remain secure, fair, and aligned with human values. The risk verticals for assessing an AI system can be summarised as follows [33, 34]:

- Robustness: the risk of an algorithm failing in unexpected circumstances or when under attack.
- Bias: the risk of an algorithm mistreating individuals or groups—significant for applications that significantly impact people's lives.
- Privacy: the potential for an algorithm to leak sensitive or personal data.

- Explainability: the danger that the system, its decisions, or its import may not be easily understandable to users and developers.
- Efficacy: The risk that the system performs poorly relative to its mandate or business case.

10.6 European Union AI Act

As already indicated, there are several good examples of effective AI regulation. The European Union's AI Act sets global benchmarks for AI regulation. This landmark legislative framework aims to regulate the development, deployment, and use of AI across the EU. Introduced in April 2021, the Act is the first comprehensive legal framework for AI globally and seeks to strike a balance between fostering innovation and addressing risks associated with AI technologies.

The Act adopts a risk-based approach (Fig. 10.2), categorising AI systems into four levels of risk: (1) unacceptable, (2) high, (3) limited, and (4) minimal. Unacceptable-risk AI systems, such as those involving social scoring by governments, are outright banned. High-risk applications, including AI in critical sectors like healthcare, law enforcement, and education, are subject to stringent transparency, accuracy, and accountability requirements. Specific transparency obligations apply for limited-risk systems, such as chatbots, while minimal-risk systems face no additional regulations. This tiered approach ensures that oversight is proportionate to the potential harm the AI system poses. It is anticipated that most AI systems will not be high-risk, as shown in Fig. 10.3. The EU process of approving a high-risk AI system is illustrated in Fig. 10.4.

The Act also emphasises ethical AI deployment, requiring safeguards to prevent discrimination, ensure human oversight, and protect fundamental rights. By mandating transparency and accountability, the legislation aims to build trust in AI technologies while addressing concerns such as bias, data privacy, and misuse. Additionally, the AI Act has provisions for creating a European AI Board to oversee implementation and ensure harmonised enforcement across member states. It supports innovation by establishing regulatory sandboxes where developers can test AI systems under controlled conditions. With the Act, the EU aims to set a global standard for responsible AI governance, encouraging other nations to adopt similar regulatory measures. By doing so, it positions itself as a leader in shaping AI's ethical and safe use while fostering a single market for trustworthy AI systems [34].

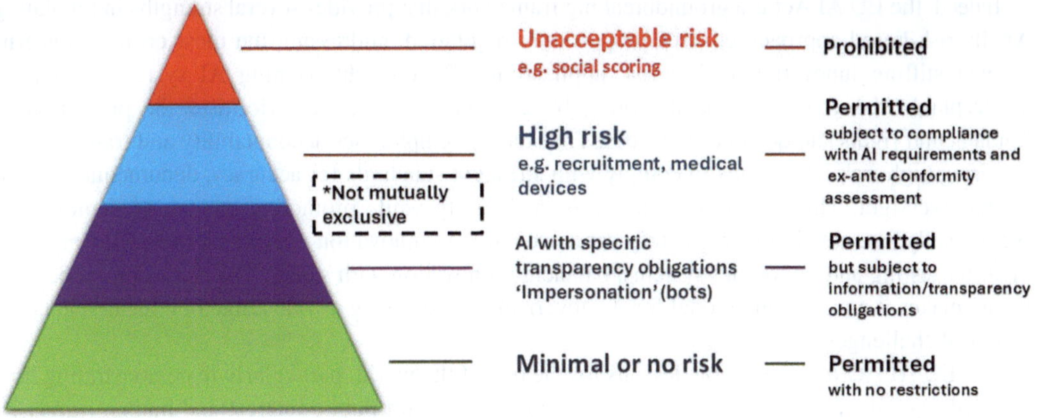

Fig. 10.2 Risk-based approach to governance [34]

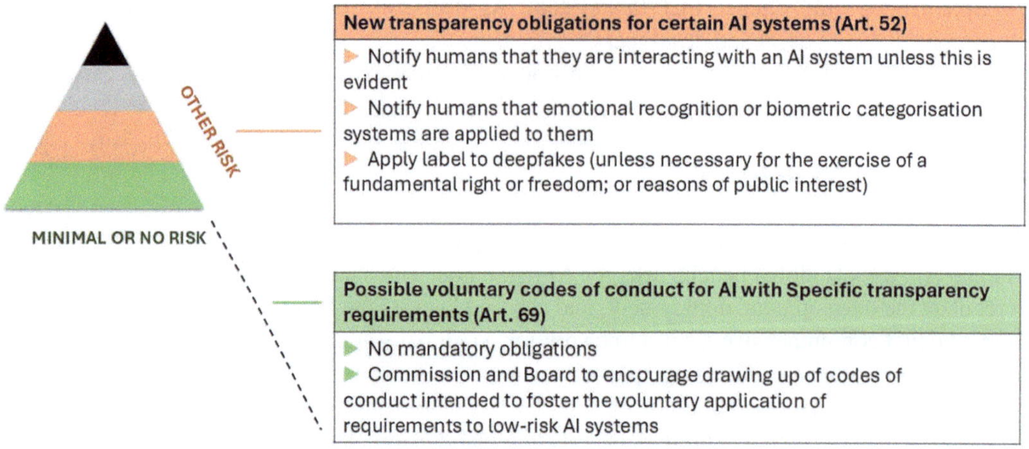

Fig. 10.3 Most AI systems will not be high-risk [34]

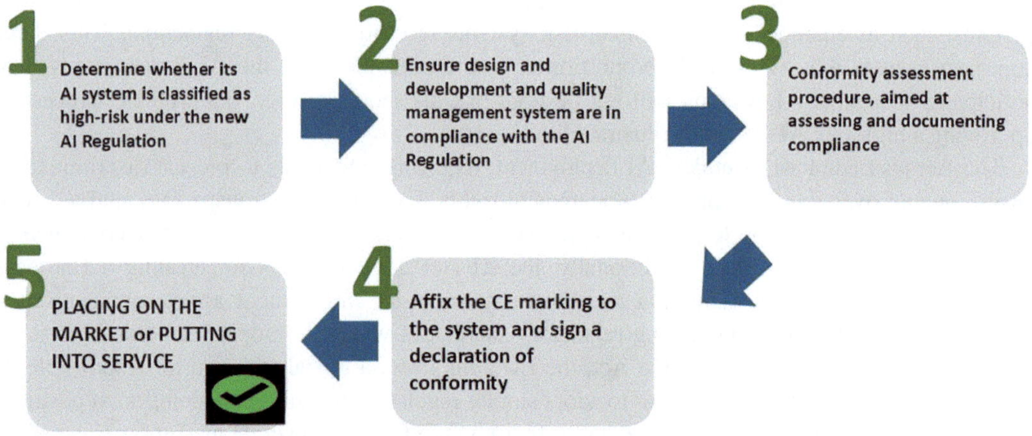

Fig. 10.4 EU process of approving a high-risk AI system [34]

Indeed, the EU AI Act is a groundbreaking framework that provides several strengths in regulating AI. Its risk-based approach ensures proportional regulation, addressing the most critical concerns without stifling innovation in low-risk applications. By outright banning AI systems that pose unacceptable risks, such as social scoring by governments, the Act prioritises the protection of fundamental rights and democratic values. The Act also emphasises accountability and transparency, requiring high-risk AI systems to comply with stringent standards for accuracy, documentation, and human oversight. This approach fosters trust and safety while promoting ethical AI deployment. Additionally, the creation of regulatory sandboxes for AI innovation provides a controlled environment for testing and improving systems, balancing innovation with safety. The Act's comprehensive nature makes it a pioneering model for AI governance, addressing a wide range of ethical, legal, and technical challenges.

The EU AI Act offers valuable lessons for the rest of the world, particularly in demonstrating how a robust regulatory framework can align AI deployment with public interest and human rights. Its emphasis on transparency and accountability serves as a template for ensuring ethical AI use, even in regions with different sociopolitical contexts. The Act also illustrates the importance of harmonising regulations within a multi-jurisdictional setting, fostering cohesion and reducing fragmentation—a challenge relevant for regions in the Global South, for example, Africa and Southeast Asia. Furthermore, the EU's focus on international cooperation, as seen in its alignment with global AI principles, underscores the need for cross-border collaboration in AI governance. However, the success of the Act depends on effective implementation and enforcement, which could provide further insights into the practicality of such comprehensive regulations. The EU AI Act offers a foundational blueprint for nations aspiring to regulate AI to ensure that technological progress aligns with societal values and global development goals [34]. As the world embarks on the ambitious mission to deploy AI to achieve the SDGs, the EU Act provides valuable lessons and emergent experiences in AI governance, legislation, and regulation [34].

Indeed, countries worldwide must leverage and learn from the legislative frameworks adopted in jurisdictions such as the EU, the United States, China, and the United Kingdom. As already explained, the most advanced examples are the EU AI Act, the OECD AI Principles, and UNESCO's AI Ethics Guidelines. Each country needs a dynamic law commission that tracks new technologies and creates new enabling laws for all new technologies (AI being just one type). The lesson from the EU legislative efforts—in particular, the development and adoption of the EU AI Act—is that three separate national Acts of Parliament must be developed:

- Digital Markets Act
- Digital Services Act
- AI Act

Each country must have these three pieces of legislation developed in conjunction with and harmonised at the regional and continental levels. For example, in Africa, this means each REC and at the AU. For example, at SADC and AU, there must be a harmonising framework consisting of:

- Digital Markets Laws and Regulations
- Digital Services Laws and Regulations
- AI Laws and Regulations

It is prudent to state that technology regulatory readiness must be linked to other developmental plans and strategies. In the case of Africa, it means aligning with existing regional and continental industrialisation plans, such as the SADC Industrialisation Strategy, SADC Infrastructure Masterplan, Programme for Infrastructure Development in Africa (PIDA), and AU Agenda 2063.

10.7 India's AI-Driven Aadhaar System

India's Aadhaar system for inclusive development, the world's most extensive biometric identification programme, has leveraged AI to enhance its inclusivity and efficiency in fostering development. It is an illustrative example of the development and implementation of AI legislation and regulations. Aadhaar, managed by the Unique Identification Authority of India (UIDAI), assigns residents a unique 12-digit identity number based on their biometric and demographic data. AI technologies have efficiently processed massive amounts of data, ensured authentication accuracy, and minimised system duplication. Using AI for facial recognition, iris scans, and fingerprint matching, Aadhaar ensures robust identity verification, which is critical for delivering government services and subsidies to marginalised communities. The integration of AI in Aadhaar has also streamlined processes for

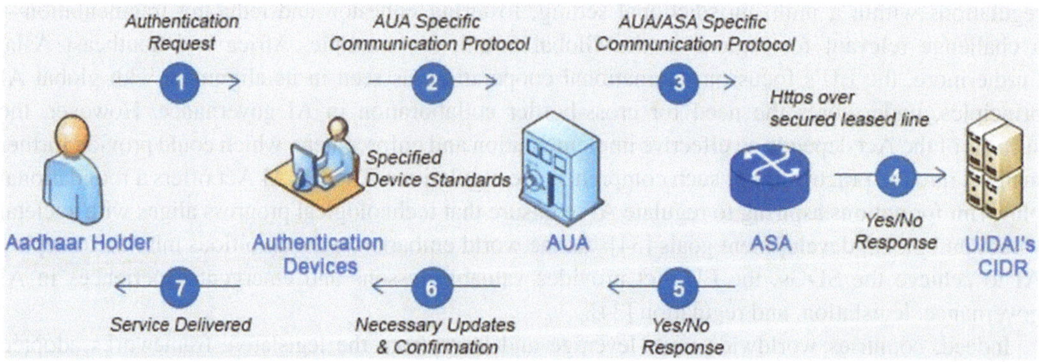

Fig. 10.5 Aadhaar identity system analysis

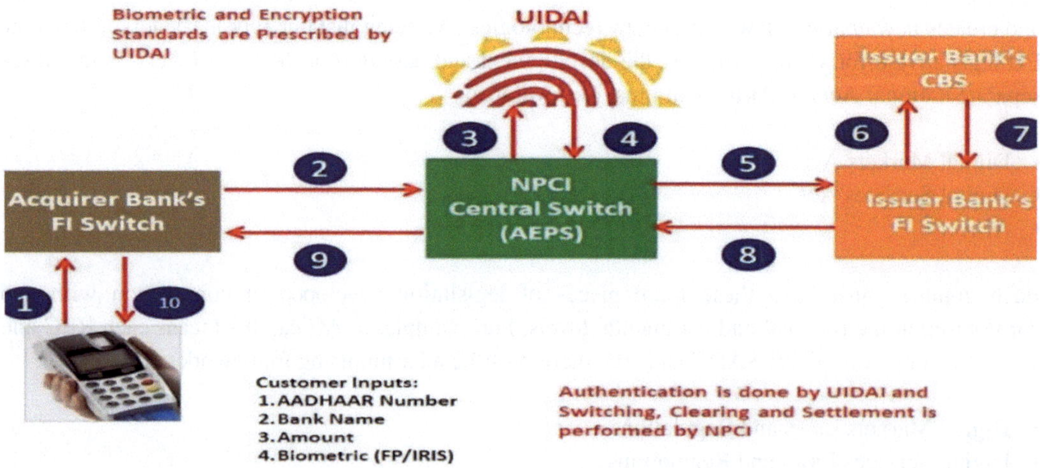

Fig. 10.6 Aadhaar-enabled payment system

opening bank accounts, accessing social welfare benefits, and availing healthcare services, making these essential services more accessible to underprivileged populations.

AI-powered Aadhaar applications have significantly contributed to India's efforts towards achieving SDGs by addressing financial inclusion, poverty alleviation, and improved access to public services. For instance, the system has facilitated the delivery of direct benefit transfers (DBT) to beneficiaries, reducing leakages and corruption in subsidy distribution. AI ensures that benefits reach the intended recipients without discrimination by linking Aadhaar to schemes such as the Public Distribution System (PDS) and the Mahatma Gandhi National Rural Employment Guarantee Act (MGNREGA). However, concerns over data privacy, surveillance, and exclusion due to technological or infrastructural gaps remain challenges. Addressing these issues while continuing to harness AI's potential in Aadhaar is essential for fostering inclusive and sustainable development in India. Figures 10.5 and 10.6 show identity system analysis by Aadhaar and an Aadhaar-enabled payment system, respectively.

Indeed, India's Aadhaar system provides valuable lessons on the governance, legislation, and regulation of AI, particularly in the context of large-scale identity management and inclusive development. One key lesson is the importance of creating robust data protection frameworks to safeguard the sensitive biometric data that underpins AI-powered systems. Effective governance requires transparency, accountability, and mechanisms to address grievances, as seen in Aadhaar's role in enabling access to public services while minimising fraud. Additionally, the Aadhaar system underscores the value of using AI to promote financial inclusion, streamline service delivery, and reduce corruption. However, it also highlights the necessity of regulatory oversight to ensure ethical AI deployment, particularly in preventing misuse of data and avoiding discrimination. The Aadhaar experience demonstrates that scaling AI to serve millions requires a balanced approach that fosters innovation while protecting individual rights and ensuring equitable access.

Despite its success, the Aadhaar system has notable limitations that offer critical lessons for future AI governance. A primary concern is the risk of exclusion, where technological or infrastructural challenges, such as unreliable Internet connectivity and faulty biometric authentication, prevent individuals from accessing essential services. This underscores the need for redundancy mechanisms and inclusive design in AI systems to account for diverse populations. Another challenge is the potential for data breaches and surveillance, as Aadhaar stores vast amounts of sensitive data, raising questions about privacy and security. The absence of comprehensive data protection legislation, such as a finalised personal data protection law in India, leaves the system vulnerable to misuse and undermines public trust.

The system's limitations highlight the importance of embedding robust legislative and regulatory safeguards into AI-powered systems, ensuring they serve as tools for empowerment rather than exclusion or exploitation. These weaknesses notwithstanding, this AI-driven biometric identification system presents many critical lessons and emerging experiences on governance, legislation, and regulation of AI as countries strive to deploy AI on the SDG journey. This is particularly the case for emerging and least industrialised economies, which share many developmental characteristics with India [95, 122].

10.8 Further Regulatory and Legislative Matters

There is a need to clarify intellectual property rights related to AI inventions, creations, and innovations. This should include issues of ownership and patentability in the context of AI-generated outputs. Another area is competition and antitrust regulations. There is a need to monitor and regulate AI applications to prevent anti-competitive practices. This can be achieved through the establishment of guidelines to ensure fair competition, prevent monopolies, and address concerns related to market concentration.

It is essential to enact laws that protect consumers from misleading or deceptive AI-driven products and services. Ensure that users should be informed about the capabilities and limitations of AI applications and have recourse in case of harm. Furthermore, issues related to cross-border data flow and data localisation must be addressed. Countries must establish regulations that balance the need for data protection with the benefits of international collaboration and data sharing.

They should also develop regulations that promote the accessibility and inclusivity of AI technologies. Ensure that AI applications are designed to accommodate individuals with diverse abilities and mitigate potential biases. Foundational to all this is digital literacy and awareness. States must implement regulations that promote digital literacy and awareness about AI among the general public. They must ensure that individuals understand the implications of AI technologies and are empowered to make informed decisions.

There is a need to integrate human rights considerations into AI regulations, ensuring that AI technologies do not violate fundamental rights. Issues related to discrimination, privacy, and freedom of expression must be addressed. However, there is a two-way street between AI and human rights—

potential violations and possible AI-driven remedies. The technology is increasingly being applied in human rights advocacy to enhance the efficiency and effectiveness of monitoring, reporting, and campaigning efforts. AI-powered tools enable the processing of large volumes of data, such as images and texts from social media and other sources, to identify potential human rights abuses in real time.

For instance, AI algorithms can detect patterns of violence in conflict zones by analysing satellite images and social media posts, providing human rights organisations with timely information to advocate for intervention and support. Additionally, AI is used to analyse legal documents and case law to assist lawyers and activists in building stronger cases for human rights violations. This technology also helps in raising public awareness and mobilising support by personalising outreach efforts and optimising advocacy campaigns. As such, while there is the potential for violation of human rights by AI systems, the technology can also act as a force multiplier in human rights advocacy, expanding the capabilities of organisations to uphold and promote human rights on a global scale.

No country, region, or continent is an island unto itself. International collaboration and partnerships on regulations and governance must be embraced. These must occur a cross both the Global South and Global North. There must be alignment of national regulations with regional, continental, and global standards and while working collaboratively to address shared challenges. However, the regional, continental, and global standards must be a product of inclusive and comprehensive consultations.

It is also instructive to emphasise that developing and implementing AI regulations should involve collaboration between government agencies, industry stakeholders, academia, and civil society. The regulatory framework should be flexible enough to accommodate technological advancements while ensuring that ethical and societal considerations are prioritised. Additionally, ongoing monitoring and updates to regulations may be necessary as the AI landscape evolves. However, this is easier said than done as explained in the next section.

10.9 Limitations of Regulations and Legislation

The law and legal systems are, by definition, conservative. Furthermore, traditionally, regulators and legislators are not technically trained or inclined. Technology is always changing, so lawmakers and regulators are always regulating and legislating the last wave of technology. It is difficult to institute anticipatory legislation when future technologies' nature and implications are unknown or knowable. For example, ChatGPT was launched in 2023; how could any jurisdiction have effectively provided its rules and regulations in 2022? No one could understand or anticipate the technology's capabilities in 2022. There is a need for a dynamic regulatory and legislative body responsible for reviewing and anticipating technological advances and innovations in general, specifically AI technologies and systems. Agile, tech-savvy (in particular with respect to AI) legislators and regulators are essential. Then, there is the challenge of bad actors. By definition, these do not obey rules and regulations. How do you legislate for bad actors?

Indeed, AI regulations and legislation aim to ensure the ethical, fair, and responsible development and deployment of AI systems. While these frameworks provide essential safeguards, they also face significant limitations. As already indicated, one major challenge is the pace of technological advancement. AI evolves far more rapidly than legislative processes, meaning that by the time laws are passed, the technology may have already outgrown the scope of the regulation. For example, while rules may address existing AI models like chatbots or recommendation systems, they might fail to anticipate future developments, such as advanced generative AI and autonomous agents with decision-making capabilities. This regulatory lag creates loopholes that bad actors can exploit, making it difficult to ensure comprehensive oversight.

Another limitation is the difficulty of enforcement. Unlike traditional technologies, AI systems often operate as "black boxes," especially in the case of Deep Learning models. It is hard to interpret how an AI system makes its decisions, which complicates accountability and the ability to determine regulatory violations. For instance, if an AI algorithm is found to be discriminatory, it can be challenging to pinpoint the specific cause or identify which party (developer, deployer, or user) is responsible. Global inconsistency in AI regulation is another major challenge. Different countries have different regulatory approaches, with the EU AI Act being stricter than frameworks in the United States, China, or India, creating a fragmented regulatory environment. Companies operating across borders must navigate multiple, and sometimes conflicting, requirements. Additionally, AI regulations often face ethical ambiguity—questions about how to define "fairness" or "bias" remain unresolved, and these definitions can differ significantly by culture, region, or sector. Finally, balancing innovation and regulation is a key limitation. Over-regulation could stifle innovation, discourage investment, and push companies to operate in less-regulated jurisdictions, while under-regulation could allow for unethical use of AI. These limitations underscore the need for flexible, adaptive, and harmonised regulatory frameworks that can evolve alongside AI technologies.

10.10 Sociology of AI Policy and Adoption

The sociology of AI policy and adoption explores the interplay between social structures, cultural values, and institutional dynamics in shaping the development and implementation of AI technologies. AI policies are not created in a vacuum; they reflect the priorities, power structures, and ideologies of the societies that create them. For instance, in highly industrialised nations, AI policies often emphasise innovation, economic competitiveness, and national security. At the same time, in developing countries, they may prioritise inclusion, equitable access, and the mitigation of societal inequalities. These differing priorities demonstrate how the societal context influences the trajectory of AI development and adoption.

At the heart of AI policy and adoption is the concept of trust. Societies vary in their levels of trust in government institutions, technology developers, and AI systems themselves. Countries with robust regulatory frameworks and transparent governance mechanisms, such as those in the European Union, have emphasised ethical AI through initiatives like the EU AI Act. These policies aim to build public trust and encourage responsible AI adoption by addressing concerns around privacy, algorithmic bias, and accountability. In contrast, in regions where trust in institutions is weaker, AI policies may struggle to gain public support, leading to slower adoption or resistance.

The sociology of AI adoption also highlights the influence of cultural norms and values. For example, collectivist cultures may prioritise AI applications that benefit communities, such as healthcare and education, over individualistic pursuits like personal convenience. On the other hand, societies with high levels of technological optimism, such as the United States and China, may embrace AI adoption more rapidly, often seeing it as a driver of progress and opportunity. These cultural attitudes shape how governments, businesses, and individuals approach AI, from policymaking to everyday use.

Economic and social inequalities significantly impact AI adoption. In many developing nations, the digital divide—disparities in access to technology and the Internet—poses a barrier to the equitable adoption of AI. Without addressing foundational issues like infrastructure, digital literacy, and affordability, these countries risk being left behind in the AI revolution. Moreover, even within wealthy nations, marginalised groups often face unequal access to AI-powered services, exacerbating existing inequalities. Therefore, policies aimed at inclusive AI adoption must address these structural disparities to ensure that AI benefits are widely shared.

Another critical sociological aspect is the role of power and influence in shaping AI policy. Large technology companies often exert significant influence over AI development, sometimes steering it in ways that align with their commercial interests rather than a public good. This raises questions about accountability and fairness, particularly in areas where governments may lack the expertise or resources to regulate these entities effectively. Public advocacy, civil society, and international cooperation play crucial roles in counterbalancing corporate power and ensuring that AI development aligns with broader societal values.

Finally, the sociology of AI adoption emphasises the need for a participatory approach to policymaking. Inclusive policy development, which incorporates diverse perspectives from stakeholders such as industry experts, academics, civil society, and marginalised groups, is critical to addressing the complex social implications of AI. By fostering open dialogue and collaboration, societies can create AI policies that advance technological innovation and promote social justice, ethical standards, and sustainability. This participatory model recognises that AI is not merely a technological tool but a deeply social phenomenon that reflects and shapes the values and priorities of the societies that use it.

10.11 Ethics, Privacy, and Moral Dilemmas

The deployment of AI to achieve SDGs presents a range of ethical considerations, privacy challenges, and moral dilemmas. On one hand, AI has the potential to revolutionise development by enabling precision interventions in healthcare, education, and environmental conservation. Conversely, its application raises concerns about equity, accountability, and unintended consequences. For instance, while AI can improve agricultural yields in developing countries, the lack of equitable access to such technologies may exacerbate existing disparities, leading to ethical questions about fairness and inclusion. Addressing these issues requires careful balancing of innovation with social responsibility.

Privacy concerns are paramount when deploying AI systems for SDG-related purposes. Many AI-driven solutions rely on vast amounts of personal data, such as health records, financial transactions, and geolocation information. Without robust data protection frameworks, individuals in vulnerable communities risk compromising their privacy. For example, while predictive analytics can help identify at-risk populations for targeted health interventions, the collection and misuse of sensitive health data could lead to stigmatisation or discrimination. Ensuring privacy in AI deployment requires strong data governance practices, including informed consent, anonymisation, and transparency in data usage.

The moral dilemmas of AI deployment are particularly pronounced in conflict-prone or authoritarian contexts. AI systems designed for development purposes, such as surveillance tools to monitor illegal deforestation or urban planning, can be repurposed for repressive practices. In these cases, the same technology that could advance SDG 13 on climate action might also be used to suppress dissent or violate human rights. This dual-use nature of AI raises significant moral questions about the responsibilities of developers, policymakers, and international organisations in safeguarding against misuse.

Bias and discrimination in AI algorithms pose another ethical challenge to achieving SDGs. Many AI systems are trained on datasets that reflect existing societal biases, which can lead to discriminatory outcomes in critical areas such as hiring, credit allocation, and access to social services. For example, an AI system deployed to optimise job matching under SDG 8 (Decent Work and Economic Growth) could inadvertently disadvantage women or minorities if the training data reflects historical patterns of exclusion. Ethical AI deployment thus requires rigorous testing, auditing, and diversification of training datasets to mitigate bias and promote fairness.

Furthermore, the increasing reliance on AI raises ethical questions about accountability and agency. When an AI system makes a critical decision, such as allocating healthcare resources or determining eligibility for welfare programmes, it can be difficult to pinpoint responsibility if something goes wrong. This "black-box" problem creates a moral dilemma around delegating decision-making to machines, especially when human lives are at stake. Policymakers must establish clear frameworks for accountability, ensuring that AI systems remain transparent and that human oversight is maintained in high-stakes applications.

Finally, there is a broader philosophical question about the role of AI in human development. While AI can accelerate progress towards the SDGs, it should not undermine the agency and dignity of the people it seeks to serve. Overreliance on AI risks disempowering communities, reducing them to passive recipients of technological solutions rather than active participants in their development. To navigate these moral dilemmas, AI deployment must be guided by human-centred design principles, emphasising empowerment, equity, and ethical governance. This approach ensures that AI serves as a tool for achieving the SDGs without compromising the core values of human rights and social justice [54].

10.12 Ethical Issues and Dilemmas Beyond SDGs

Ethical issues and dilemmas afflict all AI research, development, and applications. They are not restricted to AI deployment to achieve SDGs. AI systems can inherit biases from the data they are trained on, leading to discriminatory outcomes, particularly against marginalised groups. Ensuring fairness in AI algorithms and mitigating bias is a significant challenge. They often require large amounts of data to function effectively, raising concerns about the privacy and security of personal information. There's a risk of surveillance, data breaches, and unauthorised use of personal data.

Transparency and accountability are significant concerns. Many AI algorithms operate as "black boxes", making it difficult to understand how they arrive at their decisions. Lack of transparency can hinder accountability and make it challenging to address errors or biases in the system. As AI systems become more autonomous, there are concerns about who holds responsibility for their actions and decisions. Questions arise about the appropriate level of human oversight and control over AI technologies.

AI-driven automation has the potential to disrupt labour markets, leading to job displacement and economic inequality. Managing the societal impact of AI-driven automation is a significant ethical challenge. Then, there is the danger of manipulation and misuse. AI technologies can be used to manipulate information, spread misinformation, and influence public opinion. Ensuring the responsible use of AI to avoid harm and exploitation is crucial.

Security and safety are central. If deployed in autonomous vehicles or other high-stakes contexts, AI systems can be vulnerable to attacks and manipulation, posing risks to critical infrastructure, cybersecurity, and even physical safety. There are concerns that AI technologies could widen existing inequalities by favouring those with access to resources and expertise. Ensuring equitable access to AI and addressing digital divides are critical ethical considerations.

AI systems may face ethical dilemmas, such as autonomous vehicles deciding between different courses of action in a potential accident. Developing frameworks for ethical decision-making in AI is a pressing concern.

There are also long-term impacts and unknown risks. The full societal implications of widespread AI adoption are still uncertain. Ethical considerations must extend to the long-term effects of AI on society, the environment, and human well-being. Addressing these ethical issues requires collaboration among policymakers, technologists, ethicists, and other stakeholders to develop regulations, standards, and guidelines that promote the responsible development and use of AI technologies.

Indeed, ethical issues in AI are numerous, intractable, interrelated, and multifaceted. This reflects the challenges and complexities of developing and deploying AI systems. The ethical concerns include bias, fairness, privacy, transparency, accountability, machine autonomy vs. human control, job displacement, manipulation, misuse, inequality, accessibility, security, safety, unknown risks, machine intentionality, and ethical dilemmas.

These intractable challenges afflict all countries—with varying degrees of impact—and must be efficaciously addressed. Resolving and mitigating these ethical matters require collaboration in an AI ecosystem that includes policymakers, technologists, ethicists, theologians, pastors, seminarians, lawyers, and economists. The collective mandate of these stakeholders should be the development of regulations, standards, and guidelines that promote the ethical and responsible development and use of AI technologies.

10.13 Conclusion

AI legislative lessons can be drawn from many jurisdictions, including the efforts of the EU and India. The utility of these emerging experiences can be extended to AI deployment for SDG attainment worldwide. Effective legislation and regulation are indispensable for deploying AI to achieve the SDGs. They provide a framework for ensuring ethical use, protecting privacy, promoting sustainability, and addressing inequality. By fostering accountability and international cooperation, these frameworks align AI's transformative power with the global vision of a more equitable and sustainable future. As the world advances towards 2030, crafting and implementing these regulations will be crucial to maximising AI's contribution to the SDGs while safeguarding against potential risks.

Indeed, good AI governance, effective legislation, and comprehensive regulations are essential for deploying AI to attain the SDGs. They ensure that AI technologies are used ethically, equitably, and sustainably. Without clear frameworks, AI applications could exacerbate inequalities, infringe on human rights, and create unintended harm, undermining the very goals they are meant to support. Governance structures provide accountability, transparency, and oversight, ensuring that AI systems align with SDG priorities, such as reducing poverty, improving healthcare, and combatting climate change. Legislation safeguards data privacy, addresses algorithmic biases, and promotes fairness in AI decision-making, especially in critical areas like education and employment.

Regulations tailored to specific sectors encourage responsible innovation, minimising risks while maximising AI's transformative potential for inclusive development. Beyond national efforts, there is a need for harmonised and standardised regional, continental, and global legislation, regulations, guardrails, and governance. All these AI frameworks must be robust, balanced, dynamic, and adaptable to accommodate the rapid and relentless changes and complex dilemmas that characterise the technology. Providing anchor and context to these interventions must be a robust understanding and adept management of the sociology, ethics, privacy, and moral dilemmas of AI policy and adoption. Designing and implementing transparent, context-aware, and socially grounded governance, legislation, and regulations create a foundation for trust, collaboration, and impactful AI deployment to pursue the SDGs worldwide.

Strategic Framework for AI Deployment

11

11.1 Introduction

As discussed throughout the book, integrating AI into efforts to achieve the SDGs presents transformative opportunities to address global challenges. AI has shown immense potential in areas such as precision agriculture to address food security, predictive analytics in healthcare to improve disease prevention and treatment, and environmental monitoring to combat climate change. These innovations contribute directly to SDG goals such as eradicating hunger, improving health and well-being, and enhancing sustainability. However, a structured and comprehensive adoption framework is essential to fully harness AI's power while mitigating risks such as bias, inequality, and data misuse. Such a framework must balance ethical principles, inclusivity, sustainability, and accountability to ensure AI deployment aligns with the SDGs.

This chapter presents the principles, objectives, and activities of the adoption framework for deploying AI to achieve SDGs. It then organises them into a detailed AI Strategic Framework with six key components: Vision, Strategy, Policy, Governance, Legislation/Regulation, and Implementation Matrix. Other aspects of the adoption framework are discussed, including infrastructure, investments, global collaboration, partnerships, and capacity building. Scale and harmonisation are essential and advocated for through regional and continental economic integration and political unity. The African Union's Continental AI Strategy, launched in August 2024, was discussed and reviewed as a starting point for a conversation on establishing a more grounded, dynamic and efficacious AI Strategic Framework for the continent [3, 4, 37, 145].

11.2 AI Adoption Principles, Objectives, and Activities

An effective AI adoption framework should rest on several guiding principles. Inclusivity ensures equitable access to AI technologies, allowing all nations and communities, particularly those in developing regions, to benefit from AI-driven solutions. Sustainability minimises the environmental impact of AI systems, particularly their energy consumption and resource use. Ethics are critical in addressing biases, safeguarding human rights, and ensuring AI systems promote fairness and equality. Accountability fosters transparency and trust, requiring stakeholders to clearly explain AI processes,

A. G. O. Mutambara, *Deploying Artificial Intelligence to Achieve the UN Sustainable Development Goals*, Sustainable Development Goals Series,
https://doi.org/10.1007/978-3-031-88423-8_11

ensure proper oversight, and take responsibility for outcomes. These principles collectively ensure that AI deployment does not undermine the very goals it aims to achieve.

There is a need for contextual understanding. It is essential to begin by comprehensively understanding the local context, including socio-economic, cultural, and environmental factors, as well as specific challenges related to the SDGs in the target region or community. Diverse stakeholders, including governments, local communities, academia, civil society organisations, and the private sector, should be involved throughout the AI deployment process to ensure inclusivity, transparency, and accountability. There is a need to prioritise specific SDGs and targets based on their relevance, urgency, and potential impact. Identify clear goals and outcomes that AI interventions aim to achieve within the context of the selected SDGs.

Data governance and accessibility must be addressed by establishing robust data governance mechanisms to ensure responsible collection, storage, processing, and data sharing. Promote data accessibility and availability while addressing privacy, security, and ethical considerations. Appropriate AI technologies and methodologies must be selected based on the identified priorities, goals, and data availability. There is a need to develop or customise AI solutions tailored to local contexts and needs, considering factors such as scalability, interoperability, and resource constraints.

Adoption frameworks must integrate ethical principles, human rights standards, and AI ethics guidelines into the design, development, and deployment of AI solutions. As indicated earlier, it is essential to address issues such as fairness, transparency, accountability, bias mitigation, and privacy protection throughout the AI lifecycle.

Nevertheless, all these articulated AI adoption principles, objectives, and activities must be anchored by AI readiness. There is a need for preparation. Countries and organisations must not put the cart before the horse when developing and adopting AI. They must address critical pillars such as basic infrastructure, digital infrastructure, governance and regulatory framework, financing and investment, basic education and literacy, talent and expertise, energy/power, the semiconductor industry, compute resources infrastructure, organisational preparedness, the AI ecosystem, and the right mindset. These related, interdependent, and mutually reinforcing foundational matters must be attended to holistically and systematically.

11.3 AI Strategic Framework

Beyond addressing these critical matters, each country, organisation, or community must develop and adopt an AI Strategic Framework comprising six distinct but related components: Vision, Strategy, Policy, Governance, Legislation/Regulations, and Implementation Matrix. Figure 11.1 depicts the interplay and hierarchy among these six strategic layers. However, these frameworks must be premised on the successful resolution of the challenges outlined in the preceding section. More significantly, the six-component AI Strategic Framework must be replicated at regional and continental levels where the three levels (national, regional, and continental) continuously influence each other in refining the document. Furthermore, AI regional, continental, and global partnerships must be developed and leveraged.

Of the six components of the AI Strategic Framework, governance and legislation/regulation have been extensively defined and discussed in Chaps. 7 and 9. In socio-economic, political, and business literature, the terms vision, strategy, policy, and implementation are widely but loosely used. They are never carefully and comprehensively defined before being applied. The utility of these terms is essential in designing the adoption framework for AI deployment to achieve the SDGs. Hence, they will be carefully explained and then applied in this section.

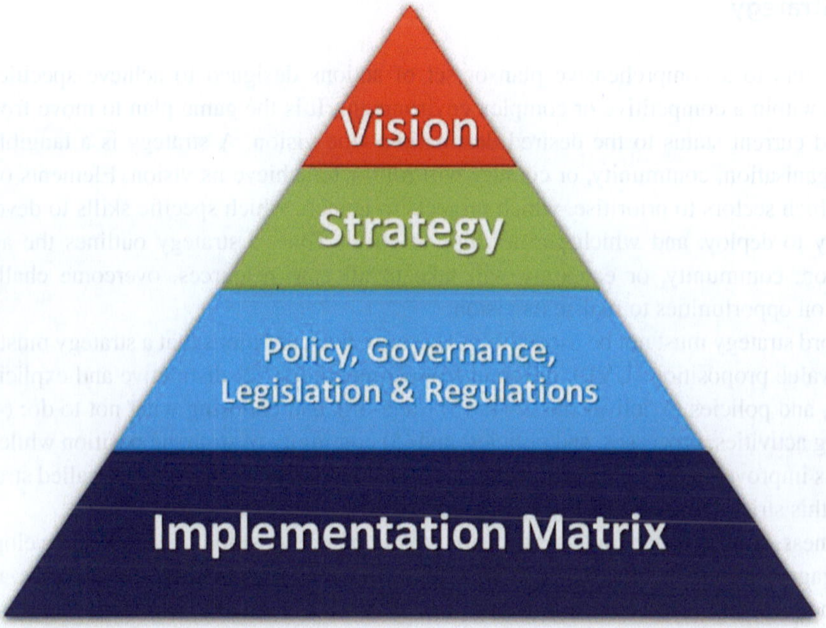

Fig. 11.1 AI adoption strategic framework

11.3.1 Vision

Vision refers to a clear and aspirational time-bound statement that defines the long-term goals and desired future state of an organisation, community, or economy. It is the overarching ambition aspired towards over a given timeframe (e.g. 5, 10, 15 years). Vision is the desired destination or the house on top of the hill. It serves as a guiding principle that shapes strategic decisions, resource allocation, and overall direction. A vision articulates what the entity aims to achieve, providing a sense of purpose and inspiration for stakeholders, including employees, investors, policymakers, and the public.

For example, in economic planning, a vision may outline goals like sustainable growth, poverty reduction, and technological innovation, helping to align policies and initiatives towards a common objective. A well-crafted vision motivates action, enhances unity, and establishes a framework for measuring progress, ensuring that short-term efforts contribute to long-term aspirations.

Having a vision for the deployment of AI to achieve the SDGs is crucial because it provides a clear, strategic direction for aligning AI technologies with global development priorities. A well-defined vision articulates the purpose, scope, and desired outcomes of AI deployment, ensuring that technological innovation is used as a force for good rather than a source of harm. Without a vision, AI initiatives risk being disjointed, misaligned, or even counterproductive, potentially exacerbating inequalities or environmental degradation. A shared vision fosters collaboration among governments, private sector actors, and civil society, enabling coordinated efforts to develop ethical, inclusive, and people-centred AI solutions. It also establishes a framework for accountability, guiding the creation of regulatory policies, ethical guidelines, and investment strategies that prioritise equity, transparency, and sustainability. Furthermore, a vision for AI in the SDG context ensures that technological advancements are directed towards critical areas such as poverty reduction, education, healthcare, and climate action, maximising their positive impact on humanity and the planet.

11.3.2 Strategy

Strategy refers to a comprehensive plan or set of actions designed to achieve specific goals and objectives within a competitive or complex environment. It is the game plan to move from a clearly understood current status to the desired destination—the vision. A strategy is a tangible approach that an organisation, community, or country will follow to achieve its vision. Elements of a strategy include which sectors to prioritise, which projects to launch, which specific skills to develop, which technology to deploy, and which partnerships to build. Thus, a strategy outlines the approach an organisation, community, or economy will take to allocate resources, overcome challenges, and capitalise on opportunities to realise its vision.

The word strategy must not be used glibly. There are five conditions that a strategy must satisfy: (1) a unique value proposition (UVP), different from competitors; (2) distinctive and explicit activities, processes, and policies to deliver the UVP; (3) trade-offs and choosing what not to do; (4) mutually reinforcing activities, processes, and policies; and (5) continuity of strategic position while embracing continuous improvement in particular technology-enabled innovations. Most so-called strategic plans do not fit this strict definition [99].

In business, strategy often includes decisions about market positioning, product development, customer engagement, and competitive differentiation. In economic planning, for instance, the strategy involves policies and programmes to foster growth, reduce inequality, and achieve sustainability. A good strategy is both adaptive and forward-looking, addressing current realities while anticipating future trends. It is a roadmap that connects long-term aspirations to actionable steps, ensuring coherence, focus, and resilience in achieving desired outcomes.

Having a strategy for the deployment of AI to achieve the SDGs is essential for ensuring that AI technologies are used effectively, ethically, and efficiently to drive sustainable development. A strategy provides a structured approach for identifying priority areas, allocating resources, managing risks, and measuring progress towards achieving the SDGs. It outlines specific actions, timelines, and roles for stakeholders, enabling the alignment of AI innovations with SDG objectives such as poverty reduction, quality education, and climate action. In designing and developing the AI deployment strategy, it is vital to ensure that the five conditions are satisfied—a unique value proposition; distinctive and explicit activities, processes, and policies to deliver; trade-offs and choosing what not to do; mutually reinforcing activities, processes, and policies; and dynamic continuity of strategic position while embracing continuous improvement.

A well-defined and anchored strategy helps to mitigate potential risks like bias, inequality, and data privacy violations, ensuring that AI deployment is ethical, inclusive, and socially responsible. By establishing clear pathways for investment, research, and capacity building, the strategy ensures that AI initiatives are scalable, adaptable, and sustainable. Moreover, a strategy facilitates partnerships among governments, private sector actors, and development organisations, promoting collaboration and knowledge sharing. Ultimately, a comprehensive AI strategy ensures that technological progress is purpose-driven and contributes to the broader goals of human well-being, environmental protection, and equitable development.

11.3.3 Policy Framework

The key difference between a policy and a policy framework lies in their scope and purpose. A policy is a specific set of rules, guidelines, or principles established by governments, organisations, or institutions to address a particular issue, and guide decision-making. It provides a clear directive on what actions should be taken and how to achieve a specific goal, such as a company's policy on data privacy or a government's policy on renewable energy adoption. Policy provides a structured approach to addressing issues, ensuring consistency, accountability, and alignment with broader objectives. Policies serve as a foundation for laws, regulations, and operational procedures, influencing behaviour, resource allocation, and strategic priorities.

On the other hand, a policy framework is a broader, overarching structure that outlines the principles, objectives, and guidelines for developing and implementing multiple related policies. It serves as a reference point for creating individual policies, ensuring that they are consistent, coherent, and aligned with the organisation's or government's overall strategic goals. While a policy focuses on specific actions, a policy framework provides the conceptual and operational foundation within which multiple policies are designed, coordinated, and implemented.

Indeed, a policy framework establishes the overarching goals, priorities, and approaches that guide the development, implementation, and evaluation of policies to address particular challenges or opportunities. In economic planning, for example, a policy framework might outline strategies for driving sustainable growth, reducing poverty, or promoting innovation. It provides coherence and consistency across individual policies, ensuring they align with broader goals and values. A well-designed policy framework also facilitates accountability, collaboration, and adaptability, serving as a reference point for policymakers, stakeholders, and implementers in achieving desired outcomes.

A policy framework for the deployment of AI to achieve the SDGs is essential for ensuring that AI technologies are developed and utilised in a way that is ethical, equitable, and aligned with sustainable development objectives. Such a framework provides clear guidelines, principles, and standards that govern the responsible use of AI, addressing critical issues like data privacy, algorithmic fairness, accountability, and human rights. Without a policy framework, AI deployment risks being inconsistent, uncoordinated, or harmful, potentially exacerbating social inequalities and environmental challenges. A well-structured framework establishes rules for oversight, defines roles and responsibilities for stakeholders, and creates mechanisms for compliance and redress. It also promotes transparency and trust, encouraging public support for AI-driven initiatives. Additionally, a policy framework enables governments, private sector actors, and development organisations to coordinate their efforts, ensuring that AI investments and innovations contribute meaningfully to SDG targets such as poverty reduction, climate action, and equitable access to healthcare and education. By fostering coherence, accountability, and shared responsibility, a policy framework ensures that AI serves as a force for positive, sustainable change.

However, in this discourse, the established distinction between a policy and a policy framework must be applied. For example, a national AI policy would provide specific rules, guidelines, or principles for AI deployment. In contrast, an AI policy framework would outline the broader goals, principles, and governance structures to guide the development of multiple AI-related policies across sectors like healthcare, education, and industry. Hence, the starting point is developing an AI policy framework, followed by a detailed and comprehensive national AI policy document.

11.3.4 Implementation Matrix

Implementation and execution refer to the processes of translating strategic plans and goals into actionable steps and ensuring their successful completion. While implementation involves setting up the necessary infrastructure, allocating resources, and defining roles and responsibilities to begin the work, execution focuses on carrying out the planned actions effectively and efficiently to achieve desired outcomes. In business, this might include launching products, marketing campaigns, or operational initiatives. Strategy in economic planning could involve rolling out policies, building infrastructure, or delivering public services. Successful implementation and execution require strong leadership, clear communication, monitoring mechanisms, and adapting to unforeseen challenges. Together, these phases are critical for bridging the gap between planning and tangible results, turning visions and strategies into measurable impact.

Implementation and execution activities must be organised into an implementation matrix—a structured tool used in planning and project management to outline the specific actions, roles and responsibilities, timelines, milestones, and resources (and their sources) needed to execute a strategy effectively. Hence, this matrix is an array of concrete plans to execute and deliver on the strategy, including prioritisation and sequencing of the activities. It serves as a detailed roadmap that breaks down goals into manageable tasks, assigning roles to stakeholders or teams and specifying deadlines for each activity. Typically presented in a tabular format, an implementation matrix includes columns for objectives, key activities, responsible parties, timelines, required resources, and performance indicators. This clarity helps track progress, identify bottlenecks, and ensure accountability at every implementation stage. By organising complex plans into actionable steps, the implementation matrix provides a practical framework for aligning efforts, optimising resource use, and achieving desired outcomes efficiently.

An implementation matrix for the deployment of AI to achieve the SDGs is essential as it provides a clear, structured, and actionable roadmap for turning strategic plans into tangible SDG outcomes. An implementation matrix enhances accountability, transparency, and efficiency by breaking down complex SDG goals into specific tasks, timelines, responsible parties, required resources, and key performance indicators. It ensures that all stakeholders, including governments, private sector actors, and civil society, have a shared understanding of their roles and responsibilities in AI deployment to achieve SDGs. This clarity minimises confusion, prevents duplication of efforts, and promotes effective resource allocation. The matrix also facilitates regular monitoring and evaluation, allowing decision-makers to track progress, identify bottlenecks, and make timely adjustments. By providing a visual, task-oriented approach, an implementation matrix ensures that AI initiatives remain on course, stay aligned with SDG targets, and deliver measurable impact in areas like poverty reduction, quality education, healthcare access, and climate action.

11.3.5 Monitoring, Measurement, Evaluation, and Feedback

A key part of the implementation matrix comprises monitoring, measurement, evaluation, and feedback. These four interventions are crucial in ensuring the successful implementation and continuous improvement of strategies, policies, and projects. They refer to the processes of systematically tracking, assessing, and analysing the progress and outcomes of a plan, project, or programme to ensure its effectiveness and alignment with objectives. Monitoring provides real-time tracking of activities and outputs, ensuring that tasks are being executed as planned and resources are used effectively. Thus, it involves the ongoing collection of data on activities and outputs to ensure tasks are being carried out as desired.

Measurement focuses on assessing progress using specific indicators, providing quantitative data to evaluate the achievement of objectives. It concentrates on quantifying results using specific metrics, such as performance or key benchmarks, to assess whether goals are being met. Evaluation involves a deeper analysis of outcomes, assessing initiatives' relevance, efficiency, impact, and sustainability to determine whether they align with intended goals. Hence, it goes beyond measurement by analysing the quality of the outcomes, often comparing them against the initial objectives and expectations. Finally, feedback integrates insights from monitoring, measurement, and evaluation into decision-making, enabling adjustments and refinements to improve performance and address challenges.

Together, these four components ensure accountability, optimise resource use, and incubate a culture of learning and adaptability, enhancing the effectiveness of initiatives across sectors. They provide critical feedback for decision-makers, enabling course corrections, optimising efficacy, and improving accountability. By integrating monitoring, measurement, evaluation, and feedback into business or economic planning, organisations, communities, and countries can improve performance and ensure meaningful, long-term results.

As already discussed, the implementation matrix enables regular monitoring and evaluation, allowing decision-makers to track progress, identify bottlenecks, and make timely adjustments. This is achieved through the matrix's monitoring, measurement, evaluation, and feedback capabilities. Having these four functions is critical for ensuring accountability, transparency, and continuous improvement. These capabilities allow stakeholders to track progress, assess the effectiveness of AI initiatives, and identify areas for improvement. Monitoring ensures that activities are being implemented as planned, while measurement quantifies outcomes using performance indicators, such as reductions in poverty, increased access to education, and enhanced healthcare delivery. Evaluation provides a deeper analysis of AI interventions' relevance, efficiency, and impact, helping stakeholders determine whether initiatives align with SDG targets. Feedback integrates lessons learned back into the decision-making process, enabling timely adjustments and adaptive management. Without these capabilities, AI projects risk inefficiencies, misalignment with development goals, and unintended negative consequences like bias and inequality. By fostering a cycle of learning and improvement, these mechanisms enhance the effectiveness, scalability, and sustainability of AI-driven initiatives aimed at accelerating SDG attainment.

11.3.6 Policy and Governance

As discussed, a key component of the AI adoption framework is strong policy anchored by effective and balanced governance. Policymakers must establish clear guidelines for the ethical use of AI, addressing issues such as data privacy, algorithmic bias, and security. Governance structures should include independent oversight bodies to enforce compliance and provide redress mechanisms. A multi-stakeholder approach that involves governments, the private sector, academia, and civil society is critical to creating balanced and inclusive policies. Complementing governance is capacity building and education, essential to developing the skills needed for AI deployment. Investments in training programmes for policymakers, technical experts, and the public are necessary to build local capacity, particularly in underrepresented and marginalised communities. These programmes should focus on building diversity in AI development, ensuring that technologies reflect varied perspectives and serve all populations equitably.

Ethical considerations form another critical pillar of the framework. AI systems must be designed with fairness in mind, ensuring they are free from biases that could exacerbate existing inequalities. Transparent and explainable AI models are crucial for fostering accountability and trust. This is especially important in sensitive areas such as healthcare and law enforcement, where AI decisions can have significant social and ethical implications. Robust ethical guidelines ensure that AI deployment respects human rights and promotes equitable outcomes across all SDGs.

Furthermore, advocacy for supportive policy environments, regulatory frameworks, and funding mechanisms that incentivise the responsible deployment of AI for SDGs must be undertaken. This can be achieved by engaging with policymakers, civil society groups, and international organisations to promote evidence-based policies and investment priorities. It is also essential to empower local communities to actively participate in decision-making processes, contribute local knowledge and insights, and take ownership of AI initiatives that address their developmental needs and priorities.

11.3.7 Infrastructure and Investment

Infrastructure and investment are equally vital. Deploying AI at scale requires significant investments in digital infrastructure, including high-speed Internet, data centres, and computational resources. These are often lacking in developing countries, where the benefits of AI are most needed. Public-private partnerships (PPPs) can help mobilise resources and foster innovation, ensuring that AI technologies are accessible and affordable. Such collaborations can bridge infrastructure gaps, enabling the deployment of AI in areas like agriculture, education, and healthcare to drive progress towards the SDGs.

Indeed, infrastructure and investment are essential for the successful deployment of AI to achieve SDGs because they provide the foundational support and resources required to develop, deploy, and scale AI technologies. Infrastructure, such as high-speed Internet, cloud computing facilities, data centres, and digital devices, is critical for enabling AI systems to function efficiently. Without adequate infrastructure, low-income countries and marginalised communities risk being excluded from the benefits of AI, thereby widening the digital divide. Conversely, investment provides the financial resources needed to build this infrastructure, support research and development (R&D), and train a skilled workforce.

Public and private sector investments are necessary to bridge funding gaps, especially in developing countries with limited access to technology. By investing in AI infrastructure, governments and development partners can create innovation hubs, foster local AI ecosystems, and promote equitable access to AI-driven solutions in healthcare, education, agriculture, and climate action. Moreover, investments in sustainable AI infrastructure, such as energy-efficient data centres and renewable energy-powered systems, ensure that AI deployment supports environmental sustainability. Infrastructure and investment accelerate progress towards the SDGs and create an enabling environment for inclusive economic growth, innovation, and long-term development.

11.3.8 Primacy of AI Infrastructure: Insights from the United States

As explained in Chap. 7, compute resources (AI infrastructure)—the hardware, software, and systems required to develop, train, and deploy AI models efficiently—will drive the AI revolution. Examples of AI infrastructure include data centres, GPUs for processing, cloud computing platforms like Amazon Web Services and Google Cloud, and specialised software frameworks like TensorFlow and PyTorch. On 14 January 2025, then US President Joe Biden signed an Executive Order on Advancing US Leadership in AI Infrastructure [137]. The order "instructs the Departments of Defence (DOD) and Energy (DOE), in coordination with other federal agencies, to lease federal sites to private-sector entities for constructing large-scale AI data centres and associated clean energy facilities". Thus, the executive order seeks to expedite the development of large-scale AI infrastructure within the United States, putting a premium on sustainable energy ambitions and national security concerns.

The United States wants to lead the AI revolution ahead of China, Russia, the European Union, and the United Kingdom. However, this supremacy is not granted. The quality of AI infrastructure will be the differentiator. Clearly, the United States does not want to be outbuilt and outperformed by competitors in this transformative AI technology, which will define and drive the economies and societies of the future. The country is also concerned about the AI carbon footprint. It does not want to undermine critical environmental standards or sacrifice efforts to protect clean water and air. Indeed, the 14 January 2025 executive order's objective is to secure US leadership of the AI revolution by ensuring that the infrastructure needed for advanced AI operations—including large-scale data centres and new clean power infrastructure—can be built quickly and at scale domestically.

The United States views building AI infrastructure as a national security imperative. The view is that as AI's capabilities grow, so do its implications for the country's safety and security. According to the presidential decree of 14 January 2025: "Domestic data centres for training and operating powerful AI models will help the United States facilitate AI's safe and secure development, harness AI in service of national security, and prevent adversaries from accessing powerful systems to the detriment of our military and national security. It will also help prevent the United States from growing dependent on other countries to access powerful AI tools". Indeed, AI has opened a new arena of superpower rivalry. The US position is that the buildout of AI infrastructure must advance US national security and competitiveness. It must support US clean energy leadership, a competitive technology ecosystem, low consumer prices, workers building AI infrastructure, and communities near it.

On 20 January 2025, just hours after his inauguration, US President Donald Trump revoked the 30 October 2023 Executive Order on the Safe, Secure and Trustworthy Development and Use of Artificial Intelligence, signed by his predecessor, Joe Biden [136]. The presidential decree sought to "reduce the risks that AI poses to consumers, workers and national security". Biden's order required developers of AI systems that pose risks to US national security, the economy, public health, or safety to "share the results of safety tests with the US government, in line with the Defence Production Act, before they are released to the public". The decree also directed agencies to "set standards for that testing and address related chemical, biological, radiological, nuclear, and cybersecurity risks". Biden's declaration came as US lawmakers had failed to pass legislation setting guardrails for AI development in the United States.

The 2024 US Republican Party election platform vowed to repeal the Biden intervention, saying: "The executive order hinders AI innovation. Republicans support AI development rooted in free speech and human flourishing". Hence, President Trump immediately rescinded the order on 20 January 2025. However, he did not revoke Biden's 14 January 2025 Executive Order on Advancing US Leadership in AI Infrastructure. This clearly shows that the commitment to the efficacy and centrality of AI infrastructure is bipartisan in the United States. Indeed, on 21 January 2025, Trump announced a US$500 billion investment in AI infrastructure in the presence of SoftBank CEO Masayoshi Son, OpenAI's Sam Altman, and Oracle's Larry Ellison [45]. Trump's AI agenda, driven by the US$500 billion private sector intervention in AI infrastructure, is anchored by the reduction in AI regulations occasioned by rescinding the sweeping Biden 2023 executive order. The US$500 billion investment, code-named "Stargate Project", will concentrate on building new AI data centres across the United States.

Stargate Project (a new company) intends to invest US$500 billion from 2025 to 2029—building new AI infrastructure in the United States. It will start by deploying US$100 billion immediately. This infrastructure will secure US leadership in AI, create hundreds of thousands of US jobs, and generate extensive economic benefits globally. The motivation is to support the re-industrialisation of the United States and provide an AI strategic capability to protect the country's national security. The initial equity funders in Stargate are SoftBank, OpenAI, Oracle, and MGX. SoftBank and OpenAI are the lead partners; the former has financial responsibility, while the latter takes charge of operations. Masayoshi Son is the chairperson of Stargate. The company's key initial technology partners are Arm, Microsoft, NVIDIA, Oracle, and OpenAI. The object is to build on sites across the United States, starting with Texas [45].

Indeed, Oracle, NVIDIA, and OpenAI will collaborate closely to build and operate this Stargate computing system, leveraging the deep collaboration between OpenAI and NVIDIA, which was initiated in 2016, and Oracle and OpenAI's recent partnership arrangements. Stargate also builds on the existing OpenAI partnership with Microsoft. The overarching objective of this Trump-backed private sector-driven US$500 billion initiative is to continue building and developing AI—in particular AGI—to turbocharge the US economy. More significantly, such massive investments in AI infrastructure have the potential to unlock social and economic value beyond the United States and benefit all of Earth's inhabitants. The challenge is for creative AI ecosystems globally to decipher how to use AI to achieve the SDGs.

The key lesson from the United States is about the primacy of AI infrastructure. Without compute resources, there will be no AI revolution. Countries must invest in AI infrastructure to leapfrog technological limitations and actively participate in the burgeoning AI-driven global economy. By developing robust AI capabilities such as data centres, GPUs, and cloud computing platforms, emerging and least industrialised economies can be producers of AI, not just users. Indeed, they can then effectively enhance their educational systems, improve healthcare delivery through bespoke predictive analytics, and drive innovation in sectors like agriculture, mining, and manufacturing. Moreover, locally developed AI infrastructure can help address local challenges, such as resource management and climate adaptation, by enabling data-driven decision-making. All these interventions will lead to the attainment of the SDGs. Ultimately, investment in AI infrastructure boosts economic growth and ensures that these regions become competitive, protect national security, and are not left behind in the global shift towards advanced technological solutions. Deploying AI to achieve the SDGs demands massive investments in AI infrastructure by all countries.

11.3.9 European Union Joins the AI Race

At the AI Global Summit in Paris (10–11 February 2025), President of the European Commission Ursula von der Leyen announced that the new EU InvestAI initiative aimed to mobilise and raise €200 billion in AI investment, with an ambition to strengthen Europe's leadership in AI. The objective is to forge European AI champions capable of competing with the United States and China. There is a clear vision to transform Europe into a leading AI powerhouse—a creator of AI systems, not just a user. The targets of this substantial European investment in AI include the following:

- Building up AI factories and Giga factories
- 12 cutting-edge AI hubs powered by EuroHPC supercomputers
- Democratising AI computing power (efforts similar to the European Organization for Nuclear Research (CERN), but focusing on AI)
- World-class infrastructure accessible to startups and scientists
- Strong focus on trustworthy AI and with a strong objective towards the industrial sector and mission-critical applications

These are great AI efforts from the European Union. Indeed, Europe is very well positioned to play a significant role in the AI revolution by leveraging its unique standing: industrial powerhouse status, rich data ecosystems, world-class talent pool, and commitment to ethical AI. The €200 billion in AI investment (rivalling the $500 billion Stargate Project in the United States) marks a significant inflection point as Europe plunges into the AI race, which the United States and China have dominated [21].

1GW Data Centre

In February 2025, President Emmanuel Macron and UAE President Sheikh Mohamed bin Zayed al-Nahyan announced that the United Arab Emirates (UAE) would be collaborating with France on an ambitious project to develop a 1GW AI data centre campus in France. The project is part of a broader joint framework agreement between the two nations that would see an injection of €50 billion by the UAE—one of the largest AI investments in Europe. A historic AI cooperation agreement was signed between the two countries. The objective is to establish a strategic partnership in AI and to develop projects and investments that support the development of the AI value chain. This includes investments in French and Emirati AI, acquiring cutting-edge chips, data centres, talent development, and establishing virtual data embassies.

The UAE investment was part of the €109 billion ($113 billion) AI investment plan for France announced by Macron at the Global AI Summit in February 2025. The balance will come from major US, Canadian, and French companies. The €109 billion intervention was a French effort to compete with the US Stargate Project—the $500-billion AI programme announced by US President Donald Trump in January 2025 [21].

What of the Global South?

All these sterling AI efforts in the United States and Europe do not imply that emerging and least industrialised economies can't play the game. The UAE AI Minister (the first in the world) has espoused some innovative ideas about the country's AI vision and strategy. These include positioning the UAE as a hub of talent, concentrating on go-to-market applications, going to market faster, unlocking MGX investment in OpenAI, investing €50 billion in the 1GW data centre in France, balancing the tension between AI safety and AI innovation, and understanding the dangers of overregulation. He posits that the AI world will not be unipolar or bipolar but multipolar. Hence, the UAE can secure a seat at the AI table by investing in AI and AI infrastructure. Indeed, the Global South must learn from the UAE [21].

Indeed, the DeepSeek success story, briefly discussed in Chap. 4, is a clear illustration of the Global South's potential. The company's product, DeepSeek-R1, is shaking Silicon Valley. In an interview in February 2025, the founder Liang Wenfeng said: "We're done following. It's time to lead". What a poignant message—an instructive lesson to the least industrialised and emerging economies. A seismic shift in AI dominance is underway, and all eyes are on China and the Global South. DeepSeek-R1—an open-source inference model from China—has sent shockwaves through the tech world by matching OpenAI and Google's top-tier LLM performance at 3.3% of the Application Programming Interface (API) cost, the price a company charges for developers to access and use their API, while embracing full openness. With a meagre $6 million, China built one of the world's finest AI models, contrasting with the billions spent by companies such as OpenAI, Meta, Google, and Microsoft. Across the world, users—especially individuals and SMEs—are switching to DeepSeek-R1, adopting it as their foundational model. Clearly, a Chinese startup dared to leapfrog industry giants and redefine the rules of AI innovation. Other Chinese AI giants, including ByteDance, Alibaba, Baidu, and Tencent, have joined the race, ignited by DeepSeek [149].

11.3.10 Regulation, Innovation, National Security, and Global Governance

There is tension between the dangers of over-regulation and the importance of allowing AI innovation to flourish. Indeed, there is a need for a balance between ensuring AI safety and embracing AI opportunities. Furthermore, to some countries, AI presents national security concerns, and they are circumspect about the nature of AI global governance. After the AI Global Summit in Paris, the United Kingdom and United States refused to sign the adopted international agreement on AI. The statement, signed by more than 60 signatories, including France, China, India, Japan, Australia, and Canada, pledged an open, inclusive, and ethical approach to AI's development. Specifically, the communiqué stated that priorities should include "ensuring AI is open, inclusive, transparent, ethical, safe, secure and trustworthy, taking into account international frameworks for all" and "making AI sustainable for people and the planet".

The refusal by the two countries to sign the declaration of the landmark Paris AI summit was a blow to hopes for a concerted and global approach to developing and regulating the technology. In a brief statement, the UK government said it had been unable to add its name to it because of concerns about national security and global governance. UK Prime Minister Keir Starmer did not attend the summit, with the Secretary of State for Science, Innovation and Technology, Peter Kyle, representing the United Kingdom. He expressed the country's reservation: "We felt the declaration didn't provide enough practical clarity on global governance, nor sufficiently address harder questions around national security and the challenge AI poses to it". The United Kingdom has previously been a champion of the idea of AI safety, with then Prime Minister Rishi Sunak holding the world's first AI Safety Summit in November 2023. Hence, by refusing to sign the Paris Declaration, the UK government risked undermining its credibility and stature as a global leader in advocating safe, ethical, and trustworthy AI innovation.

US Vice President JD Vance told delegates in Paris that too much regulation of AI could "kill a transformative industry just as it's taking off". Vance emphasised to world leaders that AI was "an opportunity that the Trump administration will not squander" and said "pro-growth AI policies" should be prioritised over safety. Vance's uncompromising and strongly worded speech in front of key political leaders, including French President Emmanuel Macron and Indian Prime Minister Narendra Modi, indicated dissatisfaction with the global approach to regulating and developing the technology. Vance warned against the EU's strident regulatory approach, stating that "excessive regulation of the AI sector could kill a transformative industry". Furthermore, he warned against cooperation with China. His comments contrasted with those of Macron, who defended the need for further regulation. "We need these rules for AI to move forward", Macron said at the summit [1, 79].

11.3.11 Implementation and Monitoring of Pilot Projects

As already articulated, implementation and monitoring are critical for translating the framework into action. Pilot projects allow stakeholders to test AI solutions in specific SDG areas, such as using AI-powered tools to optimise agricultural productivity or improve energy efficiency. These pilots provide valuable insights, enabling iterative improvements and scaling successful models. Progress must be measured using clear metrics and indicators, such as the number of people benefitting from AI-driven healthcare solutions or the reduction in greenhouse gas emissions achieved through AI-optimised energy systems. Regular monitoring and evaluation ensure that AI initiatives remain aligned with SDG objectives, fostering accountability and transparency.

Despite its transformative potential, deploying AI comes with challenges that must be addressed within the framework. Data privacy and security are significant concerns, as AI systems rely on vast amounts of sensitive information. Robust data protection laws and cybersecurity measures are essential to safeguarding this data. The framework must also address the digital divide, ensuring underserved populations can access the infrastructure and skills needed to benefit from AI technologies. Moreover, AI systems must adopt environmentally sustainable practices, such as using renewable energy and optimising computational efficiency to reduce their carbon footprint.

As already explained, the AI adoption framework should incorporate mechanisms for continuous feedback and adaptation. Stakeholder input, technological advancements, and changing global conditions necessitate periodic updates to policies and practices. Indeed, it is imperative to robust monitoring and evaluation frameworks to assess the impact, effectiveness, and sustainability of AI interventions in contributing to the selected SDGs. It is essential to continuously learn from implementation experiences, adapt strategies based on feedback, and share best practices to improve future interventions. This capacity and flexibility ensure that the framework remains relevant and effective in achieving the SDGs, as pilot projects are rolled out, evaluated, improved, and then replicated. There is a need to identify opportunities for scaling successful AI interventions and replicating best practices across different contexts, regions, or sectors. Mechanisms for sustainability, including business models, funding sources, and community ownership, should be front and centre to ensure long-term impact and scalability.

11.3.12 Global Collaboration and Partnerships

Global collaboration and partnerships are integral to the adoption framework's success. Achieving the SDGs through AI requires coordinated global efforts. The framework should encourage international collaboration on AI research, standard setting, and resource sharing. AI systems often operate across borders, making coordinated international efforts essential. There must be partnerships and collaboration among governments, academia, industry, civil society, and international organisations to leverage complementary expertise, resources, and networks. Thus, collective impact is maximised by engaging in joint initiatives, knowledge exchange platforms, and public-private partnerships.

Initiatives like the UN's AI for Good platform demonstrate the importance of partnerships in leveraging AI for sustainable development. Countries can share best practices, establish harmonised standards, and collaborate on research and innovation to maximise AI's impact. Regional organisations like the African Union can play a key role in fostering continental integration and ensuring equitable access to AI benefits. Indeed, collaborative efforts ensure that the deployment of AI leads to the attainment of SDGs equitably across regions.

11.3.13 Balance, Capacity Building, and Education

In crafting an ideal adoption framework for AI deployment to achieve the SDGs, the importance of balancing innovation with responsibility cannot be overstated. The framework must integrate robust governance, ethical principles, capacity building, and investment in infrastructure while addressing challenges such as the digital divide and environmental impact. AI can become a cornerstone of sustainable development by fostering global collaboration and inclusivity. With a clear and comprehensive framework, AI has the potential to drive transformative progress, ensuring that the global community achieves the SDGs and creates a more equitable and sustainable future for all.

AI deployment requires skilled professionals and informed policymakers. Investments in education and training programmes are essential to build local capacity in AI research, development, and governance. These programmes should focus on under-represented groups, including women and marginalised communities, to ensure diverse perspectives in AI development and decision-making. Capacity-building efforts must include public awareness campaigns to foster understanding and trust in AI technologies. There is a need to invest in capacity-building initiatives to empower local stakeholders with the skills, knowledge, and resources needed to develop, implement, and sustain AI solutions for SDGs. Furthermore, there must be the facilitation of knowledge sharing, training programmes, and collaborative learning platforms to drive innovation and expertise exchange.

11.3.14 Addressing Challenges and Risks

AI systems rely heavily on data, raising concerns about privacy and security. The framework must include robust data protection laws and cybersecurity measures to safeguard sensitive information. The framework should prioritise reducing the digital divide to prevent AI from exacerbating inequalities. Digital literacy and infrastructure investments are crucial to ensuring that underserved populations benefit from AI technologies. AI systems' energy consumption poses a challenge to environmental sustainability. The framework should encourage the use of renewable energy in AI operations and promote research into energy-efficient AI technologies.

Addressing the challenges and risks in the deployment of AI to achieve the SDGs is crucial to ensure that AI technologies are used ethically, inclusively, and sustainably. Without proper oversight, AI deployment could exacerbate inequalities, increase biases, and undermine human rights, especially in vulnerable communities. For example, biased algorithms in healthcare or recruitment could lead to discriminatory outcomes, while inadequate data privacy protections could expose sensitive information. Additionally, the environmental impact of AI systems, such as the high energy consumption of data centres, poses a threat to climate action goals (SDG 13). Addressing these challenges requires clear regulatory frameworks, robust data protection laws, and ethical guidelines to ensure fairness, transparency, and accountability.

Mitigating risks also involves promoting inclusive access to AI technologies, particularly in developing countries, to avoid widening the digital divide. By proactively identifying and addressing these challenges, stakeholders can foster public trust, strengthen the legitimacy of AI-driven initiatives, and ensure that AI deployment supports—rather than hinders—progress towards the SDGs. This approach reduces risks and maximises the long-term impact, sustainability, and equity of AI-driven development efforts.

11.4 Scale and Regulatory Harmonisation

As already stated, beyond the country, the six-component AI Strategic Framework must be dynamically replicated at regional and continental levels such that the country influences the region and vice versa, the country influences the continent and vice versa, and the region influences the continent and vice versa. The three documents must be aligned and consistent while unlocking synergies. AI deployment to achieve the SDGs requires scale and regulatory harmonisation. These will come from economic integration and political unity at regional and continental levels. This is of prime importance for economies in the Global South, such as those in Africa. It is instructive to flesh out the meaning and efficacy of terms and diction.

11.4.1 Regional Integration and Political Unity

Regional integration is the process by which neighbouring countries or states collaborate to achieve greater economic, social, and political cohesion through shared policies, agreements, and institutions. It typically involves reducing trade barriers, harmonising regulations, fostering cross-border investments, and enabling the free movement of people, goods, and services. Regional integration aims to enhance collective economic strength, improve competitiveness, and address shared challenges such as infrastructure gaps, security threats, and environmental concerns. By pooling resources and leveraging collective expertise, integrated regions can create economies of scale, stimulate development, and ensure equitable growth among member states. Examples like the

European Union (EU) and the African Continental Free Trade Area (AfCFTA) demonstrate how integration can boost trade, improve living standards, and strengthen a region's position in the global economy.

Political unity, on the other hand, involves the alignment and collaboration of nations or regions to achieve common political goals, often through shared governance structures, joint decision-making, or collective representation. It fosters peace, stability, and mutual support by prioritising dialogue and cooperation over conflict. Political unity is crucial in addressing transnational challenges like terrorism, climate change, and pandemics, which require coordinated action and shared accountability. Moreover, political unity strengthens a region's voice on the global stage, allowing it to influence international policy and advocate for its collective interests more effectively. Together, regional integration and political unity are essential for fostering development, ensuring peace, and enhancing the resilience of nations in an increasingly interconnected world.

Regional integration and political unity are essential for the successful deployment of AI to achieve the SDGs because they foster collective action, resource sharing, and harmonised regulatory frameworks. AI systems often operate across borders, requiring unified governance, ethical standards, and data-sharing protocols. Regional integration facilitates cross-border collaboration in areas like research, development, and capacity building, enabling countries to pool resources, reduce costs, and avoid duplication of efforts. Political unity ensures that governments work together to address transnational challenges like climate change, health crises, and cross-border security threats, which AI can help tackle more effectively when efforts are coordinated. Unified policies across regions, such as common AI regulations under initiatives like the African Continental Free Trade Area (AfCFTA), can create a level playing field, enhance regional competitiveness, and attract foreign investment. Moreover, regional integration promotes inclusivity and ensures that smaller or less technologically advanced nations can access AI innovations, fostering equity and reducing the digital divide. By promoting political unity, nations can present a collective voice in global AI governance, ensuring that regional interests are represented in international policy discussions. Regional integration and political unity ensure that AI deployment is efficient, fair, and aligned with the collective pursuit of sustainable development.

11.4.2 Continental Integration and Political Unity

Continental integration refers to the process of unifying countries across an entire continent through coordinated economic, social, and political frameworks to achieve shared development goals and address collective challenges. It involves harmonising policies, fostering trade, building infrastructure, and encouraging the free movement of people, goods, and services. Continental integration aims to create larger, more robust markets, attract investments, and enhance the continent's global competitiveness. For example, the African Union (AU) promotes continental integration through initiatives like the AfCFTA, which seeks to eliminate trade barriers across Africa, boosting intra-continental trade and economic growth. By pooling resources and aligning development strategies, continental integration strengthens economic resilience and fosters inclusive growth for all member nations.

Political unity at the continental level involves aligning nations around common political objectives and creating collective decision-making mechanisms to address transnational issues. It emphasises shared governance structures, mutual respect for sovereignty, and joint efforts to promote peace, stability, and sustainable development. Political unity is essential for tackling complex global challenges such as climate change, migration, and security threats, which transcend national borders and require coordinated action. For instance, political unity under the AU has facilitated peacekeeping missions and collective responses to health crises like the Ebola outbreak in Africa. Continental political unity also amplifies a region's influence in international forums, enabling member states to advocate for fairer global policies and partnerships. Together, continental integration and political unity are crucial for fostering development, safeguarding stability, and ensuring the collective prosperity of nations within a continent.

Continental integration and political unity are critical for the successful deployment of AI to achieve the SDGs as they promote a unified approach to AI governance, policy harmonisation, and resource sharing across an entire continent. Continental integration facilitates the establishment of shared regulations, ethical standards, and infrastructure, ensuring that AI technologies are developed and deployed in a way that benefits all member countries equitably. Political unity fosters collaboration among nations, enabling joint efforts to tackle transboundary challenges like climate change, health pandemics, and regional security threats using AI-driven solutions. It also enhances the continent's bargaining power in global AI governance, ensuring that collective interests are represented in international policy discussions. Initiatives like the African Union's Agenda 2063 and AfCFTA highlight the importance of leveraging AI to drive economic growth, innovation, and sustainable development. By aligning AI strategies at the continental level, nations can reduce redundancies, close technological gaps, and ensure that smaller, less technologically advanced countries are not left behind. Ultimately, continental integration and political unity enable AI's efficient, ethical, and equitable deployment, ensuring that its benefits are shared across all nations and fully aligned with the SDGs [2, 3, 86].

11.4.3 African Union's Continental AI Strategy

In July 2024, the African Union (AU) released its AI blueprint titled Continental Artificial Intelligence Strategy: Harnessing AI for Africa's Development and Prosperity. The document outlines the continent's vision, mission, guiding principles, and overall strategic objectives. The strategy is aligned with the mission and vision of the AU and designed to address Africa's problems while leveraging the continent's assets and strengths in pursuit of inclusive development and shared prosperity: "The Continental AI Strategy supports the AU Agenda 2063's vision. It is rooted in Africa's unique challenges and opportunities, as the continent can leverage its youth's digital native talent, natural resources, huge market and geopolitical position to develop and promote an ethical, responsible and inclusive AI that empowers people and contributes to the continent's economic growth, social progress and cultural renaissance".

The AI vision for the continent is articulated as "a prosperous and integrated Africa where responsible, ethical and African-centric AI is the axis of inclusive growth, resilience, socio-economic development, people empowerment, and positioning the continent as a key player in the global AI landscape". The mission "is to harness AI for accelerating social and economic transformation and promoting cultural renaissance in Africa in line with the AU Agenda 2063 and the Sustainable Development Goals, to minimise the risks that AI poses to African people and countries, and to accelerate the development of the necessary AI capabilities of the AU Member States". Figure 11.2 shows the entire membership of the African Union.

Guiding Principles, Focus, and Policy

The continental strategy's guiding principles include local first, people-centred, human rights and human dignity, peace and prosperity, inclusion and diversity, ethics and transparency, cooperation

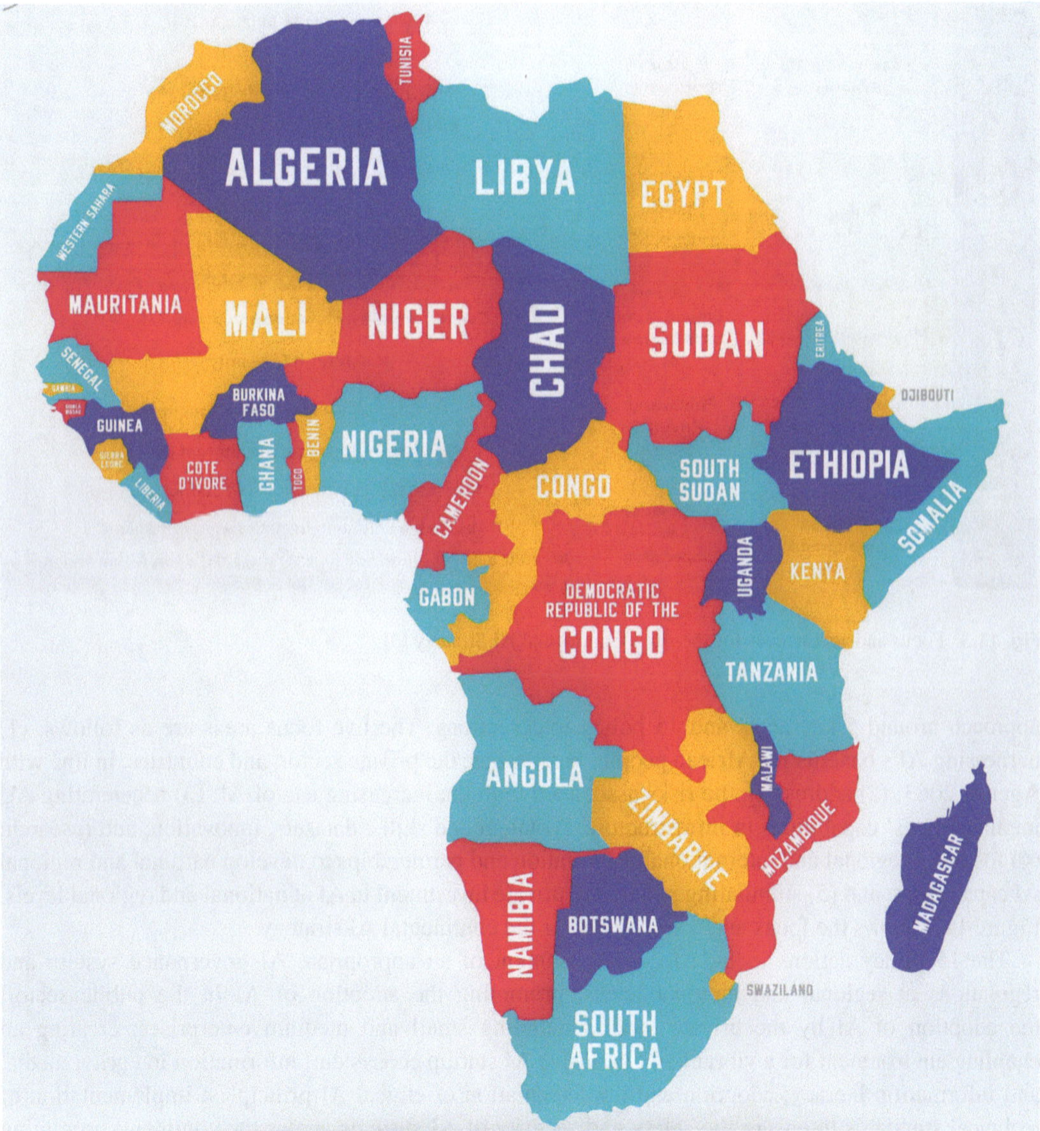

Fig. 11.2 Countries constituting the African Union

and integration, skills development, public awareness and education. The strategy acknowledges the importance of making Africa ready for AI: "To realise AI's positive and transformative potential for African development and mitigate potential risks, it is imperative to build the necessary capabilities. These range from sustained investment in infrastructure (reliable and efficient electricity, broadband connectivity, data infrastructure like data centres and cloud, and computing power) to huge sets of quality data, education and skills in AI, media and information literacy (MIL), research and innovation. Africa must also build a vibrant and inclusive AI startup enterprise ecosystem that develops and implements social and economic applications and systems".

The continental strategy prioritises regional coordination, international partnerships, and significant public and private investments. It proposes a people-centric, development-oriented, inclusive

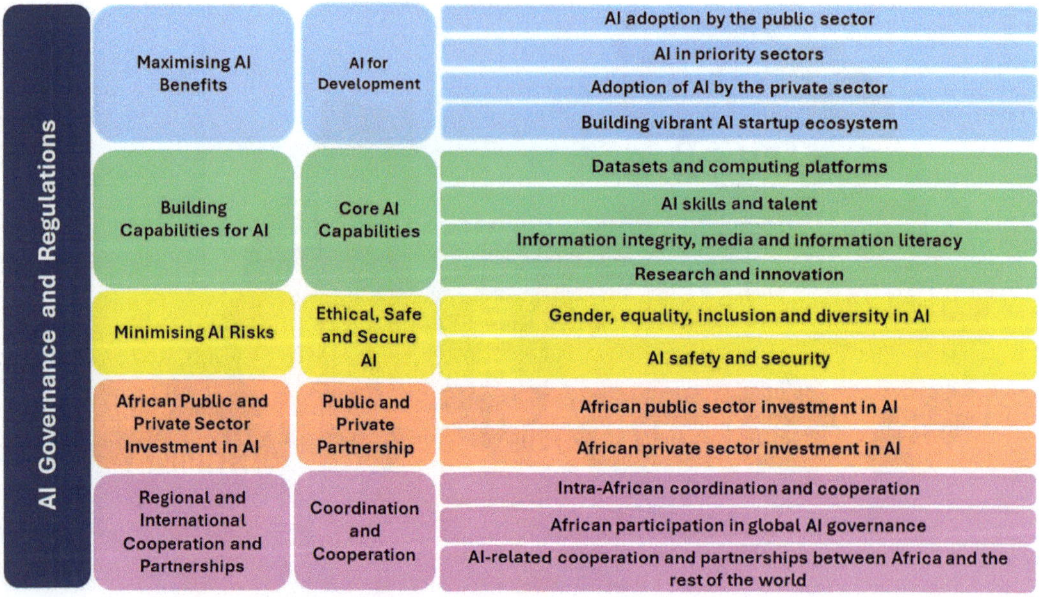

Fig. 11.3 Focus and action areas for the AU's Continental AI Strategy [3]

approach around 5 key areas and 15 policy interventions. The five focus areas are as follows: (1) harnessing AI's benefits for African people, institutions, the private sector, and countries, in line with Agenda 2063; (2) addressing the risks associated with the increasing use of AI; (3) accelerating AU member states' capabilities in infrastructure, AI talent and skills, datasets, innovation, and research; (4) fostering regional and international cooperation and partnerships to develop national and regional AI capabilities; and (5) stimulating public and private investment in AI at national and regional levels. Figure 11.3 shows the focus and action areas for the continental AI strategy.

The 15 policy actions include the establishment of an appropriate AI governance system and regulations at regional and national levels, promoting the adoption of AI in the public sector; the adoption of AI by the private sector, including small and medium enterprises; creating an enabling environment for a vibrant and inclusive AI startup ecosystem; information integrity, media, and information literacy; adoption and implementation of ethical AI principles; implementation of technical standards to ensure the safety and security of AI systems across the continent; promoting regional cooperation and coordination; accelerated African participation in global AI governance; and promotion of AI-related partnerships between Africa and the rest of the world, to mobilise financial and technical resources for AI development in Africa [3, 86].

Limitations of the AU Continental AI Strategy

The AU Continental AI Strategy document is a good starting point. It contains many pertinent issues and dimensions. The AI vision is clear and well-linked to existing AU developmental efforts. The document fairly articulates the AI implementation requirements.

Nevertheless, the AU Continental Strategy is just a laundry list without strategic coherence. There is no clarity about how the strategy will be cascaded to the RECs, much less to the member states. The review of existing country efforts is neither detailed nor substantive. Hence, lessons from current state-level efforts are neither acknowledged nor leveraged. It is important to link current RECs and national efforts to the Africa-wide strategy and vice versa. As articulated in this chapter, the six-component National AI Strategic Framework must be dynamically replicated at regional and continental levels such that the country influences the region and vice versa, the country influences the continent and vice versa, and the region influences the continent and vice versa. The AU document falls far short of this non-negotiable imperative.

More significantly, the continental strategy is decontextualised from Africa's challenges. There is a need to link AI interventions to the continent's perennial developmental ambitions. In crafting the blueprint, AI is taken as a silver bullet. No. It is just another developmental tool among many others. The strategy fails to spell out the other complementary non-AI interventions required on the continent—the critical success factors for impactful AI adoption. What must Africa do to prepare for adopting AI in terms of basic infrastructure, digital infrastructure, energy, finance and investment, governance and regulatory framework, literacy, talent, and expertise? The strategy document does not adequately address these issues, which must be attended to so that the continent is ready for AI. Furthermore, other new and transformative technologies (part of the 4IR) exist, such as quantum computing, nanotechnology, big data, 3D printing, the Internet of Things, and battery energy storage systems. They are mutually reinforcing with AI. The blueprint does not refer to these innovative technologies.

While articulating an AI strategy for Africa, the document does not acknowledge the importance of leveraging Global Africa. More importantly, the AI strategy is not contextualised within the Global South. For example, African AI efforts must leverage lessons from China, India, Singapore, and Malaysia. Furthermore, the nature and impact of the Global South-Global North technology divide are not discussed. More specifically, the role of partners from the Global North and the rest of the Global South in enabling the adoption of AI in Africa is not articulated. No clear lessons are drawn from the emergent AI experiences of jurisdictions such as China, India, Singapore, the United States, the United Kingdom, and the European Union. A review of how Africa has benefitted (or not benefitted) from previous technology revolutions and an articulated description of how AI would be different would improve the utility of the AI blueprint.

Nevertheless, the AU document is a reasonable premise for designing and developing a more grounded, dynamic, and efficacious AI Strategic Framework for the continent. African countries and regional blocs (SADC, EAC, ECOWAS, and Maghreb) can use it as a starting point or reference source as they embark on their own similarly structured AI Strategic Framework documents. Figures 11.4 and 11.5 show the countries that should develop and implement such frameworks for SADC and ECOWAS, respectively. Lessons and emerging experiences of these African efforts—at country, region, and continental levels—can be harvested and leveraged worldwide, particularly in the Global South.

11.5 Driving SDGs Through Sectorial AI Applications

AI-driven advancements across various economic and social sectors are pivotal for achieving the SDGs as they create a synergistic effect that propels holistic development. Enhancements in education, healthcare, agriculture, and industry contribute to their respective SDGs and interlink to support progress in other areas. For instance, better education leads to improved health outcomes and higher economic productivity, while advancements in healthcare reduce poverty by decreasing healthcare-related expenditures and increasing productivity through healthier populations. Similarly, improvements in infrastructure and technology enhance access to markets and information, boosting business opportunities and innovation. This integrated approach ensures that progress in one sector

Fig. 11.4 SADC AI Strategic Framework: The countries

catalyses benefits in others, leading to sustainable and inclusive development that addresses the core objectives of the SDGs. Several specific sectors can be used to illustrate the strategic value of sectoral AI applications on the SDG journey.

11.5.1 AI Applications in Agriculture and Mining

Improvements in agriculture and the unlocking of value in mining can dramatically advance the attainment of the SDGs. Specifically, enhancements in agriculture play a pivotal role in achieving SDGs related to ending hunger, achieving food security, improving nutrition, and promoting sustainable agriculture (SDGs 1, 2, and 3). Improved agricultural practices can increase food production and availability, making it more accessible and affordable, especially in emerging and least industrialised countries. This, in turn, reduces hunger and malnutrition rates and fosters socio-economic growth by boosting rural economies through increased market opportunities and job creation. Furthermore, sustainable agriculture minimises the environmental impact, conserves natural resources, and contributes to climate action (SDG 13), ensuring the longevity and health of ecosystems on which agricultural productivity depends.

Fig. 11.5 ECOWAS AI Strategic Framework: The countries

Similarly, unlocking the value in mining is essential for the progress towards multiple SDGs, including those focused on economic growth (SDG 8); industry, innovation, and infrastructure (SDG 9); and responsible consumption and production (SDG 12). The mining sector's development contributes significantly to national economies by generating employment, fostering industrial diversification, and providing critical materials for modern technologies. When managed sustainably, mining can reduce negative environmental impacts and promote the efficient use of resources. Moreover, responsible mining practices enhance local communities' resilience by ensuring fair labour conditions and equitable distribution of economic benefits, thus contributing to reducing inequalities (SDG 10) and promoting sustainable cities and communities (SDG 11).

Hence, AI applications in agriculture and mining are key ingredients in deploying AI to achieve the SDGs. Appendix A outlines examples of such interventions and their impact.

11.5.2 AI Applications in Infrastructure, Governance, and Manufacturing

Improvements in infrastructure, governance, and manufacturing can dramatically improve the attainment of the SDGs. Specifically, enhancements in infrastructure are foundational to advancing many of these goals, especially those related to poverty alleviation, urban services, and environmental sustainability. Enhanced infrastructure, such as roads, bridges, water systems, and telecommunications, can dramatically improve access to essential services, economic opportunities, and connectivity between rural and urban areas. This connectivity drives economic development by facilitating trade, attracting investment, and creating jobs, thereby tackling SDG 1 (No Poverty) and SDG 8 (Decent Work and Economic Growth). Furthermore, resilient infrastructure contributes to building sustainable cities and communities (SDG 11) and can help mitigate climate change impacts (SDG 13) by promoting energy efficiency and reducing reliance on fossil fuels.

Similarly, strong governance systems are essential for the effective implementation of the SDGs, as they ensure accountability, transparency, and inclusiveness in managing human, natural, and economic resources. Improved governance can empower communities, protect the rights of the most vulnerable, and ensure that development gains are equitably shared, directly supporting SDG 16 (Peace, Justice, and Strong Institutions). Sound governance frameworks are also essential for enhancing innovation and infrastructure development (SDG 9), ensuring sustainable industrialisation, and promoting efficient resource use (SDG 12). These elements collectively enhance the overall effectiveness of policies aimed at sustainable development across all sectors, including critical areas like health, education, and environmental protection.

Improvements in manufacturing are key to advancing the SDGs by boosting economic productivity and fostering innovation. Efficient and sustainable manufacturing processes can lead to more productive use of resources and less waste, which supports SDG 12 (Responsible Consumption and Production) and reduces environmental degradation. Additionally, enhancing manufacturing capabilities in developing countries can drive economic growth and employment (SDG 8), helping to lift populations out of poverty (SDG 1) and contributing to building resilient infrastructure (SDG 9). Moreover, modernising manufacturing through technological advancements can lead to more sustainable practices, significantly reducing the industry's carbon footprint and aligning with SDG 13 (Climate Action). Thus, investing in and upgrading the manufacturing sector holds profound potential to impact various aspects of economic, social, and environmental development globally.

Hence, AI applications in infrastructure, governance, and manufacturing are key ingredients in deploying AI to achieve the SDGs. Appendix B outlines examples of such interventions and their impact.

11.5.3 AI Applications in Financial Services

Financial service improvements are critical to advancing the SDGs, particularly in promoting inclusive economic growth and poverty reduction. Enhanced financial services ensure broader access to capital and financial resources, enabling entrepreneurs and small businesses to initiate or expand their operations, which fosters job creation (SDG 8: Decent Work and Economic Growth). Additionally, accessible financial services such as savings accounts, insurance, and credit can empower individuals, especially in underserved and rural areas, by safeguarding their earnings and reducing their vulnerability to economic shocks, directly contributing to SDG 1: No Poverty.

Moreover, integrating sustainable practices within the financial sector can accelerate progress towards environmental goals (SDG 13: Climate Action). Financial institutions prioritising green finance, such as funding for renewable energy projects or sustainable agriculture, play a crucial role in mitigating the impact of climate change and promoting sustainable practices across industries. This shift supports the transition towards a low-carbon economy and aligns investors with broader global commitments to sustainability. Therefore, the evolution of financial services to support economic, social, and environmental development is essential for the holistic achievement of the SDGs.

Hence, AI applications in financial services are key to deploying AI to achieve SDGs. Appendix C outlines examples of such interventions and their impact.

11.5.4 Sectorial Efforts as Part of a National AI Strategy

These sectorial AI efforts must not be executed in a vacuum. As discussed earlier, there must be a National AI Strategy anchored by a National AI Vision, which is clearly articulated at the highest level of government. For example, on 13 January 2025, UK Prime Minister Keir Starmer announced an ambitious AI plan to make the United Kingdom a world leader in AI through five key strategic initiatives: (1) a sovereign AI drive so that the country can produce its own AI models—becoming an AI *maker*, not *taker*; (2) encouraging the private sector through *AI growth zones*; (3) embedding AI in the public sector by *rapidly piloting* AI-powered services; (4) creation of a new AI energy council to drive the use of clean energy in the AI growth zones; and (5) overhaul of the UK's copyright law with respect to AI. This level of clarity and detail will enhance AI's efficacy as an SDG driver. Other economies with well-designed and effectively articulated national AI strategic frameworks include the United States, European Union, United Arab Emirates, China, India, Singapore, Malaysia, Rwanda, Kenya, and South Africa.

On 14 January 2025, the country published a draft of its Kenya National AI Strategy 2025–2030. This ambitious strategy aims to harness AI's transformative power to drive sustainable development, enhance public services, and improve social and economic equity. Kenya's explicit AI vision is to be: "An African leader in AI R&D, innovation and commercialisation for inclusive socio-economic development". Hence, the blueprint envisions the country as the leading AI hub for "model innovation, driving sustainable development, economic growth, and social inclusion while positioning itself as an AI research and application leader in Africa". At the core of the strategy is "creating solutions tailored to Kenya's unique needs and those of the African continent". The country's AI ambitions are anchored by three key pillars (AI digital infrastructure, data, and AI research and innovation) and supported by four enablers (governance, talent development, private and public investments, and ethics, equity, and inclusion) [29, 62, 86].

11.6 Conclusion

This chapter explains the importance of designing, developing, and executing an AI adoption framework to address the 17 UN SDGs. The starting point is laying out the framework's principles, objectives, and activities. The key contribution is a detailed AI Strategic Framework with six key components: Vision, Strategy, Policy, Governance, Legislation/Regulation, and Implementation Matrix. Emphasis is also placed on the centrality of infrastructure, investments, global collaboration, partnerships, and capacity building. Scale and harmonisation are emphasised by discussing regional and continental economic integration and political unity. The strengths and limitations of the African Union's Continental AI Strategy, launched in August 2024, are also discussed and reviewed as a foundational source in crafting a more grounded, dynamic, and efficacious AI Strategic Framework for Africa. The example of deploying AI in a holistic, integrated, and strategic approach that involves the African continent, its regions, and its countries can be replicated in other regions of the world as they pursue the SDG agenda. AI-driven advancements across various economic and social sectors are presented as pivotal for achieving the SDGs. However, these must be anchored by a national AI strategic framework.

Indeed, an ideal adoption or deploying AI to achieve the SDGs must balance innovation with responsibility, inclusivity with efficiency, and local priorities with global collaboration. By incorporating strong governance, ethical principles, capacity building, and robust infrastructure, such a framework ensures that AI serves as a powerful tool for socio-economic and political transformation by assisting in delivering the SDGs. Addressing challenges such as the digital divide, environmental impact, and data privacy is essential to realising AI's transformative potential. AI can become a key SDG enabler through coordinated efforts, a commitment to shared values, and an ecosystem approach. By integrating all the elements and initiatives presented in this chapter into a comprehensive AI adoption framework, stakeholders can enhance the effectiveness, inclusivity, and sustainability of AI deployments on the SDG journey.

Deploying AI to Achieve UN SDGs: Details of the 17 Goals 12

12.1 Introduction

Chapter 5 demonstrated the efficacy of AI in driving the SDG agenda and explored illustrative examples in addition to discussing specific cases from a range of countries. That material presents the foundation of this chapter, which presents detailed and comprehensive outlines of the proposed deployment of AI to achieve each of the 17 SDGs. These recommendations are a product of applying the concepts and ideas developed and advocated for in this book, in particular, leveraging emerging global experiences (Chap. 6); enablers (Chap. 7); drivers (Chap. 8); further critical success factors (Chap. 9); governance, legislation, and ethics (Chap. 10); and strategic framework (Chap. 11). Figure 12.1 shows the broader context in which AI can assist in achieving the SDGs.

12.2 AI and SDG 1 (No Poverty)

UN SDG 1 (No Poverty) aims to eradicate extreme poverty in all its forms by 2030. Despite global progress, over 700 million people still live below the international poverty line, facing challenges related to food security, healthcare, education, and economic opportunities. AI has emerged as a transformative tool in the fight against poverty, offering innovative solutions to improve resource allocation, enhance social protection systems, expand financial inclusion, and empower marginalised communities. This section explores the application of AI in achieving SDG 1 by addressing poverty through data-driven approaches in economic development, public health, education, agriculture, and social safety nets. Additionally, it examines the ethical considerations and challenges of deploying AI in poverty alleviation.

12.2.1 Introduction

Eradicating poverty is the cornerstone of sustainable development. Poverty is a multidimensional issue, encompassing inadequate access to food, education, healthcare, and economic opportunities. Addressing these challenges requires a combination of targeted policies, efficient resource man-

A. G. O. Mutambara, *Deploying Artificial Intelligence to Achieve the UN Sustainable Development Goals*, Sustainable Development Goals Series,
https://doi.org/10.1007/978-3-031-88423-8_12

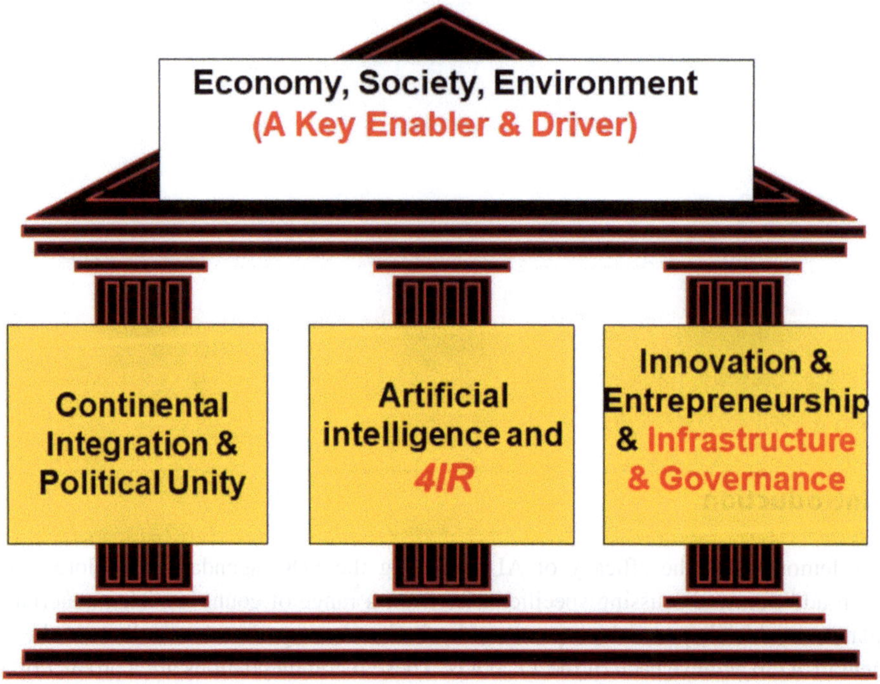

Fig. 12.1 Context in which AI can be used to achieve UN SDGs

agement, and innovative technologies. With its ability to analyse vast datasets, identify trends, and optimise decision-making, AI offers unprecedented opportunities to tackle poverty at scale. From predicting food insecurity to enabling access to financial services, AI-driven solutions can potentially transform the fight against poverty.

This section explores how AI contributes to achieving SDG 1 by supporting poverty alleviation initiatives across various sectors. It highlights AI's role in improving social protection, expanding access to essential services, enhancing agricultural productivity, and fostering inclusive economic growth. By leveraging AI's capabilities, we can create data-driven strategies to empower vulnerable populations and reduce poverty worldwide.

12.2.2 Economic Empowerment and Job Creation

Economic empowerment is critical for lifting individuals and communities out of poverty. AI has the potential to enhance economic opportunities by creating new industries, optimising resource use, and enabling more efficient labour markets. AI is driving the creation of new industries, such as data science, AI engineering, and technology-driven entrepreneurship. These fields generate employment opportunities that offer higher wages and economic mobility. Educational institutions and organisations are leveraging AI-driven platforms to provide training and upskilling programmes, enabling workers to acquire in-demand skills and transition to roles in the digital economy. For instance, platforms like Coursera and edX use AI to personalise learning paths, making education accessible and relevant to marginalised communities.

In addition to job creation, AI improves labour market efficiency by matching job seekers with opportunities based on their skills, experience, and preferences. AI-powered recruitment platforms analyse resumes and job descriptions, identifying suitable matches and reducing hiring times. This benefits workers in low-income regions by connecting them with employers who may otherwise overlook their talent.

Moreover, AI supports entrepreneurship by providing small- and medium-sized enterprises (SMEs) with affordable tools for business analytics, customer engagement, and market research. AI-powered platforms help SMEs optimise their operations, reach new markets, and improve profitability, fostering economic empowerment and reducing poverty.

12.2.3 Enhancing Social Protection Systems

Social protection systems, such as cash transfers, food subsidies, and unemployment benefits, are essential for protecting vulnerable populations from falling into poverty. AI enhances the efficiency and effectiveness of these systems by improving beneficiary identification, fraud detection, and resource allocation.

AI-driven data analysis can identify individuals and households most in need of assistance by analysing demographic, economic, and social indicators. For example, Machine Learning models can process data from national surveys, census records, and geospatial imagery to locate areas with high poverty rates. This targeted approach ensures that social protection programmes reach those who need them most, reducing exclusion errors and maximising impact.

Fraud detection is another area where AI improves social protection systems. AI algorithms analyse transaction patterns to identify irregularities, such as duplicate claims or misuse of funds. By detecting and preventing fraud, AI ensures that resources are allocated fairly and efficiently, protecting the integrity of social safety nets.

In addition, AI supports dynamic resource allocation by predicting future demand for social protection services. AI models can forecast needs by analysing economic trends, population growth, and environmental factors and help governments allocate resources proactively. This ensures that social protection systems are resilient and capable of responding to economic shocks, natural disasters, and other crises.

12.2.4 Expanding Financial Inclusion

Financial inclusion is a critical enabler of poverty reduction, as it allows individuals to save, invest, and manage financial risks. AI has the potential to expand access to financial services for unbanked and underserved populations, promoting economic empowerment and stability.

One of the most significant applications of AI in financial inclusion is alternative credit scoring. Traditional credit scoring models rely on formal financial histories, excluding individuals without access to banking services. AI-driven models, however, analyse alternative data sources, such as mobile phone usage, utility payments, and social media activity, to assess creditworthiness. This enables financial institutions to provide loans and other financial products to individuals in low-income and rural areas, empowering them to start businesses, invest in education, and improve their livelihoods.

AI also enhances digital payment systems, making financial transactions more accessible and affordable. Mobile payment platforms powered by AI enable individuals to transfer money, pay bills, and save securely without needing a traditional bank account. These systems promote financial independence and reduce reliance on cash, particularly in regions with limited banking infrastructure.

Moreover, AI-powered financial literacy tools educate users about budgeting, saving, and investing. These tools provide personalised recommendations and support, helping individuals make informed financial decisions and build long-term economic resilience.

12.2.5 Improving Public Health

Health disparities are a significant driver of poverty, as inadequate access to healthcare reduces productivity and increases financial strain. AI can enhance public health systems by improving disease prevention, diagnosis, and treatment, particularly in low-resource settings.

AI-driven disease prediction models analyse data from electronic health records, environmental sensors, and social determinants of health to identify populations at risk of illness. For example, AI algorithms can predict outbreaks of infectious diseases, such as malaria or dengue fever, by analysing weather patterns and population density. This enables governments and health organisations to implement preventive measures, reducing the disease burden in vulnerable communities.

In diagnosis, AI-powered tools improve access to quality healthcare by enabling non-specialist healthcare workers to identify illnesses accurately. For instance, AI-based apps like Ada Health analyse symptoms to provide preliminary diagnoses and treatment recommendations, helping patients in remote areas receive timely care. Similarly, AI-driven imaging tools analyse medical scans for conditions like tuberculosis or cancer, reducing diagnostic errors and improving outcomes.

AI also supports equitable healthcare delivery by optimising resource allocation. Machine Learning models predict demand for medical supplies, personnel, and infrastructure, ensuring that resources are directed to areas with the greatest need. By improving healthcare access and quality, AI addresses a critical dimension of poverty and promotes social well-being.

12.2.6 Boosting Agricultural Productivity

Agriculture is the primary livelihood for many people living in poverty, yet smallholder farmers often face challenges such as low yields, pests, and climate variability. AI offers innovative solutions to improve agricultural productivity, enhance food security, and increase farmers' incomes.

AI-enabled precision agriculture optimises resource use and improves crop management. AI-powered platforms analyse data from soil sensors, weather forecasts, and satellite imagery to provide farmers with recommendations on planting, irrigation, and fertilisation. For example, AI algorithms can predict the optimal time for planting and harvesting based on environmental conditions, maximising yields and reducing waste. By increasing productivity, AI helps farmers earn higher incomes and achieve greater economic stability.

AI also supports pest and disease management by detecting infestations early and recommending targeted interventions. Machine Learning models analyse images of crops to identify signs of pests or diseases, enabling farmers to take preventive action. This reduces crop losses and minimises the need for chemical pesticides, promoting sustainable agricultural practices. In addition, AI facilitates access to markets by connecting farmers with buyers and providing real-time price information. AI-driven platforms help farmers negotiate fair prices and reduce post-harvest losses by optimising supply chains. By improving market access, AI empowers farmers to participate in the broader economy and escape the cycle of poverty.

12.2.7 Addressing Urban Poverty

Urban poverty is a growing challenge as cities worldwide face issues such as housing shortages, unemployment, and inadequate infrastructure. AI can address urban poverty by enhancing social services, improving infrastructure, and creating economic opportunities in urban areas.

AI-powered urban planning tools analyse data on population growth, housing demand, and transportation needs to optimise infrastructure development. For example, AI models can simulate different urban development scenarios, helping city planners design affordable housing, efficient public transit systems, and accessible public spaces. By creating inclusive urban environments, AI reduces inequalities and improves the quality of life for low-income residents. In addition to infrastructure, AI enhances social services in urban areas by enabling targeted interventions. Machine Learning models analyse data on unemployment, education, and healthcare access to identify neighbourhoods in need of support. This allows governments and NGOs to implement programmes that address specific challenges, such as job training or healthcare outreach, in vulnerable communities.

AI also promotes entrepreneurship and innovation in urban areas by providing tools for business analytics, market research, and customer engagement. By supporting small businesses and startups, AI fosters economic growth and creates job opportunities, reducing urban poverty.

12.2.8 Challenges and Ethical Considerations

While AI offers powerful tools for poverty alleviation, its deployment raises ethical considerations and challenges. One primary concern is the risk of exacerbating inequalities if AI systems are not designed to be inclusive. For example, biased algorithms may exclude marginalised populations from social protection programmes or financial services. Ensuring fairness and equity in AI design requires diverse and representative datasets, regular audits, and transparency.

Data privacy is another critical issue, particularly in applications involving sensitive information such as health records or financial data. Robust data protection frameworks are essential to safeguard individuals' privacy and maintain public trust in AI-driven solutions.

The cost of implementing AI technologies can also be a barrier, particularly in low-income regions. Governments and international organisations must prioritise investments in affordable and scalable AI solutions to ensure that all communities benefit from these technologies.

12.2.9 Policy Recommendations

To maximise AI's potential for achieving SDG 1, governments, organisations, and technology companies must collaborate to create ethical frameworks, promote innovation, and invest in inclusive AI research. Policies should prioritise transparency, fairness, and accountability, ensuring AI solutions align with poverty alleviation goals.

Promoting public-private partnerships can drive innovation and scale AI-driven solutions for poverty reduction. These partnerships can also provide funding and technical support for implementing AI in low-income regions. Investing in digital literacy and capacity-building programmes is essential for equipping marginalised communities with the skills needed to benefit from AI technologies. By fostering digital inclusion, we can ensure that AI's benefits are equitably distributed.

12.2.10 Conclusion

Artificial Intelligence has the potential to transform the fight against poverty by enhancing economic opportunities, improving social protection systems, expanding access to essential services, and empowering marginalised communities. From precision agriculture and financial inclusion to public health and urban development, AI offers innovative solutions to address the complex dimensions of poverty. However, ethical considerations such as data privacy, algorithmic bias, and affordability must be carefully managed to ensure that AI serves as a force for social good. By fostering collaboration, investing in digital education, and promoting inclusive AI development, we can leverage AI's capabilities to achieve SDG 1 and create a world where no one is left behind.

This section highlights AI's role in achieving SDG 1 by addressing the economic, social, and environmental dimensions of poverty, offering recommendations for ethical AI implementation, and promoting inclusive and sustainable solutions for poverty alleviation.

12.3 AI and SDG 2 (Zero Hunger)

UN SDG 2 (Zero Hunger) seeks to end hunger, achieve food security, improve nutrition, and promote sustainable agriculture by 2030. Achieving these objectives requires addressing the challenges of food production, distribution, and access, particularly in the face of climate change, population growth, and resource limitations. Artificial Intelligence (AI) offers transformative solutions to enhance agricultural productivity, optimise food distribution systems, reduce waste, and improve nutritional outcomes. This section explores how AI contributes to achieving SDG 2 through applications in precision agriculture, food supply chain optimisation, climate resilience, and personalised nutrition. It also discusses the challenges and ethical considerations associated with deploying AI in the fight against hunger.

12.3.1 Introduction

Hunger and malnutrition remain pressing global issues, affecting over 800 million people despite advancements in agriculture and food systems. SDG 2 aims to eradicate hunger and malnutrition while ensuring that food production systems are sustainable and resilient. However, achieving these goals is increasingly complex due to factors such as climate change, urbanisation, and inequitable resource distribution. AI has emerged as a powerful tool to address these challenges, providing data-driven insights, predictive analytics, and automation capabilities that enhance food systems at every stage. From optimising crop yields to reducing post-harvest losses, AI-driven innovations play a critical role in pursuing Zero Hunger.

This section examines AI's applications in advancing SDG 2, focusing on its role in sustainable agriculture, food distribution, waste reduction, climate adaptation, and nutrition improvement. By understanding how AI can transform food systems, we can leverage its potential to ensure a world where everyone has access to sufficient, nutritious, and sustainably produced food.

12.3.2 Precision Agriculture and Sustainable Farming

Agriculture is central to food security, but traditional farming methods often lead to resource inefficiencies, environmental degradation, and low productivity. AI-driven precision agriculture addresses these challenges by enabling data-informed decision-making that optimises resource use and enhances crop yields.

One of AI's most significant applications in agriculture is using Machine Learning algorithms to analyse data from sensors, drones, and satellite imagery. These tools provide farmers with real-time information about soil health, crop growth, and pest activity. For example, AI-powered platforms like FarmLogs and Climate FieldView offer recommendations on planting, irrigation, and fertilisation, ensuring that resources are used efficiently. Precision agriculture increases productivity by tailoring agricultural practices to specific conditions while minimising waste, water use, and chemical inputs.

AI also supports pest and disease management by detecting infestations early and recommending targeted interventions. Machine Learning models analyse images of crops to identify signs of pests or diseases, enabling farmers to take preventive action before damage becomes severe. This reduces reliance on chemical pesticides, lowers production costs, and promotes sustainable farming practices. For instance, companies like Plantix and PEAT use AI-driven tools to provide farmers with diagnostic insights and solutions for managing crop health.

Furthermore, AI-powered robotics and automation systems are transforming labour-intensive tasks such as planting, weeding, and harvesting. Autonomous tractors and drones equipped with AI algorithms perform these tasks with precision, reducing labour costs and increasing efficiency. By integrating AI into farming operations, farmers can achieve higher yields and greater sustainability, supporting SDG 2s goal of ending hunger through increased agricultural productivity. Figure 12.2 shows a swarm of cooperating drones spraying fertiliser in a maize field, while Fig. 12.3 illustrates the automation of tomato harvesting using an AI-driven robot.

Fig. 12.2 A swarm of cooperating drones spraying fertiliser in a maize field. (Image credit: iStock kinwun.)

Fig. 12.3 Automation of tomato harvesting using an AI-driven robot. (Image credit: iStock kinwun.)

12.3.3 Optimising Food Supply Chains

Food insecurity is not solely a result of inadequate production; inefficiencies in food supply chains also contribute significantly to hunger and malnutrition. AI has the potential to optimise supply chains by improving logistics, reducing food waste, and ensuring that food reaches those who need it most.

AI-driven supply chain management systems analyse data on production, transportation, and storage to identify inefficiencies and recommend improvements. For example, Machine Learning algorithms can predict demand for perishable goods, enabling producers and retailers to adjust inventory levels accordingly. This reduces overproduction and prevents food from spoiling before it reaches consumers. In addition, AI-powered route optimisation tools minimise transportation delays and fuel consumption, ensuring that food is delivered quickly and sustainably.

AI also enhances traceability in food supply chains, enabling stakeholders to monitor the journey of food from farm to table. Blockchain technology, integrated with AI, provides a secure and transparent record of transactions, helping to identify and address disruptions in the supply chain. For instance, IBM Food Trust uses AI and Blockchain to track food items, ensuring that they are safe, fresh, and ethically sourced. By improving traceability, AI promotes consumer confidence and reduces food loss at various stages of the supply chain.

Moreover, AI-powered platforms facilitate market access for smallholder farmers by connecting them directly with buyers and reducing reliance on intermediaries. These platforms use AI to match farmers with markets based on factors such as price, demand, and location, ensuring fair compensation and reducing post-harvest losses. By streamlining food supply chains, AI contributes to SDG 2s objective of ensuring that everyone has access to sufficient and affordable food.

12.3.4 Reducing Food Waste

Food waste is a significant barrier to achieving Zero Hunger, with an estimated one-third of all food produced globally going to waste. AI offers innovative solutions to reduce food waste at every stage, from production to consumption. In production, AI-driven predictive analytics help farmers and food processors align supply with demand, minimising surplus. For example, AI algorithms analyse historical sales data, market trends, and weather forecasts to predict crop yields and consumer demand. This enables producers to plan harvests and inventory levels more accurately, reducing the likelihood of excess production.

In retail and distribution, AI-powered inventory management systems monitor stock levels and expiration dates, alerting retailers to prioritise the sale of perishable items. Companies like Winnow Solutions use AI to analyse food waste patterns in restaurants and cafeterias, providing insights to reduce waste and optimise menu planning. AI helps businesses reduce waste and save costs by identifying inefficiencies in food handling and storage.

On the consumer side, AI applications encourage sustainable consumption by providing personalised recommendations for meal planning and food storage. For example, apps like Too Good To Go and Olio connect consumers with surplus food from restaurants and grocery stores, promoting redistribution and reducing waste. By addressing food waste across the value chain, AI supports SDG 2s mission to ensure sustainable food systems and reduce hunger.

12.3.5 Building Climate Resilience

Climate change poses a significant threat to global food security, as rising temperatures, changing precipitation patterns, and extreme weather events disrupt agricultural systems. AI plays a crucial role in building climate resilience by providing tools to predict, adapt to, and mitigate the impacts of climate change on food production.

AI-driven climate models analyse historical weather data, satellite imagery, and environmental factors to predict the impacts of climate change on specific regions and crops. These models enable policymakers and farmers to develop adaptive strategies, such as shifting planting schedules, selecting climate-resilient crop varieties, and implementing water conservation measures. For example, AI-powered tools like ClimateAI provide farmers with actionable insights to optimise their operations in the face of changing climate conditions.

In addition to prediction, AI supports disaster preparedness by identifying areas at risk of droughts, floods, or other extreme events. By analysing geospatial data, AI algorithms can recommend infrastructure improvements, such as irrigation systems or flood barriers, to protect agricultural lands and ensure food security. These proactive measures reduce vulnerability and promote resilience, particularly in regions that are heavily dependent on agriculture.

AI also contributes to sustainable water management, a critical component of climate-resilient agriculture. Machine Learning models analyse data on rainfall, soil moisture, and crop water requirements to optimise irrigation practices. By ensuring that water is used efficiently, AI helps farmers maintain productivity while conserving a vital resource. Through these applications, AI enhances the ability of food systems to withstand climate challenges, supporting SDG 2s goal of sustainable agriculture.

Furthermore, by combining AI and greenhouse innovation, farming in dry climates can be revolutionised. The smart farming infrastructure comprises a controlled environment (greenhouse), drones, measurements, data analytics, and precision agriculture. The result is energy-efficient, profitable, and sustainable food production through low-cost innovative inputs and high crop yields. This strategy has turned a desert into a green belt in the UAE's Al Ain region.

12.3.6 Improving Nutrition Outcomes

Hunger is not only about the quantity of food but also its quality. Malnutrition affects millions of people, particularly in low-income regions with limited access to nutritious food. AI offers innovative solutions to improve nutrition outcomes by enabling personalised dietary recommendations, fortifying food systems, and addressing micronutrient deficiencies. AI-powered platforms analyse individual health data, dietary preferences, and nutritional needs to provide personalised meal plans and recommendations. For instance, apps like Nutrino and EatLove use AI algorithms to create customised diets that promote balanced nutrition. These tools are particularly valuable for addressing specific nutritional needs in vulnerable populations, such as children, pregnant women, and the elderly.

AI also supports the development of nutrient-enriched crops through precision breeding and genetic analysis. Machine Learning models identify genetic traits associated with higher nutrient content, enabling researchers to develop biofortified crops such as iron-rich beans or vitamin A-enriched rice. These innovations improve the nutritional value of staple foods, addressing micronutrient deficiencies and enhancing public health.

In addition, AI facilitates the monitoring and evaluation of nutrition programmes by analysing data on food distribution, consumption patterns, and health outcomes. By providing insights into the effectiveness of interventions, AI helps governments and organisations design more targeted and impactful nutrition strategies. Through these applications, AI contributes to SDG 2s mission to end all forms of malnutrition and ensure that everyone has access to nutritious food.

12.3.7 Challenges and Ethical Considerations

While AI offers significant potential for advancing SDG 2, its implementation in food systems raises ethical considerations and challenges. One primary concern is data accessibility and inclusivity. AI-driven solutions often rely on high-quality data, which may be limited in low-income regions with poor infrastructure. Addressing this challenge requires investment in data collection and capacity building to ensure that AI benefits are equitably distributed.

Another concern is algorithmic bias, which can lead to unequal access to resources or services. For example, biased data in predictive models may overlook the needs of marginalised communities, exacerbating inequalities in food access. Ensuring fairness in AI design requires diverse and representative datasets and regular audits to identify and address biases. Data privacy is also critical, particularly in applications involving personal health or agricultural data. Robust data protection frameworks are essential to safeguard individuals' privacy and maintain trust in AI-driven solutions. Additionally, the environmental impact of AI itself, such as the energy required for training large models, must be considered to ensure that AI applications align with sustainability goals.

12.3.8 Policy Recommendations

To maximise AI's potential for achieving SDG 2, governments, international organisations, and private sector stakeholders must collaborate to create ethical frameworks and invest in inclusive

AI research. Policies should prioritise transparency, accountability, and equity in AI deployment, ensuring that benefits reach underserved populations.

Promoting public-private partnerships can drive innovation in AI for food systems, enabling the development of cost-effective and scalable solutions for smallholder farmers and low-income communities. These partnerships can also foster knowledge-sharing and capacity building, ensuring that all stakeholders have the tools and expertise to implement AI technologies effectively.

Investing in digital infrastructure and data collection is critical for enabling AI applications in regions with limited resources. By improving access to technology and data, governments and organisations can ensure that AI-driven solutions are accessible to all, supporting global efforts to end hunger.

12.3.9 Conclusion

Artificial Intelligence offers the transformative potential to advance SDG 2 by enhancing agricultural productivity, optimising food supply chains, reducing waste, building climate resilience, and improving nutrition outcomes. From precision farming and predictive analytics to personalised nutrition and supply chain transparency, AI provides innovative solutions to address the complex challenges of hunger and malnutrition. However, ethical considerations such as data privacy, algorithmic bias, and accessibility must be carefully managed to ensure that AI serves as a force for equitable and sustainable development. By fostering collaboration, investing in digital education, and promoting inclusive AI policies, we can harness AI's capabilities to achieve Zero Hunger and create a world where everyone has access to sufficient, nutritious, and sustainably produced food. This section outlines AI's role in achieving SDG 2, focusing on its applications in agriculture, food systems, and nutrition while addressing challenges and recommendations for ethical and inclusive AI implementation in the fight against hunger.

12.4 AI and SDG 3 (Good Health and Well-Being)

UN SDG 3 (Good Health and Well-Being) aims to ensure healthy lives and promote well-being for all ages. Despite significant progress in global health, challenges such as communicable and non-communicable diseases, limited healthcare access, and emerging pandemics persist. Artificial Intelligence (AI) has emerged as a transformative force in healthcare, offering innovative solutions for disease prevention, diagnosis, treatment, and health system optimisation. This section explores the role of AI in achieving SDG 3 by enhancing early disease detection, personalising treatment, improving healthcare delivery, supporting public health surveillance, and promoting equitable access to health services. It also discusses the challenges, ethical considerations, and future directions for AI in advancing global health.

12.4.1 Introduction

Health is a cornerstone of sustainable development, directly influencing economic growth, education, and overall well-being. However, achieving universal health coverage and addressing health disparities remain significant challenges. SDG 3 underscores the importance of equitable access to healthcare, the prevention and treatment of diseases, and the reduction of premature mortality. With its ability to process vast datasets, identify patterns, and make predictions, AI offers powerful tools to transform healthcare systems and improve health outcomes worldwide.

This section examines AI's applications in global health, focusing on its contributions to disease prevention, diagnosis, treatment, public health monitoring, and healthcare accessibility. By leveraging AI's potential, we can address systemic health challenges and accelerate progress towards achieving SDG 3.

12.4.2 Early Disease Detection and Prevention

Early detection of diseases is critical for improving health outcomes, as it allows for timely intervention and treatment. AI excels in this area by analysing medical data, such as imaging, genetic information, and patient histories, to accurately identify early warning signs of diseases.

One prominent application of AI is in medical imaging. Machine Learning algorithms analyse X-rays, MRIs, and CT scans to detect abnormalities that may indicate conditions such as cancer, cardiovascular disease, or neurological disorders. For example, AI systems like Google's DeepMind Health have demonstrated the ability to identify breast cancer in mammograms with greater accuracy than human radiologists. By enabling earlier and more accurate diagnoses, AI reduces the burden of late-stage disease and improves survival rates.

AI also supports disease prevention by identifying individuals at risk of developing chronic conditions. Predictive models analyse lifestyle factors, genetic predispositions, and environmental exposures to assess risk levels and recommend preventive measures. For instance, AI-driven platforms like HeartFlow assess cardiovascular risk based on imaging and patient data, guiding clinicians in designing personalised prevention plans. These tools empower healthcare providers and patients to take proactive steps to mitigate health risks, aligning with SDG 3s focus on preventive care.

In the context of infectious diseases, AI is instrumental in outbreak prediction and surveillance. Machine Learning models analyse data from environmental sensors, social media, and travel patterns to predict the spread of diseases such as malaria, dengue fever, or COVID-19. For example, BlueDot, an AI-powered platform, identified the emergence of COVID-19 before it was officially reported. By providing early warnings, AI enables governments and health organisations to implement preventive measures, reducing the impact of epidemics.

12.4.3 Personalised Medicine and Treatment

Personalised medicine tailors treatment to individual patients based on their unique characteristics, such as genetic makeup, lifestyle, and medical history. AI is at the forefront of this revolution, offering tools to analyse complex data and guide clinicians in developing targeted therapies.

AI-powered genomic analysis is a key driver of personalised medicine. Machine Learning algorithms identify genetic variations associated with diseases, enabling the development of precision therapies. For example, AI tools have been used to identify mutations in cancer genomes, guiding oncologists in selecting treatments that target specific genetic abnormalities. Platforms like Tempus and Foundation Medicine leverage AI to provide actionable insights from genomic data, improving treatment efficacy and reducing adverse effects. Figure 12.4 represents a visualisation of a robot android hand holding cyber DNA.

In addition to genomics, AI supports personalised treatment by analysing patient data to predict drug responses and optimise dosages. Machine Learning models assess how individual patients metabolise medications, enabling clinicians to avoid underdosing or overdosing. This approach is

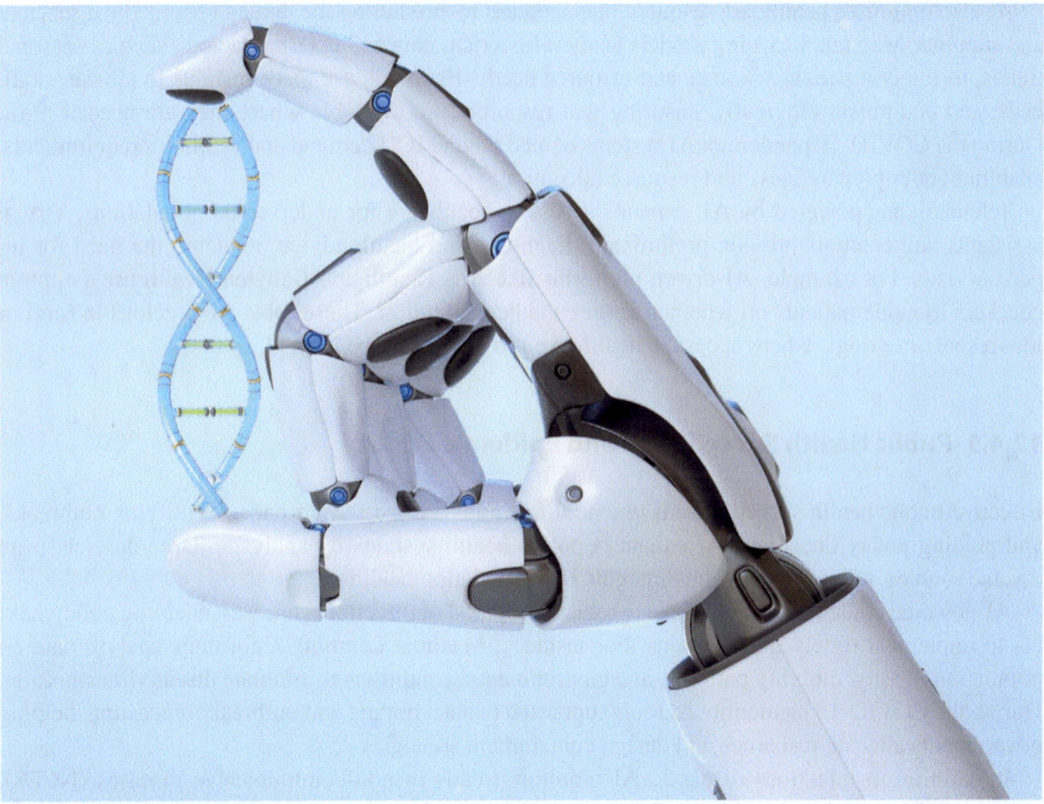

Fig. 12.4 Robot android hand holding cyber DNA. (Image credit: iStock ilexx.)

particularly valuable in managing conditions like diabetes or hypertension, where precise dosing is critical for achieving optimal outcomes.

AI-driven clinical decision support systems further enhance treatment by providing evidence-based recommendations. These systems analyse medical literature, clinical guidelines, and patient data to assist clinicians in selecting the most appropriate interventions. For example, IBM Watson Health uses AI to suggest cancer treatment options based on individual patient profiles, supporting personalised care.

12.4.4 Improving Healthcare Delivery

AI has the potential to address inefficiencies in healthcare delivery by streamlining workflows, reducing administrative burdens, and optimising resource allocation. These improvements enhance the quality of care while expanding access to health services.

One significant application of AI is automating administrative tasks, such as scheduling, billing, and documentation. AI-powered systems handle repetitive tasks, freeing healthcare providers to focus on patient care. For example, Natural Language Processing (NLP) tools transcribe and summarise medical notes, reducing the time clinicians spend on documentation. This increases productivity and allows for more patient-centred care.

AI also optimises healthcare resource management by predicting the demand for medical services and supplies. Machine Learning models analyse historical data and external factors, such as seasonal trends, to forecast patient volumes and resource needs. Hospitals use these insights to allocate staff, beds, and equipment efficiently, ensuring that resources are available where they are needed most. During the COVID-19 pandemic, AI systems helped predict ICU demand and ventilator requirements, enabling better preparedness and resource allocation.

Telemedicine, powered by AI, expands access to healthcare for underserved populations. Virtual assistants and chatbots provide preliminary diagnoses and health advice, reducing the need for in-person visits. For example, AI-driven platforms like Ada Health and Babylon Health use symptom checkers to guide patients on whether to seek medical attention. These tools are precious in rural or low-resource settings, where access to healthcare providers may be limited.

12.4.5 Public Health Surveillance and Epidemic Control

Effective public health surveillance is essential for monitoring disease trends, identifying outbreaks, and guiding policy decisions. AI enhances public health systems by analysing large datasets from diverse sources, providing real-time insights into population health.

AI-powered epidemiological models predict the spread of infectious diseases, enabling policymakers to implement timely interventions. For instance, Machine Learning algorithms analyse data on population density, mobility patterns, and environmental conditions to simulate disease transmission. During the COVID-19 pandemic, AI tools supported contact tracing and outbreak forecasting, helping governments allocate resources and design containment strategies.

In addition to infectious diseases, AI monitors trends in non-communicable diseases (NCDs), such as diabetes and heart disease. Predictive models analyse data from electronic health records, wearable devices, and social determinants of health to identify high-risk populations. These insights inform public health campaigns and resource allocation, ensuring that interventions target the most vulnerable groups.

AI also supports vaccine development and distribution by accelerating research and optimising logistics. Machine Learning models analyse clinical trial data to identify promising vaccine candidates, reducing the time required for growth. In vaccine distribution, AI-powered systems manage cold chain logistics, ensuring that vaccines reach their destinations safely and efficiently. These innovations contribute to global immunisation efforts, aligning with SDG 3s goal of preventing and controlling diseases.

12.4.6 Advances in Biological Research

AI is increasingly recognised as a transformative tool in biological research, offering unprecedented capabilities in data analysis and predictive modelling. Advances in this field will dramatically and directly impact SDG 3 (Good Health and Well-Being), as well as assist in the attainment of other related goals such as SDG 2 (Zero Hunger), SDG 9 (Industry, Innovation, and Infrastructure), and SDG 15 (Life on Land). Consequently, it is prudent to flesh out the various applications of AI across different areas of biological research, demonstrating the significant potential enhancements

The first significant application is in genomic sequencing and analysis. AI algorithms are adept at processing vast amounts of genomic data, identifying genetic variations and mutations. This capability is crucial for unravelling the genetic underpinnings of diseases and can expedite the discovery of new therapeutic targets, ultimately advancing personalised medicine.

In the realm of drug discovery and design, AI proves invaluable by accelerating the identification of viable drug candidates through predictive modelling of drug-target interactions and analysis of biological pathways. This application speeds up the drug discovery process and reduces the costs associated with pharmaceutical research, making the development of new treatments more efficient.

Protein structure prediction is another area where AI makes a critical impact. Deep Learning models assist in predicting the three-dimensional structures of proteins, which is essential for understanding their functions and interactions. Such insights are key to drug design, as they allow for the development of molecules that can interact precisely with specific proteins. Indeed, there has been a revolutionary application of AI in protein structure prediction using Deep Learning models. This is essential for understanding protein functions, thus enabling the design of drugs that target specific proteins. The leading AI innovation in this endeavour is called AlphaFold. DeepMind, a subsidiary of Alphabet Inc. (Google's parent company), developed this AI system.

AlphaFold is specifically designed to predict the structure of proteins based on their amino acid sequences. The technology gained significant attention and acclaim for its groundbreaking performance in the field of protein folding. In particular, it demonstrated remarkable accuracy in predicting the 3D structures of proteins during the 2020 Critical Assessment of Protein Structure Prediction competition, outperforming other methods by a substantial margin. AlphaFold has since been used to predict 3D shapes of 200 million proteins, creating a vast protein structure database. It also developed a novel molecular syringe to deliver drugs at the molecular level.

AI also plays a pivotal role in disease diagnosis and biomarker discovery. By analysing diverse datasets, AI can identify patterns and biomarkers that might not be visible to human analysts. This capability enhances the diagnosis and prognosis of diseases, facilitating early intervention and tailored treatment plans.

Clinical trial optimisation benefits from AI through more efficient patient recruitment and the prediction of treatment outcomes. AI models can analyse historical data and ongoing trial results to refine recruitment criteria and trial protocols, enhancing the efficacy and efficiency of clinical studies.

In high-throughput experimental settings, AI aids in high-throughput screening by automating the analysis of large datasets to identify promising compounds. This automation speeds up the screening process and increases its accuracy, reducing the likelihood of false positives and negatives.

The use of AI in image analysis in microscopy transforms the way biological images are processed and analysed. AI algorithms can perform complex tasks such as segmenting cells, quantifying phenotypic characteristics, and identifying morphological patterns in tissue samples, which are critical for both basic research and clinical diagnostics.

AI's application extends to metagenomics and microbiome analysis, where it helps decipher complex data from microbial ecosystems. AI tools analyse sequences and functional data to understand microbial diversity and its role in health and disease, providing insights crucial for developing interventions to modify the microbiome.

Lastly, AI facilitates the analysis of biological networks by helping researchers map and understand complex interactions within biological systems, such as metabolic pathways, genetic regulation networks, or protein-protein interaction networks. This analysis is fundamental in revealing new therapeutic targets and understanding disease mechanisms.

Overall, the integration of AI into biological research has not only enhanced the scope and accuracy of scientific studies but has also accelerated the pace of discoveries, opening up new possibilities for medical and environmental advancements. By leveraging AI, the biological research community is better equipped to tackle complex biological questions, leading to deeper insights and more effective interventions. These advancements can be leveraged to achieve SDG 3, with an indirect impact on SDGs 2, 9, and 15.

12.4.7 Promoting Health Equity

Health disparities remain a significant barrier to achieving SDG 3, as access to healthcare often varies based on geographic, socio-economic, and demographic factors. AI has the potential to bridge these gaps by enabling equitable access to health services and addressing social determinants of health.

Telemedicine, supported by AI, expands access to healthcare in remote and underserved areas. Virtual consultations and AI-powered diagnostic tools allow patients to receive care without travelling long distances. For example, AI-driven mobile health units equipped with diagnostic devices bring healthcare services to rural communities, addressing disparities in access. AI also identifies and addresses social determinants of health, such as poverty, education, and housing. Machine Learning models analyse data on socio-economic factors to predict health outcomes and guide interventions. For instance, AI algorithms can identify neighbourhoods with high rates of preventable diseases, enabling governments to implement targeted public health programmes. By addressing the root causes of health disparities, AI promotes health equity and supports SDG 3s focus on universal health coverage.

Furthermore, AI enables cost-effective healthcare delivery by automating processes and reducing inefficiencies. This lowers the cost of care, making health services more affordable for low-income populations. AI-driven platforms also facilitate access to affordable medications by analysing supply chain data and recommending cost-saving measures.

12.4.8 Challenges and Ethical Considerations

While AI offers transformative potential in healthcare, its implementation raises ethical considerations and challenges. Data privacy is a primary concern, as AI systems rely on sensitive medical information. Ensuring compliance with data protection regulations, such as GDPR and HIPAA, is essential to safeguard patient confidentiality and maintain trust in AI-driven healthcare.

Algorithmic bias is another significant issue, as biased data can lead to unequal outcomes. For example, AI models trained on data from predominantly high-income populations may not perform well in low-resource settings. Addressing bias requires diverse and representative datasets and rigorous testing to ensure fairness and inclusivity.

Transparency and accountability in AI decision-making are also critical. Clinicians and patients must understand how AI systems generate recommendations, particularly in high-stakes situations. Developing explainable AI models and establishing clear guidelines for AI use in healthcare are essential for building trust and ensuring ethical practice.

Finally, the digital divide poses a challenge to equitable AI adoption, as low-income and rural communities may lack the infrastructure needed to access AI-driven healthcare services. Addressing this issue requires investment in digital infrastructure and training programmes to ensure that all populations benefit from AI advancements.

12.4.9 Policy Recommendations

To maximise AI's potential for advancing SDG 3, governments, healthcare organisations, and technology developers must collaborate to establish ethical frameworks and invest in inclusive AI research. Policies should prioritise transparency, accountability, and equity in AI deployment, ensuring that innovations benefit all populations.

Promoting public-private partnerships can drive innovation and scale AI solutions for global health. These partnerships can fund research, develop affordable technologies, and foster knowledge-sharing to address global health challenges. Investing in digital literacy and workforce training is essential for equipping healthcare providers with the skills needed to use AI technologies effectively. By integrating AI into medical education and public health curricula, we can prepare the next generation of health professionals to leverage AI for improved outcomes.

Finally, fostering global cooperation in AI research and deployment can ensure that AI-driven healthcare solutions address the needs of diverse populations. Collaborative initiatives can promote data sharing, ethical standards, and inclusive innovation, accelerating progress towards SDG 3.

12.4.10 Conclusion

Artificial Intelligence offers transformative potential to achieve SDG 3 by enhancing disease prevention, personalising treatment, improving healthcare delivery, and promoting health equity. AI provides innovative solutions to address global health challenges, from early diagnosis and predictive analytics to telemedicine and public health surveillance. However, ethical considerations such as data privacy, algorithmic bias, and equitable access must be carefully managed to ensure that AI serves as a force for good. By fostering collaboration, investing in digital education, and promoting inclusive AI policies, we can harness AI's capabilities to improve health outcomes and achieve universal well-being, bringing us closer to the vision of SDG 3.

This section highlights AI's transformative role in achieving SDG 3. It offers insights into its applications in healthcare, addresses ethical challenges, and provides recommendations for inclusive and sustainable AI implementation in global health.

12.5 AI and SDG 4 (Quality Education)

The UN SDG 4 aims to ensure inclusive and equitable quality education and promote lifelong learning opportunities for all. Education is fundamental to the well-being of individuals, communities, and societies at large. However, despite significant progress, global educational challenges persist, including unequal access to quality education, teacher shortages, outdated pedagogical models, and lack of infrastructure, especially in the Global South. The advent of AI has the potential to address these challenges by revolutionising teaching and learning processes, improving educational outcomes, and scaling access to quality education. This section explores how AI can contribute to achieving SDG 4, focusing on the various applications of AI in education, including personalised learning, accessibility, teacher training, curriculum development, and the challenges and ethical considerations of AI in education.

12.5.1 Personalised Learning

One of the most significant applications of AI in education is the development of personalised learning systems. Personalised learning tailors the educational experience to individual students, helping them learn at their own pace and according to their unique needs and preferences. AI enables personalised learning by analysing large datasets of student performance and behaviour, allowing it to adapt content and assessments to the needs of each learner.

Fig. 12.5 3D rendering robot learning anchored in problem-solving. (Image credit: Shutterstock phonlamai)

AI-powered adaptive learning platforms, such as DreamBox and Knewton, use algorithms to adjust the content and difficulty of lessons based on real-time student performance. For example, if a student struggles with a particular concept, the system provides additional resources and practice to help them master the topic before moving on to more advanced material. This helps ensure that no student is left behind or forced to move ahead before they have fully understood the material. AI-based tutoring systems, such as Squirrel AI, provide students with personalised one-on-one tutoring sessions, helping to bridge learning gaps. These systems use Machine Learning algorithms to assess a student's strengths and weaknesses, providing targeted exercises, explanations, and feedback that are tailored to the student's learning style and progress. This individualised attention can supplement traditional classroom instruction and provide more effective learning experiences, particularly in resource-limited settings. Indeed, there is scope for using chatbots and humanoid robots to deliver quality education. Figure 12.5 illustrates 3D rendering robot learning anchored in problem-solving.

AI enables data-driven insights for teachers. AI can also help teachers personalise their teaching methods by providing them with detailed insights into student performance. AI-driven platforms can analyse students' responses to quizzes, tests, and assignments, allowing educators to identify patterns and predict which students might need additional support. Teachers can use this information to provide targeted interventions and adjust their teaching strategies to meet the needs of individual students.

12.5.2 Scaling Access to Education

Access to quality education remains a significant challenge, especially in rural and underserved areas. AI can play a critical role in overcoming barriers to education and providing scalable solutions for reaching marginalised communities and remote regions.

AI can enable remote learning for students in areas without access to physical schools or teachers. Platforms like Khan Academy, Coursera, and edX offer online courses that are accessible to anyone with an Internet connection. AI helps enhance these platforms by providing personalised learning experiences and recommending courses based on the learner's interests and goals. In regions with limited access to quality teachers, AI can provide a solution by delivering high-quality educational content through virtual classrooms.

Mobile phones are widely used in many developing countries, and AI can help leverage this infrastructure to provide learning opportunities on the go. AI-powered mobile applications like Duolingo for language learning and Google Classroom for organising and delivering educational content enable students to access learning materials, collaborate with peers, and receive feedback directly through their smartphones. These applications can be particularly useful in low-resource environments lacking traditional educational infrastructure.

AI-driven language translation tools, such as Google Translate, can break down language barriers by providing instant translation services for educational content. This is particularly important in multilingual societies, where students may not speak the language of instruction. AI can automatically translate textbooks, articles, and instructional materials into various languages, ensuring students in diverse linguistic contexts can access quality learning resources.

12.5.3 Enhancing Teacher Training and Development

Teachers are the cornerstone of education systems, but there is a global shortage of qualified teachers, particularly in underserved regions. AI can help address this shortage by improving teacher training and professional development. AI has proven to be a powerful tool for professional advancement. AI-powered platforms can provide teachers with personalised professional development opportunities. These platforms assess teachers' knowledge, teaching style, and classroom challenges to recommend training materials and resources most relevant to their needs. For example, Teachers Pay Teachers and Edmodo offer professional development courses that use AI to tailor content to individual teachers, enabling them to improve their skills at their own pace.

AI can also assist in preparing teachers for real classroom scenarios. Virtual reality (VR) and AI-based simulations allow teachers to practise classroom management, lesson delivery, and student engagement strategies in a controlled environment. Using AI to simulate different classroom situations, teachers can build confidence and develop the necessary skills to handle various challenges before entering real-world classrooms.

AI can help reduce teachers' administrative burden by automating routine tasks such as grading, attendance tracking, and lesson planning. For instance, AI-driven grading systems can quickly and accurately grade multiple-choice exams and essays, giving teachers more time to focus on instruction and student interaction. Automating administrative tasks also enables teachers to spend more time on personalised instruction and student support.

12.5.4 Supporting Lifelong Learning

As the global workforce evolves, lifelong learning becomes increasingly important. AI can play a key role in supporting individuals throughout their careers by providing access to continuous education and skill development. AI-powered platforms like LinkedIn Learning, Udacity, and Coursera offer a wide range of courses for adults seeking to upgrade their skills or switch careers. AI tailors course recommendations based on an individual's learning history, career goals, and the current job market, ensuring learners receive the most relevant content for personal and professional growth.

In addition to traditional academic education, AI can support vocational education and skill development. AI-driven platforms can offer personalised training in industries such as healthcare, technology, agriculture, and manufacturing. By providing customised learning paths and real-time feedback, AI ensures that individuals acquire practical, job-relevant skills that meet the needs of the labour market.

12.5.5 Promoting Inclusivity and Accessibility

AI has the potential to address educational inequalities and promote inclusivity by providing support to marginalised groups, including students with disabilities and girls and women in underserved regions.

AI-powered assistive technologies can help students with disabilities participate more fully in education. For example, speech-to-text and text-to-speech technologies, such as Microsoft's Seeing AI, help visually impaired students access written materials. AI-driven tools can also provide real-time captioning and sign language interpretation, making learning more accessible to students with hearing impairments.

AI can enhance gender equality in education. It can help bridge gender gaps by identifying barriers that prevent girls and women from accessing quality education. AI-powered analytics can track enrolment, retention, and graduation rates by gender, enabling policymakers and educators to implement targeted interventions that support girls' education. AI-based applications, such as SheTrades by the International Trade Centre, help empower women by providing educational resources and skill-building tools to improve their employability and entrepreneurial opportunities.

12.5.6 Ethical Considerations in AI for Education

While AI presents significant opportunities for transforming education, its implementation raises several ethical considerations that must be addressed to ensure fairness, equity, and transparency in education systems.

As discussed earlier, AI systems are only as good as the data on which they are trained. If the data is biased, the resulting AI algorithms may perpetuate or exacerbate existing inequalities in education. For example, AI-based admissions systems or grading algorithms may unintentionally favour certain demographic groups over others. To mitigate this risk, developers must ensure that AI systems are trained on diverse and representative data, and algorithms should be regularly audited for fairness and transparency.

AI systems in education rely heavily on student data, including personal information, academic performance, and learning habits. This data must be handled securely and in compliance with data privacy regulations, such as the General Data Protection Regulation (GDPR). Educational institutions must prioritise protecting student privacy while leveraging AI to enhance learning outcomes.

As AI solutions become increasingly integrated into education systems, students in low-income or rural areas could be left behind due to a lack of access to the necessary technology. To ensure that AI benefits all students, governments and international organisations must work to provide equitable access to digital tools and infrastructure.

12.5.7 Conclusion

AI has the potential to transform education by making learning more personalised, scalable, inclusive, and efficient. From adaptive learning platforms to teacher training and lifelong learning, AI can contribute significantly to achieving SDG 4 by improving access to quality education and promoting lifelong learning opportunities for all. However, to fully realise the potential of AI in education, it is essential to address the ethical challenges of data privacy, algorithmic bias, and digital inequality. By carefully implementing AI in education, we can create more equitable, inclusive, and high-quality learning environments for all. As we move towards 2030, AI will be a key driver in achieving the goal of inclusive and equitable quality education for all.

12.6 AI and SDG 5 (Gender Equality)

12.6.1 Introduction

UN SDG 5 aims to achieve gender equality and empower all women and girls. It recognises the persistent gender disparities that exist globally in areas such as access to education, economic participation, healthcare, political representation, and protection from violence. Despite progress, these gaps remain deeply entrenched, particularly in developing regions, where women and girls face cultural, economic, and institutional barriers that restrict their ability to participate in society fully.

With its transformative capabilities, AI presents an opportunity to accelerate progress towards SDG 5. By harnessing AI in areas like data collection, resource allocation, decision-making, and advocacy, we can empower women and girls, tackle gender inequalities, and create more inclusive societies. This section explores the ways AI can contribute to achieving gender equality, focusing on the key areas of education, healthcare, economic empowerment, violence prevention, and political participation while also addressing the challenges and ethical concerns of AI in gender equality.

12.6.2 AI for Gender Equality in Education

Education is a fundamental human right and key to achieving gender equality. However, girls and women face significant barriers to accessing quality education, particularly in rural and conflict-affected regions. AI has the potential to remove these barriers by improving access to educational content, personalising learning experiences, and providing data-driven insights to inform policy decisions.

AI-powered personalised learning platforms can tailor educational content to students' specific needs and learning styles. AI systems can provide individualised learning support in regions with limited access to qualified teachers. For example, platforms like Khan Academy and Duolingo use AI to adapt lessons based on students' performance, enabling girls to learn at their own pace. This is especially beneficial in overcoming barriers such as overcrowded classrooms or cultural norms that may restrict girls' access to formal education.

AI can be leveraged to identify and address gender biases in textbooks, curricula, and learning materials. By analysing large datasets of educational content, AI systems can detect language or imagery that perpetuates stereotypes or marginalises women and girls. Tools like Reverso Context and Gender Decoder can assist educators and content creators in developing inclusive learning materials that reflect diverse and positive portrayals of both genders. This helps to foster gender-neutral

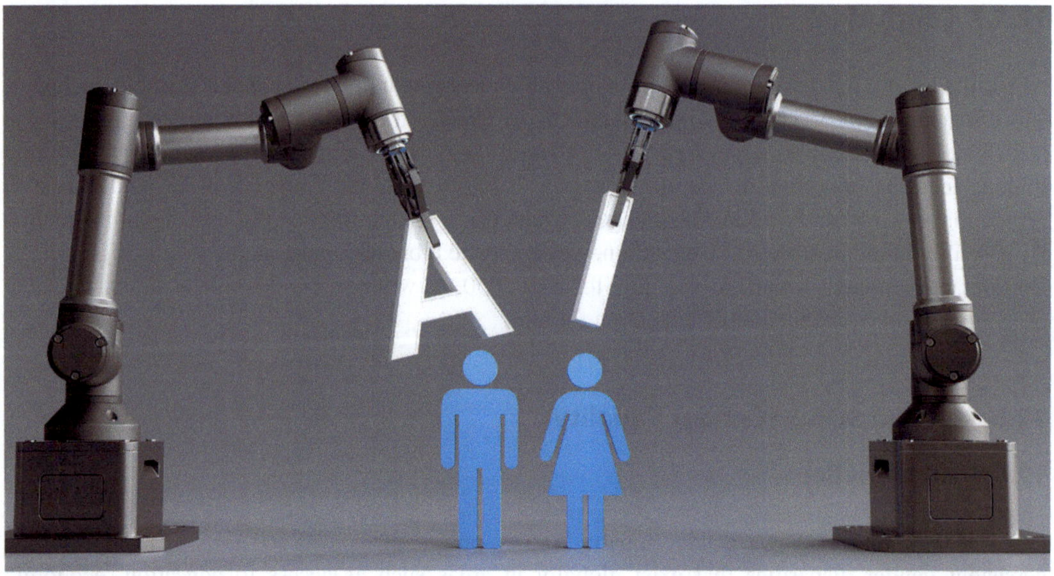

Fig. 12.6 AI can reduce gender inequality. (Image credit: iStock sarawuth702)

education environments that promote equality from an early age. Indeed, AI deployed effectively can reduce gender inequality as symbolically illustrated in Fig. 12.6.

AI-driven e-learning platforms are instrumental in breaking down geographical and logistical barriers that hinder girls' education in remote or underdeveloped areas. With mobile penetration rising globally, AI-powered mobile applications can offer education on the go. For example, AI tutors such as Squirrel AI and Cognii provide personalised guidance, assessments, and tutoring in subjects such as math, science, and literacy, ensuring that girls who cannot attend traditional schools still have access to quality learning resources.

12.6.3 Healthcare for Women and Girls

Healthcare is another area where AI can significantly contribute to gender equality by improving access to quality health services, particularly for women and girls who face systemic barriers to healthcare access. AI can help identify gender-specific health risks, provide personalised treatments, and offer better maternal and reproductive health services.

AI technologies can improve maternal health outcomes by predicting and managing risks during pregnancy and childbirth. AI-powered tools like PathAI and Aidoc analyse medical imaging to detect complications such as preeclampsia, gestational diabetes, and foetal distress, allowing for earlier interventions. In regions where access to skilled healthcare providers is limited, these tools can provide crucial support in diagnosing and managing high-risk pregnancies, ultimately reducing maternal mortality.

AI can also enhance access to sexual and reproductive health information, empowering women and girls to make informed decisions about their health. Platforms like Clue and Flo Health use AI to help women track menstrual cycles, predict ovulation, and monitor reproductive health. These tools provide important insights into women's health, offering tailored advice on contraception, pregnancy, and fertility.

AI can be used to better understand and address gender-specific health issues by analysing large datasets of health records and genetic information. Many medical studies have historically excluded women or failed to account for gender differences in health outcomes. AI-driven analytics can ensure that research is more inclusive by identifying health trends and disparities that disproportionately affect women, such as autoimmune diseases or mental health disorders.

12.6.4 Economic Empowerment and Workforce Participation

Economic empowerment is central to achieving gender equality. AI can drive economic inclusion for women by improving access to jobs, creating new business opportunities, and promoting financial independence. However, to fully realise its potential, AI must be designed to address the unique challenges women face in the labour market, such as the gender pay gap, employment discrimination, and lack of financial literacy.

AI can assist women in finding job opportunities and improving their skills. For example, platforms like LinkedIn use AI algorithms to match women with job openings based on their skills, experiences, and career goals. AI-driven recruitment tools can also reduce bias in hiring by providing employers with data-driven insights and helping them assess candidates based on their qualifications rather than gender. Additionally, AI can help women access training and reskilling programmes that align with current labour market needs, such as tech and digital skills courses offered through platforms like Coursera and Udemy.

Financial exclusion is a significant barrier for women, particularly in developing regions, where women often lack access to formal financial services. AI-driven financial technologies, or Fintech, can enhance financial inclusion by providing women with access to banking, credit, and insurance services. AI systems can analyse alternative data, such as mobile phone usage patterns or social media activity, to assess the creditworthiness of women who may not have traditional credit histories. Services like Kiva and Grameen Bank use AI to offer microloans to women entrepreneurs, enabling them to start or expand businesses and achieve economic independence.

AI tools also help women entrepreneurs by providing personalised business advice, market insights, and access to funding. AI-driven platforms like Shopify and Wix help women launch and manage online businesses by offering tools for product recommendation, inventory management, and customer engagement. Additionally, AI platforms that analyse market trends and consumer behaviour, such as IBM Watson Analytics, can provide women entrepreneurs with valuable insights into market demand and competition, helping them make informed decisions to grow their businesses.

12.6.5 Preventing Violence Against Women and Girls

One of the key priorities of SDG 5 is to eliminate violence against women and girls. AI technologies can support this goal by helping prevent, detect, and respond to violence in real time while also providing resources for survivors of abuse. AI can play a pivotal role in preventing violence by identifying early warning signs and enabling timely interventions. For instance, AI systems can analyse social media posts, phone calls, and other digital footprints to detect patterns that may indicate domestic violence, harassment, or trafficking. AI-powered platforms like Benevolent and SafeCity collect data from public spaces and online platforms to identify high-risk areas for sexual harassment or violence, allowing law enforcement and social services to intervene proactively.

AI-powered chatbots, such as Ada and Wysa, provide confidential and immediate support to survivors of violence. These AI systems offer emotional support, information about legal rights, and guidance on accessing healthcare and counselling services. AI can also help survivors by connecting them to emergency services and shelters, offering anonymity and privacy during a critical time.

AI can play a significant role in gender policy and advocacy. The technology can assist organisations in understanding the scale and scope of gender-based violence by analysing large datasets of reported incidents. By identifying trends and hotspots, AI enables policymakers and NGOs to design more effective interventions and allocate resources where they are most needed. AI-driven analytics can help monitor the effectiveness of laws and policies aimed at preventing violence against women, ensuring that these measures are evidence-based and responsive to the needs of survivors.

12.6.6 Political Participation and Leadership

Achieving gender equality in political participation is another critical aspect of SDG 5. AI can enhance women's political engagement by providing them with platforms for advocacy, improving their access to political resources, and supporting their leadership ambitions.

AI can help amplify women's voices in politics by identifying potential candidates, increasing voter engagement, and promoting gender-inclusive political campaigns. AI-powered platforms like Revolutio use data analytics to assess public opinion, enabling female politicians to tailor their messaging to resonate with voters and address their concerns effectively. Furthermore, AI can help reduce gender bias in political discourse by monitoring media coverage and social media platforms for gender stereotypes and misrepresentation.

AI can also support women's rights organisations in advocating for policy changes by providing insights into public attitudes, tracking legislation, and identifying opportunities for intervention. AI-powered tools that analyse social media data, such as Brandwatch and Crimson Hexagon, can help organisations gauge public sentiment and mobilise support for gender equality initiatives.

12.6.7 Challenges and Ethical Considerations

While AI presents immense opportunities for achieving gender equality, its deployment must be approached with caution to avoid exacerbating existing biases or creating new forms of discrimination.

AI systems are only as unbiased as the data on which they are trained. If the data used to develop AI algorithms reflects historical gender biases, the resulting systems may perpetuate these biases, leading to discriminatory outcomes. For example, AI algorithms used in hiring or lending may inadvertently favour male candidates or exclude women from financial opportunities. AI developers must use diverse, representative datasets to ensure fairness and regularly audit their algorithms for bias.

AI applications often rely on sensitive personal data, which raises concerns about privacy and security. In the context of gender equality, women may be particularly vulnerable to exploitation or misuse of their data. Therefore, it is crucial to implement strong data protection policies and ensure that women's personal information is handled securely and with their consent.

12.6.8 Conclusion

AI has the potential to accelerate progress towards achieving SDG 5, fostering gender equality and empowering women and girls in all spheres of life. From education and healthcare to economic empowerment and violence prevention, AI can help address gender disparities by providing personalised solutions, improving access to resources, and supporting the empowerment of women and girls worldwide. However, to fully harness the power of AI for gender equality, it is essential to address the challenges of algorithmic bias, privacy, and digital exclusion. By prioritising ethical AI development and inclusive design, we can create a future where AI is a powerful tool for advancing gender equality and empowering women and girls across the globe.

12.7 AI and SDG 6 (Clean Water and Sanitation)

12.7.1 Introduction

Access to clean water and sanitation is fundamental to human well-being, health, and development. UN SDG 6 aims to ensure the availability and sustainable management of water and sanitation for all by 2030. Despite significant advancements in water infrastructure, over two billion people worldwide still lack access to safely managed drinking water services, and nearly half the global population lacks access to safely managed sanitation. Furthermore, the impacts of climate change, population growth, and urbanisation have exacerbated water scarcity and environmental pollution, making it increasingly challenging to meet the demands of a growing global population.

AI, with its capacity for data analysis, predictive modelling, automation, and optimisation, offers transformative solutions for addressing the challenges associated with water management and sanitation. AI can help improve water use efficiency, reduce waste, monitor water quality, detect leaks, optimise resource allocation, and support decision-making. This section explores how AI can contribute to achieving SDG 6 by examining its applications in water management, wastewater treatment, water quality monitoring, leak detection, and the management of water-related disasters.

12.7.2 Water Resource Management

Water resource management is a critical component of achieving SDG 6. Efficient and sustainable water use is essential to meet the needs of agriculture, industry, and domestic consumption while preserving water ecosystems. AI can enhance water management by improving decision-making, optimising water distribution, and predicting future water demands. AI-based predictive modelling tools can forecast water availability by analysing historical data, climatic conditions, and hydrological patterns. By using Machine Learning algorithms to process data from weather stations, satellites, and sensors, AI can predict changes in water supply and demand in real time. These tools allow policymakers to plan water allocation more effectively, prepare for droughts or floods, and reduce the over-extraction of water resources. For instance, IBM's Water Management Solution uses AI to predict water availability and optimise irrigation, helping to ensure water resources are used efficiently, especially in water-scarce regions.

AI can revolutionise agricultural water use by enabling smart irrigation systems that optimise water usage. Traditional irrigation methods are often inefficient, leading to excessive water use or poor crop yield. AI-powered irrigation systems, such as those developed by companies like Rachio and Hydropoint, use Machine Learning to adjust irrigation schedules based on factors such as soil moisture, weather forecasts, and crop water needs. By reducing water waste and improving crop yield, these systems promote more sustainable water use in agriculture, which accounts for around 70% of global freshwater consumption.

AI can optimise the management and distribution of water within municipal systems. AI algorithms can analyse real-time data from sensors placed in water networks to identify inefficiencies and recommend improvements. For example, Veolia's AI-driven platform uses data from smart meters, sensors, and weather forecasts to predict water demand and adjust supply accordingly, ensuring that water is distributed efficiently and sustainably to urban areas, especially during peak demand periods.

Wastewater treatment is a critical aspect of SDG 6, as the safe disposal and treatment of wastewater is essential for maintaining public health and protecting water ecosystems. Traditional wastewater treatment methods are energy-intensive and often inefficient. AI has the potential to optimise wastewater treatment processes, reducing energy consumption and improving the quality of treated water.

AI can enhance the efficiency of wastewater treatment plants by optimising processes such as filtration, chemical dosing, and aeration. By continuously analysing data from sensors monitoring key parameters like pH, dissolved oxygen, and turbidity, AI can adjust treatment processes in real time to ensure that wastewater meets safety standards while minimising energy use. For example, Xylem's AI-powered solutions are designed to monitor and control various components of the wastewater treatment process, ensuring optimal performance, reducing operational costs, and improving water quality.

AI can be used to predict equipment failures and schedule maintenance in waste-water treatment plants. Machine Learning algorithms can analyse historical maintenance data, sensor readings, and operational performance to identify potential issues before they cause system breakdowns. This helps reduce downtime, optimise plant performance, and extend the lifespan of equipment. For instance, GE's Predix platform uses AI to predict failures in critical infrastructure, ensuring that wastewater treatment plants can operate without unexpected disruptions.

AI can enable real-time monitoring of water quality in wastewater treatment systems. Integrating AI with sensors and data collection systems makes it possible to continuously monitor parameters such as chemical concentrations, pathogens, and contaminants in treated water. AI algorithms can detect anomalies and alert operators to potential risks, allowing them to take timely action to correct issues. This improves the efficiency of treatment processes and ensures that treated wastewater meets health and environmental standards.

12.7.3 Leak Detection and Water Loss Prevention

Water loss due to leaks in urban water distribution systems is a significant challenge for achieving SDG 6. According to the UN, around 30% of the global water supply is lost due to leakage, which wastes valuable water resources and increases the cost of water provision. AI can help detect leaks more efficiently, saving water and reducing the cost of repairs. AI-powered leak detection systems can identify water leaks in real time by analysing data from sensors placed throughout the water distribution network. Machine Learning algorithms can identify abnormal patterns in flow rates, pressure levels, and water consumption to detect leaks before they escalate. For example, Aquasave and UtilityAI use AI to monitor and analyse water distribution systems, providing utilities with early alerts about leaks, enabling faster response times, and reducing the volume of water lost.

AI can also predict when pipelines are likely to fail by analysing historical data, sensor readings, and environmental conditions. By assessing factors such as the age and material of pipes, pressure levels, and maintenance history, AI can predict potential weaknesses in the system and recommend preventive actions. This predictive approach helps prevent leaks and breaks, ensuring the long-term sustainability of water distribution systems.

12.7.4 Water Quality Monitoring

Ensuring the quality of water is essential for public health and environmental protection. Contaminants such as heavy metals, pathogens, and chemicals pose significant risks to both drinking water and aquatic ecosystems. AI can enhance water quality monitoring by providing real-time data analysis and early warning systems to prevent water contamination.

AI-enabled sensors can continuously monitor water quality in rivers, lakes, and reservoirs, detecting contaminants such as bacteria, nitrates, and industrial pollutants. These sensors, integrated with AI algorithms, can analyse water samples in real time and provide instant feedback on the water's safety for consumption. For example, Aquasend provides AI-based sensors that detect contaminants and send alerts when water quality deteriorates. This helps authorities take immediate action to address pollution sources before they pose a risk to public health.

Machine Learning algorithms can also be used to predict water quality changes based on environmental factors such as rainfall, temperature, and industrial activity. By analysing historical data from water quality sensors, AI can forecast potential pollution events and recommend preventive measures. For example, Google's AI and Earth Engine is used for real-time monitoring of water quality, helping authorities detect harmful algal blooms or oil spills in oceans and freshwater sources before they spread.

12.7.5 Climate Change Adaptation for Water Management

Climate change seriously threatens water resources by altering precipitation patterns, increasing the frequency of extreme weather events, and exacerbating water scarcity in certain regions. AI can support climate change adaptation by helping governments and organisations better manage water resources under changing climatic conditions.

AI can be used to develop predictive models that assess the impact of climate change on water resources, enabling governments and water utilities to make informed decisions about water management. By analysing climate data, AI can help predict droughts, floods, and other extreme events that affect water availability. This information can then be used to design more resilient water systems and adapt water usage patterns in anticipation of climate change impacts.

AI can assist in forecasting future water demand based on population growth, urbanisation, and climate scenarios. By integrating data from various sources, including demographic trends, meteorological data, and historical consumption patterns, AI can predict changes in water demand and help water utilities plan for future needs. This forecasting ability helps ensure that water systems are prepared for both current and future challenges.

12.7.6 Ethical Considerations and Challenges

While AI holds significant potential to advance SDG 6, its implementation must be cautiously approached to address various challenges and ethical concerns. To start with, there are data privacy and security issues. AI systems in water and sanitation rely on collecting and analysing vast amounts of data, including information on water consumption, quality, and infrastructure. This raises concerns about the privacy and security of sensitive data, particularly in regions with limited data protection regulations. It is essential to ensure that data is handled ethically, with proper safeguards in place to protect the privacy of individuals and communities.

AI models are only as good as the data they are trained on. If the data used to train AI systems is incomplete, unrepresentative, or biased, the resulting models may not accurately reflect the needs of all populations. For example, AI-driven water management systems may not account for the specific needs of marginalised communities or rural populations. It is crucial to ensure that AI models are designed to be inclusive and equitable, taking into account the diverse needs of different populations.

Access to AI technology may be limited in developing countries and rural areas, where resources for implementing advanced technologies may be scarce. Promoting equitable access to AI tools and infrastructure is essential to ensure that AI benefits all populations, particularly in regions where water scarcity and sanitation challenges are most acute.

12.7.7 Conclusion

Artificial Intelligence holds immense potential to advance the achievement of SDG 6 by improving the management of water resources, optimising wastewater treatment, detecting leaks, enhancing water quality monitoring, and adapting to the challenges posed by climate change. By leveraging AI, we can develop smarter, more efficient systems for water distribution and wastewater treatment, reduce water wastage, and improve access to safe and clean water for all. However, to fully realise the potential of AI in water and sanitation, it is essential to address challenges related to data privacy, bias, and equitable access to technology. With careful planning and ethical implementation, AI can be a powerful tool in building a more sustainable and equitable future for all.

12.8 AI and SDG 7 (Affordable and Clean Energy)

UN SDG 7 (Affordable and Clean Energy) seeks to ensure universal access to affordable, reliable, and modern energy services, increase the share of renewable energy in the global energy mix, and improve energy efficiency. AI has emerged as a powerful tool to achieve these goals in a world facing rising energy demands and pressing environmental concerns. By optimising energy generation, enhancing grid management, increasing renewable energy adoption, improving energy efficiency, and expanding energy access, AI can play a transformative role in making clean energy more accessible and sustainable. This section explores how AI applications contribute to achieving SDG 7 through innovations in energy management, grid optimisation, renewable energy forecasting, smart energy consumption, and expanding access to clean energy in underserved regions. Additionally, it examines the ethical considerations and challenges associated with implementing AI in the energy sector.

12.8.1 Introduction

The starting point is acknowledging the complex two-way interplay between AI and energy. Developing and running AI systems such as LLMs require massive amounts of energy. Hence, AI is the reason why there is a need for more energy globally. In fact, the AI energy requirement is so huge and critical that some AI companies are procuring or building nuclear power plants to improve their access to energy. On the other hand, AI systems are part of the solution by enabling increased energy generation, efficient distribution, demand-side management, and the transition to renewable energy. More specifically, AI can be used in (1) smart grid management, (2) energy storage optimisation, (3) microgrid development in underserved areas, (4) demand forecasting and load management, and (5) optimisation of renewable energy systems. Hence, while the AI revolution demands large amounts of energy, the technology can also contribute to creating more efficient, sustainable, and equitable energy systems globally. Indeed, with respect to energy, AI is both a problem and a solution.

Sustainable energy access is fundamental for economic development, environmental sustainability, and social progress. However, achieving universal access to clean energy remains challenging, particularly in regions with limited infrastructure and high reliance on fossil fuels. SDG 7 emphasises the need for affordable and clean energy solutions, focusing on three main targets: universal energy access, increased use of renewable energy, and improved energy efficiency. AI offers unique capabilities that can help address these challenges, providing insights that support energy management, efficiency, and sustainability. Through predictive analytics, real-time data processing, and advanced automation, AI enables energy systems to operate more effectively, reducing environmental impact and supporting a transition to renewable energy sources.

As already intimated, this section examines how AI can help achieve SDG 7 by optimising energy grids, improving renewable energy generation, enhancing energy efficiency, and expanding energy access to underserved regions. By understanding how AI supports clean energy goals, we can leverage its potential to create a more sustainable and resilient global energy system.

12.8.2 Optimising Energy Generation and Distribution

Efficient energy generation and distribution are essential for achieving universal access to clean and reliable energy. AI has the potential to optimise these processes, enhancing the performance of both traditional and renewable energy sources while reducing waste and improving reliability. Indeed, AI can improve the efficacy of various clean energy sources such as solar, wind, geothermal, biofuel, and hydroelectric power. This is pictorially displayed in Fig. 12.7.

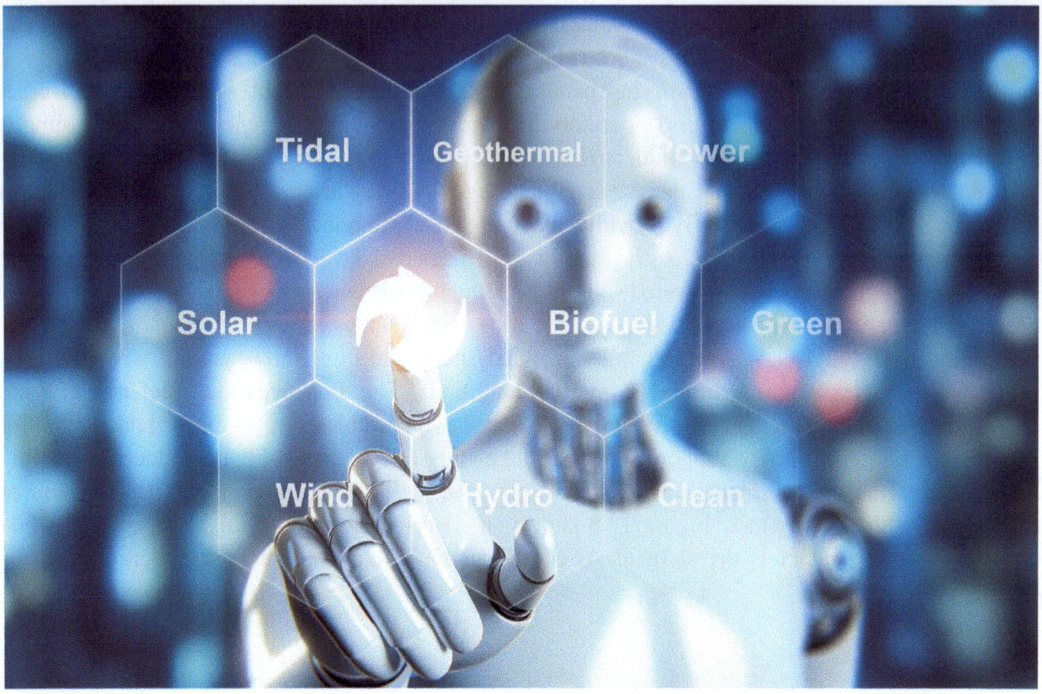

Fig. 12.7 AI can improve the efficacy of various clean energy sources. (Image credit: iStock scarfsinn86)

In traditional power plants, AI-driven predictive maintenance systems monitor equipment conditions in real time, identifying potential issues before they lead to failures. By analysing data from sensors embedded in turbines, generators, and other machinery, AI algorithms can detect signs of wear and tear, allowing operators to conduct maintenance proactively. This reduces downtime, minimises repair costs, and extends the lifespan of critical equipment, ensuring a reliable energy supply. In renewable energy systems, AI enhances efficiency by predicting and managing the variability inherent in sources like solar and wind power. AI-powered predictive analytics tools forecast energy production based on weather data, historical performance, and environmental conditions. For example, Google's DeepMind has used AI to improve wind energy efficiency by predicting wind patterns 36 hours in advance, allowing grid operators to integrate wind power more reliably. By stabilising renewable energy generation, AI contributes to a more balanced and efficient energy grid.

Moreover, AI plays a significant role in optimising energy distribution within smart grids. Smart grids are electricity networks that use digital technology to monitor and manage the flow of electricity, enabling real-time adjustments to meet demand. AI algorithms analyse data on electricity consumption patterns, grid load, and weather conditions to balance supply and demand effectively. This prevents grid overload, reduces transmission losses, and enhances overall efficiency, supporting SDG 7s reliable and resilient energy systems goal.

12.8.3 Improving Renewable Energy Forecasting

Renewable energy sources like solar and wind are essential for reducing greenhouse gas emissions, but their intermittency presents challenges for consistent energy supply. AI-driven forecasting models have the potential to mitigate these challenges by accurately predicting energy generation from renewable sources, allowing operators to integrate renewable energy sources more effectively into the grid.

AI-based forecasting models use Machine Learning algorithms to analyse historical weather data, satellite imagery, and sensor readings. These models can predict energy generation levels hours, days, or even weeks in advance by identifying patterns in temperature, sunlight, wind speed, and other factors. This capability enables grid operators to plan for variability in renewable energy supply, ensuring a stable and reliable flow of electricity.

In solar energy, AI algorithms analyse satellite images and weather patterns to predict how much sunlight will reach solar panels. By providing accurate solar irradiance forecasts, AI helps optimise solar power generation and supports grid stability. For instance, the Solar Energy Research Institute of Singapore has developed AI-powered solar forecasting models that improve the integration of solar energy in urban environments.

In wind energy, AI can predict changes in wind speed and direction, enabling wind farms to maximise energy production during favourable conditions. This improves the efficiency of wind energy systems, reduces reliance on fossil fuels, and makes renewable energy a more viable option for meeting energy demands. By supporting renewable energy forecasting, AI helps reduce uncertainty and promotes the use of clean energy sources, contributing to SDG 7s goal of a larger share of renewables in the energy mix. Figure 12.8 shows an AI-driven robot cleaning solar panels, while Fig. 12.9 illustrates a factory driven by an AI-coordinated hybrid of solar and wind energy sources.

Fig. 12.8 AI-driven robot cleaning solar panels. (Image credit: iStock sarawuth702.)

12.8.4 Energy Efficiency in Buildings and Industries

Energy efficiency is a critical component of SDG 7, as reducing energy consumption minimises environmental impact and conserves resources. AI offers significant opportunities to improve energy efficiency in buildings, industrial facilities, and transportation systems by enabling real-time monitoring, predictive analytics, and automated energy management. In buildings, AI-powered energy management systems monitor and adjust heating, ventilation, and air conditioning (HVAC) systems based on occupancy patterns, weather conditions, and user preferences. For example, smart thermostats equipped with AI algorithms learn residents' schedules and adjust temperature settings accordingly, minimising energy wastage. Similarly, AI-driven lighting systems control lighting based on occupancy, turning lights on or off as needed to conserve energy. These AI systems help reduce electricity bills, lower greenhouse gas emissions, and promote sustainable building practices by optimising energy use.

In industrial facilities, AI enhances energy efficiency by monitoring machinery performance, optimising production processes, and reducing resource consumption. For instance, AI-powered predictive maintenance systems analyse data from equipment sensors to identify energy inefficiencies, enabling companies to address issues proactively. By minimising energy waste, AI contributes to more sustainable industrial practices and supports SDG 7s goal of energy efficiency.

AI also plays a role in optimising energy use in transportation, particularly in electric vehicles (EVs) and public transportation systems. AI algorithms analyse traffic patterns, energy consumption, and charging needs to determine optimal routes and charging schedules for EVs, reducing energy consumption and improving battery efficiency. In public transit, AI-powered systems optimise bus and train schedules to minimise idle time and fuel consumption, promoting energy-efficient transportation options.

Fig. 12.9 AI-coordinated hybrid renewable energy sources. (Image credit: iStock andreus.)

12.8.5 Access to Clean Energy in Underserved Regions

Achieving universal access to affordable and reliable energy remains a significant challenge, especially in remote and underserved regions. AI offers innovative solutions to expand clean energy access in these areas by enabling decentralised energy systems, optimising microgrids, and supporting off-grid renewable energy projects.

AI-driven microgrids, which are localised energy grids that operate independently from the central grid, provide a sustainable solution for remote communities with limited access to conventional energy infrastructure. AI algorithms optimise microgrid performance by balancing energy generation, storage, and consumption based on real-time data. For example, AI can predict energy demand patterns and adjust the flow of electricity to prevent shortages or excesses. By ensuring efficient energy distribution, AI-powered microgrids offer a reliable and affordable energy source for communities without access to traditional energy grids.

In addition, AI supports off-grid renewable energy projects by managing distributed energy resources, such as solar panels and battery storage systems. Off-grid systems, often powered by solar energy, provide electricity to remote regions that lack centralised infrastructure. AI algorithms optimise the charging and discharging cycles of battery storage, ensuring that energy generated during the day is stored and used efficiently at night. This enables off-grid systems to provide continuous electricity, supporting SDG 7s goal of universal energy access.

Furthermore, AI-powered mapping tools use satellite data to identify regions with limited energy access, enabling governments and organisations to prioritise investments in clean energy infrastructure. By identifying high-impact areas for solar or wind installations, AI helps expand clean energy access and promotes energy equity, ensuring that all communities can benefit from modern energy services.

12.8.6 Smart Energy Consumption and Demand Response

Smart energy consumption and demand response are essential for balancing supply and demand in energy systems, especially as more renewable energy sources are integrated into the grid. AI supports these processes by enabling real-time monitoring, automated control, and predictive analytics that help manage energy consumption more effectively. In demand response programmes, AI algorithms analyse data on electricity prices, grid load, and consumer behaviour to encourage users to shift energy consumption to off-peak hours. For example, AI-powered demand response platforms send notifications to consumers to adjust their energy use when grid demand is high, reducing the strain on the grid. In return, consumers receive incentives for participating in demand response, helping to stabilise the grid and lower energy costs.

AI also supports smart metering systems, which provide consumers with real-time information about their energy consumption patterns. Smart meters use AI to analyse usage data, identify energy-saving opportunities, and provide personalised recommendations to reduce energy use. By promoting conscious energy consumption, smart meters empower consumers to make sustainable choices, supporting SDG 7s emphasis on energy efficiency.

In addition, AI-enabled home energy management systems allow households to monitor and control appliances remotely, optimising energy use based on factors such as time of day, weather conditions, and electricity rates. These systems ensure that energy is used only when needed, reducing waste and lowering energy bills. By fostering smart energy consumption, AI contributes to a more sustainable energy system that supports both consumers and the environment.

12.8.7 Challenges and Ethical Considerations

While AI presents numerous benefits for achieving SDG 7, its application in the energy sector raises several ethical considerations and challenges. One primary concern is data privacy, as AI-driven energy management systems collect and process data on individual energy consumption patterns. Ensuring that AI systems comply with data protection regulations is essential to protect consumers' privacy and maintain public trust.

Another challenge is AI's environmental impact, as training AI models and maintaining data centres require significant energy. As the demand for AI-driven energy solutions grows, it is essential to develop energy-efficient algorithms and use renewable energy sources to power data centres, minimising AI technologies' carbon footprint.

Bias in AI algorithms is also a concern, particularly in applications related to energy access and distribution. Biased data can result in unfair resource allocation or limit access to energy services for certain populations. Ensuring that AI models are trained on diverse and representative datasets and conducting regular audits is crucial for promoting fairness and equity in AI-driven energy systems.

Finally, the cost of implementing AI technology can be a barrier for low-income regions, limiting access to the benefits of AI-driven energy solutions. To address this, governments and international organisations should prioritise investment in affordable AI-based solutions and provide financial support to underserved communities, ensuring that the benefits of AI in energy are equitably distributed.

12.8.8 Policy Recommendations

To fully leverage AI's potential for achieving SDG 7, governments, energy providers, and technology companies must collaborate to establish ethical AI frameworks, promote innovation, and invest in inclusive research. Policies should prioritise data privacy, transparency, and accountability, ensuring that AI-driven solutions align with the goals of sustainable and equitable energy access.

Promoting public-private partnerships can drive innovation in AI-driven energy systems, enabling the development of cost-effective and scalable solutions for clean energy access. These partnerships can also foster knowledge-sharing and encourage the adoption of best practices in AI-driven energy management.

Investing in digital literacy and capacity-building programmes is essential for equipping energy providers, policymakers, and communities with the skills needed to implement and benefit from AI technologies. By providing training in AI and energy management, we can create a workforce capable of supporting a sustainable energy transition.

In addition, encouraging open-access data and research in AI for energy applications can accelerate innovation and ensure that AI-driven energy solutions are accessible to a broad range of stakeholders. Collaborative research initiatives can help address the unique energy challenges of different regions and promote the use of AI for sustainable energy solutions worldwide.

12.8.9 Conclusion

Artificial Intelligence holds transformative potential for achieving SDG 7 by enhancing energy generation, improving renewable energy forecasting, increasing energy efficiency, and expanding access to clean energy. AI provides innovative solutions to address global energy challenges, from optimising smart grids and microgrids to promoting smart energy consumption and demand response. However, ethical considerations such as data privacy, environmental impact, and algorithmic fairness must be carefully managed to ensure that AI serves as a force for good in the energy sector. AI can contribute to a sustainable and resilient energy system through responsible governance, investment in digital education, and collaboration. This brings us closer to realising SDG 7s vision of affordable, reliable, and clean energy for all.

This section explores how AI contributes to SDG 7 by supporting energy optimisation, renewable energy integration, energy efficiency, and clean energy access. It addresses the ethical challenges and offers recommendations for sustainable AI implementation in the energy sector.

12.9 AI and SDG 8 (Decent Work and Economic Growth)

UN SDG 8 (Decent Work and Economic Growth) emphasises the need for inclusive economic growth, productive employment, and decent work for all. In today's rapidly changing global economy, achieving these goals requires innovative strategies to enhance productivity, improve working conditions, and promote sustainable economic practices. AI offers transformative potential for addressing the complexities of economic development by automating processes, enhancing labour market efficiency, fostering innovation, supporting financial inclusion, and improving workplace safety. This section explores how AI applications contribute to achieving SDG 8 by creating new job opportunities, enhancing economic productivity, supporting fair labour practices, and promoting inclusive growth. Additionally, it addresses the ethical considerations and challenges associated with AI deployment in economic contexts.

12.9.1 Introduction

Sustainable Development Goal 8 calls for sustained, inclusive, sustainable economic growth, productive employment, and decent work. In an era characterised by rapid technological advancement, AI has emerged as a key driver of economic development, offering tools to optimise processes, improve efficiency, and create new opportunities. While AI's capacity for data processing, prediction, and automation provides numerous benefits, its application in economic development raises questions about job displacement, data privacy, and economic inequality. This section examines the potential of AI to advance SDG 8 by focusing on key areas such as job creation, productivity improvement, fair labour practices, financial inclusion, and workplace safety. By understanding how AI can support inclusive and sustainable growth, we can harness its potential to create a more equitable economic future.

12.9.2 Job Creation and Productivity

One of AI's primary benefits is its potential to increase economic productivity by automating repetitive tasks, optimising workflows, and enabling data-driven decision-making. This increased productivity supports economic growth and opens the door for job creation in new, technology-driven industries.

AI-driven automation can free workers from mundane, repetitive tasks, allowing them to focus on higher-value activities that require creativity, problem-solving, and interpersonal skills. For example, AI applications in manufacturing enable robots to handle routine assembly tasks, while human workers can oversee production lines, manage quality control, and innovate new solutions. This shift towards more meaningful work increases productivity and enhances job satisfaction and engagement, supporting SDG 8s emphasis on decent work.

Additionally, AI is fuelling the creation of entirely new industries, such as data science, Machine Learning engineering, and AI ethics consulting. These fields require a range of skills, from technical expertise in algorithms to ethical considerations in AI deployment. As demand for AI-skilled professionals grows, educational institutions are developing programmes to train individuals for these emerging roles. By creating new job opportunities, AI contributes to economic growth and provides a pathway for workers to transition into high-demand roles in the digital economy.

Moreover, AI-driven analytics can identify productivity bottlenecks, enabling businesses to streamline operations and improve resource use. In retail, for example, AI-powered supply chain management systems analyse sales data to forecast demand accurately, reducing overproduction and inventory costs. By optimising resource allocation and minimising waste, AI enhances economic efficiency and promotes sustainable production practices that contribute to inclusive growth.

In an article published by Axios on 28 May 2025, titled "Behind the Curtain: A White-Collar Bloodbath," the CEO of Anthropic (a key producer of AI systems), Dario Amodei, articulates a pessimistic view regarding the impact of AI on jobs. He opines that AI could eliminate 50% of all entry-level white-collar jobs and increase unemployment by 10–20% over the next 1–5 years. He insists that AI companies and the US government need to stop "sugar-coating what is coming." Amodei envisions a potential mass elimination of jobs across various sectors, including technology, finance, law, consulting, and other white-collar professions, particularly in entry-level roles. This is quite a dim outlook, but he claims he is speaking out in "hopes of jarring government and fellow AI companies into preparing and protecting the United States." On 3 June 2025, DeepMind (a major AI research laboratory) CEO and Nobel Laureate Demis Hassabis dismissed Amodei's extreme views. He argues that AI will create many new jobs within the next five years and that people should not fear AI but rather prepare to work alongside it. Specifically, he posits that the younger generation should focus on learning AI skills, as well as the foundational principles of STEM.

Indeed, the impact of AI on jobs has generated a broad spectrum of views and perspectives. The key issue is to emphasise that there are three ways AI will impact all jobs. Without a doubt, there will be job losses, an occurrence also known as job displacement. However, the majority of the jobs will be modified, a process known as human augmentation. Lastly, there will be the creation of entirely new jobs. The task of policymakers and the entire AI ecosystem is to ensure that new jobs, along with modified Jobs, are more numerous than the ones destroyed. This job equation must be satisfied. This will involve reskilling job losers, building competencies in preparation for modified jobs, and developing new skills and capabilities required for the new AI jobs. All these job retention activities should be anchored by continuous education and reskilling at every level of employment.

12.9.3 Enhancing Labour Market Efficiency

The labour market is a crucial area where AI can significantly impact by improving efficiency, facilitating job matching, and promoting fair hiring practices. Traditional recruitment processes often rely on subjective criteria, leading to biased hiring decisions and limiting access to opportunities for marginalised groups. AI has the potential to transform these processes by making recruitment more objective, efficient, and inclusive.

AI-powered recruitment platforms streamline hiring by automating resume screening, assessing candidate skills, and matching job seekers with relevant roles. Machine Learning algorithms analyse applicants' skills, experience, and preferences to identify potential job matches, reducing hiring times and ensuring that candidates are well-suited for their positions. By improving job matching, AI can help companies attract diverse talent and create more inclusive workplaces, aligning with SDG 8s commitment to productive employment and equal opportunities.

Additionally, AI algorithms can help reduce hiring bias by focusing on objective criteria rather than demographic factors. For example, some AI-driven hiring platforms anonymise candidate information, ensuring that hiring decisions are based solely on qualifications and experience. This approach promotes fair hiring practices, provides equal access to job opportunities, and supports workforce diversity.

Beyond recruitment, AI can facilitate workforce development by identifying skills gaps and recommending targeted training programmes. AI-driven learning platforms like LinkedIn Learning and Coursera offer personalised skill-building recommendations, enabling workers to acquire skills that are in demand in the labour market. By providing access to skill development, AI supports workers in advancing their careers and improving their employability, contributing to economic resilience and inclusive growth.

12.9.4 Financial Inclusion and Economic Empowerment

Access to financial services is essential for economic growth, as it enables individuals to save, invest, and manage risks. However, financial exclusion remains a significant challenge, particularly in low-income regions with limited access to banking services. AI can expand financial inclusion by providing affordable, accessible financial services to underserved populations, promoting economic empowerment and reducing inequality.

AI-driven credit assessment models can evaluate creditworthiness based on alternative data sources, such as mobile phone usage, transaction history, and social media activity. This approach enables financial institutions to assess individuals who lack traditional credit histories, such as those in low-income or rural areas. By expanding access to credit, AI empowers individuals to start businesses, invest in education, and improve their economic well-being. Companies like Tala and Branch use AI to provide microloans to underserved populations in developing countries, demonstrating how AI can promote economic inclusion.

AI-powered financial platforms also offer affordable, user-friendly tools for managing personal finances. For instance, AI-driven chatbots can provide financial advice, help users set savings goals, and offer budget recommendations. By improving financial literacy and enabling better money management, these tools empower individuals to make informed financial decisions, supporting SDG 8s goal of inclusive economic growth.

Furthermore, AI supports the development of digital payment systems that facilitate transactions for unbanked populations. Mobile payment platforms powered by AI enable individuals to receive and transfer funds securely, reducing reliance on cash and improving economic stability. These systems promote economic participation for individuals who previously lacked access to formal banking services, contributing to inclusive growth and poverty reduction.

12.9.5 Fair Labour Practices and Decent Work

SDG 8s central objective is to promote decent work, encompassing fair wages, safe working conditions, and protection against exploitation. AI can contribute to fair labour practices by enhancing transparency in supply chains, monitoring workplace safety, and detecting labour violations.

AI-powered supply chain analytics enable companies to monitor working conditions across their supply networks, identifying risks of forced labour, underpayment, and unsafe practices. By analysing data from suppliers, factories, and distribution centres, AI can detect anomalies that may indicate labour violations. For example, AI-driven systems can track worker hours, wages, and labour contracts to ensure compliance with labour laws. Companies like Patagonia and H&M use AI to monitor supplier practices, promoting transparency and ethical sourcing.

In addition to supply chain monitoring, AI-driven safety systems can improve workplace conditions by identifying potential hazards and reducing the risk of accidents. For example, computer vision technology can detect safety violations, such as workers not wearing protective gear, and alert supervisors in real time. AI-powered safety monitoring systems can also analyse environmental data, such as temperature, noise levels, and air quality, to ensure that workplaces adhere to safety standards. By promoting a safe work environment, AI supports the goal of decent work and reduces workplace risks.

AI also enables workers to report labour violations anonymously through digital platforms, empowering employees to hold employers accountable for unfair practices. These reporting tools can analyse complaints and prioritise cases that require immediate attention, ensuring that labour violations are addressed promptly. By promoting transparency and accountability, AI-driven platforms support fair labour practices and protect workers' rights.

12.9.6 Boosting Workplace Safety and Efficiency

Workplace safety is a fundamental aspect of decent work, as it ensures that workers are protected from hazards and able to perform their duties in a secure environment. AI can enhance workplace safety by monitoring work environments, detecting potential risks, and providing predictive insights that help prevent accidents.

AI-driven safety monitoring systems use sensors and computer vision to track workplace conditions in real time. For instance, in construction sites, AI-powered cameras can detect safety violations, such as workers not wearing helmets or harnesses. These systems alert supervisors immediately, enabling swift corrective action and reducing the risk of accidents. Additionally, AI algorithms can monitor environmental factors, such as air quality, noise levels, and temperature, ensuring that workplaces comply with safety regulations and protect employees' health.

Predictive analytics powered by AI can also identify trends in workplace accidents, helping organisations develop targeted safety interventions. By analysing historical incident data, Machine Learning models can identify patterns and predict areas of high risk. For example, AI-driven safety systems in mining operations analyse equipment performance, worker behaviour, and environmental conditions to predict and prevent accidents. This proactive approach enhances workplace safety, supports employee well-being, and aligns with SDG 8s emphasis on decent work.

Furthermore, AI-enabled robotics can perform hazardous tasks in industries such as manufacturing, mining, and agriculture, reducing the risk of injury for human workers. Robots equipped with AI algorithms can handle dangerous chemicals, operate heavy machinery, and perform tasks in extreme environments, protecting workers from harm and ensuring a safer workplace.

12.9.7 Challenges and Ethical Considerations

While AI offers significant potential for advancing SDG 8, its deployment in economic development raises ethical considerations and challenges. One primary concern is job displacement, as AI-driven automation may replace certain roles in industries such as manufacturing and retail. To address this challenge, investing in reskilling and upskilling programmes that equip workers with skills needed for emerging roles in AI-integrated industries is essential.

Data privacy and security are also critical considerations, particularly in applications involving financial information, personal data, and workplace monitoring. Ensuring that AI systems comply with data protection regulations is essential to protect individuals' privacy and maintain trust in AI-driven services. Transparent data practices and user consent are key to fostering a responsible AI ecosystem.

Algorithmic bias is another concern, as biased data can result in unfair outcomes, such as discriminatory hiring decisions or unequal access to financial services. Addressing bias in AI algorithms requires diverse and representative data, regular audits, and transparent model development. Ensuring fairness in AI-driven economic applications is essential to promoting inclusive growth and equal opportunity.

12.9.8 Policy Recommendations

To fully harness AI's potential for achieving SDG 8, governments, organisations, and educational institutions must collaborate to create ethical frameworks and invest in inclusive AI research. Policies should prioritise transparency, fairness, and accountability, ensuring that AI-driven solutions support sustainable economic growth and social equity.

Promoting digital literacy and workforce development programmes is essential for preparing workers for an AI-integrated economy. Public-private partnerships that provide access to education, reskilling, and digital skills training can help workers transition to high-demand roles in the digital economy. Additionally, fostering cross-sector collaboration between industry, academia, and government can drive innovation in AI applications that support inclusive growth.

Establishing regulatory frameworks for ethical AI use in the workplace is critical for protecting workers' rights and ensuring that AI-driven safety monitoring, labour practices, and recruitment processes align with the principles of decent work. These frameworks should address issues such as data privacy, algorithmic bias, and the responsible use of automation.

12.9.9 Conclusion

Artificial Intelligence offers powerful tools for advancing SDG 8 by enhancing economic productivity, promoting inclusive labour practices, supporting financial inclusion, and improving workplace safety. By leveraging AI to drive innovation, create new job opportunities, and support decent work, we can create an economy that fosters sustainable growth and equal opportunity. However, ethical considerations such as data privacy, job displacement, and algorithmic bias must be carefully managed to ensure that AI serves as a force for good in economic development. AI can contribute to a more inclusive and resilient economy through responsible governance, investment in digital education, and collaboration. This brings us closer to achieving SDG 8s vision of sustainable and equitable growth for all.

This section explores AI's potential contribution to SDG 8 by enhancing productivity, supporting fair labour practices, promoting financial inclusion, and ensuring workplace safety. It also addresses the ethical challenges and recommendations for responsible AI implementation in economic development.

12.10 AI and SDG 9 (Industry, Innovation, and Infrastructure)

UN SDG 9 (Industry, Innovation, and Infrastructure) emphasises the need for resilient infrastructure, inclusive and sustainable industrialisation, and the promotion of innovation. As economies evolve and global competition intensifies, the need for advanced infrastructure and technology becomes crucial to achieving sustainable growth and addressing global challenges. AI is emerging as a transformative tool that can enhance productivity, optimise resource use, and drive innovation across industries. This section explores the application of AI in achieving SDG 9 through advancements in manufacturing, infrastructure management, sustainable resource utilisation, innovation ecosystems, and technological inclusion. It also discusses the challenges and ethical considerations associated with implementing AI in these areas.

12.10.1 Introduction

Sustainable Development Goal 9 emphasises the role of industry, innovation, and infrastructure in building sustainable economies and supporting inclusive development. Industrialisation has long been a driver of economic growth, yet its environmental impact and resource demands necessitate a shift towards sustainable practices. AI offers powerful tools to support this transition, enabling data-driven decision-making, predictive analytics, and automation across industries. With applications in smart manufacturing, infrastructure resilience, resource management, and digital inclusion, AI has the potential to address the complexities of modern industrialisation in a sustainable way.

This section examines how AI contributes to achieving SDG 9 by supporting efficient manufacturing processes, enhancing infrastructure durability, promoting sustainable resource use, fostering innovation ecosystems, and bridging technological divides. Leveraging AI in these critical areas can create a foundation for sustainable growth and industrial advancement.

12.10.2 Smart Manufacturing and Process Optimisation

Manufacturing is a cornerstone of industrialisation, yet traditional manufacturing processes often result in waste, inefficiencies, and environmental degradation. AI has the potential to revolutionise manufacturing by enabling automation and smart production systems that are efficient, sustainable, and adaptive to market changes. Figure 12.10 shows the use of AI-powered robots for heavy industrial automation in a smart factory, whereas Fig. 12.11 shows an AI-powered automated robot assembly line producing manufacturing electric vehicles.

Fig. 12.10 AI-powered robots for heavy industrial automation in a smart factory. (Image credit: iStock gorodenkoff.)

Fig. 12.11 AI-powered automated robot assembly line manufacturing electric vehicles. (Image credit: iStock gorodenkoff.)

One significant application of AI in manufacturing is predictive maintenance, which uses Machine Learning algorithms to predict when equipment is likely to fail, allowing for proactive maintenance. Predictive maintenance minimises costly downtime, reduces material wastage, and extends the life of machinery. By analysing data from sensors embedded in machines, AI algorithms can identify patterns that signal potential issues, enabling manufacturers to address problems before they escalate. This approach optimises production and supports the sustainable use of resources by reducing equipment turnover and waste.

AI-driven quality control is another critical application in manufacturing. Machine vision and Deep Learning algorithms can accurately inspect products for defects, ensuring that only quality products reach consumers. By reducing defective products, AI minimises waste and enhances resource efficiency. For instance, AI-powered cameras can inspect items on assembly lines, detecting minute flaws and enabling quick corrective action. This ensures consistent product quality and reduces the need for rework, saving resources and energy.

In addition, AI-powered supply chain optimisation enhances manufacturing efficiency by accurately predicting demand and adjusting production accordingly. By analysing historical data and market trends, AI models can forecast fluctuations in demand, allowing manufacturers to produce just the right amount of goods. This helps avoid overproduction, reduces inventory waste, and ensures efficient resource use, aligning with SDG 9s goal of sustainable industrialisation.

12.10.3 Enhancing Infrastructure Resilience and Maintenance

Infrastructure resilience is a fundamental aspect of SDG 9, as resilient infrastructure ensures economic stability, supports quality of life, and protects communities from disruptions. AI has significant potential to improve infrastructure resilience through predictive analytics, real-time monitoring, and automated maintenance systems. Infrastructure includes transportation networks, water, energy, telecommunication, and public works.

Predictive analytics in infrastructure management allows city planners and engineers to anticipate potential failures in infrastructure components, such as bridges, roads, and buildings. By analysing historical data and environmental conditions, AI algorithms can forecast infrastructure deterioration and suggest maintenance schedules that prevent costly failures. For example, AI-powered monitoring systems in Japan's transportation infrastructure collect data on structural conditions and environmental stressors, predicting when and where repairs are needed. This proactive approach extends the lifespan of infrastructure, reduces repair costs, and enhances safety for urban populations. AI also assists in developing and managing sophisticated transportation networks. Figure 12.12 illustrates an AI-enabled road transportation network that leverages traffic big data analytics and IoT.

Real-time monitoring powered by AI is another valuable tool in infrastructure management. Sensors placed on infrastructure components continuously collect data on vibrations, temperature, humidity, and other factors that impact structural integrity. Machine Learning algorithms process this data to detect anomalies and alert engineers to potential issues. For instance, AI-driven systems in smart cities can monitor water pipelines, identify leaks, and prevent water waste. By enabling timely maintenance, AI minimises resource consumption and supports the sustainability of urban infrastructure.

Moreover, AI-driven urban planning tools enable cities to design infrastructure that is adaptable to future changes. By analysing demographic, environmental, and economic data, AI models can simulate how urban growth, climate change, and technological advancements might impact infrastructure needs. This ensures that cities invest in infrastructure that is resilient, efficient, and capable of meeting future demands, supporting the development of sustainable, inclusive cities.

Fig. 12.12 AI-enabled road transportation networks. (Image credit: iStock nanostockk.)

12.10.4 Sustainable Resource Management

Efficient resource management is a critical component of sustainable industrialisation, as resource scarcity and environmental concerns pose challenges to industrial growth. AI-driven solutions for sustainable resource management can optimise energy, water, and material use, reducing waste and supporting the circular economy. A circular economy is an economic model designed to minimise waste and maximise the utility of resources by keeping products, materials, and energy in use for as long as possible. Unlike the traditional linear economy, which follows a "take, make, dispose" approach, a circular economy is based on the following principles: (1) designing out waste and pollution, products are designed for longevity, repairability, and recyclability, minimising waste generation; (2) keeping products and materials in use, this involves reusing, repairing, refurbishing, re-manufacturing, and recycling products to extend their lifecycle; (3) regenerating natural systems, the aim is to use renewable resources and processes that restore, rather than degrade, natural ecosystems. Figure 12.13 shows the difference between a linear economy and a circular one, while Fig. 12.14 illustrates the main elements of a circular economy.

AI plays a vital role in optimising power grids, integrating renewable energy sources, and enhancing energy efficiency in the energy sector. AI algorithms can predict energy demand patterns, enabling grid operators to adjust supply and minimise waste. By analysing weather data and historical consumption patterns, AI systems can forecast energy production from renewable sources like solar and wind, ensuring a stable and efficient energy supply. For example, in Germany, AI-powered grid management systems support the integration of renewable energy, reducing reliance on fossil fuels and promoting a greener industrial base.

AI also enhances water resource management by predicting usage patterns, detecting leaks, and optimising water distribution. Machine Learning models analyse data from sensors in water

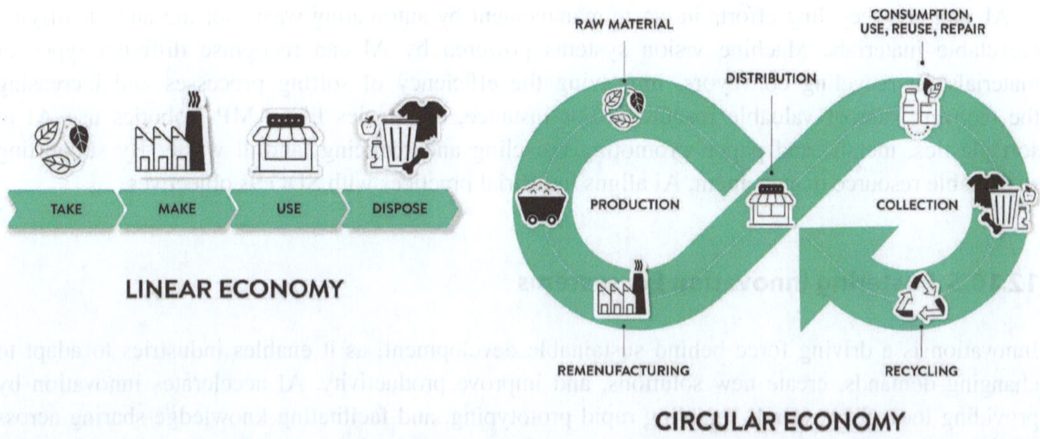

Fig. 12.13 Comparing linear and circular economies. (Image credit: iStock whaledesign.)

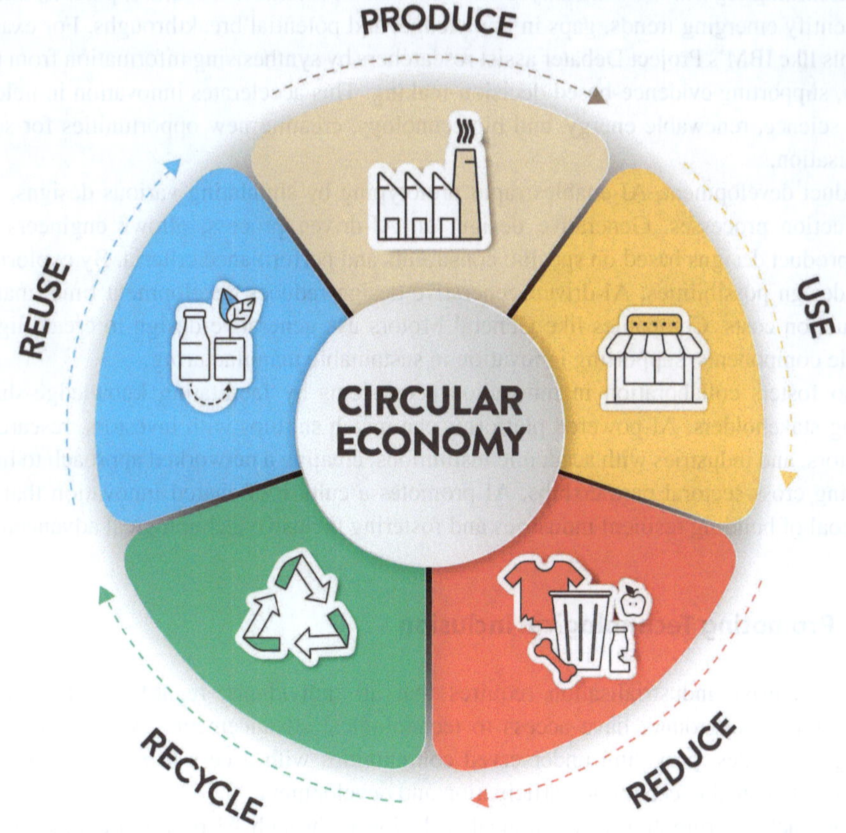

Fig. 12.14 Key elements and characteristics of a circular economy. (Image credit: iStock whaledesign.)

infrastructure to monitor usage trends and identify inefficiencies. AI-driven irrigation systems optimise water use in agriculture by providing crops with the precise amount of water needed based on soil moisture, weather forecasts, and crop types. This approach conserves water, supports agricultural productivity, and promotes sustainable practices in water-scarce regions.

AI supports recycling efforts in waste management by automating waste sorting and identifying recyclable materials. Machine vision systems powered by AI can recognise different types of materials on recycling conveyors, improving the efficiency of sorting processes and increasing the recovery rate of valuable resources. For instance, companies like AMP Robotics use AI to sort plastics, metals, and paper, promoting recycling and reducing landfill waste. By supporting sustainable resource management, AI aligns industrial practices with SDG 9s objectives.

12.10.5 Fostering Innovation Ecosystems

Innovation is a driving force behind sustainable development, as it enables industries to adapt to changing demands, create new solutions, and improve productivity. AI accelerates innovation by providing tools for research, enabling rapid prototyping, and facilitating knowledge-sharing across sectors.

One significant application of AI in fostering innovation is in research and development (R&D). Machine Learning algorithms can analyse vast amounts of scientific literature, patents, and research data to identify emerging trends, gaps in knowledge, and potential breakthroughs. For example, AI-driven tools like IBM's Project Debater assist researchers by synthesising information from thousands of studies, supporting evidence-based decision-making. This accelerates innovation in fields such as materials science, renewable energy, and biotechnology, creating new opportunities for sustainable industrialisation.

In product development, AI enables rapid prototyping by simulating various designs, materials, and production processes. Generative design, an AI-driven process, allows engineers to create multiple product designs based on specific constraints and performance criteria. By exploring a wide range of design possibilities, AI-driven generative design reduces development time, material use, and production costs. Companies like General Motors use generative design to create lightweight, sustainable components, supporting innovation in sustainable manufacturing.

AI also fosters collaboration in innovation ecosystems by facilitating knowledge-sharing and connecting stakeholders. AI-powered platforms can match startups with investors, researchers with collaborators, and industries with academic institutions, creating a networked approach to innovation. By enabling cross-sectoral partnerships, AI promotes a culture of shared innovation that advances SDG 9s goal of building resilient industries and fostering inclusive technological advancement.

12.10.6 Promoting Technological Inclusion

Achieving inclusive industrialisation requires that all individuals, regardless of geographic or socio-economic background, have access to technological advancements. AI has the potential to bridge digital divides, providing underserved communities with access to resources, training, and opportunities that enable economic participation and development.

One approach to promoting technological inclusion is through AI-powered education platforms that provide remote learning opportunities and digital skills training. These platforms use adaptive learning algorithms to tailor content to individual skill levels, enabling learners from diverse backgrounds to acquire the technical skills needed for the modern workforce. For example, Coursera and edX offer AI-driven courses in data science, programming, and engineering, providing accessible education that empowers individuals to participate in tech-driven industries. This democratisation of education supports the development of an inclusive workforce aligned with SDG 9s objectives.

AI also facilitates access to digital financial services, promoting economic inclusion for unbanked and underserved populations. AI-powered mobile banking platforms enable individuals in rural areas to save, borrow, and make transactions without physical access to banks. By using alternative data sources, such as mobile phone usage and transaction history, AI-driven credit scoring models assess creditworthiness, allowing individuals without formal credit histories to access loans. Companies like Tala and Branch leverage AI to provide financial services to underserved communities in developing countries, promoting economic empowerment and reducing inequality.

In addition to education and finance, AI can support small and medium-sized enterprises (SMEs) by providing affordable access to advanced technology. AI-driven tools for business analytics, inventory management, and customer engagement enable SMEs to compete with larger corporations. By enhancing the capabilities of SMEs, AI promotes inclusive industrialisation, supports local economies, and contributes to a balanced economic landscape.

12.10.7 Challenges and Ethical Considerations

While AI offers transformative potential for achieving SDG 9, its application in industry and infrastructure presents challenges and ethical concerns. One major issue is the risk of job displacement, as automation may replace certain roles in manufacturing and other labour-intensive sectors. To address this, investing in reskilling and upskilling programmes is essential, ensuring that workers can transition to roles that complement AI technologies rather than compete with them. Privacy and data security are also critical considerations, particularly in applications that involve personal or sensitive data. AI-driven systems in infrastructure and resource management rely on large datasets that, if mishandled, could result in privacy breaches. Establishing robust data governance frameworks is essential for protecting individuals' privacy and maintaining public trust in AI applications.

Algorithmic bias is another concern, as biased data can result in unfair or discriminatory outcomes. Ensuring that AI models are trained on diverse and representative datasets and conducting regular audits is crucial for promoting fairness and equity in AI-driven systems. Transparent AI design and inclusive data practices are key to ensuring that AI contributes to sustainable and fair industrialisation.

12.10.8 Policy Recommendations

To maximise AI's potential for achieving SDG 9, governments, industries, and technology developers must collaborate to establish ethical AI frameworks and invest in inclusive AI research. Policies should emphasise transparency, fairness, and accountability in AI applications, ensuring that AI-driven solutions support sustainable development and social equity. Investing in partnerships between private companies, research institutions, and governments can drive innovation and accelerate the adoption of AI in industry and infrastructure. These partnerships can create open-access datasets, share best practices, and promote cross-sectoral learning, facilitating a collaborative approach to sustainable industrialisation.

In addition, promoting digital literacy and workforce development programmes is essential for preparing workers for an AI-integrated economy. By providing reskilling opportunities and access to digital education, we can ensure that individuals are equipped to succeed in tech-driven industries, supporting inclusive growth.

12.10.9 Conclusion

Artificial Intelligence offers powerful tools for advancing SDG 9 by transforming industries, optimising infrastructure, and promoting sustainable resource use. From predictive maintenance and quality control in manufacturing to resilient infrastructure management, AI provides solutions that enhance productivity, reduce waste, and support sustainable industrial growth. While ethical considerations such as data privacy, job displacement, and algorithmic bias must be managed carefully, responsible AI governance can ensure that AI serves as a positive force in sustainable industrialisation. By fostering collaboration, investing in digital literacy, and promoting equitable access to technology, AI can help create resilient, inclusive, and sustainable industries and infrastructure, bringing us closer to achieving SDG 9.

This section explores AI's transformative role in achieving SDG 9, focusing on applications in smart manufacturing, infrastructure resilience, sustainable resource management, innovation ecosystems, and technological inclusion. The section offers insights into ethical challenges and recommendations for fostering sustainable industrial growth.

12.11 AI and SDG 10 (Reduced Inequalities)

UN SDG 10 (Reduced Inequalities) seeks to address disparities in income, opportunity, and access to resources across populations. Despite global advancements, inequality continues to pose challenges in terms of social and economic development, impacting marginalised and underserved communities worldwide. AI has emerged as a transformative tool to tackle these issues by enhancing access to education, improving healthcare, promoting financial inclusion, supporting labour market equity, and facilitating inclusive policymaking. This section explores the potential of AI to help achieve SDG 10 by addressing these critical areas, and it also discusses the ethical considerations, challenges, and future directions involved in implementing AI for social equity.

12.11.1 Introduction

Sustainable Development Goal 10 underscores the need to reduce inequalities within and among countries by 2030. In an interconnected world, addressing inequality goes beyond income disparities; it encompasses issues such as access to healthcare, education, technology, financial services, and representation in decision-making processes. AI holds significant promise in addressing these inequities by offering solutions that can bridge gaps in resources, provide personalised services to underserved communities, and support data-driven decision-making that promotes social equity. By leveraging AI's capacity to analyse data, predict trends, and automate processes, we can create more inclusive systems that empower marginalised groups and promote equal opportunities for all.

This section examines the application of AI in key areas related to inequality: education, healthcare, financial inclusion, the labour market, and policymaking. By focusing on these domains, AI can make meaningful contributions towards achieving SDG 10, fostering a world where resources, opportunities, and benefits are distributed more equitably.

12.11.2 Inclusive Education

Education is a fundamental tool for reducing inequality, as it provides individuals with the skills and knowledge necessary to improve their social and economic status. However, barriers such as location, economic background, and language limitations can prevent access to quality education. AI has the potential to overcome these obstacles by making learning more accessible, personalised, and inclusive.

One application of AI in education is adaptive learning, where Machine Learning algorithms adjust educational content based on individual student performance. AI-powered platforms, such as Khan Academy and Coursera, use adaptive learning techniques to personalise lessons, ensuring that each student learns at their own pace. By catering to individual learning needs, these platforms help students who may struggle in traditional classroom settings to catch up and succeed. Additionally, AI-driven educational tools can support students with disabilities, providing personalised support through text-to-speech, sign language translation, and screen-reading software. These tools ensure that learning materials are accessible to all, promoting inclusive education.

AI can also address language barriers in education, a significant obstacle for marginalised communities. AI-based language translation tools, such as Google Translate, enable students to access educational content in their native languages, expanding their learning opportunities. This application is particularly valuable in multilingual countries and refugee communities where language barriers often hinder educational progress. By making learning resources accessible to a diverse range of students, AI contributes to SDG 10s goal of reducing inequalities in education.

12.11.3 Promoting Health Equity

Healthcare disparities are a critical aspect of inequality, as access to quality healthcare varies widely based on income, geographic location, and social status. AI can enhance health equity by improving access to medical services, enabling early disease detection, and supporting personalised treatment plans, particularly in underserved communities.

Telemedicine platforms powered by AI offer virtual consultations, connecting patients in remote or low-income areas with medical professionals. AI-based diagnostic tools can analyse symptoms and provide recommendations, helping healthcare providers offer accurate diagnoses even without physical presence. For instance, apps like Ada Health and Babylon Health use AI algorithms to evaluate symptoms and suggest potential diagnoses, allowing people with limited healthcare access to receive timely medical guidance. This approach reduces barriers to healthcare access and enables underserved populations to receive quality care.

AI is also instrumental in analysing healthcare data to identify health disparities and prioritise interventions for vulnerable populations. Machine Learning models can detect patterns in health records to identify at-risk populations based on factors such as socio-economic status, geographic location, and underlying health conditions. This information allows public health officials to target resources where they are needed most, ensuring equitable healthcare delivery. Furthermore, AI-powered predictive models can forecast disease outbreaks, enabling governments to allocate resources proactively in underserved areas, preventing widespread health crises.

Another area where AI promotes health equity is personalised medicine. By analysing genetic, environmental, and lifestyle data, AI algorithms can develop individualised treatment plans that consider a person's unique health needs. While personalised medicine has traditionally been limited to high-income patients, AI-driven platforms can democratise this approach, making personalised treatment available to more people and improving health outcomes across different socio-economic groups.

12.11.4 Financial Inclusion

Financial inclusion is essential for reducing economic inequality, as access to financial services enables individuals to save, borrow, and invest, empowering them to improve their financial well-being. However, many people, particularly in developing countries, remain excluded from the formal financial system due to a lack of credit history, documentation, or geographic access to banking facilities. AI has the potential to expand financial inclusion by providing alternative credit assessments, streamlining financial services, and enhancing access to affordable financial products.

AI-powered credit scoring models can assess creditworthiness based on alternative data sources, such as mobile phone usage, transaction history, and social media activity. This approach enables financial institutions to evaluate individuals who lack traditional credit histories, such as those in low-income or rural areas. Companies like Tala and Branch use AI algorithms to analyse mobile data to determine the creditworthiness of applicants, allowing people without formal credit histories to access loans. By expanding access to credit, AI promotes economic empowerment and reduces financial disparities. In addition to credit assessment, AI can streamline financial services for underserved populations.

AI-powered chatbots provide 24/7 customer support and assist users with account management, loan applications, and financial planning. These chatbots operate in multiple languages, making financial services accessible to a broader audience, including those with limited literacy skills. AI-driven financial inclusion initiatives also extend to remittance services, enabling individuals in low-income regions to receive funds from family members abroad and promoting economic stability in marginalised communities.

Furthermore, AI-driven fintech solutions offer affordable and accessible insurance products for low-income populations. AI models can assess risks and provide customised insurance plans by analysing data from mobile devices, environmental factors, and health indicators. For example, AI-driven microinsurance products in Africa provide affordable health and crop insurance to low-income farmers, helping them recover from financial losses and reduce economic vulnerability. Through these applications, AI enhances financial inclusion, empowers individuals, and reduces economic inequalities.

12.11.5 Supporting Labour Market Equity

The labour market is a significant area of inequality, as job opportunities, wages, and working conditions often vary based on gender, race, and socio-economic background. AI has the potential to address labour market disparities by promoting fair hiring practices, improving workforce training, and supporting inclusive employment policies.

AI-driven recruitment platforms can help eliminate bias in hiring by anonymising applicant data and focusing on skills and qualifications rather than demographic information. For example, Pymetrics uses AI-based assessments to evaluate candidates' cognitive and emotional attributes, reducing bias in hiring decisions. By promoting objective assessments and reducing reliance on traditional resumes, AI helps create more inclusive hiring practices prioritising merit over background. Additionally, AI can identify diverse talent pools, providing companies with insights into underrepresented candidates and supporting diversity in the workforce.

AI also facilitates workforce training and upskilling by providing personalised learning pathways based on individuals' skill levels and career goals. AI-powered platforms such as LinkedIn Learning and Coursera offer tailored training recommendations that help individuals acquire skills aligned with job market demands. This approach is particularly valuable for workers in low-income communities who may lack access to traditional education and training resources. By promoting skill development and enhancing employability, AI reduces inequality in job opportunities and promotes economic mobility. In addition to supporting fair hiring and training, AI can assist policymakers in identifying labour market trends and developing inclusive employment policies. By analysing labour force data, AI models can identify wage gaps, employment disparities, and demographic trends, providing insights that inform policy interventions. Governments can use this data to implement policies that support equal pay, protect workers' rights, and create inclusive labour markets that reduce economic inequality.

12.11.6 Inclusive Policymaking and Governance

Effective policymaking is essential for reducing inequality, as equitable policies ensure that resources, opportunities, and services are distributed fairly. AI has the potential to enhance policymaking by providing data-driven insights, enabling predictive analysis, and improving transparency in government operations.

One application of AI in policymaking is data analysis, which enables policymakers to identify social and economic disparities within populations. By analysing demographic, economic, and health data, AI models can pinpoint regions and communities that require targeted interventions. For example, AI-powered data analysis can help governments assess poverty levels, educational disparities, and access to healthcare services, allowing policymakers to prioritise resources where they are most needed. This data-driven approach ensures that policy decisions are based on evidence, promoting social equity.

AI also facilitates predictive analysis, which can help policymakers anticipate the impact of policies on different population groups. By simulating various policy scenarios, AI models can predict potential outcomes and assess the effectiveness of interventions before they are implemented. For instance, predictive models can evaluate the impact of social welfare programmes on poverty reduction, enabling governments to refine policies for maximum effectiveness. This approach supports inclusive policymaking that addresses the unique needs of marginalised communities, aligning with SDG 10s goal of reducing inequality.

In addition to policy analysis, AI can improve government transparency by detecting and preventing corruption. Machine Learning algorithms can analyse financial records, procurement data, and transaction histories to identify irregularities and potential corruption. By promoting transparency and accountability, AI-driven anti-corruption measures build public trust in government institutions and ensure that resources are allocated fairly.

12.11.7 Ethical Considerations and Challenges

While AI offers powerful tools for reducing inequality, its application in socially sensitive areas presents ethical considerations and challenges. One major concern is the risk of algorithmic bias, which can reinforce existing inequalities rather than mitigate them. AI models trained on biased data may produce unfair outcomes, such as biased hiring recommendations or discriminatory credit assessments. To address this issue, it is essential to ensure that AI algorithms are trained on diverse datasets and undergo regular audits to identify and mitigate biases.

Privacy is another critical concern, particularly in applications involving personal data related to health, finances, or employment. Ensuring that AI systems comply with data protection regulations and prioritise user consent is essential for building public trust and preventing misuse of sensitive information.

Additionally, there is a risk that reliance on AI could exacerbate digital divides, as marginalised populations may lack access to the digital infrastructure necessary to benefit from AI-driven services. Addressing this challenge requires investment in digital literacy and infrastructure, particularly in low-income and rural areas, to ensure that all individuals can access the benefits of AI.

12.11.8 Policy Recommendations

To maximise AI's potential in achieving SDG 10, governments, organisations, and tech companies must collaborate to establish ethical frameworks and invest in inclusive AI research. Policies should promote transparency, fairness, and accountability in AI applications, ensuring that AI-driven solutions align with the goals of social equity and inclusivity. Investing in public-private partnerships and fostering cross-sector collaboration can drive innovation in AI for social good, enabling the development of solutions that address inequality at a global scale. Furthermore, promoting open-access datasets and AI tools for public use can encourage research and innovation in AI-driven social impact.

To bridge digital divides, governments and organisations should prioritise digital literacy programmes and provide resources for marginalised communities to access AI-driven services. By investing in digital inclusion, we can ensure that AI benefits are equitably distributed across all segments of society.

12.11.9 Conclusion

Artificial Intelligence can significantly contribute to achieving SDG 10 by reducing inequalities in education, healthcare, finance, the labour market, and policymaking. From improving access to quality education and healthcare to enhancing financial inclusion and supporting fair labour practices, AI provides powerful tools for creating a more equitable world. However, ethical considerations such as algorithmic bias, privacy concerns, and digital divides must be carefully managed to ensure that AI supports inclusive, sustainable development. Through responsible governance, collaboration, and commitment to ethical standards, AI can serve as a transformative force in reducing inequality and empowering individuals and communities to realise their full potential.

This section discusses AI's transformative impact on achieving SDG 10 by addressing critical areas of inequality, offering recommendations for responsible AI implementation, and highlighting the importance of collaboration for promoting social equity.

12.12 AI and SDG 11 (Sustainable Cities and Communities)

UN SDG 11 (Sustainable Cities and Communities) seeks to make cities inclusive, safe, resilient, and sustainable. As urbanisation accelerates worldwide, achieving sustainable urban development has become essential to ensure high quality of life for citizens while minimising environmental impacts. AI has emerged as a transformative technology for addressing the complex challenges of urbanisation, such as managing resources efficiently, reducing pollution, improving public transportation, enhancing public safety, and promoting affordable housing. This section explores how AI can support the achievement of SDG 11 by fostering sustainable urban planning, optimising transportation networks, enhancing energy efficiency, monitoring air quality, and supporting resilient infrastructure. It also examines the challenges and ethical considerations associated with the application of AI in urban development.

12.12.1 Introduction

The world is rapidly urbanising, with more than half of the global population currently living in cities, a figure projected to rise significantly in the coming decades. As urban areas expand, challenges related to pollution, traffic congestion, housing shortages, and infrastructure resilience become increasingly complex. SDG 11, Sustainable Cities and Communities, underscores the need for cities that are hubs of economic opportunity and safe, inclusive, and environmentally responsible spaces. The potential of AI to analyse complex data, recognise patterns, and make real-time decisions has made it an invaluable tool in urban development. AI applications, ranging from smart transportation systems to real-time air quality monitoring, can play a vital role in making cities more sustainable and improving the lives of urban residents.

This section explores how AI can help achieve SDG 11 through innovative solutions in urban planning, transportation, energy management, environmental monitoring, and infrastructure resilience, creating the foundations for sustainable urban development.

12.12.2 Smart Urban Planning

Urban planning is essential to shaping cities to promote quality of life and environmental sustainability. Traditional urban planning approaches, however, often struggle to address the dynamic and complex nature of modern urban spaces. AI has become a powerful tool in urban planning, enabling data-driven decision-making that anticipates the needs of growing populations, optimises land use, and promotes sustainable growth.

One of the most valuable applications of AI in urban planning is predictive modelling, which uses historical data and Machine Learning algorithms to forecast urban growth patterns. For example, AI algorithms can analyse demographic trends, migration data, and economic indicators to predict population density, enabling city planners to design infrastructure that meets future demands. By forecasting where population growth is likely to occur, AI helps planners allocate resources more effectively, ensuring that neighbourhoods have adequate access to transportation, schools, healthcare, and green spaces.

In addition, AI-driven geographic information systems (GIS) allow urban planners to analyse land use patterns and simulate different development scenarios. These tools provide insights into the potential impact of new infrastructure projects on the environment and urban communities. For instance, an AI-powered GIS system could analyse how the construction of a new highway might impact nearby residential areas, helping planners minimise disruptions and ensure equitable development. Through data-driven insights, AI enables more inclusive and sustainable urban planning, aligning with SDG 11s goals of creating inclusive and sustainable cities. Figure 12.15 illustrates the smart city concept, where AI effectively manages urban infrastructure.

Fig. 12.15 Smart city concept: AI effectively manages urban infrastructure. (Image credit: iStock andreysuslov.)

12.12.3 Optimising Transportation Systems

Transportation is a critical component of sustainable urban development, affecting mobility, air quality, and quality of life for city residents. AI-powered solutions are transforming urban transportation systems by optimising traffic flow, reducing congestion, and improving public transportation services.

One prominent application of AI in transportation is traffic management—optimising transportation systems and reducing traffic congestion. AI algorithms can analyse data from traffic sensors, GPS devices, and cameras to monitor real-time traffic conditions and identify patterns of congestion. AI systems can reduce bottlenecks and improve traffic flow by predicting traffic flow and adjusting traffic signals accordingly. For example, cities like Los Angeles and London use AI-driven traffic management systems that dynamically adjust signal timings based on current traffic conditions, reducing delays and lowering vehicle emissions. These systems improve commute times, reduce pollution, and contribute to the development of cleaner, more efficient cities.

AI is also enhancing public transportation systems by improving scheduling, route planning, and demand prediction. Machine Learning algorithms can analyse historical ridership data and factors such as weather and events to predict peak travel times. This enables transportation authorities to optimise bus, train, and metro schedules, reducing wait times for commuters and increasing ridership. Additionally, AI-powered route optimisation ensures that public transportation vehicles operate on the most efficient routes, reducing fuel consumption and emissions. These improvements in public transportation support SDG 11s objectives of making cities more accessible and reducing the environmental impact of urban mobility.

The emergence of autonomous vehicles, supported by AI, presents further opportunities for sustainable transportation. Autonomous shuttles, for instance, can provide first- and last-mile connectivity in urban areas, bridging the gap between public transportation and people's destinations. While still in the experimental phase, autonomous vehicles have the potential to reduce traffic congestion, enhance road safety, and create more sustainable transportation systems in the long term.

12.12.4 Enhancing Energy Efficiency in Urban Areas

Energy efficiency is essential for sustainable urban development, as reducing energy consumption lowers greenhouse gas emissions and conserves resources. AI can play a crucial role in enhancing energy efficiency in cities by optimising energy use in buildings, managing electricity grids, and integrating renewable energy sources.

In the building sector, AI-driven energy management systems are becoming increasingly popular. These systems use data from sensors to monitor energy use in real time and adjust heating, ventilation, and lighting based on occupancy patterns. For example, smart thermostats powered by AI can learn residents' schedules and adjust temperature settings accordingly, minimising energy wastage. Similarly, AI can control lighting systems in commercial buildings, turning lights on or off based on occupancy. By optimising energy consumption, these AI systems help reduce energy bills and contribute to lower carbon emissions, supporting the goal of creating energy-efficient cities.

AI also plays a significant role in managing electricity grids, particularly in integrating renewable energy sources such as solar and wind. Machine Learning models can analyse weather forecasts and energy demand patterns to predict fluctuations in renewable energy supply and adjust grid operations accordingly. This capability allows cities to rely more on clean energy sources, reducing dependence on fossil fuels and lowering their carbon footprint. For instance, AI-driven grid management systems in cities like San Diego and Amsterdam have enabled more efficient integration of renewable energy, supporting SDG 11s vision of sustainable urban energy systems.

Moreover, AI can support energy efficiency in transportation by optimising electric vehicle (EV) charging infrastructure. By analysing EV usage patterns and location data, AI systems can determine where to place charging stations to maximise accessibility and convenience for EV owners. This ensures that urban areas can support the growing demand for electric vehicles, contributing to sustainable urban transportation networks.

12.12.5 Environmental Monitoring and Air Quality Management

Environmental health is a critical component of SDG 11, as cities must address issues such as pollution, waste management, and climate resilience. AI technologies are being deployed to monitor air quality, track pollution sources, and assess environmental conditions, enabling cities to take proactive steps towards a cleaner and healthier urban environment. AI-powered air quality monitoring systems use data from sensors placed throughout urban areas to track pollutants like carbon dioxide, nitrogen dioxide, and particulate matter. By analysing this data in real-time, AI algorithms can identify pollution hotspots, predict air quality trends, and alert residents and authorities to dangerous conditions. For example, cities like Beijing and New Delhi, which face severe air pollution issues, use AI-driven air quality forecasting systems that provide early warnings and help mitigate health risks.

Machine Learning models can also identify pollution sources by analysing patterns in emissions data. For instance, AI systems can determine whether pollution spikes are due to traffic, industrial activities, or construction work, enabling targeted interventions. These insights are invaluable for urban policymakers seeking to reduce pollution levels, as they provide the data needed to implement more effective regulations and promote cleaner industries.

In addition to air quality, AI supports urban resilience by monitoring environmental risks such as floods, landslides, and heat waves. By analysing meteorological data, satellite imagery, and historical records, AI systems can predict the likelihood of extreme weather events and alert city authorities in advance. This capability is essential for protecting vulnerable communities and building resilient infrastructure, contributing to SDG 11s sustainable, resilient cities goal.

12.12.6 Building Resilient Urban Infrastructure

Resilient infrastructure is crucial for creating cities that can withstand natural disasters, climate change, and population growth. AI plays a vital role in designing, managing, and maintaining resilient urban infrastructure, ensuring that cities are prepared for both current and future challenges.

One application of AI in resilient infrastructure is predictive maintenance, which uses Machine Learning algorithms to monitor the condition of critical infrastructure, such as bridges, roads, and buildings. By analysing data from sensors embedded in infrastructure components, AI can detect signs of wear and tear, structural weaknesses, and other potential issues. This proactive approach enables city authorities to perform maintenance before problems escalate, preventing costly repairs and minimising disruptions. Predictive maintenance is particularly valuable in earthquake-prone regions, where infrastructure integrity is essential for public safety.

AI also supports sustainable construction practices by optimising building designs for durability, energy efficiency, and resource conservation. Machine Learning models can analyse construction materials, structural designs, and environmental conditions to recommend sustainable building practices. For example, AI can simulate the impact of climate change on building structures, guiding architects to design buildings that can withstand extreme weather events. This ensures that urban infrastructure remains resilient and adaptable to changing environmental conditions, supporting the long-term sustainability of cities.

Furthermore, AI contributes to the development of smart water management systems, which are essential for cities facing water scarcity. By analysing data from water meters, weather forecasts, and consumption patterns, AI systems can optimise water distribution, reduce leaks, and encourage conservation. These systems are particularly valuable in arid urban regions, where efficient water management is vital for sustainable living.

12.12.7 Sustainable Housing Solutions

Affordable and sustainable housing is a core aspect of SDG 11, ensuring that all city residents have access to safe and secure living conditions. AI can help address housing shortages by analysing urban demographics, economic conditions, and housing demand to guide the development of affordable housing solutions.

One application of AI in housing development is demand forecasting, which uses Machine Learning algorithms to predict housing needs based on population growth, income levels, and migration patterns. This insight enables urban planners to prioritise affordable housing projects and allocate resources where they are most needed. AI can also support the design and construction of low-cost, energy-efficient housing by recommending sustainable materials and construction methods.

AI-driven platforms are also being used to streamline housing allocation processes, making it easier for residents to find affordable housing options. These platforms analyse data on available housing, income brackets, and eligibility criteria to match residents with suitable housing opportunities. AI supports SDG 11s commitment to inclusive urban development by improving access to affordable housing.

12.12.8 Challenges and Ethical Considerations

While AI holds significant potential for advancing SDG 11, it also raises ethical considerations and challenges that must be addressed. Privacy concerns are particularly relevant in applications such as smart surveillance, environmental monitoring, and traffic management, where large amounts of data on residents' movements and behaviours are collected. Ensuring data protection and respecting individuals' privacy is essential to building public trust in AI-driven urban solutions.

Another challenge is algorithmic bias, which can result in unfair outcomes for certain groups. For instance, biased data in predictive policing or housing allocation algorithms could disproportionately impact marginalised communities. Addressing bias in AI models requires careful data selection, diverse representation, and regular audits to ensure fairness and equity in urban applications.

Moreover, AI systems' energy consumption, especially in large-scale applications, contributes to environmental impacts. As cities adopt AI technologies to promote sustainability, it is important to consider AI's ecological footprint and prioritise energy-efficient algorithms and renewable energy sources.

12.12.9 Policy Recommendations

To harness AI's full potential for SDG 11, governments, urban planners, and technology companies must collaborate to create ethical frameworks for AI deployment in cities. Policies should emphasise data privacy, transparency, and fairness, ensuring that AI applications align with the goals of social equity and environmental sustainability. Establishing guidelines for AI ethics, promoting community involvement, and investing in public education on AI are essential for building inclusive smart cities. Furthermore, fostering partnerships between cities, technology providers, and research institutions can drive innovation and knowledge sharing in AI-driven urban development. Open-access data and shared resources can enable cities worldwide to benefit from AI advancements, promoting global progress towards SDG 11.

12.12.10 Conclusion

Artificial Intelligence has the potential to revolutionise urban development, making cities more sustainable, inclusive, and resilient. AI provides innovative solutions to modern cities ' complex challenges, from smart urban planning and optimised transportation to enhanced energy efficiency and resilient infrastructure. While ethical considerations such as data privacy, algorithmic bias, and energy consumption must be managed carefully, responsible governance and collaboration can ensure that AI serves as a catalyst for sustainable urban transformation. By leveraging AI's capabilities, we can move closer to realising SDG 11s vision of cities that promote quality of life, environmental stewardship, and social inclusion for all.

This section outlines how AI can contribute to achieving SDG 11 through applications that promote sustainable urban planning, transportation efficiency, environmental monitoring, infrastructure resilience, and affordable housing.

12.13 AI and SDG 12 (Responsible Consumption and Production)

UN SDG 12 (Responsible Consumption and Production) aims to ensure sustainable management and efficient use of resources, reduce waste generation, and promote sustainable practices across industries. In today's global economy, meeting the demands of a growing population while reducing environmental impact poses significant challenges. AI offers powerful solutions for addressing these challenges by optimising resource management, reducing waste, and encouraging sustainable practices in both production and consumption. This section explores the application of AI to achieve SDG 12 through various sectors, including energy, manufacturing, agriculture, waste management, and consumer behaviour. It also discusses the ethical considerations, challenges, and future directions of leveraging AI for sustainable consumption and production.

12.13.1 Introduction

Sustainable Development Goal 12 focuses on promoting responsible consumption and production to reduce the ecological footprint of economic growth. Global consumption patterns and production processes have historically contributed to resource depletion, environmental degradation, and climate change, making sustainable practices essential for a balanced and healthy future. With its ability to analyse complex data, predict outcomes, and automate decision-making, AI has emerged as a transformative tool in this mission. By enabling more efficient resource use, minimising waste, and driving sustainable business practices, AI can significantly contribute to achieving SDG 12.

This section will examine how AI applications support SDG 12 in key areas such as sustainable manufacturing, energy optimisation, efficient agriculture, waste reduction, and promoting conscious consumer behaviour. Integrating AI across these sectors can pave the way for a more sustainable and resource-efficient global economy.

12.13.2 Sustainable Manufacturing

Manufacturing is one of the most resource-intensive sectors, accounting for substantial energy use, material waste, and greenhouse gas emissions. AI can play a crucial role in optimising manufacturing processes, reducing waste, and promoting resource efficiency, making it a powerful ally in achieving SDG 12.

One of the most impactful applications of AI in manufacturing is predictive maintenance, which uses Machine Learning algorithms to predict equipment failures and schedule maintenance before breakdowns occur. This approach prevents costly downtime, reduces waste, and extends the lifespan of machinery. For example, AI models can analyse data from sensors embedded in manufacturing equipment to detect anomalies, alerting operators to potential issues before they become critical. By optimising maintenance schedules and minimising machine failures, AI reduces resource wastage and ensures more efficient production.

AI also contributes to sustainable manufacturing by optimising supply chain processes. Machine Learning algorithms can forecast demand more accurately, helping manufacturers to produce the right amount of goods and avoid overproduction. Furthermore, AI can streamline inventory management, reducing excess stock and minimising waste. For instance, AI-powered demand forecasting tools analyse historical sales data, market trends, and external factors such as weather conditions to predict demand fluctuations accurately. This insight enables manufacturers to adjust production levels dynamically, leading to more efficient resource use and reduced waste generation.

Another application of AI in sustainable manufacturing is quality control. AI-driven image recognition systems can detect product defects during production, ensuring only high-quality items reach consumers. By identifying defective products early, AI minimises the need for rework, reduces material wastage, and enhances the sustainability of production processes.

12.13.3 Energy Efficiency and Resource Optimisation

Energy efficiency is a key aspect of responsible consumption and production, as reducing energy use mitigates greenhouse gas emissions and conserves valuable resources. AI can enhance energy efficiency in various sectors by optimising energy consumption, managing demand, and integrating renewable energy sources.

In industrial settings, AI-powered energy management systems can monitor and optimise energy use by analysing data from sensors and adjusting equipment settings in real time. For instance, AI algorithms can analyse energy usage patterns across production lines, identifying inefficiencies and suggesting adjustments to minimise waste. By optimising energy consumption at each stage of production, AI helps companies reduce their environmental footprint and lower operating costs.

AI is increasingly used to optimise heating, ventilation, and air conditioning (HVAC) systems in the building and construction sector. Smart buildings equipped with AI-based HVAC controls can adjust temperature settings based on occupancy, weather, and time of day, ensuring energy-efficient climate control. For example, Google's DeepMind applied AI to optimise the cooling system in its data centres, achieving a 40% reduction in energy used for cooling. This application demonstrates AI's potential to enhance energy efficiency in both residential and commercial buildings, contributing to sustainable resource use.

AI also plays a role in integrating renewable energy sources, such as solar and wind, into power grids. By analysing weather patterns and energy demand, AI can predict fluctuations in renewable energy availability and balance supply accordingly. This capability helps utilities manage energy distribution more effectively, reducing the need for fossil fuel-based energy and supporting a transition to cleaner energy sources.

12.13.4 Agriculture for Sustainable Food Production

Agriculture is one of the primary contributors to environmental degradation, accounting for significant water consumption, greenhouse gas emissions, and land use. AI has the potential to transform agriculture by promoting more sustainable food production practices, optimising resource use, and reducing waste.

Precision agriculture, enabled by AI, uses data-driven insights to optimise agricultural practices, ensuring that resources such as water, fertiliser, and pesticides are used efficiently. For instance, AI algorithms analyse data from soil sensors, weather forecasts, and crop health imagery to determine the exact amount of water and nutrients each plant requires. This targeted approach minimises resource wastage, reduces environmental impact, and increases crop yield. By applying AI-driven insights to manage irrigation and fertilisation, farmers can produce more food with fewer resources, aligning with SDG 12s goal of sustainable food production.

AI also supports sustainable agriculture through pest and disease management. Machine Learning models can analyse data from satellite imagery and drones to detect signs of crop diseases or pest infestations early. By identifying these issues in their initial stages, AI allows farmers to take preventive measures, minimising crop loss and reducing the need for chemical pesticides. This proactive approach reduces the environmental impact of agriculture while ensuring food security and productivity.

In addition to optimising resource use, AI can reduce food waste along the supply chain. Predictive analytics help food producers, retailers, and suppliers match production with demand more accurately, minimising excess inventory and reducing waste. For instance, AI-powered demand forecasting tools can predict consumer demand for perishable goods, enabling supermarkets and food suppliers to adjust their orders accordingly. By aligning supply with demand, AI reduces food wastage, conserving resources and promoting sustainable consumption.

12.13.5 Reducing Waste and Promoting Circular Economy

Waste management is a critical component of responsible consumption and production, as reducing waste generation and promoting recycling are essential for sustainability. AI can support waste reduction efforts by enhancing recycling processes, enabling the circular economy, and helping consumers make more sustainable choices. As already explained, a circular economy is an economic model designed to minimise waste and maximise the utility of resources by keeping products, materials, and energy in use for as long as possible.

AI-driven sorting technologies are revolutionising recycling by automating the identification and separation of recyclable materials. Machine vision systems powered by AI can recognise different types of waste—such as plastics, glass, and metals—on recycling conveyor belts. This automated sorting process improves the efficiency of recycling facilities, reducing contamination and increasing the recovery rate of recyclable materials. For example, AMP Robotics uses AI to sort materials from mixed recycling streams, ensuring that valuable resources are recycled and repurposed rather than sent to landfills. AI also promotes a circular economy by enabling manufacturers to design products with sustainability in mind. By analysing product lifecycle data, AI can identify opportunities to use recycled materials, extend product lifespans, and design for disassembly. This approach encourages companies to adopt circular business models that reduce resource extraction and waste generation. For instance, companies like HP and Philips use AI to design products that can be refurbished, reused, or recycled, supporting a shift towards sustainable production.

In consumer sectors, AI-based recommendation systems can encourage sustainable purchasing decisions by suggesting eco-friendly products. For instance, e-commerce platforms use AI algorithms to recommend products with minimal environmental impact, helping consumers make informed choices that align with SDG 12s objectives.

12.13.6 Shaping Sustainable Consumer Behaviour

Consumer behaviour plays a significant role in achieving responsible consumption. AI can influence sustainable consumer practices by providing personalised recommendations, raising awareness, and empowering individuals to make more sustainable choices.

Personalised recommendation engines are an example of how AI can promote sustainable consumption. By analysing user preferences and purchasing history, AI algorithms can suggest eco-friendly alternatives to conventional products, making it easier for consumers to choose sustainable options. E-commerce platforms like Amazon and Alibaba leverage AI to recommend environmentally friendly products, contributing to more sustainable consumption patterns.

AI can also raise awareness of consumption impacts by providing insights into products' environmental footprints. For instance, apps like GoodGuide and CoGo use AI to inform consumers about product sustainability, such as carbon footprint, material sources, and ethical practices. By making this information easily accessible, AI empowers consumers to make choices that reflect their values and contribute to environmental sustainability.

Moreover, AI is used in sustainable behaviour tracking, encouraging individuals to reduce waste and energy consumption. Apps like JouleBug and My Green Butler use AI-driven insights to encourage users to adopt sustainable habits, such as conserving water, reducing energy use, and recycling more effectively. These platforms help people visualise the environmental impact of their daily actions, promoting responsible consumption and supporting the goals of SDG 12.

12.13.7 Ethical Considerations and Challenges

While AI holds great promise for achieving SDG 12, its application in sustainable consumption and production presents ethical considerations and challenges. One key concern is data privacy, as AI-driven insights often rely on personal data related to consumption habits, energy use, and purchasing behaviour. Ensuring that AI systems adhere to data protection standards is essential to protect consumer privacy and maintain public trust.

Another challenge is the risk of bias in AI algorithms, particularly in applications such as product recommendations and demand forecasting. Bias in training data can result in biased outcomes, which may limit access to sustainable options for certain populations. Addressing algorithmic bias requires diverse data representation and regular audits to ensure fairness and inclusivity. Additionally, the environmental impact of AI itself should not be overlooked. Training complex AI models requires significant computing power, which can result in high energy consumption. As AI becomes more widely adopted, it is essential to develop energy-efficient algorithms and promote the use of renewable energy in data centres to minimise AI's ecological footprint.

12.13.8 Policy Recommendations

To fully leverage AI's potential for SDG 12, governments, businesses, and international organisations must collaborate to create regulatory frameworks that promote ethical AI use in sustainable consumption and production. Policies should prioritise data privacy, transparency, and inclusivity, ensuring that AI applications align with the goals of environmental sustainability and social equity.

Investing in AI research focused on sustainability is crucial for developing innovative solutions to global resource challenges. Supporting public-private partnerships and fostering knowledge-sharing between countries can drive advancements in AI-driven sustainability practices. Furthermore, promoting open-access datasets and AI tools for sustainable production can encourage innovation and enable smaller businesses to adopt responsible practices.

As AI continues to evolve, interdisciplinary collaboration among AI researchers, environmental scientists, and policymakers will be essential to address the complex challenges of sustainable consumption and production. By integrating AI into a comprehensive framework of responsible resource management, we can create a more sustainable and equitable global economy.

12.13.9 Conclusion

Artificial Intelligence offers transformative potential for achieving SDG 12 by promoting responsible consumption, reducing waste, and optimising resource use across industries. From sustainable manufacturing and energy efficiency to circular economy initiatives and consumer behaviour, AI enables data-driven approaches to address global environmental challenges. While ethical considerations such as data privacy, algorithmic bias, and AI's energy footprint must be carefully managed, responsible governance and innovation can ensure that AI serves as a force for good in advancing SDG 12. By fostering partnerships, investing in sustainable AI research, and encouraging responsible consumption, AI can help create a sustainable future where economic growth is harmonised with environmental stewardship.

This section outlines AI's transformative impact on achieving SDG 12, focusing on its applications, challenges, and recommendations for promoting responsible consumption and production.

12.14 AI and SDG 13 (Climate Action)

As already presented, SDG 13 (Climate Action) aims to "take urgent action to combat climate change and its impacts". Climate change poses a significant threat to global development, disproportionately affecting the poorest and most vulnerable populations. Addressing climate change is essential for the successful implementation of all Sustainable Development Goals, as it impacts food security, water availability, health, and biodiversity.

Achieving SDG 13 requires collective action from governments, businesses, and civil society to implement sustainable practices, invest in green technologies, and promote policies that mitigate climate change and adapt to its impacts. More significantly, AI has emerged as a pivotal tool in addressing the multifaceted challenges posed by climate change, aligning directly with SDG 13. By leveraging AI's capabilities in data analysis, predictive modelling, and optimisation, stakeholders can enhance mitigation and adaptation strategies, thereby fostering a more resilient and sustainable future. Indeed, AI technology, with its ability to process vast amounts of data and identify patterns, is a powerful tool in the fight against climate change and environmental degradation, as visualised and illustrated in Fig. 12.16.

12.14.1 Enhancing Climate Data Analysis and Monitoring

AI's proficiency in processing vast datasets enables more accurate climate modelling and monitoring. Machine Learning algorithms can analyse complex climate data to identify patterns and predict future climate scenarios. For instance, AI-driven models have been utilised to improve weather forecasting accuracy, providing critical information for disaster preparedness and resource management.

Moreover, AI facilitates real-time monitoring of environmental parameters. Satellite imagery combined with AI algorithms can detect deforestation, track ice cap melting, and monitor ocean health, offering timely insights for policymakers and conservationists. These applications are instrumental in assessing the effectiveness of climate policies and identifying areas requiring immediate intervention.

Fig. 12.16 AI as a powerful tool in the fight against climate change and environmental degradation. (Image credit: Shutterstock tonton.)

12.14.2 Optimising Energy Systems

The transition to renewable energy sources is central to mitigating climate change. AI contributes to this transition by optimising energy systems for efficiency and reliability. Smart grids powered by AI can balance energy supply and demand, integrate diverse energy sources, and reduce transmission losses. For example, AI algorithms have been employed to predict energy consumption patterns, enabling utilities to adjust generation and distribution accordingly.

Additionally, AI aids in the maintenance and operation of renewable energy infrastructure. Predictive maintenance models can foresee equipment failures, minimising downtime and enhancing the longevity of assets like wind turbines and solar panels. This proactive approach ensures a steady energy supply and reduces operational costs and resource wastage.

12.14.3 Advancing Sustainable Agriculture

Agriculture is both a contributor to and a victim of climate change. AI offers solutions to make agricultural practices more sustainable and resilient. Powered by AI, precision agriculture enables farmers to optimise resource use—such as water and fertilisers—thereby reducing greenhouse gas emissions and enhancing crop yields. Machine Learning models can analyse soil health, weather patterns, and crop conditions to provide tailored recommendations for farmers.

Furthermore, AI-driven early warning systems can predict pest infestations and extreme weather events, allowing farmers to take preventive measures. These systems contribute to food security and reduce the vulnerability of agricultural communities to climate-induced disruptions.

12.14.4 Facilitating Climate Research and Policy Development

AI accelerates climate research by processing and interpreting complex datasets, leading to new insights and discoveries. Natural Language Processing (NLP) tools can analyse scientific literature to identify emerging trends and knowledge gaps, guiding future research directions. For instance, AI has been used to synthesise information from numerous studies to assess the impacts of climate change on biodiversity.

In policy development, AI models can simulate the outcomes of various policy scenarios, assisting decision-makers in crafting effective climate strategies. By evaluating the potential impacts of different interventions, AI provides a data-driven foundation for policies aimed at emission reductions and adaptation measures.

12.14.5 Promoting Climate Adaptation and Resilience

AI enhances resilience planning and supports communities in adapting to climate change. Predictive analytics can identify regions most susceptible to climate hazards, informing infrastructure development and disaster preparedness initiatives. For example, AI models have been employed to predict flood risks, guiding the construction of resilient infrastructure and evacuation planning.

Additionally, AI-driven platforms can disseminate climate information and adaptation strategies to vulnerable populations. Mobile applications equipped with AI provide farmers with weather forecasts and agricultural advice, enabling them to adjust practices in response to changing climatic conditions. These tools empower communities to make informed decisions, reducing their vulnerability to climate impacts.

12.14.6 Enhancing Climate Finance and Carbon Markets

Financial resources are required to support activities aimed at mitigating and adapting to the impacts of climate change. This is the essence of climate finance. These funds are critical for achieving SDG 13 and related objectives outlined in international agreements such as the 2015 Paris Agreement. Climate finance is provided by governments, private sector entities, multilateral development banks, and other institutions and can be sourced domestically or internationally.

The central disagreement between the Global North and the Global South at COP29 in Baku, Azerbaijan, in November 2024, was on climate finance—resources allocated to support activities aimed at mitigating and adapting to the impacts of climate change. Nearly 200 nations agreed to a climate finance deal, committing developed countries (the main historical polluters) to provide US$300 billion annually by 2035 to assist developing nations (the primary victims of Loss and Damage) in addressing climate change impacts. This agreement aims to support these countries in transitioning to clean energy and adapting to climate-related challenges.

However, many Global South countries expressed extreme dissatisfaction with the deal. They argued that the pledged amount fell significantly short of their actual needs. Emerging and least industrialised economies had advocated for a minimum of US$500 billion annually immediately, emphasising that the agreed-upon US$300 billion by 2035 was insufficient to effectively combat the escalating climate crisis. Their ambition was for a commitment (from the historical polluters) of US1$ trillion annually by 2035. Furthermore, concerns were raised at COP29 about the vagueness of the agreement regarding the specific sources of funding and the mechanisms for its distribution, leading to frustration among nations most vulnerable to climate impacts.

Buttressing and igniting the Global South's scepticism is the history of failed promises and lack of follow-through by the Global North on climate finance. In 2009, during the Copenhagen Climate Summit, developed nations pledged to mobilise a paltry US$100 billion annually by 2020 to assist developing countries in addressing climate change through mitigation and adaptation efforts. This commitment was reaffirmed in the 2015 Paris Agreement, emphasising the importance of financial support for vulnerable nations. However, this financial target was not met by the 2020 deadline. It wasn't until 2022 that (the inadequate) US$100 billion target was reportedly achieved, 2 years later than promised.

AI can be used to streamline climate finance by assessing the environmental impact of investments and facilitating carbon trading. Machine Learning algorithms can evaluate the sustainability of projects, aiding investors in making environmentally responsible decisions. For instance, AI has been used to analyse companies' carbon footprints, informing investment portfolios aligned with climate goals.

In carbon markets, AI enhances transparency and efficiency. Blockchain technology combined with AI ensures accurate tracking of carbon credits, preventing fraud and double-counting. This integration fosters trust in carbon trading systems, encouraging participation and investment in emission reduction initiatives.

12.14.7 Conclusion

AI serves as a catalyst in the global effort to combat climate change, offering innovative solutions across various sectors. By enhancing data analysis, optimising energy and agricultural systems, facilitating research and policy development, promoting adaptation, and streamlining climate finance, AI aligns with and advances the objectives of SDG 13. However, the deployment of AI must be approached with consideration of ethical implications, data privacy, and inclusivity to ensure that its benefits are equitably distributed and contribute effectively to sustainable development.

12.15 AI and SDG 14 (Life Below Water)

The world's oceans and marine ecosystems are threatened by pollution, overfishing, and climate change. The danger is unprecedented. UN SDG 14 (Life Below Water) highlights the global commitment to conserve and sustainably use oceans, seas, and marine resources. AI has emerged as a powerful tool in these efforts, enabling innovative solutions to address complex environmental challenges. This section explores the applications of AI in marine conservation, from monitoring ecosystems and combatting pollution to managing fisheries and predicting climate impacts on marine biodiversity, ultimately presenting AI as an indispensable asset in achieving SDG 14.

12.15.1 Introduction

Sustainable Development Goal 14 embodies the commitment to protect and restore the health of the world's oceans. Oceans cover more than 70% of the Earth's surface and are essential for life, influencing global weather patterns, providing critical habitats, and supporting countless species. However, human activities have caused rapid degradation of these ecosystems, threatening the delicate balance of marine life. Innovative technologies are required to address these pressing issues. AI stands out as an up-and-coming solution because it can analyse massive datasets, recognise patterns, and generate predictive insights. This section examines how AI applications can drive progress towards SDG 14 by safeguarding marine biodiversity, managing resources sustainably, and addressing the impact of climate change on marine environments.

12.15.2 Marine Ecosystem Monitoring

Monitoring marine ecosystems effectively is essential for understanding the health of the oceans and detecting changes that could signal environmental threats. AI has revolutionised this domain by enhancing our ability to analyse data from remote sensing and satellite imagery. Advanced Machine Learning algorithms can process high-resolution images to track ocean temperatures, identify marine species, and assess coral reef health. These technologies offer real-time insights that would otherwise take years to gather and process manually.

For instance, the CoralNet platform uses AI to monitor coral reefs, classifying and quantifying coral and other organisms in underwater images. Machine Learning can detect changes in coral health over time, allowing conservationists to intervene before degradation becomes irreversible. Additionally, autonomous underwater vehicles (AUVs) and drones equipped with AI have enabled researchers to explore and monitor marine habitats that are difficult for humans to reach. These technologies help map ocean floors, track fish populations, and gather extensive data in fragile ecosystems, facilitating informed conservation efforts.

12.15.3 Combatting Marine Pollution

Marine pollution, particularly from plastics, severely threatens ocean health. AI has become instrumental in identifying, tracking, and mitigating pollution sources in oceans and coastal areas. Some AI-driven systems, such as The Ocean Cleanup, use Machine Learning algorithms to detect floating plastic and predict its movement across the ocean. This enables efficient planning and deployment of cleanup operations in areas where plastic waste accumulates, such as the Great Pacific Garbage Patch.

In addition to plastic, oil spills and other chemical pollutants require immediate intervention to prevent widespread ecological damage. AI-based predictive models simulate the spread of oil spills, providing responders with crucial data on how and where the spill may disperse. This capability helps teams mobilise resources to contain the spill before it causes extensive harm to marine life. Furthermore, real-time monitoring systems powered by AI analyse water quality through sensors deployed in oceans. These sensors detect and report levels of pollutants, pH changes, and chemical composition, enabling stakeholders to respond swiftly to pollution incidents.

12.15.4 Sustainable Fisheries Management

Sustainable fisheries are vital for protecting marine biodiversity and ensuring that fishing communities have access to resources over the long term. AI contributes significantly to sustainable fishing practices by predicting fish population dynamics based on environmental and historical data. Machine Learning algorithms analyse variables such as water temperature, ocean currents, and migration patterns, allowing fishery managers to set fishing limits that prevent overexploitation.

AI also addresses illegal, unreported, and unregulated (IUU) fishing, a major issue that threatens fish stocks and marine ecosystems worldwide. Satellite tracking and image recognition technologies powered by AI can monitor fishing vessels and detect suspicious activities. Platforms like Global Fishing Watch use AI to track and visualise fishing activities across the globe, enabling authorities to identify and take action against illegal fishing practices. By ensuring adherence to sustainable practices, AI helps protect fish populations and supports the livelihood of communities dependent on marine resources.

12.15.5 Adapting to Climate Change Marine Environments

The ocean is profoundly affected by climate change, with rising temperatures, shifting currents, and acidification all impacting marine ecosystems. AI plays a crucial role in climate adaptation strategies by forecasting these changes and helping conservationists implement proactive measures. By analysing historical and real-time data, AI models can predict shifts in oceanic climate patterns, including rising sea levels and temperature anomalies. These insights allow scientists to anticipate and respond to events that threaten marine ecosystems.

AI also supports the preservation of coral reefs, which are particularly vulnerable to warming waters and acidification. AI models can predict coral bleaching events by analysing factors like sea surface temperature and sunlight exposure. Conservationists can use this information to prioritise protection efforts and focus on resilient areas, enhancing coral survival rates. The application of AI in climate adaptation is essential for SDG 14 and complements broader climate change initiatives aimed at protecting ecosystems worldwide.

12.15.6 Challenges and Ethical Considerations

While AI holds enormous potential for marine conservation, it also presents certain challenges and ethical concerns. Collecting extensive marine data raises privacy and security issues, particularly for coastal communities that rely on the ocean for their livelihoods. Ensuring that these communities have a voice in AI-driven conservation efforts is essential for ethical stewardship. Additionally, AI models are only as accurate as the data on which they are trained. In regions with limited infrastructure for monitoring and data collection, AI predictions can suffer from inaccuracies, potentially leading to misguided interventions.

Interdisciplinary collaboration is also necessary to address these ethical concerns. AI researchers, marine scientists, government agencies, and local communities must work together to develop and deploy AI systems in ways that respect environmental and social considerations. Through responsible collaboration, AI's full potential can be harnessed for the benefit of both marine ecosystems and human communities.

12.15.7 Policy Recommendations

To unlock AI's potential in marine conservation, governments and private organisations must invest in AI research focused on ocean sustainability. Funding AI-driven marine research can accelerate the development of new technologies for ecosystem monitoring, pollution management, and climate adaptation. Additionally, creating regulatory frameworks that govern the use of AI in marine environments will ensure responsible development and usage, promoting transparency and accountability. International partnerships are crucial for sharing AI tools, data, and best practices in marine conservation. Global collaboration can strengthen collective efforts towards achieving SDG 14 by facilitating knowledge exchange and technology transfer. In an interconnected world, a cooperative approach allows countries to address shared environmental challenges more effectively.

12.15.8 Conclusion

Artificial Intelligence represents a transformative tool in our quest to achieve Sustainable Development Goal 14. By enabling real-time ecosystem monitoring, enhancing pollution detection, supporting sustainable fisheries management, and aiding in climate adaptation, AI provides innovative solutions for conserving and sustainably using ocean resources. While challenges remain, including data limitations and ethical concerns, interdisciplinary collaboration and responsible governance can maximise AI's positive impact. As we continue to explore and develop AI applications in marine conservation, it is evident that this technology has the potential to play a pivotal role in creating a sustainable future for the world's oceans.

12.16 AI and SDG 15 (Life on Land)

UN SDG 15 (Life on Land) aims to protect, restore, and promote sustainable use of terrestrial ecosystems, manage forests sustainably, combat desertification, and halt and reverse land degradation and biodiversity loss. As anthropogenic activities continue to degrade natural environments, AI has emerged as a potent tool for addressing these challenges. Through applications in biodiversity monitoring, deforestation control, habitat restoration, and climate adaptation, AI presents innovative approaches for achieving SDG 15. This section explores the various ways in which AI can support land conservation and management efforts alongside the ethical and operational challenges it poses.

12.16.1 Introduction

Sustainable Development Goal 15 addresses the urgent need to protect and preserve life on land, covering issues such as deforestation, desertification, and the loss of biodiversity. Terrestrial ecosystems, including forests, grasslands, wetlands, and deserts, are essential for human survival and ecological balance. They provide critical services such as clean air, water, climate regulation, and habitats for numerous species. Yet, these ecosystems face significant degradation due to expanding agriculture, urban development, and industrialisation. AI offers a unique opportunity to mitigate these impacts by enhancing our capacity for data collection, analysis, and action planning. This section examines how AI applications can advance conservation efforts, focusing on its role in biodiversity monitoring, forest management, land restoration, and tackling climate impacts on terrestrial ecosystems.

12.16.2 Biodiversity Monitoring and Species Conservation

Monitoring biodiversity and protecting endangered species are critical components of SDG 15. Traditionally, biodiversity monitoring has been labour-intensive and time-consuming, relying on field observations and manual data collection. AI, however, can process vast amounts of ecological data, significantly improving our ability to monitor wildlife populations, track species, and identify biodiversity hotspots.

Machine Learning models, for example, can analyse images from camera traps placed in natural habitats to identify animal species and estimate their population size. With AI-driven image recognition, researchers can automate the identification process, distinguishing between species based on physical characteristics. Platforms like Wildbook employ AI to identify individual animals based on unique features such as skin patterns or fur markings, facilitating the study of individual animal behaviours, movement patterns, and population health.

In addition to images, bioacoustic monitoring utilises AI to analyse sounds captured in nature to detect species, particularly elusive animals like birds and amphibians, that are harder to monitor visually. Algorithms trained to recognise specific animal calls can provide valuable information on species presence and distribution. These AI systems allow conservationists to monitor remote and sensitive habitats with minimal human disturbance, offering a more sustainable way to track biodiversity and protect endangered species.

12.16.3 Forestry and Deforestation Control

Forests play a crucial role in maintaining ecological stability by storing carbon, supporting biodiversity, and providing resources for livelihoods. However, illegal logging, land conversion for agriculture, and forest fires pose significant threats to forests globally. AI has emerged as a powerful tool in the battle against deforestation, helping to monitor forests, detect illegal activities, and manage forest resources more effectively.

AI-powered satellite imagery analysis can track deforestation in near real time. Machine Learning algorithms can process high-resolution satellite images to identify changes in forest cover, enabling rapid detection of illegal logging and land-clearing activities. For example, Global Forest Watch, powered by AI, provides a platform where governments and conservation organisations can monitor forest loss globally and respond to deforestation hotspots. These systems help prevent illegal logging and support the development of sustainable forest management policies by providing accurate data on forest health and carbon sequestration.

Furthermore, drones equipped with AI can survey forested areas, offering granular data on tree species, health, and growth patterns. AI analyses this data to generate insights on forest dynamics and ecosystem health. Additionally, in fire-prone regions, AI-driven fire detection models can predict potential forest fires by analysing environmental factors like temperature, humidity, and wind speed, providing authorities with early warnings to prevent fire outbreaks.

12.16.4 Combatting Desertification and Land Degradation

Desertification and land degradation pose serious challenges to ecosystems and communities, particularly in arid and semi-arid regions. Land degradation affects soil fertility, reduces agricultural productivity, and contributes to food insecurity. AI offers innovative solutions to detect and combat land degradation by enabling more accurate land monitoring, predictive modelling, and sustainable land management practices.

Remote sensing and satellite imagery, analysed with AI, are vital for detecting signs of desertification and soil degradation. Machine Learning models can process vast amounts of data from satellite images, identifying areas with declining vegetation, soil erosion, and other degradation indicators. For instance, AI can detect subtle changes in soil colour or vegetation cover that signal the early stages of desertification, enabling timely intervention.

In addition to detection, AI-driven predictive models help forecast land degradation trends based on environmental factors, rainfall patterns, and human activities. By anticipating areas at risk of degradation, conservationists and policymakers can implement land management strategies that promote soil health and sustainable agriculture. Additionally, AI can support precision agriculture, which uses data to optimise planting, irrigation, and fertilisation practices. Precision agriculture minimises soil depletion and helps maintain productive landscapes, reducing the risk of land degradation.

12.16.5 Habitat Restoration and Reforestation Efforts

Restoring degraded ecosystems and reforesting deforested lands are vital actions for achieving SDG 15. Habitat restoration enhances biodiversity and improves carbon sequestration, supporting climate goals. AI aids in reforestation and habitat restoration by optimising the selection of tree species, predicting successful growth patterns, and monitoring restoration progress. Figure 12.17 compares deforestation and reforestation—what a dramatic difference.

AI-based decision support systems can analyse environmental conditions, such as soil composition, rainfall, and temperature, to determine which tree species are best suited for specific reforestation projects. These insights help conservationists design restoration plans that maximise ecological resilience and biodiversity. AI-guided drones can also be used to plant seeds in degraded areas, increasing the efficiency and accuracy of reforestation efforts. This approach, known as precision reforestation, enables large-scale tree planting in a relatively short time, addressing deforestation at scale.

Fig. 12.17 Comparison of deforestation and reforestation. (Image credit: iStock martins.)

Once reforestation projects are underway, AI assists in tracking their success. Satellite imagery and remote sensing data analysed by Machine Learning algorithms monitor tree growth, survival rates, and overall forest health. These insights allow organisations to evaluate the effectiveness of restoration initiatives, make data-informed adjustments, and ensure the long-term sustainability of reforested areas.

12.16.6 Climate Adaptation and Resilience in Terrestrial Ecosystems

Climate change significantly impacts terrestrial ecosystems, with effects such as shifting habitats, altered species distribution, and increased frequency of extreme weather events. AI plays a critical role in climate adaptation efforts by enabling detailed climate modelling and resilience planning for vulnerable ecosystems.

AI models can simulate the effects of climate change on specific ecosystems, predicting how species and habitats might respond to temperature changes, precipitation patterns, and other climate variables. Conservationists use these predictions to identify climate-resilient species, prioritise conservation areas, and design protected zones that support biodiversity under changing conditions. In some cases, AI models recommend assisted migration of species from habitats threatened by climate change to areas where conditions are more favourable.

AI-driven tools can predict crop yield under various climate scenarios for agriculture, which is highly dependent on stable terrestrial ecosystems, enabling farmers to adapt their practices to anticipated climate impacts. This adaptability supports the sustainability of land use and mitigates the risks posed by climate variability, contributing to SDG 15 by preserving productive and resilient ecosystems.

12.16.7 Challenges and Ethical Considerations

Despite its promise, the application of AI in environmental conservation presents certain challenges. Data availability is a primary concern, as AI models rely on extensive, high-quality datasets to make accurate predictions. In regions with limited infrastructure for data collection, conservation efforts may struggle to gather the necessary information, resulting in less effective AI applications.

Ethical concerns also arise regarding data privacy and AI's impact on Indigenous communities and local stakeholders. For instance, surveillance technologies used to monitor forests might inadvertently affect the privacy of local populations. Ensuring that AI-driven conservation efforts respect these communities' rights and cultural values is essential for ethical and inclusive environmental management.

Furthermore, there is a risk that reliance on AI could create a gap between technology developers and field-based conservationists, who possess valuable ecological knowledge. Interdisciplinary collaboration and community involvement are crucial for balancing AI capabilities with on-the-ground expertise and promoting sustainable and ethical use of AI in conservation.

12.16.8 Policy Recommendations

To fully leverage AI's potential for achieving SDG 15, governments, NGOs, and the private sector must invest in AI research focused on environmental applications. Developing open-access datasets and sharing AI tools across conservation networks can foster innovation and encourage collaboration. Additionally, establishing regulatory frameworks for AI in conservation ensures responsible deployment, protecting both ecosystems and communities.

International partnerships can facilitate technology transfer, enabling countries to benefit from AI advancements regardless of their technological infrastructure. By promoting cross-border cooperation, we can address global environmental challenges more effectively, creating a united effort towards sustainable land management and biodiversity conservation.

12.16.9 Conclusion

Artificial Intelligence offers powerful solutions to some of the most pressing issues related to life on land. By enabling biodiversity monitoring, managing forests sustainably, combatting land degradation, and supporting climate adaptation, AI provides innovative pathways towards achieving SDG 15. While challenges and ethical considerations persist, responsible governance, interdisciplinary collaboration, and community engagement can ensure that AI serves as a positive force in environmental conservation. As we move forward, the integration of AI into conservation efforts will be vital for creating resilient terrestrial ecosystems and a sustainable future for generations to come.

This section comprehensively explores AI applications that support SDG 15, demonstrating AI's transformative.

12.17 SDG 16 (Peace, Justice, and Strong Institutions)

UN SDG 16 (Peace, Justice, and Strong Institutions) aims to foster peaceful and inclusive societies, provide access to justice for all, and build accountable, inclusive institutions. The challenges in achieving these goals are immense, ranging from corruption and crime to human rights violations and ineffective governance. AI is emerging as a powerful tool to address these issues, enhancing governance, improving access to justice, and fostering transparency and accountability. This section explores the potential of AI to contribute to SDG 16 by examining its applications in various domains, such as crime prevention, judicial reform, transparency enhancement, human rights protection, and challenges and ethical considerations associated with AI deployment in these sensitive areas.

12.17.1 Introduction

SDG 16 is central to the mission of achieving a peaceful, just, and inclusive world. A foundation for social progress, SDG 16 underscores the importance of reducing violence, ending corruption, ensuring transparent governance, and promoting the rule of law. However, widespread challenges—ranging from bureaucratic inefficiency and judicial backlogs to human trafficking and cybercrime—pose substantial barriers to realising these goals. In response, AI technologies are increasingly being utilised as solutions. AI can process vast datasets, identify trends, predict outcomes, and automate tasks, making it invaluable for addressing complex governance and justice issues. This section discusses how AI can help achieve SDG 16, focusing on applications that support law enforcement, judicial reform, anti-corruption initiatives, human rights, and institutional transparency.

12.17.2 Crime Prevention and Law Enforcement

One of the most promising applications of AI in supporting peace and justice is in crime prevention and law enforcement. By analysing large datasets, AI can identify crime patterns, predict potential incidents, and assist law enforcement in deploying resources effectively. Predictive policing is an area where AI has shown significant promise, using Machine Learning algorithms to analyse historical crime data and anticipate where future crimes are likely to occur. For example, AI-powered systems like PredPol (Predictive Policing) analyse data from past incidents—such as location, time, and type of crime—to provide real-time recommendations on where police should focus their efforts. This proactive approach can help law enforcement agencies prevent crime and allocate resources more efficiently. Additionally, AI can assist with facial recognition and video analysis, identifying suspects more accurately and aiding investigations. By enhancing the precision and speed of investigative processes, AI can reduce crime rates and improve public safety, supporting the goal of creating peaceful and secure communities.

However, it is essential to consider the ethical implications and risks associated with AI-driven predictive policing. Issues like algorithmic bias, privacy concerns, and the potential for AI misuse by law enforcement need to be carefully managed to ensure that AI applications do not lead to unjust outcomes or undermine public trust in law enforcement. Indeed, if carefully and effectively applied, AI can improve the delivery of justice, crime prevention, and law enforcement. This is illustrated in Fig. 12.6 (Fig. 12.18).

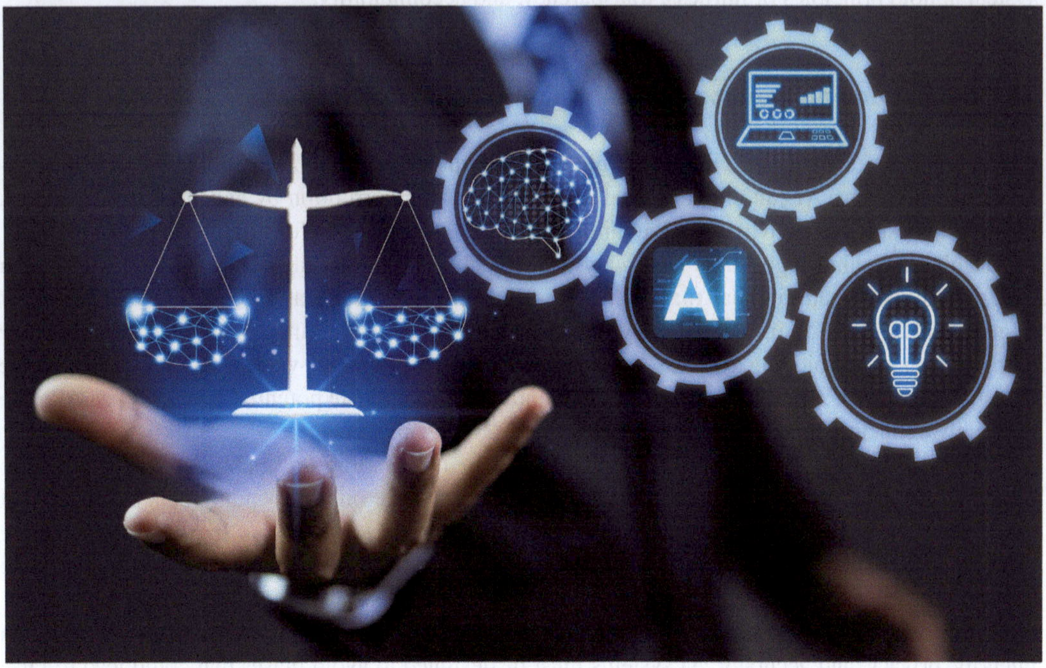

Fig. 12.18 AI and the law: Enhance justice, crime prevention, and law enforcement. (Image credit: iStock areesarak.)

12.17.3 Judicial Reform and Access to Justice

The judicial system is a cornerstone of justice and peace, but many countries face challenges such as case backlogs, limited access to legal resources, and procedural inefficiencies. AI has the potential to transform judicial processes by streamlining case management, improving access to legal information, and supporting fair trial practices.

One of the most promising AI applications in judicial reform is Natural Language Processing (NLP), which can analyse and summarise legal documents, helping judges, lawyers, and court clerks process cases more efficiently. For example, AI systems can analyse previous rulings, relevant statutes, and case details to suggest probable outcomes or assist judges in decision-making. By reducing the time spent on document review and research, AI can alleviate case backlogs, enabling courts to handle more cases within a shorter time frame and improving access to justice.

Moreover, AI-powered chatbots and virtual legal assistants are being used to provide legal information and support to the public, especially in regions with limited legal resources. These tools can answer basic legal questions, guide individuals through court processes, and even help prepare legal documents, making legal assistance more accessible. For instance, platforms like DoNotPay use AI to help people contest parking tickets and handle small claims, offering affordable legal support and empowering individuals to seek justice.

While AI offers promising solutions for judicial reform, challenges such as the accuracy and fairness of AI-driven legal recommendations, data privacy concerns, and the need for human oversight remain. To build trust in these systems, it is essential to ensure that AI applications in the judicial system uphold principles of fairness, transparency, and accountability.

12.17.4 Enhancing Transparency and Combatting Corruption

Corruption is a pervasive issue that undermines public trust in institutions and hinders social and economic development. AI can play a pivotal role in combatting corruption by enhancing transparency, identifying suspicious activities, and monitoring financial transactions for illicit behaviour.

Machine Learning algorithms can analyse financial records, procurement data, and transaction histories to detect patterns that may indicate corrupt practices, such as bribery, embezzlement, or money laundering. For instance, AI systems can flag irregularities in procurement processes by identifying unusual bidding patterns or sudden increases in contract values. By automating these detection processes, AI enables authorities to identify and investigate potential corruption cases more efficiently, reducing opportunities for illicit activities to go unnoticed.

Additionally, AI is being used to monitor government spending and public sector activities to ensure accountability. For example, AI-driven tools can analyse public spending data to ensure that funds are allocated and used according to budgetary guidelines. In Kenya, the Open Contracting Data Standard initiative uses AI to analyse procurement data and identify corruption risks in public contracts, helping authorities make procurement processes more transparent and accountable.

However, deploying AI in anti-corruption efforts requires robust data governance to ensure data integrity, privacy, and accuracy. AI systems must be designed to avoid biases that could lead to false positives or disproportionately target specific individuals or groups. Ensuring that AI systems in anti-corruption initiatives are transparent and accountable is critical to building public trust and achieving SDG 16 goals.

12.17.5 Protecting Human Rights

AI is also proving to be a valuable tool for human rights protection, as it can help monitor and document violations, identify at-risk populations, and raise awareness of human rights abuses. AI-driven image and video analysis tools are being used to detect instances of human rights violations, such as violence against civilians or forced labour. For example, satellite imagery analysis can detect evidence of conflict, such as destroyed infrastructure or mass graves, which human rights organisations can then use to document abuses and bring perpetrators to justice.

AI can also help identify vulnerable populations and prevent human trafficking. By analysing data from social media, immigration records, and transportation data, AI can identify patterns that may indicate trafficking networks or identify individuals at risk of exploitation. For instance, AI algorithms are used by NGOs to monitor social media platforms for suspicious job postings or recruitment tactics that are often associated with human trafficking schemes. This proactive approach enables authorities to intervene before exploitation occurs, protecting individuals from harm.

While AI can enhance human rights protection, ethical considerations such as data privacy and the risk of AI misuse for surveillance are crucial. AI systems used in human rights monitoring must operate under strict ethical guidelines to protect individuals' privacy and prevent the misuse of data by repressive regimes or other actors.

12.17.6 Strengthening Institutions

Strong, transparent institutions are essential for a functioning society, fostering trust between governments and citizens. AI has significant potential to strengthen institutional performance by improving public service delivery, enhancing decision-making, and promoting transparency in government operations.

For example, AI-powered data analytics can help policymakers make more informed decisions by analysing demographic, economic, and social data to identify areas of need and assess the impact of policies. By providing data-driven insights, AI enables governments to allocate resources more effectively and respond to societal challenges more efficiently. Additionally, AI can enhance public service delivery by automating administrative tasks, reducing bureaucratic inefficiencies, and minimising delays. For instance, AI chatbots can handle routine citizen inquiries, reducing wait times and improving access to information.

AI also plays a role in election monitoring and fraud prevention, ensuring the integrity of electoral processes. Machine Learning algorithms can analyse voting patterns and detect irregularities, such as vote manipulation or coercion, providing transparency in the electoral process. By strengthening institutional integrity and transparency, AI supports the establishment of robust, accountable governance structures necessary for peace and justice. However, the use of AI in governance raises questions about accountability, data privacy, and potential biases in algorithmic decision-making. Ensuring that AI applications in public administration are transparent, fair, and subject to oversight is crucial to maintaining public trust and preventing misuse.

12.17.7 Challenges and Ethical Considerations

While AI presents numerous opportunities to advance SDG 16, its application in sensitive areas such as justice, law enforcement, and human rights poses challenges and ethical dilemmas. One of the main concerns is the risk of algorithmic bias, which can lead to unfair treatment or reinforce existing inequalities. For example, biased data in predictive policing algorithms can disproportionately target marginalised communities, leading to discriminatory outcomes. Ensuring fairness in AI requires careful data selection, diverse representation in datasets, and regular audits to identify and correct biases.

Privacy is another critical concern, particularly in applications involving surveillance, human rights monitoring, or public administration. AI systems that analyse personal data must adhere to strict data protection regulations to safeguard individual privacy and prevent misuse. Additionally, transparency and accountability in AI deployment are essential for building trust in these technologies. Institutions must ensure that AI systems are subject to human oversight, with clear mechanisms for challenging or appealing AI-driven decisions when necessary.

12.17.8 Policy Recommendations

To harness AI's full potential for achieving SDG 16, governments, international organisations, and the private sector must invest in ethical AI research, promote data-sharing initiatives, and establish regulatory frameworks for AI in governance. Investing in open-access datasets and AI tools for public use can enhance transparency and encourage innovation in peace, justice, and governance sectors. International cooperation is also necessary to address global challenges, such as cross-border crime and human rights abuses, and to establish ethical standards for AI deployment.

Creating legal frameworks that ensure responsible AI use, protect data privacy, and uphold human rights is essential. These frameworks should include ethical AI design guidelines, accountability mechanisms, and measures to prevent misuse. Building public trust in AI requires a commitment to transparency, fairness, and inclusivity in AI development and deployment.

12.17.9 Conclusion

Artificial Intelligence holds significant potential to advance SDG 16 by supporting peace, justice, and strong institutions. AI provides innovative solutions for building a more just and inclusive society, from crime prevention and judicial reform to transparency enhancement and human rights protection. However, ethical challenges such as algorithmic bias, privacy concerns, and the need for transparency must be addressed to ensure that AI applications serve the public good. With responsible governance, interdisciplinary collaboration, and adherence to ethical principles, AI can play a pivotal role in creating peaceful, just, and strong institutions, contributing to a sustainable future for all.

This section explores AI's transformative role in achieving SDG 16, detailing applications, challenges, and recommendations for harnessing AI effectively and responsibly to foster a more peaceful and just society.

12.18 AI and SDG 17 (Partnerships for the Goals)

UN SDG 17 (Partnerships for the Goals) emphasises the need for global cooperation, resource mobilisation, and capacity building to achieve sustainable development across all sectors and nations. As the world becomes increasingly interconnected, AI presents unique opportunities to strengthen global partnerships, improve data and resource-sharing mechanisms, enhance policy implementation, and bridge gaps between developed and developing regions. This section explores how AI can drive progress towards SDG 17 by facilitating partnerships, optimising resource allocation, advancing data analysis and transparency, enhancing financial inclusion, and supporting capacity-building efforts. It also discusses the ethical considerations, challenges, and future directions of integrating AI into global partnership frameworks.

12.18.1 Introduction

Sustainable Development Goal 17 addresses the critical need for international collaboration and resource-sharing to achieve the 2030 Agenda for Sustainable Development. This goal highlights the importance of partnerships that span across borders and sectors, focusing on areas such as finance, technology, trade, and capacity building. In an era of rapid digital transformation, AI has emerged as a powerful tool for bridging divides, fostering innovation, and maximising the impact of development efforts. With its ability to analyse massive datasets, identify trends, and create predictive models, AI can support SDG 17 by enhancing coordination and strengthening partnerships that drive sustainable development.

This section examines the role of AI in realising SDG 17s vision of strong, collaborative partnerships by addressing key areas where AI can make a significant impact. From facilitating data-sharing initiatives and improving transparency to optimising resource distribution and enabling financial inclusion, AI has the potential to transform how nations and organisations work together to address global challenges. In doing so, AI can contribute to building more inclusive, equitable, and resilient societies worldwide.

12.18.2 Facilitating Data Sharing and Transparency

Data sharing is fundamental to effective partnerships, enabling stakeholders to make informed decisions, assess impact, and respond to emerging challenges. AI's capacity for data processing and analysis supports the development of data-sharing platforms that enhance transparency, accountability, and information exchange across sectors and borders. For example, AI-driven platforms can centralise and standardise data on environmental indicators, health statistics, and economic metrics, providing partners with real-time access to relevant information and enabling them to monitor progress towards the SDGs.

Machine Learning algorithms can also process data across languages, making insights accessible to stakeholders globally and removing language barriers that often hinder cross-border cooperation. The UN's Global Pulse initiative leverages AI and big data to analyse information from various sources, including social media and satellite imagery, to support development objectives and provide actionable insights for policymakers. By facilitating the rapid dissemination and interpretation of data, AI strengthens transparency and fosters trust among partners, empowering them to make coordinated, evidence-based decisions.

However, effective data-sharing mechanisms require robust data governance to protect privacy, ensure data accuracy, and prevent misuse. For AI to support SDG 17 effectively, data-sharing initiatives must include clear guidelines on data protection, ownership, and accountability, ensuring that all partners can benefit from AI-driven insights without compromising ethical standards.

12.18.3 Financial Inclusion and Resource Mobilisation

Financial inclusion is essential to achieving SDG 17, as it enables individuals, especially in developing countries, to access financial services that can improve their quality of life and economic stability. AI has proven to be instrumental in promoting financial inclusion by facilitating access to financial services, optimising credit assessments, and enhancing resource mobilisation efforts. For example, AI-driven algorithms can analyse alternative data, such as mobile phone usage and transaction history, to assess the creditworthiness of individuals who lack formal financial records. This approach enables financial institutions to extend credit and other financial services to underserved populations, supporting economic empowerment and poverty reduction.

In resource mobilisation, AI can assist governments and organisations in identifying funding opportunities, optimising tax revenue collection, and managing public finances more effectively. Machine Learning models can analyse economic data to predict tax revenue, monitor financial flows, and detect anomalies that may indicate potential revenue loss. By maximising resource mobilisation, AI enables governments to allocate funds towards sustainable development initiatives and invest in programmes that support SDG 17s partnership goals.

One example of AI's impact on financial inclusion is in mobile banking and digital payment systems. These AI-driven platforms enable people in remote and underserved regions to participate in the global economy, making it easier to transfer funds, save money, and access loans. Financial technology (Fintech) companies are using AI to streamline transactions, reduce fraud, and improve accessibility for those previously excluded from traditional banking systems. As AI continues to advance, it has the potential to unlock new sources of financing and mobilise resources that support sustainable development partnerships worldwide.

12.18.4 Capacity Building and Knowledge Sharing

Capacity building is a key component of SDG 17, as it empowers countries, especially developing nations, to implement sustainable development strategies independently. AI plays a significant role in capacity building by facilitating knowledge sharing, enhancing education, and developing technical skills. Through AI-powered online learning platforms, individuals in developing regions can access high-quality educational resources, upskilling themselves in fields such as data science, environmental management, and public health. AI-based adaptive learning systems personalise educational content to meet learners' needs, maximising the effectiveness of capacity-building efforts.

Moreover, AI can support knowledge sharing by translating and disseminating scientific research, policy analyses, and best practices across languages and cultural contexts. By breaking down language barriers, AI-driven translation tools make valuable knowledge accessible to a global audience, allowing countries to share insights and learn from each other's experiences. Organisations such as UNESCO and the World Bank are leveraging AI to facilitate knowledge exchange, providing decision-makers with the resources they need to address local challenges and enhance their capacity for sustainable development.

AI also supports the development of public sector capabilities by streamlining administrative processes and improving service delivery. For example, governments can use AI to analyse demographic data, monitor program effectiveness, and allocate resources efficiently. By providing public officials with data-driven insights, AI enables them to make informed decisions and enhance institutional capacities, contributing to stronger, more resilient institutions.

12.18.5 Optimising Resource Allocation and Disaster Response

The efficient allocation of resources is vital for achieving sustainable development, particularly in the face of natural disasters, public health crises, and humanitarian emergencies. AI-driven resource allocation models can analyse data on population needs, resource availability, and logistical constraints to optimise the distribution of food, water, medical supplies, and other essential resources. This capability is especially valuable for international organisations and humanitarian agencies, as it enables them to respond swiftly and effectively to crises.

In the context of disaster response, AI can analyse satellite images, weather patterns, and social media data to predict and assess the impact of natural disasters. Machine Learning algorithms can identify regions at high risk of floods, earthquakes, or wildfires, enabling governments and international partners to deploy resources proactively and mitigate potential damage. For instance, during the COVID-19 pandemic, AI models helped allocate medical supplies, identify infection hotspots, and predict healthcare demands, demonstrating the potential of AI to support global health and disaster preparedness.

AI's predictive capabilities are also valuable for long-term resource planning, enabling stakeholders to anticipate and prepare for future needs. By providing a data-driven approach to resource allocation, AI enhances resilience and ensures that resources are directed where they are needed most, reinforcing the impact of SDG 17s partnership initiatives.

12.18.6 Improving Policy Implementation and Monitoring

Effective policy implementation is essential to achieving SDG 17, as it ensures that SDGs are translated into actionable initiatives. AI can improve policy implementation by providing policymakers with data-driven insights that inform decision-making, monitor progress, and evaluate the effectiveness of interventions. For instance, AI models can analyse data on poverty rates, education outcomes, and healthcare access to assess the impact of policies and identify areas where further action is needed.

AI-powered dashboards and monitoring systems offer real-time tracking of SDG indicators, allowing governments and organisations to monitor progress towards the SDGs and adjust their strategies accordingly. By providing a comprehensive view of development outcomes, these tools enable stakeholders to identify bottlenecks, allocate resources more effectively, and make evidence-based policy adjustments. For example, in India, the NITI Aayog SDG India Index uses data and AI analytics to assess the country's progress on the SDGs, providing a transparent and data-driven framework for monitoring sustainable development.

AI can also enhance policy enforcement by detecting irregularities or inefficiencies in government operations. Machine Learning algorithms can identify patterns of corruption, mismanagement, or fraud, allowing authorities to take corrective action and ensure that public resources are used for their intended purposes. Through these applications, AI strengthens policy implementation and promotes accountability, supporting the establishment of strong institutions aligned with SDG 17.

12.18.7 Ethical Considerations and Challenges

While AI offers numerous benefits for SDG 17, its application in global partnerships raises ethical considerations and challenges. One key concern is data privacy, as collecting and analysing sensitive information can infringe on individual rights and create security vulnerabilities. To mitigate these risks, AI-driven partnerships must adhere to robust data governance frameworks that protect privacy, ensure data accuracy, and promote transparency.

Bias in AI algorithms is another significant challenge, particularly in applications related to resource allocation, financial inclusion, and policy implementation. Biased data can result in unfair outcomes, reinforcing inequalities and compromising the effectiveness of development initiatives. Ensuring that AI systems are transparent, inclusive, and regularly audited for bias is essential for achieving equitable outcomes.

Additionally, AI's potential to replace jobs poses a challenge to capacity building, as automation may reduce demand for certain types of labour. To address this issue, capacity-building programmes must prioritise digital literacy and technical skills development, enabling individuals to adapt to a changing labour market. As AI technologies continue to evolve, fostering an inclusive, ethical approach to AI in global partnerships will be essential for realising SDG 17s vision of sustainable and equitable development.

12.18.8 Policy Recommendations

To harness AI's full potential for SDG 17, governments, international organisations, and the private sector must invest in AI research, develop ethical standards, and promote data-sharing initiatives. Creating open-access datasets and AI tools for public use can enhance transparency, enable innovation, and encourage collaboration in sustainable development efforts. Additionally, establishing global regulatory frameworks for AI ensures that AI applications in development adhere to ethical standards and protect human rights.

International partnerships are crucial for addressing global challenges and sharing best practices in AI-driven sustainable development. Countries can leverage AI's capabilities to address shared challenges, promote equitable resource distribution, and strengthen institutional resilience by fostering cross-border cooperation. AI can play a transformative role in driving SDG 17's partnership goals in this collaborative environment, ultimately contributing to a sustainable and inclusive future.

12.18.9 Conclusion

Artificial Intelligence presents transformative opportunities for achieving SDG 17 by enhancing global partnerships, promoting transparency, enabling financial inclusion, and supporting capacity building. From facilitating data-sharing initiatives to optimising resource allocation, AI strengthens international cooperation and empowers stakeholders to address complex global challenges. However, ethical considerations such as data privacy, algorithmic bias, and job displacement must be carefully managed to ensure that AI supports inclusive, sustainable development. With responsible governance, international collaboration, and a commitment to ethical principles, AI can play a pivotal role in realising the vision of SDG 17, creating a world where global partnerships drive progress towards a more sustainable, just, and resilient future.

This section outlines AI's potential to advance SDG 17 through partnerships that promote transparency, enhance resource distribution, foster inclusion, and support sustainable development efforts worldwide.

12.19 Conclusion

This chapter has articulated the deployment of AI to achieve the SDGs for each of the 17 goals. Figure 12.19 illustrates the impact of AI on SDGs (direct, indirect, minor, and major), while Figs. 12.20 and 12.21 summarise the application of AI to achieve SDGs 1 to 17. In all these interventions, regulations and governance are critical. Figure 12.22 outlines the hierarchy of the key AI governance functions.

Using AI to achieve the UN SDGs is challenging due to several technical, ethical, and systemic issues. However, it can be done. Indeed, AI systems rely on large amounts of high-quality data, often unavailable or fragmented in regions where the SDGs are most critical. Data disparities and lack of infrastructure, especially in low-income countries, create significant barriers to AI deployment.

Additionally, algorithmic bias and inequality in AI systems can exacerbate existing disparities, undermining the inclusive nature of the SDGs. Ethical concerns such as data privacy, surveillance, and the misuse of AI further complicate its application, particularly in sensitive areas like healthcare and governance. The energy-intensive nature of AI technologies also poses environmental challenges, potentially conflicting with some sustainability goals.

The pursuit of each SDG using AI must take cognisance of these challenges. More specifically, there must be leveraging of emerging global experiences, understanding and deploying of the enablers, drivers, and all critical success factors. The appropriate AI governance, legislation, and ethical guidelines must be developed and put in place. More significantly, the AI interventions for the 17 SDGs can only be successful within a well-structured strategic framework. As proposed in this chapter, AI can become a powerful tool for achieving all 17 SDGs through collaboration, integration, political unity, capacity building, and strategic thinking. Both the Global North and Global South can deploy AI to achieve the SDGs. However, it cannot be business as usual. [4, 38, 60, 112, 117, 125, 128, 139]

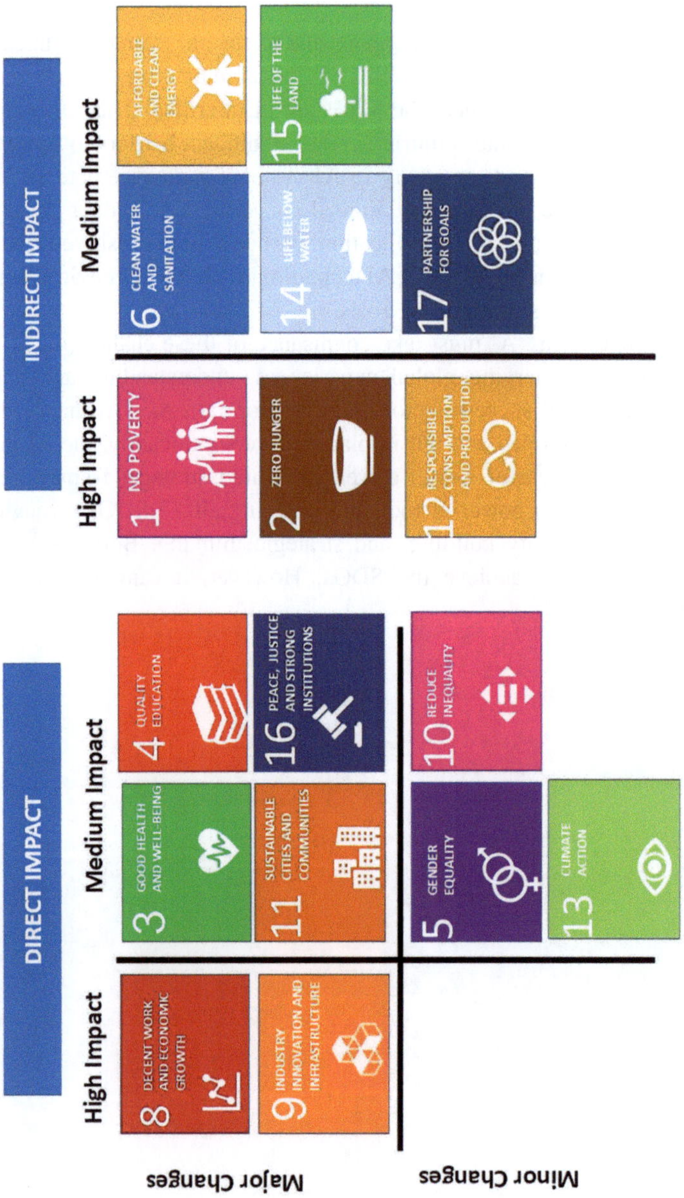

Fig. 12.19 Assessing the impact of AI on SDGs

SDG	Application of AI	Examples
1. No Poverty	AI-driven data analysis identifies poverty patterns, enabling targeted interventions.	AI models predict areas at risk of poverty, guiding resource allocation.
2. Zero Hunger	AI optimizes agricultural practices through precision farming, enhancing crop yields.	Machine learning algorithms analyze soil data to recommend optimal planting strategies.
3. Good Health and Well-being	AI assists in disease diagnosis and personalized treatment plans.	AI algorithms detect early signs of diseases from medical imaging.
4. Quality Education	AI-powered platforms provide personalized learning experiences.	Adaptive learning systems tailor educational content to individual student needs.
5. Gender Equality	AI analyzes data to identify and address gender disparities.	Natural language processing tools monitor media for gender bias.
6. Clean Water and Sanitation	AI monitors water quality and predicts contamination events.	Sensors with AI capabilities detect pollutants in water sources.
7. Affordable and Clean Energy	AI optimizes energy consumption and integrates renewable energy sources.	Smart grids use AI to balance energy supply and demand efficiently.
8. Decent Work and Economic Growth	AI enhances productivity and creates new job opportunities in tech sectors.	Automation of routine tasks allows workers to focus on complex problem-solving.
9. Industry, Innovation, and Infrastructure	AI drives innovation in manufacturing and infrastructure development.	Predictive maintenance systems reduce downtime in industrial equipment.

Fig. 12.20 Application of AI to achieve SDGs 1 to 9

SDG	Application of AI	Examples
10. Reduced Inequalities	AI identifies and addresses social and economic disparities.	Data analysis highlights underserved communities, guiding policy decisions.
11. Sustainable Cities and Communities	AI improves urban planning and management.	Traffic management systems use AI to reduce congestion and emissions.
12. Responsible Consumption and Production	AI optimizes supply chains, reducing waste and promoting sustainability.	AI forecasts demand accurately, minimizing overproduction.
13. Climate Action	AI models predict climate patterns and assess environmental impacts.	Climate simulations inform policy decisions on emission reductions.
14. Life Below Water	AI monitors marine biodiversity and detects illegal fishing activities.	Drones equipped with AI identify and track marine species populations.
15. Life on Land	AI aids in wildlife conservation and forest management.	AI analyzes satellite imagery to detect deforestation.
16. Peace, Justice, and Strong Institutions	AI enhances transparency and combats corruption.	AI systems analyze financial transactions to identify fraudulent activities.
17. Partnerships for the Goals	AI facilitates global collaboration through data sharing and analysis.	Platforms powered by AI connect stakeholders to share best practices.

Fig. 12.21 Application of AI to achieve SDGs 10 to 17

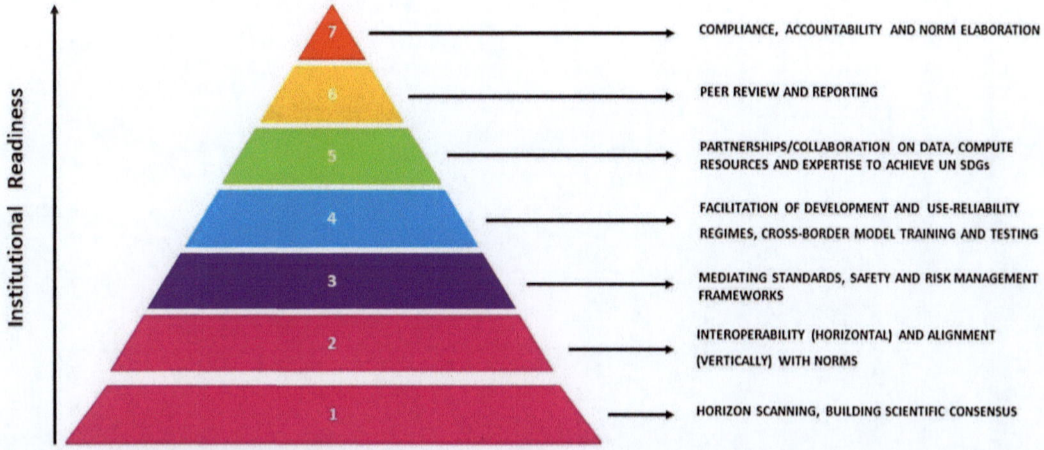

Fig. 12.22 Hierarchy of AI governance functions in pursuit of SDGs

Conclusion 13

13.1 Leveraging AI to Achieve Inclusive GDP Growth

As AI enhances the attainment of SDGs, it has also emerged as a transformative force driving GDP growth in most countries. AI is reshaping industries and redefining economic opportunities by enhancing productivity, fostering innovation, and enabling new business models. In manufacturing, AI-powered automation reduces costs and increases output, while predictive analytics optimise supply chains. In the service sector, AI-driven chatbots, virtual assistants, and recommendation engines enhance customer experiences, boosting consumption and revenue generation. These advances collectively accelerate economic activity, contributing to higher GDP growth rates globally.

Appreciating the nature and scale of AI's economic opportunity is instructive. The economics of AI is about assessing how AI impacts economic systems, markets, labour, productivity, and overall societal welfare. As AI technologies develop, they influence various economic activities, from production and consumption to trade and labour markets. Understanding the economics of AI is essential to assessing its potential benefits and challenges and formulating policies that maximise SDG attainment while mitigating any negative impacts.

A key driver of the economics of AI is productivity growth. AI can significantly enhance productivity by automating routine tasks, optimising processes, improving efficiency, and enabling better decision-making. AI technologies like Machine Learning and predictive analytics can lead to innovations across many industries, such as agriculture, mining, manufacturing, infrastructure, and finance. However, the distribution of productivity gains may not be equitable, benefitting larger firms or specific sectors while others may lag behind. As already intimated, AI can boost economic growth by enhancing efficiency, reducing costs, and fostering innovation. Countries investing heavily in AI research and adoption, like the United States and China, will likely see a competitive advantage in the global economy. Nevertheless, the benefits of AI-driven growth may not be evenly distributed, potentially increasing inequality between countries, firms, or social groups.

Productivity is a critical driver of GDP growth. By automating repetitive and mundane tasks, AI enables human workers to focus on higher-value activities requiring creativity, problem-solving, and emotional intelligence. For example, in the healthcare sector, AI-assisted diagnostics and personalised treatment plans enhance the efficiency of medical professionals, reducing workload and improving

A. G. O. Mutambara, *Deploying Artificial Intelligence to Achieve the UN Sustainable Development Goals*, Sustainable Development Goals Series,
https://doi.org/10.1007/978-3-031-88423-8_13

patient outcomes. Across sectors, this reallocation of human effort drives economic productivity, facilitating faster GDP growth.

Innovation is another essential avenue through which AI impacts GDP. By unlocking new insights from large datasets, AI facilitates the creation of novel products, services, and business models. For instance, AI accelerates drug discovery in the pharmaceutical industry, enabling faster development of life-saving medications. Similarly, in agriculture, AI-driven precision farming techniques optimise crop yields, reducing waste and enhancing food security. These technological advancements expand existing markets and create new economic frontiers, further stimulating GDP growth.

However, the impact of AI on GDP is not uniform across regions. Advanced economies with strong digital infrastructure and access to skilled labour are better positioned to leverage AI's benefits. In contrast, developing countries may face barriers such as limited technological infrastructure, inadequate education systems, and the absence of enabling policies. Addressing this digital divide is critical to ensuring that all nations can participate in the AI-driven economic transformation. International cooperation, capacity building, and targeted investments are essential to bridging these gaps.

AI's deployment poses certain risks to GDP growth despite its economic potential. Automation could lead to job displacement, particularly in sectors reliant on routine tasks such as manufacturing, logistics, and customer service. Without adequate reskilling programmes, displaced workers may face prolonged unemployment, reducing household consumption and slowing GDP growth. Policymakers must design social safety nets and reskilling initiatives to ensure smooth labour market transitions.

As already stated, AI is significantly driving global GDP by enhancing productivity, promoting innovation, and creating new economic opportunities through automating tasks, optimising operations, and enabling data-driven decision-making across various industries. In 2024, AI is projected to add US$11 trillion to global GDP (US$105 trillion)—a 10.4% impact! Of the US$11 trillion, US$4.5 trillion will come from generative AI alone. This means that generative AI alone will be equivalent to the German economy, which is the number three economy in the world. Staggering! The financial contribution of generative AI to the world economy will be greater than the individual sizes of the vast economies of Japan, India, the United Kingdom, France, Russia, and Canada. It is bigger than the collective GDP of Africa (US$3.1 trillion). Just one type of AI! The overall AI contribution will be more than double the German economy and larger than all national GDPs except those of the United States and China. That is the transformative economics of AI [77, 144].

AI's impact on the global economy is expected to intensify in the next 5 years. As discussed, the stunning GDP impact stems from AI's ability to revolutionise sectors such as healthcare, finance, manufacturing, infrastructure, and agriculture, leading to increased efficiency, new products and services, and new markets. However, realising these gains will require addressing challenges related to workforce displacement, ethical considerations, and ensuring equitable access to AI technologies across different regions.

Indeed, AI profoundly impacts GDP growth through enhanced productivity, innovation, and labour reallocation. While it presents opportunities for economic expansion, its benefits must be shared equitably across regions and social groups. By addressing associated risks and ensuring inclusive AI adoption, societies can harness AI's full potential for sustainable and inclusive economic growth. This will be dramatically manifested through phenomenal national and global GDP growth rates. By focusing on achieving the SDGs and shared economic prosperity in every country, stakeholders can ensure that the AI-driven GDP growth is inclusive worldwide—no individual, community, or state should be left behind [16, 17, 86].

13.2 Deploying AI to Achieve SDGs: Key Imperatives

Indeed, deploying AI to achieve the SDGs is feasible and can be highly effective for several reasons advanced in the book. A key enabler is data-driven decision-making. AI technologies enable data analysis at unprecedented scales and speeds. By harnessing large volumes of structured and unstructured data, AI can provide valuable insights and inform evidence-based decision-making processes crucial for addressing complex development challenges. AI systems can optimise resource allocation, streamline processes, and automate repetitive tasks, increasing efficiency and cost-effectiveness in implementing SDG-related initiatives. This efficiency allows for the maximisation of impact with limited resources.

Predictive analytics and early warning systems are key ingredients. AI-powered predictive analytics can anticipate emerging trends, identify risks, and provide early warning systems for potential crises such as natural disasters, disease outbreaks, or food shortages. This proactive approach enables timely interventions to mitigate risks and minimise negative impacts on SDG progress. AI technologies enable the development of personalised solutions tailored to specific contexts, demographics, and needs. By targeting interventions more effectively, AI can enhance the relevance, accessibility, and effectiveness of SDG-related programmes and services.

AI promotes innovation and collaboration across sectors by enabling interdisciplinary approaches, data sharing, and co-creation of solutions. Through partnerships between governments, academia, industry, and civil society, AI can unlock new opportunities for addressing SDG challenges through novel technologies and approaches. Successful AI interventions can be scaled rapidly and replicated across different contexts, regions, or sectors, leveraging economies of scale and maximising impact. This scalability enables the widespread adoption of effective solutions and accelerates progress towards achieving the SDGs on a global scale.

AI facilitates real-time monitoring and evaluation of SDG-related programmes and initiatives, allowing stakeholders to track progress, identify bottlenecks, and adjust strategies accordingly. This adaptive approach improves accountability, transparency, and learning for continuous improvement.

AI can empower marginalised communities by providing access to information, services, and opportunities that were previously inaccessible. By addressing digital divides and promoting inclusivity, AI contributes to leaving no one behind, a core principle of the SDGs. AI technologies have cross-cutting applications across various SDGs and sectors, ranging from healthcare and education to agriculture, environmental conservation, and infrastructure development. This versatility enables holistic approaches to sustainable development and fosters synergies between different goals and targets.

AI interventions, designed with sustainability as a key anchor, can have long-lasting impacts beyond short-term gains. By building local capacities, strengthening institutions, and fostering innovation ecosystems, AI contributes to building resilient communities capable of achieving sustainable development over the long term.

The main ingredients for the successful deployment of AI to achieve the SDGs can be summarised as follows:

- Addressing the basics such as general infrastructure, digital infrastructure, financing and investment, basic education, talent and expertise, energy/power, compute resources infrastructure, organisational preparedness, AI ecosystem, and the right mindset.
- Harvesting and studying examples of effective deployment for AI for SDGs.
- Sharing the obtained lessons and emerging experiences worldwide.
- Putting in place and leveraging key enablers and drivers (critical success factors).
- Developing and adopting enabling legislation and safety regulations.

- Development and adoption by each country, organisation, sector and community of an AI Strategic Framework comprising six distinct but related components: vision, strategy, policy, governance, legislation/regulations, and implementation matrix.
- Adopting AI as part of a broader digital and technology revolution.

Indeed, deploying AI to achieve the UN SDGs offers immense potential for accelerating progress towards a more sustainable, inclusive, and prosperous future for all. By leveraging AI technologies strategically and responsibly, stakeholders can unlock new opportunities and overcome existing challenges in pursuit of the SDGs.

13.3 Are the SDGs Achievable in 2030?

The United Nations set an ambitious target of achieving all the SDGs by 2030. Whether this target will be fully achieved depends on various factors, including political will, global cooperation, financial investment, technological advancements, and societal changes. As of the current trajectory, it's unlikely that all goals will be fully achieved by 2030. However, significant progress has been made in many areas since the SDGs were adopted in 2015.

While there has been progress in certain areas, such as poverty reduction, improvements in healthcare, and advancements in renewable energy adoption, challenges remain in other areas, such as inequality, climate change mitigation, and biodiversity conservation. Addressing inequalities within and among countries remains a significant challenge. Disparities in income, access to education, healthcare, clean water, and sanitation persist, particularly for marginalised and vulnerable populations.

Climate change poses a significant threat to achieving many of the SDGs. Without urgent and decisive action to reduce greenhouse gas emissions, mitigate climate impacts, and enhance resilience, progress towards SDGs could be undermined. Achieving the SDGs requires strong international cooperation, partnerships, and solidarity among countries, as many of the challenges are transnational in nature. Strengthening global governance mechanisms and commitments to multilateralism are essential for addressing shared challenges effectively.

Adequate financial resources, investment, and technology transfer are crucial for implementing SDG-related projects and programmes, particularly in developing countries. Enhancing domestic resource mobilisation, attracting foreign investment, and ensuring the effective use of development assistance are critical for achieving the SDGs. Integrating the SDGs into national development plans, policies, and strategies is essential for mainstreaming sustainable development principles across sectors and governance levels. Coordinated and coherent policy approaches are needed to address interconnected challenges and promote synergies among different goals.

Improving data collection, monitoring, and reporting mechanisms are essential for tracking progress towards the SDGs, identifying gaps and challenges, and informing evidence-based decision-making. Strengthening statistical capacities, data disaggregation, and accountability mechanisms are critical for ensuring transparency and accountability in implementing the SDGs.

While achieving all the SDGs by 2030 may be challenging, it is essential to maintain momentum, scale up efforts, and redouble commitments to sustainable development. Even if all goals are not fully achieved by the target date, progress towards the SDGs can still significantly impact people's lives, the environment, and the planet's future.

13.4 Resolving the Risks of AI to SDGs

Continued dedication, innovation, and collaboration are needed to advance the global agenda for sustainable development. One of the major tasks is resolving the risks that AI poses to achieving SDGs. These dangers are discussed in Chap. 5. While AI has the potential to significantly contribute to the achievement of the SDGs, it also poses several challenges that could undermine progress towards these goals. AI systems involve extensive computing facilities, huge financial investments, consumption of a lot of energy, use of critical minerals, and generation of significant greenhouse gas emissions. A lot of water is also used to cool the hardware. The AI energy requirements are so massive that some AI companies are busy procuring nuclear plants to satisfy their internal needs. AI's colossal carbon footprint can potentially negate the SDG agenda.

Indeed, the growing adoption of AI, particularly large models like GPT-4 and Deep Learning systems, has led to a significant increase in energy consumption and a larger carbon footprint. Training and deploying these models require vast computational power, storage, and continuous processing, often relying on energy-intensive cloud data centres. Globally, there must be a concerted effort to reduce AI's carbon footprint by developing less energy-intensive, cheaper, and greener AI. Several strategies and solutions must be developed and deployed to mitigate AI's environmental impact. They range from optimising models' efficiency to leveraging renewable energy sources.

Model optimisation is one of the most effective ways to reduce AI's carbon footprint. Techniques such as pruning, quantisation, and transfer learning enable AI models to operate with less computational power. Pruning removes unnecessary neurons and connections, while quantisation reduces the precision of mathematical calculations, significantly lowering the energy required for training and inference. Transfer learning allows models to be fine-tuned on existing pre-trained models, avoiding the need for full-scale retraining. Additionally, federated learning allows AI models to be trained directly on decentralised edge devices, like smartphones, without requiring large data transfers to centralised servers. This approach reduces the dependency on energy-hungry data centres and cuts emissions associated with cloud-based training.

Another critical approach is the use of renewable energy to power the data centres where most AI computation takes place. Tech giants like Google, Microsoft, and Amazon are committed to powering their data centres with 100% renewable energy. Cloud providers are also investing in green infrastructure, including hyperscale data centres that use more efficient cooling systems and energy storage solutions to reduce reliance on fossil fuels. For instance, data centres now employ liquid cooling and natural air cooling instead of traditional air conditioning, significantly cutting energy use. Companies can substantially reduce emissions by running energy-hungry AI models on servers powered by wind, solar, or hydroelectric power.

Another emerging solution is the adoption of green AI development practices. This approach focuses on tracking and reducing the energy consumption of AI models throughout their lifecycle. Tools like Carbontracker and Machine Learning CO2 Impact Calculator enable developers to measure and report the carbon emissions associated with training models, fostering greater accountability. Research conferences and AI competitions also encourage researchers to report their AI models' environmental impact as part of responsible publishing practices. Additionally, companies are increasingly relying on specialised, energy-efficient hardware like Google's TPUs (Tensor Processing Units), application-specific integrated circuits (ASICs), and other low-power AI accelerators that reduce computational overhead.

Government regulations and ethical AI development are critical in addressing AI's carbon footprint. Regulatory frameworks like the EU AI Act propose stricter sustainability criteria, encouraging companies to design AI systems that are both socially and environmentally responsible. Governments may also introduce carbon taxes and impose energy-efficiency standards on data centres, requiring

Strategy	How it Works	Impact on Carbon Reduction
Model Optimisation	Reduce model size pruning quantisation	Major Reduction
Green Cloud Computing	Use renewable energy powered clouds services	Major Reduction
Renewable Energy	Power data centres with wind solar or hydro	Major Reduction
Edge AI	Run AI models on local devices not in the cloud	Significant Reduction
Low-Power AI Chips	Use energy-efficient chips like TPUs and ASICs	Significant Reduction
Transparency and Reporting	Report carbon impact and train models ethically	Moderate Reduction
Government Regulations	Impose rules on carbon reporting and energy use	Moderate Reduction

Fig. 13.1 Key strategies for reducing the AI carbon footprint

tech companies to disclose the emissions associated with AI models. Meanwhile, ethical AI certifications—similar to green building certifications like LEED—are being explored as a means to hold companies accountable for sustainable AI development.

Finally, decentralised AI and Edge AI offer game-changing potential for sustainability. Instead of relying on large, centralised data centres, companies can deploy AI models locally on edge devices like smartphones, IoT sensors, and autonomous vehicles. By processing AI workloads on-device, organisations reduce the need for constant cloud connectivity, saving energy and cutting emissions. Companies like Apple have already embraced this approach with their Neural Engine, which enables on-device AI processing.

Indeed, AI's enormous carbon footprint can be addressed through a combination of model optimisation, renewable energy adoption, hardware efficiency, green AI development practices, and regulatory oversight. Figure 13.1 summarises the strategies, how they work, and their corresponding impact. From energy-efficient AI models to green data centres, the path towards sustainable AI development is clear. By embracing these strategies, companies, governments, and researchers can ensure that AI's risks to SDGs are mitigated and that the technology becomes a positive tool for achieving the goals in 2030 [26, 31].

13.5 Addressing AI Inequity

There is AI inequity between the Global North and the Global South, as discussed in Chap. 5. This requires undivided, razor-sharp attention, as it undermines the latter's effectiveness in deploying AI to achieve the SDGs. There is a need to democratise AI within the Global North and between the Global South and Global North. The rapid rise of AI has created a significant divide between highly industrialised countries and the least industrialised ones. Wealthier nations have the infrastructure, technical expertise, and financial resources to develop and deploy AI systems. At the same time, least industrialised countries often lack access to the tools, skills, and data necessary to leverage AI for economic and social development. If left unaddressed, this divide will deepen existing inequalities, hindering efforts to achieve the SDGs, particularly in areas such as education, healthcare, and economic growth.

Strategies and solutions must be developed and deployed to bridge this gap, focusing on capacity building, infrastructure development, regulatory reform, and international cooperation. Building local capacity and skills is one of the most critical steps in addressing AI inequity. AI development requires specialised skills in Machine Learning, data science, and software engineering—skills often in short supply in the least industrialised countries (LICs). To close this gap, education and training programmes must be prioritised. National governments, universities, and international development agencies can collaborate to introduce AI-focused curricula at schools and universities, offer professional certification programmes, and support the development of AI research hubs. Online education platforms like Coursera, edX, and Udacity offer accessible, low-cost courses on AI and Machine Learning, but local in-person bootcamps and coding workshops should complement these efforts. Moreover, partnerships with tech giants like Google, Microsoft, and IBM can facilitate mentorship programmes and knowledge exchange, giving students and professionals in LICs direct access to AI expertise. Ensuring gender inclusion in AI education is equally vital, as women remain underrepresented in the global AI workforce.

Infrastructure development is another critical component in closing the AI divide. AI relies on computational power, high-speed Internet, and access to cloud-based platforms—resources that are limited in many LICs. One solution is to build affordable, sustainable digital infrastructure, such as satellite-based Internet (like Starlink) or 5G networks. Partnerships with cloud providers like Amazon Web Services (AWS), Google Cloud, and Microsoft Azure can grant LICs access to cloud credits, allowing universities, startups, and research institutions to use cloud-based platforms at reduced costs. For AI development to thrive, LICs must also reduce the cost of AI hardware like GPUs, edge devices, and IoT sensors, which are essential for processing AI models. Instead of building large data centres (which are expensive and energy-intensive), LICs can leverage cloud-based AI platforms and edge computing to run models locally on small devices. Decentralised AI approaches, such as federated learning, can also reduce reliance on large cloud systems by allowing AI models to be trained directly on edge devices like smartphones, thereby saving energy and infrastructure costs.

Data sovereignty is another essential issue in addressing AI inequity. Data is the fuel for AI, but many LICs have little control over their own data. Multinational corporations often extract and store data from LICs to train models abroad. To counter this, LICs should adopt data localisation policies and create regional data repositories where data is stored and processed locally. This would ensure that the benefits of AI development are retained within the country and that privacy regulations are respected. Countries like India, South Africa, and Nigeria have begun to develop data protection regulations modelled on frameworks like the European Union's General Data Protection Regulation (GDPR). By creating data governance frameworks, LICs can retain ownership of their own data, enabling them to train AI models tailored to local needs. Additionally, open data initiatives can encourage the use of anonymised, non-sensitive datasets for research, enabling universities and startups in LICs to access vital training data.

To bridge the AI divide, it is also essential to promote local AI innovation. While highly industrialised countries focus on AI for automation, robotics, and consumer technology, LICs have distinct healthcare, agriculture, education, and climate resilience priorities. AI solutions in these areas can have transformative effects. For instance, AI-powered weather prediction tools can help farmers make better planting decisions, while AI-based medical diagnostic tools can support rural clinics with limited access to doctors. Supporting local innovation hubs and AI incubators can foster the growth of homegrown startups. International organisations like AI for Good provide platforms for showcasing AI-driven solutions aimed at addressing development challenges. Public-private partnerships are also essential. Large technology companies can co-invest with development finance institutions, offering funding, mentorship, and technical resources to local AI startups working on projects that support the SDGs.

Funding and investment play a central role in addressing AI inequity. Many of the least industrialised countries lack the financial capacity to invest in AI development independently. International donors, development finance institutions (DFIs), and multilateral organisations like the World Bank, African Development Bank (AfDB), and International Finance Corporation (IFC) can provide concessional loans, grants, and technical assistance to support AI-related projects. These funds can be directed towards building local AI capacity, supporting AI research, and promoting AI-based startups. Impact investors also play a role by supporting social-impact-driven startups that use AI for sustainable development. One effective financing strategy is results-based financing, where governments and private funders provide financial support only if AI-driven projects achieve tangible, measurable development outcomes—such as improving healthcare outcomes or increasing agricultural yields.

Technology transfer and knowledge sharing are also crucial to ensuring AI equity. Wealthy nations and international bodies like the United Nations and the World Economic Forum can facilitate the transfer of AI-related technologies, tools, and intellectual property to LICs. Companies like Google, Microsoft, and Facebook have democratised access to AI development tools by making open-source libraries like TensorFlow, PyTorch, and Keras freely available. Technology transfer initiatives can also include access to patents and intellectual property, enabling LICs to use, modify, and improve existing AI technologies. Tech-sharing platforms allow LICs to develop AI models without incurring licensing fees. Public-private partnerships can further support this effort, as corporations can help fund local AI development hubs and provide hands-on training to local AI developers.

Another crucial aspect of reducing the AI divide is regulatory reform. Countries with clear national AI strategies are better positioned to attract investment, foster local innovation, and build public trust. The European Union (EU) has established the EU AI Act, prioritising fairness, transparency, and ethical AI development. LICs can take inspiration from such frameworks to develop their own national AI regulations. These regulations should cover data privacy, ethical AI development, and inclusive access to AI resources. Clear regulations will enable LICs to participate in cross-border AI collaborations and establish partnerships with foreign investors and development organisations.

Finally, to address the power imbalance with Big Tech, LICs must reduce their reliance on a handful of dominant corporations like Google, Microsoft, Amazon, and OpenAI. Instead of relying on proprietary cloud platforms, LICs can support regional AI consortia like AI4D (AI for Development) in Africa or Data Science Nigeria. Such initiatives build local capacity for AI development and ensure that technological innovation is centred around local priorities, not the profit motives of foreign corporations. Regional partnerships also strengthen collective bargaining power, enabling LICs to negotiate better deals on cloud usage, patents, and intellectual property.

AI holds tremendous potential to accelerate economic and social development, but without deliberate intervention, the gap between highly industrialised and least industrialised countries will continue to widen. Capacity building, data sovereignty, infrastructure development, regulatory reform, and technology transfer are critical components of the solution. To achieve AI equity, the Global South must prioritise investments in AI education, build resilient infrastructure, and foster local innovation that addresses local challenges. With support from international donors, public-private partnerships, and development finance institutions, LICs can access the tools, skills, and resources needed to close the AI divide. If done successfully, AI can become a force for inclusive, sustainable development globally. More specifically, the strategies and solutions outlined in this section will mitigate AI inequity and thus enable the technology to be effectively deployed to achieve the SDGs without leaving any country or community behind.

13.6 Rethinking the SDG Framework

Critics of the SDG approach have raised significant concerns regarding the framework's efficacy. Objections include the non-binding nature of the goals, the lack of coherent enforcement mechanisms, the ambitious scope of the goals (17 goals and 169 targets), and sovereignty concerns from major world powers such as the United States and Russia. The SDG nomenclature has also become a source of discord. The term "goals" implies optional aspirations or distant ideals that countries may choose to pursue at their discretion. This framing has undermined the essential nature of what the SDGs truly represent—a comprehensive risk assessment of humanity's most pressing threats in the 21st century. The UN SDGs are not merely desirable objectives but rather identify critical systemic risks that, if left unaddressed, threaten global stability and human well-being, regardless of national boundaries or economic development.

For example, Climate Action (SDG 13) should not be viewed merely as an environmental aspiration. Climate change serves as an existential risk multiplier that worsens resource scarcity, extreme weather events, and mass displacement. The World Bank estimates that without decisive action, climate change could force more than 140 million people to migrate within their countries by 2050, creating instability that crosses borders. Similarly, Clean Water and Sanitation (SDG 6) is far more than a development target. Water scarcity presents a security threat, with UN research indicating that water-related tensions impact regions accounting for 45% of global GDP. Furthermore, Reduced Inequalities (SDG 10) must be regarded as an existential concern. Indeed, inequality functions as a significant risk accelerator and multiplier. IMF research shows that high inequality reduces economic growth by 0.8 percentage points annually in affected nations while increasing political instability.

Consequently, it is prudent to recast the SDG framework as a global risk mitigation strategy, where addressing systemic risks is not optional but essential for the planet's survival. Reframing the SDGs away from aspirational targets leads to a clearer understanding of their true significance. They are not optional or aspirational "good deeds" but existential interventions to guarantee the preservation of the earth while ensuring inclusive development and economic transformation leading to shared prosperity globally.

13.7 Beyond 2030

As 2030 beckons, it is clear that the SDGs are desperately and critically behind schedule. Urgent, innovative, and radical interventions are required from governments to treat the SDGs as the key developmental strategy informing and driving all national policies. They must translate the goals into national legislation and policies, develop elaborate implementation plans, and reallocate public budgets to anchor the execution. Razor-focused attention must be given to poverty reduction, education, gender equality, and sustainable resource use. Indeed, there is a foundational need to rethink and reimagine global development frameworks post-2030. The SDGs constitute a universal and holistic framework for global progress, not a menu for selective adoption in pursuit of narrow political agendas by national elites. This instrumentalisation of the SDGs is counter-productive. It leads to fragmented implementation, where goals that advance the political or economic interests of the few are given precedence over those addressing broad issues of poverty eradication, economic empowerment, climate change, environmental sustainability, governance, and equity.

Indeed, the implementation of SDGs must be context-aware, culturally sensitive, anchored by national history, and integrated into all developmental programmes. It must address both local needs and global imperatives. One-size-fits-all goals must give way to adaptable frameworks that are informed and driven by national, regional, and continental socioeconomic aspirations and environmental challenges. To crown it all, there must be effective SDG monitoring, evaluation, and feedback systems at the three levels—country, regional bloc, and continent. As the adage goes, if you can't measure it, you can't manage it. More significantly, that which is measured gets done.

Beyond 2030, there is a need for innovative and futuristic thinking. For example, the paper "Beyond 2030: Structures for Achieving Sustainable Development" by Tom Cernev and Richard Fenner proposes a forward-looking framework for international development goals to succeed the SDGs post-2030 [20]. Drawing on lessons from the Millennium Development Goals and SDGs, and recognising global challenges such as climate change, planetary boundaries, and global catastrophic risks, the authors advocate a streamlined set of seven interconnected goals categorised under environmental, physical, social, and collaboration themes. These goals emphasise sustainability within planetary limits, equitable access to basic services, poverty and inequality reduction, and institutional strengthening. The paper introduces a novel system of five-year risk assessments to evaluate the consequences of unmet targets, aiming to create a dynamic and accountable framework aligned with long-term objectives like the 2050 net-zero target. There must be urgency of international cooperation, realistic goal-setting, and improved financing mechanisms to ensure sustainable development in an increasingly complex global environment. Figure 13.2 shows the lessons learnt from both the MDGs and SDGs. These insights must be used to inform the post-2030 SDG strategy.

Summary of Lessons Learnt from the MDGs	Summary of Lessons Learnt from the SDGs
• No obligations or binding conditions • Insufficient data/progress tracking • Insufficient progress achieved • Progress is undermined by climate change, environmental degradation, and conflict • Wording and ambition of goals/targets means they were always unlikely to be achieved • Approached social issues with a one-size-fits-all mentality • Did not sufficiently address environmental/sustainability issues • Insufficient initial baselines for progress to be measured against	• Insufficient data/progress tracking • Too many targets are unachievable and require too many resources • Contradictions between goals/targets should be avoided • Targets are black and white and do not consider the spectrum of possibilities • Too much influence from international foundations and non-governmental organisations • Overuse as a marketing strategy for companies • No obligations or binding conditions • Easily derailed by world events

Fig. 13.2 Summary of lessons learnt from MDGs and SDGs [20]

13.8 Final Reflection

On 25 April 2023, UN Secretary-General António Guterres delivered remarks during the launch of the Special Edition of the Sustainable Development Goals (SDGs) Progress Report. In his address, he stated, "Unless we act now, the 2030 Agenda will become an epitaph for a world that might have been". This statement underscores the critical need for immediate and transformative action to achieve the SDGs by their 2030 deadline. The report highlighted that only 12% of SDG targets were on track, with progress on 50% being weak and insufficient, and over 30% of targets either stalled or regressed [130,131].

Guterres emphasised that without prompt and collective efforts, the aspirations outlined in the 2030 Agenda risked remaining unfulfilled, serving as a poignant reminder of missed opportunities for global development and cooperation. The Secretary-General's remarks reflect the urgency and gravity of humanity's global challenges in achieving the UN 2030 SDG agenda. The statement underscores the growing recognition that progress on key SDGs is either stagnating or regressing due to a convergence of global crises, such as climate change, economic inequality, pandemics, and geopolitical conflicts. Indeed, without immediate and collective action, the vision for a sustainable and just world by 2030 could remain unfulfilled. The phrase "epitaph for a world that might have been" evokes a powerful metaphor, symbolising the death of a shared dream if humanity fails to act. It is now July 2025, and Guterres's trepidation is still accurate and unresolved. There is a need for action.

Governments, businesses, civil society, and individuals must accelerate efforts in tackling critical global issues. Specific areas that require urgent attention include (1) reducing greenhouse gas emissions to avoid surpassing the 1.5°C global warming threshold; (2) bridging the gaps in education, healthcare, and access to technology between the Global North and the Global South; (3) supporting post-pandemic recovery efforts, especially in vulnerable and least-developed countries (LDCs); and (4) mobilising the necessary financial resources to support sustainable development, especially for emerging and least industrialised economies burdened by debt crises.

The United Nations Pact for the Future, adopted by the UN General Assembly at the UN Summit of the Future on 22 September 2024, addressed sustainable development and financing for development; international peace and security; science, technology and innovation and digital cooperation; youth and future generations; and transformation of global governance. The Pact incorporates a Global Digital Compact and a Declaration on Future Generations. Recognising that 85 percent of the Sustainable Development Goals (SDGs) defined in Agenda 2030 in 2015, were off track, stagnating or in recession, at the end of 2024, and that the financing gaps for the SDGs and climate action under the Paris Agreement on Climate Change had reached trillions of US dollars, while fiscal constraints and debt distress marked financing conditions for developing countries, the UN Pact for the Future prioritised the need for urgent reforms to the global financial architecture to enable effective means for financing for development, and a just energy transition.

The world must understand the moral imperative of solidarity and shared responsibility. Collective humanity and shared dignity demand nothing less. The potential future envisioned in the 2030 SDG agenda—a world free from poverty, hunger, and environmental degradation—is slowly becoming elusive, if not illusionary. That desired future—complete attainment of the SDGs—is not inevitable. It is contingent on immediate and transformative action. Political and business leaders, policymakers, academics, civil society activists, and ordinary citizens must reignite momentum towards the SDGs, ensuring that 2030 becomes a milestone of achievement rather than a moment of regret.

This agenda demands visionary leadership at organisational, national, regional, continental and global levels—characterised by the ability to create and articulate a clear, compelling future vision that inspires and motivates others to achieve the shared SDGs. These leaders must have a unique blend of foresight, passion, and innovation, enabling them to see beyond the current SDG implementation challenges, and anticipate potential redemptive interventions. All this must be anchored by our shared common humanity and global interests, not narrow national, sectarian, or hegemonic interests. Global cooperation, regional/continental integration, moving up global value chains, inclusive economic transformation, addressing the climate crisis, and use of advanced technology (in particular AI) can play a significant role in the arduous journey to 2030.

Of course, there is the danger that AI will widen global inequality. Left unchecked, AI can intensify global disparities by consolidating power and wealth in affluent nations while exploiting labour and resources in emerging and least industrialised countries. Without deliberate interventions such as those proposed in this publication, AI will entrench existing inequities, leading to heightened political instability, environmental degradation, and cultural dominance by a select few. This book seeks to mitigate and prevent this potential calamity. The UN Secretary-General, António Guterres, is right: "We must never allow AI to stand for Advancing Inequality." Indeed, this transformative technology can be leveraged to achieve the SDGs in every country. However, there is a need for clarity regarding the critical enablers, drivers, and strategic framework.

AI Applications in Agriculture and Mining

A

Improvements in agriculture and the unlocking of value in mining can dramatically advance the attainment of several SDGs. Specifically, agriculture directly affects SDGs 1, 2, 3, and 13, while mining influences SDGs 8, 9, 10, 11, and 12. Hence, AI applications in these two sectors are foundational in deploying AI to achieve the SDGs. This appendix outlines examples of such interventions and their impact.

A.1 Precision Agriculture: Kenya

IBM's Watson-based precision agriculture initiative in Kenya is a compelling case study on the application of AI in agriculture. This initiative leverages AI to improve crop yield, manage resources, and reduce agricultural risk in unpredictable climates. This approach has brought significant advances for smallholder farmers in Kenya, particularly in regions where traditional agriculture faces challenges due to climate variability, limited access to technology, and financial constraints.

A.1.1 Background

Kenya's agriculture is critical to its economy, with smallholder farmers producing the bulk of its food supply. However, issues such as erratic rainfall, outdated farming practices, and limited access to modern technologies have led to low crop yields and economic instability. Precision agriculture with IBM's Watson Decision Platform for Agriculture combines weather data, remote sensing, and Machine Learning models to offer precision agriculture insights tailored for each farmer:

- Weather Prediction and Crop Scheduling: Watson uses AI to analyse weather patterns, rainfall, and temperature forecasts to guide planting and harvesting schedules. Farmers receive tailored recommendations on when to plant, irrigate, or fertilise crops based on real-time weather insights. This data-driven approach helps mitigate losses due to climate unpredictability.
- Soil Health Monitoring and Fertiliser Optimisation: Watson provides insights into optimal fertiliser usage and soil conditioning by integrating soil health data with AI. Farmers can precisely adjust soil treatment to enhance crop growth, prevent nutrient depletion, and increase sustainability.
- Pest and Disease Control: AI models in Watson detect patterns of pest infestations and predict outbreaks based on environmental factors. This allows farmers to apply pesticides only when needed, reducing chemical usage and costs while protecting crops more effectively.

A. G. O. Mutambara, *Deploying Artificial Intelligence to Achieve the UN Sustainable Development Goals*, Sustainable Development Goals Series,
https://doi.org/10.1007/978-3-031-88423-8

- Yield Prediction and Market Insights: Watson predicts potential crop yields by analysing historical and current data on soil health, crop type, and weather patterns. It also provides market insights, helping farmers make informed decisions about when to sell crops for maximum profit.

A.1.2 Outcomes and Impact

- Increased Crop Yield: Farmers using the Watson platform reported improved crop yields due to better crop timing, optimised fertiliser use, and reduced losses from pests.
- Resource Efficiency: AI-driven recommendations reduced water usage and fertiliser application, contributing to more sustainable farming practices.
- Economic Stability: With data-backed predictions and insights, farmers gained financial stability through improved yields and better timing for crop sales.
- Climate Resilience: The technology allowed farmers to adapt to climate variability, which is increasingly vital in regions vulnerable to climate change.

A.1.3 Key Findings

- AI and Accessibility: IBM's solution is accessible through mobile phones, making it suitable even for rural farmers with limited technology access.
- Sustainability: Precision agriculture helps reduce waste and conserve resources, aligning with sustainable agricultural practices.
- Scalability: AI models can be adapted for various crops and regions, showing potential for widespread application across sub-Saharan Africa.

A.1.4 Conclusion

This case demonstrates AI's transformative potential in agriculture, especially for the Global South, where climate resilience, resource management, and yield optimisation are crucial. Using AI, farmers can make data-driven decisions that improve productivity, sustainability, and resilience.

A.2 Mining: Rio Tinto

A.2.1 Background

Rio Tinto, a global mining corporation, implemented AI-driven autonomous mining operations in the Pilbara region of Western Australia. The company aimed to boost productivity and reduce operational costs by adopting AI technology for vehicle automation, remote operations, and predictive maintenance. These advancements became part of their "Mine of the Future" programme, which leverages AI and Machine Learning to revolutionise traditional mining processes.

A.2.2 Problem

Mining operations are labour-intensive, high-risk, and costly. Some specific challenges include:

- Safety Risks: Mining is hazardous, with risks from heavy machinery, blasting, and working in remote areas.
- Operational Inefficiency: Relying on human operators limits efficiency, especially in large and remote mines.
- High Maintenance Costs: Mining equipment suffers from significant wear and tear, leading to high maintenance costs.
- Labour Shortages in Remote Regions: Attracting a skilled workforce to remote mining areas is challenging.

A.2.3 AI Solution

Rio Tinto developed a range of AI applications for mining, with a particular focus on Autonomous Haulage Systems (AHS) for mining trucks, predictive maintenance for equipment, and operational optimisation using data analytics.

- Autonomous Haulage System: Rio Tinto implemented an autonomous haulage fleet in its Pilbara iron ore mines, using AI-powered trucks that operate without human drivers. AI algorithms allow trucks to navigate mining routes, avoid obstacles, and make real-time decisions. Remote operators can monitor the fleet from a control centre hundreds of miles away, further reducing the need for onsite personnel.
- Predictive Maintenance: Using Machine Learning, Rio Tinto monitors and predicts equipment health based on data from sensors installed on machinery. AI algorithms analyse patterns in equipment use, wear, and environmental conditions, predicting when failures are likely to occur. This reduces downtime and extends equipment life, significantly lowering maintenance costs and improving operational efficiency.
- Data-Driven Decision Making: AI and data analytics optimise operations by analysing extensive data generated from mine sites. AI models identify bottlenecks and inefficiencies, allowing Rio Tinto to adjust schedules, optimise routes, and maximise ore extraction.

A.2.4 Impact

- Safety: Autonomous vehicles reduce human involvement in high-risk zones, minimising accidents and fatalities. Rio Tinto has reported a significant reduction in onsite incidents.
- Increased Productivity: Autonomous trucks operate 24/7, unaffected by human limitations, which increases productivity. The Pilbara mines achieved a 15% improvement in productivity within the first few years of deploying the autonomous fleet.
- Cost Savings: Rio Tinto saves millions in operational costs each year by reducing the reliance on human operators and optimising maintenance. Predictive maintenance reduces downtime by 30%, allowing equipment to remain operational longer.
- Environmental Benefits: Autonomous operations minimise fuel consumption by optimising routes and reducing idle times, lowering greenhouse gas emissions.

A.2.5 Future Applications and Scaling Potential

Rio Tinto is expanding its autonomous operations and plans to integrate AI for even more tasks, such as drilling and blasting. The success in Pilbara has set a precedent for other mining companies, leading to increased interest in AI-driven automation across the industry.

A.2.6 Lessons Learnt

- Safety and Productivity Synergy: AI reduces safety risks while improving productivity, showing how technology can align business and safety goals.
- Cost Efficiency through Predictive Maintenance: Machine learning helps predict failures, saving costs on repairs and preventing downtime.
- Scalability of Autonomous Systems: Autonomous systems can be scaled to cover large areas and support multiple operations remotely, making them ideal for extensive mining sites.

A.2.7 Conclusion

Rio Tinto's experience shows how AI-driven automation and predictive analytics can transform mining, making it safer, more efficient, and more sustainable.

AI Applications in Infrastructure, Governance, and Manufacturing

B

Improvements in infrastructure, governance, and manufacturing can dramatically improve the attainment of the SDGs. Infrastructure directly influences SDGs 1, 8, 9, 11, and 13; governance affects the implementation of all SDGs, particularly SDGs 9, 12, and 16, while manufacturing drives SDGs 1, 9, 12, and 13. Hence, AI applications in these sectors are critical in deploying AI to achieve the SDGs. This appendix outlines examples of such interventions and their impact.

B.1 Infrastructure Management: Barcelona

B.1.1 Background

One significant example is the application of AI in predictive maintenance and smart city infrastructure management. Barcelona, a leading smart city, has heavily invested in infrastructure innovation to improve public services, manage resources efficiently, and enhance the quality of life for its residents. The city implemented AI as part of its broader digital transformation strategy to address infrastructure challenges such as ageing assets, energy efficiency, and real-time management of public services.

B.1.2 Problem

Barcelona faced several infrastructure management challenges:

- Ageing Infrastructure: With an older infrastructure system, the city needed a way to prevent costly and disruptive failures.
- Resource Management: Efficient use of water, electricity, and other resources was crucial to reduce costs and environmental impact.
- Crowd and Traffic Management: As a popular tourist destination, Barcelona faced challenges managing high foot and vehicle traffic, leading to congestion and maintenance strain on public infrastructure.

A. G. O. Mutambara, *Deploying Artificial Intelligence to Achieve the UN Sustainable Development Goals*, Sustainable Development Goals Series,
https://doi.org/10.1007/978-3-031-88423-8

B.1.3 AI Solution

Barcelona integrated AI across its infrastructure management systems, focusing on predictive maintenance, resource optimisation, and traffic management. The solution drivers are predictive maintenance, smart sensors, and data analytics for city infrastructure.

Predictive Maintenance for Infrastructure

- The city installed IoT sensors across critical infrastructure components, including bridges, roads, and public buildings, to monitor structural health and performance.
- AI algorithms analyse data from these sensors to detect signs of wear, corrosion, and strain, predicting potential failures before they occur.
- This proactive approach allows maintenance teams to address issues before they escalate, reducing unplanned downtime and extending the lifespan of assets.

Energy and Resource Management

- Barcelona uses AI to optimise energy usage in public lighting and buildings. For instance, streetlights are equipped with smart sensors that adjust brightness based on pedestrian traffic and ambient lighting conditions.
- AI-driven water management systems monitor water usage in public parks, fountains, and green spaces. The system can adjust irrigation schedules by analysing real-time weather data, conserving water and reducing operational costs.
- These AI-enabled systems help the city reduce unnecessary energy and water consumption, promoting sustainability and reducing utility expenses.

Traffic and Crowd Management

- Barcelona implemented an AI-powered traffic management system that uses data from cameras and sensors to monitor vehicle and pedestrian flow in real time.
- The system predicts congestion patterns, enabling traffic lights to adjust timing automatically based on traffic density, reducing delays and managing crowd flow more effectively.
- During large events, such as festivals and sports events, the AI system provides real-time data to city authorities, enabling them to reroute traffic and manage public transport availability to prevent overcrowding.

Smart Waste Management

- Barcelona's waste collection system incorporates AI to determine optimal collection routes and schedules. Sensors in waste bins monitor fill levels, alerting waste collection teams when bins are near capacity.
- AI algorithms then calculate the most efficient collection routes based on real-time data, reducing fuel consumption and labour costs.

B.1.4 Impact

- Reduced Maintenance Costs and Downtime: By implementing predictive maintenance, Barcelona reduced infrastructure repair costs by 25% and extended the lifespan of key infrastructure assets. With AI-powered alerts and preventative measures, unplanned infrastructure failures decreased by 30%, minimising disruptions for residents and businesses.

- Improved Energy Efficiency and Sustainability: The AI-driven lighting system has led to energy savings of over 30%, helping Barcelona lower its carbon footprint and utility expenses. Similarly, the AI-enabled water management system has cut water usage in parks by 25%, contributing to the city's environmental sustainability goals.
- Enhanced Traffic Flow and Reduced Congestion: Dynamic traffic management through AI improved traffic flow by approximately 20% in high-density areas, reducing commuting times and emissions from idling vehicles. During major events, AI-based traffic solutions have reduced congestion, allowing residents and tourists to navigate the city more easily.
- Higher Quality of Public Services: With optimised waste collection schedules, the city improved cleanliness in public spaces, leading to higher satisfaction among residents and tourists. Real-time infrastructure management also contributes to the city's reputation as a smart and sustainable destination, attracting global attention and boosting tourism.

B.1.5 Future Applications and Scaling Potential

Barcelona plans to expand AI applications to other areas, such as public safety and healthcare infrastructure. The city is exploring AI-based air quality monitoring systems and emergency response optimisation to improve urban resilience and public health outcomes.

B.1.6 Lessons Learnt

- Proactive management reduces costs. It extends asset life. Indeed, predictive maintenance enabled by AI helps cities prevent costly repairs and extend the longevity of public assets, saving taxpayer money and reducing service interruptions.
- Resource efficiency drives sustainability. AI's role in optimising energy and water use demonstrates that technology can contribute significantly to sustainability goals, providing both environmental and economic benefits.
- Real-time data improves decision-making and public satisfaction. AI allows for data-driven decisions, leading to improved services that enhance residents' quality of life and city resilience.

B.1.7 Conclusion

Barcelona's application of AI in infrastructure management showcases the power of AI to make cities more sustainable, efficient, and responsive. By integrating AI across key urban systems, Barcelona has established itself as a global leader in smart infrastructure, creating a model for other cities seeking to modernise their infrastructure management.

B.2 Energy Sector: M-KOPA Solar (Kenya)

A notable example of AI applications in the energy sector in the Global South is the work of M-KOPA Solar. This Kenyan company utilises AI to provide affordable solar energy solutions. M-KOPA Solar leverages AI to manage financing, optimise energy distribution, and ensure sustainable energy access for off-grid communities in sub-Saharan Africa.

B.2.1 Background

In Kenya, as in many parts of the Global South, access to reliable electricity is limited, especially in rural and underserved areas. Traditional grid expansion is often economically and logistically challenging, leading to significant energy poverty. To address this issue, M-KOPA Solar provides affordable solar home systems with a pay-as-you-go model. By integrating AI, M-KOPA enhances the affordability, accessibility, and efficiency of renewable energy distribution.

B.2.2 AI Application in Operations

- Credit Scoring and Financing with AI: M-KOPA Solar's business model relies on a pay-as-you-go financing plan, where users make small daily payments through mobile money platforms to own the solar system. AI algorithms analyse historical payment data and customer usage patterns to create individual credit scores for each customer. This enables M-KOPA to determine flexible payment plans, making it easier for low-income households to access clean energy.
- Demand Forecasting and Energy Distribution Optimization: AI models predict demand patterns based on user data and seasonal factors, optimising the distribution of energy resources. This forecasting allows M-KOPA to manage supply chains more effectively, ensuring that solar systems and maintenance support reach high-demand areas promptly.
- Predictive Maintenance and Customer Support: By analysing usage data from solar systems, AI detects anomalies that could indicate equipment issues or the need for maintenance. This predictive maintenance approach helps prevent system failures, reduce downtime, and ensure reliable energy access for users. AI-powered chatbots and support systems also provide customer assistance, improving service delivery without requiring extensive human resources.
- Product Customization and Expansion of Services: AI-driven insights help M-KOPA design products that better fit the needs of different communities. For example, they can tailor products for users with higher energy requirements, such as small businesses. Additionally, AI allows M-KOPA to expand its services beyond solar energy to include other products, like TVs and smartphones, using the same financing model.

B.2.3 Outcomes and Impact

- Increased Energy Access: M-KOPA Solar has connected over a million households in sub-Saharan Africa to clean, affordable solar power, providing reliable energy for lighting, cooking, and business activities.
- Improved Affordability: The AI-powered credit scoring system has made solar energy financially accessible to low-income households, allowing them to access energy without significant upfront costs.
- Economic Empowerment: Access to reliable energy has enabled small businesses to extend their operating hours, increasing local economic productivity and community well-being.
- Environmental Impact: M-KOPA's renewable energy solutions help reduce reliance on kerosene lamps and diesel generators, reducing carbon emissions and improving home indoor air quality.
- Scalability and Adaptability: The success of M-KOPA in Kenya has led to expansion into other African countries, including Uganda and Tanzania, demonstrating the scalability of AI-enabled solutions in the energy sector.

B.2.4 Lessons Learnt

- AI and Financial Inclusion: M-KOPA's AI-driven credit scoring system provides financial inclusion by making energy financing accessible to underserved communities.
- Sustainability: By offering affordable, renewable energy alternatives, M-KOPA aligns with global sustainability goals and addresses the energy needs of communities sustainably.
- Economic Development: AI-driven solar energy distribution has enabled community growth by supporting small businesses and reducing reliance on fossil fuels.
- Potential for Widespread Adoption: M-KOPA's model can be replicated across other countries in the Global South with similar challenges, showing the power of AI in addressing energy poverty and supporting sustainable development.

B.2.5 Conclusion

This case study illustrates how AI in the energy sector can foster economic empowerment, support environmental sustainability, and offer scalable solutions to energy poverty in the Global South. By enabling affordable solar energy access, AI-powered solutions like M-KOPA Solar play a crucial role in advancing clean energy and improving the quality of life for underserved populations.

B.3 Mobile Network Optimisation: Vodafone

B.3.1 Background

Vodafone, a global telecommunications leader, operates extensive mobile networks across Europe and other regions, handling millions of users daily. Vodafone UK adopted AI to optimise its network performance, automate customer service, and provide a better user experience. The initiative aimed to address common challenges such as network congestion, maintenance costs, and improving customer satisfaction.

B.3.2 Problem

Vodafone UK faced multiple challenges typical of large telecom providers:

- Network Congestion and Downtime: Growing data usage led to increased congestion, which degraded the quality of service, particularly during peak hours.
- Reactive Maintenance: Network issues were often addressed reactively, leading to prolonged downtime and increased operational costs.
- Customer Service Inefficiencies: Handling millions of customer interactions presented challenges, with long wait times and difficulty addressing diverse issues promptly.

B.3.3 AI Solution

Vodafone implemented AI-driven solutions for predictive network maintenance, real-time network optimisation, and an AI-based customer service assistant. Thus, the key drivers of the solution are predictive maintenance, network optimisation, and AI-powered customer service:

- Predictive Maintenance: AI algorithms analyse network data to predict when components are likely to fail.

  Vodafone uses Machine Learning models to monitor data from cell towers and network components, identifying patterns that indicate potential failures. By predicting outages and performance drops, the system schedules maintenance proactively, preventing issues before they impact users.
- Real-Time Network Optimisation: Vodafone uses AI algorithms for real-time network management. These algorithms analyse user activity, traffic patterns, and data usage to dynamically allocate bandwidth, particularly during peak hours. For instance, when AI detects a high concentration of data traffic in one area, it reallocates resources from low-traffic areas to maintain service quality. This optimisation ensures a smoother user experience, reduces network congestion, and minimises dropped calls and slowdowns.
- AI-Powered Customer Service (TOBi Chatbot): Vodafone implemented an AI chatbot named TOBi to handle customer queries. TOBi uses Natural Language Processing (NLP) to understand and respond to customer inquiries across multiple channels (app, web, SMS). TOBi handles common issues like billing questions, service upgrades, and technical troubleshooting, reducing the need for human agents. The AI system can also transfer complex queries to live agents with relevant information, improving service efficiency and reducing wait times.

B.3.4 Impact

- Reduced Downtime and Maintenance Costs: Predictive maintenance reduced network failures by nearly 30%, significantly lowering unplanned downtime and enhancing user experience. Maintenance has also reduced operational costs, as repairs are scheduled strategically and prevent more costly emergency fixes.
- Improved Network Quality: Real-time network optimisation resulted in a 15% reduction in congestion-related service issues, leading to fewer dropped calls and better data speeds. During high-demand events (e.g., concerts and sports events), Vodafone uses AI to manage traffic surges effectively, providing stable connectivity even in densely packed areas.
- Enhanced Customer Experience: TOBi handles over 60% of customer interactions, significantly reducing the load on customer service representatives and cutting customer wait times by 30%. The chatbot's self-learning ability improves response accuracy, and customer satisfaction with TOBi has remained high, with most users able to resolve their issues without needing live assistance.
- Increased Customer Retention: By improving network reliability and reducing wait times, Vodafone has observed higher customer satisfaction and a lower churn rate, as users experience fewer service-related frustrations.

B.3.5 Future Applications and Scaling Potential

Vodafone aims to scale these AI-driven solutions across other regions and expand its capabilities, including more sophisticated network analytics, personalised customer services, and further automation in network troubleshooting.

B.3.6 Lessons Learnt

- Proactive Problem Prevention: Predictive maintenance improves network reliability and customer satisfaction by reducing reactive maintenance and costly emergency repairs.
- Resource Optimisation for Real-Time Demands: AI-driven network optimisation effectively manages resources dynamically, especially during peak hours and high-demand events, ensuring consistent service quality.
- Enhanced Customer Support Through AI: AI chatbots improve service efficiency by handling high-volume, repetitive queries, allowing human agents to focus on more complex cases and improving customer satisfaction.

B.3.7 Conclusion

Vodafone's use of AI in network management and customer service highlights how AI-driven maintenance, optimisation, and automation can transform mobile telephony, ensuring reliable connectivity, reduced costs, and an improved customer experience.

B.4 Smart Governance: Amsterdam

B.4.1 Background

Amsterdam, one of Europe's most digitally advanced cities, has used AI to address urban challenges and improve city governance. The city faces issues common to large urban areas: increasing population density, traffic congestion, and rising demands for efficient public services. Amsterdam's city council implemented AI-driven solutions as part of its Amsterdam Smart City initiative, aiming to improve urban management, citizen engagement, and policy decision-making.

B.4.2 Problem

As Amsterdam's population and tourism grew, city management became more complex, with challenges including:

- Traffic Congestion: The city's narrow streets and heavy traffic cause frequent congestion and pollution, impacting residents' quality of life.
- Resource Allocation: Public services needed more efficient resource management to meet the demands of a growing population.

- Public Engagement and Transparency: The city sought to increase citizen engagement in governance and make data-driven decisions more transparent.

B.4.3 AI Solution

Amsterdam used AI across various governance functions, focusing on traffic management, predictive infrastructure maintenance, and citizen engagement. Thus, the drivers of the solution are traffic management, predictive maintenance, and citizen engagement.

AI-Driven Traffic Management

- Amsterdam implemented AI to monitor and manage real-time traffic data. Sensors and cameras around the city collect data on traffic flow, congestion, and environmental conditions.
- AI algorithms analyse this data, predicting traffic patterns and suggesting optimal routes to reduce congestion.
- Additionally, AI-powered systems manage traffic signals dynamically, adjusting the timing to alleviate bottlenecks and prioritise public transportation during peak hours.

Predictive Maintenance for Public Infrastructure

- AI models monitor the condition of public infrastructure, including roads, bridges, and water management systems, by analysing data from sensors.
- Using predictive maintenance, Amsterdam can identify when infrastructure components are likely to fail or require repairs, allowing for timely maintenance and reducing the likelihood of costly, unexpected repairs.
- This proactive approach improves public safety and minimises disruptions caused by sudden infrastructure failures.

Citizen Engagement Through AI-Powered Chatbots

- The city uses AI-powered chatbots on its website to handle enquiries from residents, ranging from local services to administrative tasks.
- AI helps automate responses to common questions, reducing wait times and improving access to information.
- Additionally, Amsterdam uses AI-driven data analytics to gather feedback from residents, understand citizen concerns, and make data-informed policy adjustments.

Transparency and Ethical AI

- Amsterdam developed an open-source AI registry, which lists all AI systems used by the city, providing transparency regarding how data is used in city governance.
- This registry addresses public concerns about data privacy and bias, detailing each AI tool's purpose, data sources, and potential implications on citizens' privacy.

B.4.4 Impact

Reduced Traffic Congestion and Emissions

- By using AI to dynamically manage traffic, Amsterdam reduced congestion by 10–20% in key areas, contributing to improved air quality and a more efficient transport system.
- This efficiency supports the city's goal to reduce carbon emissions and encourages the use of public transportation.

Cost Savings and Infrastructure Reliability

- Predictive maintenance reduced unplanned infrastructure failures, allowing the city to save significantly on emergency repair costs and extend the life of public assets.
- These savings enable better resource allocation, allowing Amsterdam to invest in other public projects and services.

Improved Citizen Satisfaction

- AI-powered chatbots reduced response times for citizen inquiries by 30%, allowing residents to access information more easily and interact more efficiently with local government.
- The transparent AI registry and open data initiatives increased public trust in the city's governance and its approach to data privacy and ethical AI use.

Enhanced Data-Driven Decision-Making

- AI allows Amsterdam's city council to make more informed decisions by analysing real-time data on traffic, environmental impact, and citizen feedback, improving urban planning and policy adjustments.

B.4.5 Future Applications and Scaling Potential

Amsterdam plans to expand its use of AI into additional areas, including waste management, environmental monitoring, and health services. By using AI to predict and optimise resource allocation, the city aims to create an even more sustainable and responsive urban environment.

B.4.6 Lessons Learnt

- Efficiency and Sustainability: AI helps address urban challenges by improving traffic flow, infrastructure maintenance, and resource allocation, making cities more efficient and environmentally sustainable.
- Citizen-Centric AI: AI solutions that enhance public service access, transparency, and responsiveness increase public satisfaction and trust.
- Ethical and Transparent Use of AI: Open-source registries and clear communication about data use build public trust and create a model for ethical AI use in governance.

B.4.7 Conclusion

Amsterdam's use of AI demonstrates how smart governance can use technology to improve urban living. It shows that transparent, data-driven decision-making can create a responsive, efficient, citizen-centred city government.

B.5 Manufacturing: Siemens (Germany)

Siemens has been a leader in the 4IR, integrating AI, IoT, and automation technologies to transform manufacturing. The Amberg Electronics Plant, a "smart factory" in Germany, produces programmable logic controllers (PLCs) used globally in industrial automation. Siemens aimed to increase production quality and efficiency, reduce downtime, and enhance customisation options to meet changing customer demands.

B.5.1 Problem

Traditional manufacturing faces several challenges that Siemens wanted to overcome:

- Quality Control: Ensuring consistent product quality is complex, particularly with high production volumes and small tolerances in electronics manufacturing.
- Maintenance Costs: Equipment failures and unplanned downtimes result in lost productivity and high maintenance costs.
- Customization and Flexibility: Adapting production lines for small-batch, custom orders is traditionally time-consuming and costly.
- Data Utilisation: Vast amounts of manufacturing data often remain underutilised, missing the opportunity for real-time insights.

B.5.2 AI Solution

Siemens implemented AI-based solutions across the Amberg plant, focusing on quality inspection, predictive maintenance, and process optimisation.

- Automated Quality Control: Siemens uses machine vision powered by deep learning algorithms for real-time quality control. Cameras capture images of each component and assembly on the production line. AI algorithms analyse these images for defects such as misalignments, soldering issues, or surface damage. The system flags faulty products immediately, allowing for corrective action before they advance in the production line, resulting in a near-zero defect rate.
- Predictive Maintenance: Sensors on machines monitor critical parameters like temperature, vibration, and wear in real time. Machine Learning models analyse this data to predict when parts will likely fail, allowing Siemens to proactively schedule maintenance. By implementing predictive maintenance, Siemens has reduced unplanned downtimes by nearly 20%, improving overall equipment efficiency and reducing maintenance costs.
- Process Optimization with Digital Twins: Siemens uses digital twins (virtual models of physical assets) to simulate and optimise production processes. The digital twin collects data from sensors and processes it in real time, allowing AI to simulate different production scenarios, optimise machine settings, and predict outcomes without disrupting production. This allows for quick adjustments in response to product variations or customer-specific requirements, enhancing production flexibility.

B.5.3 Impact

- Enhanced Quality Control: The plant achieved a 99.999% quality rate, effectively eliminating most defects and drastically reducing waste. Real-time quality control has also lowered the need for manual inspections, increasing throughput and decreasing costs.
- Reduced Downtime and Maintenance Costs: Predictive maintenance improved machine availability by 10–15% and reduced unexpected machine stoppages, significantly lowering maintenance costs and boosting productivity.
- Improved Operational Efficiency: AI-driven process optimisation and automated adjustments allow the Amberg plant to respond to changing production requirements quickly. The plant achieved a 1400% productivity increase over 20 years, largely attributed to AI-driven innovations.
- Sustainability: AI-driven quality and efficiency improvements reduced material waste and energy consumption, helping Siemens meet sustainability goals while lowering production costs.

B.5.4 Future Applications and Scaling Potential

Siemens aims to expand its AI-driven smart manufacturing model to other plants and apply these innovations to multiple sectors. The company also plans to enhance its AI-powered digital twin technology for even greater adaptability in production processes, from electronics to heavy machinery.

B.5.5 Lessons Learnt

- Quality and Efficiency Synergy: AI ensures quality and efficiency without sacrificing one for the other, making manufacturing more sustainable and cost-effective.
- Predictive Maintenance as a Cost Saver: Proactive maintenance extends equipment life and minimises disruptions, proving essential for high-volume manufacturing.
- Flexibility through Digital Twins: Digital twins allow factories to adapt quickly to changing demands, offering a level of flexibility traditionally challenging in manufacturing.

B.5.6 Conclusion

Siemens' success at the Amberg plant demonstrates the transformative potential of AI in manufacturing. It showcases how AI-driven quality control, predictive maintenance, and process optimisation can lead to world-class manufacturing efficiency and adaptability.

AI Applications in Financial Services

C

Improvements in financial services are critical to advancing the SDGs, particularly in promoting inclusive economic growth and poverty reduction. Indeed, financial services are central to achieving SDGs 1, 2, 8, and 13. Hence, AI applications in this sector are key to deploying AI to achieve the SDGs. This appendix outlines examples of such interventions and their impact.

C.1 Banking Sector: JPMorganChase

JPMorganChase, one of the largest banks in the United States, has implemented AI across its operations to improve efficiency, customer experience, and risk management. Their AI initiatives cover a range of areas, from automating back-office functions to enhancing fraud detection and risk assessment.

C.1.1 Key Applications of AI

- Fraud Detection and Risk Management: JPMorgan uses AI to detect fraud and assess risks in real time. Their AI systems analyse vast amounts of transactional data to identify unusual patterns and flag potential fraud. For example, if a customer typically makes small transactions and suddenly there's a large transaction in a foreign country, the AI model can flag this for review. The AI systems also assess credit risk by analysing the financial history and behaviour of individuals and businesses, helping the bank make more accurate lending decisions.
- Contract Review (COiN—Contract Intelligence): JPMorgan's COiN platform uses Machine Learning to analyse and extract important information from legal documents. The system can process thousands of loan agreements and contracts in seconds, a task that previously took hundreds of thousands of human hours annually. COiN improves accuracy and frees up human capital for more complex decision-making.
- Customer Service (Chatbots and Virtual Assistants): JPMorgan's virtual assistant "Erica" interacts with customers via text and voice, helping them with routine tasks like checking balances, paying bills, and transferring money. The AI-powered assistant uses Natural Language Processing (NLP) to understand and respond to customer queries, improving response times and customer satisfaction.
- Algorithmic Trading: AI plays a critical role in JPMorgan's trading strategies. Using Machine Learning models, the bank analyses market trends, predicts price movements, and executes trades

A. G. O. Mutambara, *Deploying Artificial Intelligence to Achieve the UN Sustainable Development Goals*, Sustainable Development Goals Series,
https://doi.org/10.1007/978-3-031-88423-8

automatically. AI-driven algorithms help identify trading opportunities faster than human traders, giving JPMorgan a competitive edge in the market.
- Personalised Banking Experience: AI helps JPMorgan offer personalised recommendations to customers, such as customised financial products, investment advice, and spending insights. By analysing customer data (transaction history, financial behaviour, etc.), AI provides tailored solutions to meet individual customer needs, enhancing engagement and loyalty.
- Loan Approval Processes: JPMorgan uses AI to streamline loan approval processes. AI models can assess applicants' creditworthiness more accurately by analysing various data points, including credit scores, income, employment history, and even alternative data like social media behaviour. This allows for faster loan approval decisions and reduces defaults by identifying riskier applicants.

C.1.2 Impact of AI Implementation

- Efficiency Gains: AI applications like COiN and algorithmic trading have significantly reduced manual workload and human error. Contract review processes, for example, were reduced from thousands of hours to seconds.
- Cost Savings: Automating back-office processes and risk assessment tasks has saved JPMorgan millions of dollars annually in operational costs.
- Enhanced Customer Experience: AI-driven personalisation and customer service tools like virtual assistants have improved response times and customer satisfaction.
- Risk Mitigation: The use of AI in fraud detection and credit risk assessment has strengthened the bank's ability to mitigate potential losses, ensuring better customer security.

C.1.3 Challenges Faced

- Data Privacy Concerns: Implementing large-scale AI requires managing and protecting vast amounts of sensitive customer data. JPMorgan has had to invest heavily in cybersecurity and data privacy measures.
- Regulatory Compliance: AI systems, especially those involved in risk management and trading, must comply with strict financial regulations. Ensuring that AI models are transparent and unbiased is critical to meeting regulatory standards.
- Integration with Legacy Systems: Like many large financial institutions, JPMorgan faced challenges integrating AI into existing legacy systems, requiring significant investment in new infrastructure and employee training.

C.1.4 Conclusion

JPMorganChase's adoption of AI has revolutionised many of its operations, leading to substantial efficiency gains, cost savings, and improved risk management. While there are challenges regarding data privacy and regulatory compliance, the long-term benefits of AI in enhancing customer service and operational efficiency are evident. This case highlights the transformative potential of AI in the banking sector.

C.2 Banking Sector: HSBC

The Hong Kong and Shanghai Banking Corporation (HSBC), one of the world's largest financial institutions, has embraced AI to enhance customer experience, improve operational efficiency, and strengthen security. The bank's AI initiatives span across multiple facets of its business, including fraud detection, customer service, risk management, and compliance.

C.2.1 Key Applications of AI

- Fraud Detection and Risk Management: HSBC uses AI-powered systems to identify and prevent real-time fraudulent activities. These systems analyse vast amounts of transactional data to detect anomalies or unusual patterns that may indicate fraud. The AI models continuously learn from past transactions, improving accuracy and reducing false positives. AI algorithms also help HSBC in detecting money laundering activities by scanning large datasets for suspicious behaviour. These AI systems can process huge volumes of data much faster than human analysts, identifying potential risks more effectively.
- Chatbots and Virtual Assistants: HSBC has implemented AI-driven chatbots to handle customer inquiries and provide support through various channels, including its mobile app and website. The chatbots use Natural Language Processing (NLP) to understand and respond to customer queries in real time. By leveraging AI, the bank has significantly improved customer service efficiency, providing 24/7 support without the need for human intervention in many cases. These virtual assistants can handle routine tasks like balance inquiries and transaction histories and even help with lost card reporting.
- Loan Application Processing: AI has transformed how HSBC processes loan applications. Machine Learning algorithms assess an applicant's creditworthiness by analysing various data points, such as credit scores, income, spending patterns, and even alternative data like social media activity (where permitted). Using AI, HSBC has expedited the loan approval process, reduced paperwork, and made faster, more accurate lending decisions. The AI models are also better at predicting loan default risks, which reduces non-performing loans (NPLs).
- Compliance and Regulatory Adherence: AI helps HSBC manage compliance with strict banking regulations by monitoring transactions and ensuring they adhere to legal requirements. AI-based systems can flag suspicious transactions or actions that may violate anti-money laundering (AML) laws, know your customer (KYC) guidelines, or other regulations. These AI models can analyse a vast amount of structured and unstructured data to identify compliance risks, offering better oversight than traditional rule-based systems.
- Personalised Banking Services: HSBC uses AI to deliver personalised financial products and services to its customers. AI algorithms analyse customer behaviour, financial habits, and preferences to offer tailored recommendations on savings plans, investment options, or insurance products. By personalising banking services, HSBC has improved customer engagement and satisfaction. AI-driven personalisation enables the bank to proactively suggest solutions that meet the specific needs of individual customers.
- Predictive Analytics for Customer Insights: Predictive analytics, powered by AI, allows HSBC to forecast customer needs and behaviour. By analysing transaction data, spending patterns, and customer interactions, AI helps the bank predict future actions, such as whether a customer may require a new loan or investment product. This predictive capability allows HSBC to enhance customer retention by anticipating needs and providing timely, relevant offers or advice, boosting customer loyalty.

C.2.2 Impact of AI Implementation

- Improved Fraud Detection: HSBC's AI systems have reduced the number of fraudulent transactions and enhanced the speed and accuracy of fraud detection processes. Machine learning allows quicker pattern identification, reducing the bank's exposure to fraud-related losses.
- Increased Operational Efficiency: Automating customer service, loan processing, and compliance checks have resulted in significant cost savings for HSBC. The bank has been able to scale its operations without proportionally increasing headcount, improving its cost-to-income ratio.
- Enhanced Customer Experience: AI-powered chatbots and personalised banking services have improved customer experience by providing faster and more accurate services, increasing overall customer satisfaction. The ability to offer round-the-clock support and tailored financial products has boosted customer loyalty.
- Regulatory Compliance: HSBC has strengthened its regulatory adherence through AI, reducing the risk of penalties related to non-compliance. The ability to analyse massive datasets for potential regulatory breaches has also increased transparency and accountability.

C.2.3 Challenges Faced

- Data Privacy and Security: The large-scale use of customer data to power AI systems raises concerns about data privacy. HSBC must ensure that AI systems comply with data protection regulations such as GDPR and adequately protect sensitive customer information.
- Bias in AI Models: AI models can sometimes exhibit bias if the data used to train them is not diverse or representative of the wider population. This can be particularly problematic in areas like loan approvals, where biased models could lead to discriminatory outcomes. HSBC continuously monitors and updates its AI systems to avoid such biases.
- Legacy System Integration: Integrating AI into HSBC's legacy systems has been challenging. AI requires modern infrastructure and scalable technology platforms, so the bank has had to invest heavily in upgrading its IT systems to leverage AI capabilities fully.
- Employee Training and Adaptation: Implementing AI requires employees to learn new skills and adapt to new working methods. HSBC has invested in upskilling its workforce to ensure they can work alongside AI systems and manage more complex tasks beyond what AI can handle.

C.2.4 Conclusion

HSBC's application of AI has driven significant improvements in fraud detection, customer service, risk management, and operational efficiency. The bank has benefitted from cost savings, enhanced customer experience, and stronger regulatory compliance. While there are challenges related to data privacy, biases in AI, and integration with legacy systems, HSBC's AI initiatives demonstrate how transformative AI can be for the banking sector. This case highlights AI's broad and deep potential to reshape banking operations and customer interactions, offering a model for other financial institutions looking to harness the power of AI.

C.3 Banking Sector: BBVA

Banco Bilbao Vizcaya Argentaria (BBVA), a multinational Spanish banking group, has been a leader in applying AI to enhance banking services. Over the past few years, BBVA has integrated AI into multiple facets of its operations, focusing on improving customer interactions, risk management, fraud detection, and internal process efficiency. This case study highlights how BBVA has successfully leveraged AI to transform its business model and drive significant competitive advantages.

C.3.1 Challenges Before AI Implementation

- Personalised Customer Experience: Customers demanded more personalised banking services, but BBVA struggled to meet these expectations with traditional banking methods. Manual data analysis to understand customer behaviour and offer personalised services was time-consuming and lacked precision.
- Risk Assessment in Credit and Loan Applications: BBVA needed to manage loan risk more effectively, as their manual and rule-based systems for credit scoring and risk assessment were limited in scope and prone to human error.
- Fraud Detection and Security: Traditional systems were reactive rather than predictive, making preventing fraudulent transactions in real time difficult. The need for a more dynamic and scalable approach to fraud detection became critical as digital banking services expanded.
- Operational Efficiency: The bank sought to streamline its internal processes, particularly in areas like compliance, data entry, and customer onboarding, where manual methods were resource-intensive and prone to delays.

C.3.2 AI Solutions Implemented

- AI-Driven Personalized Banking: BBVA implemented an AI-powered recommendation engine called Bconomy, which analyses customer data to offer personalised financial advice and services. By leveraging Machine Learning algorithms, the system can predict customer needs, suggest appropriate financial products, and help customers manage their finances more effectively.
  Impact: Bconomy helped BBVA deliver personalised experiences by analysing spending habits, income patterns, and savings behaviour. Customers received tailored suggestions for improving their financial health, such as recommendations on saving strategies, loan products, or investment opportunities.
- AI in Credit Scoring and Loan Risk Assessment: BBVA introduced AI-powered risk models to evaluate loan applications. These models use Machine Learning to analyse customer data beyond traditional credit scores, incorporating behavioural and transactional data, social media activity, and more to provide a holistic view of risk.
  Impact: This approach significantly improved the bank's ability to make accurate lending decisions (a 15% increase in loan approvals). AI-enabled models reduced loan default rates by 25% and expanded access to credit for underbanked populations who traditionally had poor or limited credit histories.
- Fraud Detection Using AI: BBVA implemented AI-based fraud detection systems that monitor real-time transactions and customer behaviour to identify suspicious activities. The Machine Learning models are continuously trained to detect new fraud patterns and behaviours, enabling the system to predict and prevent potential fraud.

Impact: BBVA's AI-based fraud detection system reduced fraud incidents by 30% in the first year of implementation. The system's accuracy in detecting fraud with minimal false positives increased customer trust and loyalty.

- Process Automation and Operational Efficiency: BBVA adopted Robotic Process Automation (RPA) combined with AI to automate routine tasks such as customer onboarding, regulatory compliance, and data entry. AI-enhanced RPA enabled the bank to streamline complex workflows and eliminate bottlenecks in processes that traditionally required manual intervention.
  Impact: Automating internal processes helped BBVA reduce operational costs by approximately 22%. Processes that previously took days, such as loan approvals and compliance checks, were reduced to just hours, improving overall efficiency.

C.3.3 Key Outcomes

- Enhanced Customer Engagement and Satisfaction: Implementing Bconomy significantly increased customer satisfaction, with a 20% rise in customer engagement. Personalised recommendations led to higher uptake of financial products such as savings accounts, mortgages, and insurance policies.
- Improved Credit Decisioning: The AI-driven risk models enhanced BBVA's ability to make more informed credit decisions, with a 15% increase in loan approvals and a 25% decrease in loan defaults, especially among traditionally underserved customers.
- Reduced Fraud and Enhanced Security: The AI fraud detection system helped prevent approximately €150 million in fraudulent activities, ensuring stronger customer protection without compromising user experience. False positives were reduced by 35%, allowing legitimate transactions to be processed smoothly.
- Cost Reduction and Operational Efficiency: BBVA's investment in AI and automation led to a 22% reduction in operational costs. Labour-intensive tasks were streamlined, freeing employees to focus on value-added services such as customer relationship management.
- AI as a Strategic Differentiator: BBVA's successful integration of AI positioned it as a technology-driven leader in the European banking industry. By adopting AI across various functions, the bank improved customer-facing services and back-end operations, strengthening its market position.

C.3.4 Challenges and Learnings

- Data Privacy and Ethical AI: One of the key challenges BBVA faced was ensuring compliance with strict European data protection regulations, particularly the GDPR. The bank had to invest heavily in ensuring its AI systems were secure, transparent, and complied with privacy laws.
- Change Management: Implementing AI across an organisation as large as BBVA required a cultural shift. Employees had to be trained to work with AI-driven systems, and there were initial concerns about job displacement. The bank managed this by focusing on AI as an enabler of new roles rather than replacing existing jobs.

C.3.5 Conclusion

BBVA's experience with AI demonstrates how banking institutions can harness technology to drive growth, reduce risks, and enhance operational efficiency. By using AI to personalise services, enhance credit decision-making, and detect fraud, BBVA improved customer experiences and strengthened its competitive edge in the banking industry. Its AI transformation is a clear example of how traditional banks can innovate and stay relevant in a rapidly changing digital landscape.

This case is an excellent model for how AI can be successfully integrated across banking operations to address customer needs, improve efficiency, and drive sustainable growth.

C.4 Banking Sector: Reserve Bank of India

C.4.1 Introduction: Role of AI in Central Banking

- Overview of Central Banks: The Reserve Bank of India (RBI), like other central banks, plays a pivotal role in monetary policy, financial regulation, and ensuring the stability of the country's banking system.
- Motivation for AI Adoption: With financial systems' increasing complexity, AI and Machine Learning have become valuable tools in managing economic forecasting, financial supervision, and fraud detection.

C.4.2 Context: Bank's Journey Towards AI Adoption

- Background on the Indian Economy: India, being one of the fastest-growing economies in the world, faces challenges related to inflation control, financial inclusion, and regulatory oversight of its vast banking sector.
- Digital Transformation: RBI has embarked on a digital transformation journey to address these issues, adopting AI and big data analytics to improve decision-making, economic forecasting, and fraud detection.

C.4.3 AI Applications in Bank's Functions

Monetary Policy and Economic Forecasting

- AI for Economic Data Analysis: RBI has integrated AI to analyse vast amounts of economic data, ranging from inflation rates to unemployment statistics. Machine learning algorithms have been deployed to predict economic trends more accurately, using a combination of traditional economic indicators and alternative data sources (social media, real-time market data).
- Improved Forecasting Models: AI-enhanced models have helped the RBI in making better projections for inflation and economic growth, allowing it to make more informed decisions regarding interest rates and monetary supply adjustments.

Financial Supervision and Risk Management

- AI for Regulatory Oversight: The RBI uses AI to monitor the financial health of banks and other financial institutions. AI systems help flag irregularities in financial reporting or behaviour that may indicate instability, thereby enabling earlier intervention.
- Stress Testing: AI models have been integrated into RBI's stress testing frameworks, simulating economic crises and assessing how the banking system would cope under different scenarios. This helps in improving policy responses to potential economic shocks.

C.4.4 Fraud Detection and Anti-money Laundering

- AI for Transaction Monitoring: RBI has implemented AI to detect patterns in financial transactions that indicate money laundering, fraud, or other illicit activities. In real time, Natural Language Processing (NLP) is employed to analyse unstructured data, such as transaction narratives.
- Collaboration with Banks: RBI collaborates with commercial banks to implement AI-driven fraud detection systems to spot anomalies and suspicious behaviour faster than traditional methods.

C.4.5 Challenges Faced in AI Implementation

- Data Quality and Availability: The availability of high-quality data is crucial for effective AI systems. RBI initially faced challenges related to fragmented data across different sectors and financial institutions. However, it has made strides in building integrated data lakes to pool information from various sources.
- Skills and Expertise Gaps: RBI faced challenges in terms of technical expertise in AI. To bridge this gap, it has initiated partnerships with academic institutions and technology firms to build AI capabilities. To overcome the skills gap, RBI launched internal training programmes and workshops to upskill its employees on AI technologies and data science.
- Ethical and Governance Concerns: One of the challenges with AI is the opacity of decision-making processes, especially in high-stakes areas like monetary policy. RBI has been developing AI governance frameworks to ensure that AI systems remain transparent and accountable to stakeholders.

C.4.6 Impact of AI Adoption

- Improved Decision-Making: AI has allowed RBI to improve its decision-making processes, particularly in monetary policy, where AI-driven economic forecasts have proven more accurate than traditional models. This has resulted in more responsive policy adjustments, such as interest rate changes.
- Enhanced Financial Regulation: AI has strengthened RBI's regulatory oversight capabilities. The early detection of financial risks has enabled more proactive interventions, reducing the likelihood of systemic crises.
- Reduction in Financial Crime: AI systems have significantly enhanced the efficiency of detecting fraudulent transactions and AML activities. The number of flagged suspicious transactions has increased, and investigations are more targeted and timely.
- Increased Financial Inclusion: RBI's use of AI has indirectly supported financial inclusion by enabling more data-driven policies. For instance, AI models analysing financial exclusion patterns have informed initiatives to extend banking services to underserved populations.

C.4.7 Lessons for Other Central Banks in the Global South

- Scalability of AI Solutions: Other central banks in the Global South can learn from RBI's experience by adopting scalable AI solutions, starting with small pilot projects focused on specific tasks like fraud detection or monetary forecasting.

- Importance of Collaboration: RBI's partnerships with technology firms, academic institutions, and commercial banks were instrumental in overcoming technical and skill-related challenges. Other countries can replicate this collaborative approach to build AI capabilities.
- Policy and Regulatory Adjustments: Developing a robust regulatory framework for AI deployment in central banking is critical to mitigating risks and ethical concerns. RBI's efforts to establish AI governance frameworks provide a model for others.

C.4.8 Conclusion

RBI's integration of AI in central banking is a powerful example of how AI can transform financial governance in the Global South. As AI technologies evolve, central banks must continue to innovate, develop human capital, and build robust ethical frameworks to maximise the benefits of AI in their financial systems. This case study on the Reserve Bank of India highlights how central banks in emerging economies leverage AI to enhance economic forecasting, regulatory oversight, and fraud detection. It also provides valuable lessons for other central banks looking to adopt AI.

C.5 Investment Management: BlackRock

One illustrative case study highlighting the applications of AI in investment management is BlackRock's Aladdin Platform, a comprehensive risk management and investment management system that leverages AI and Machine Learning to enhance decision-making, risk assessment, and portfolio optimisation. BlackRock, the world's largest asset manager, created Aladdin (Asset, Liability, Debt, and Derivative Investment Network) to enhance investment decision-making, manage risks more effectively, and streamline operations for BlackRock and its clients. Aladdin combines sophisticated data analytics, machine learning, and Natural Language Processing (NLP) to assist in investment management.

C.5.1 Key AI Applications

Risk Management and Analysis

- Portfolio Risk Analytics: Aladdin uses AI algorithms to analyse massive amounts of market data and simulate scenarios to evaluate the risks associated with different investment choices. For example, machine learning models are employed to identify patterns in market behaviour, enabling Aladdin to assess the impact of potential market events on portfolio risk.
- Stress Testing and Scenario Analysis: Aladdin employs predictive analytics and machine learning to simulate stress tests, assessing how portfolios might perform under various extreme market conditions, such as interest rate changes, political instability, or economic downturns.

Asset Allocation and Portfolio Optimisation

- Dynamic Asset Allocation: Using machine learning, Aladdin can process historical and real-time data to predict future asset performance, recommending an optimal allocation strategy that balances risk and reward for specific portfolio goals.
- Rebalancing Algorithms: Aladdin's AI-powered algorithms automatically recommend portfolio adjustments based on market changes and the client's investment objectives. This ensures that portfolios stay aligned with their risk tolerance, time horizon, and strategic goals.

Predictive Analytics for Market Trends

- Market Sentiment Analysis: Aladdin uses NLP to process and analyse news articles, financial reports, and social media sentiment, identifying trends that may impact market performance. By understanding sentiment changes, Aladdin helps predict short-term price movements and longer-term market trends, providing additional data to inform investment decisions.
- Economic and Market Forecasting: AI-driven predictive models analyse economic indicators and market data to forecast potential market scenarios, helping portfolio managers make proactive decisions based on likely economic outcomes.

Trade Execution and Operational Efficiency

- Automated Trading Strategies: Aladdin's AI capabilities help execute trades efficiently by leveraging real-time market data to find optimal entry and exit points. This can reduce trading costs and increase returns by optimising trade execution.
- Compliance and Regulatory Reporting: Aladdin's AI-based compliance tools monitor trades, transactions, and portfolio activities in real time, flagging any actions that may breach regulatory requirements. This ensures all investment activities comply with legal standards, reducing compliance risks.

Personalized Client Experience and Advisory Services

- Customized Investment Strategies: Aladdin analyses individual client profiles, including investment goals, risk tolerance, and financial history, to recommend personalised investment strategies. This personalised approach enables BlackRock to serve a wide range of clients, from retail investors to large institutions, better.
- AI-Driven Client Communication: By using NLP, Aladdin generates insights and reports that are easy to understand for clients, helping advisors communicate complex financial data effectively and keep clients informed about their investments.

C.5.2 Impact of AI Intervention

- Enhanced Decision-Making: Aladdin's AI tools have enhanced decision-making for portfolio managers by providing timely insights into risk exposure, market trends, and portfolio performance. This data-driven approach has made BlackRock's investment strategies more responsive to market changes.
- Risk Mitigation and Increased Resilience: Through AI-powered risk management tools, Aladdin has significantly improved BlackRock's ability to manage market volatility and identify potential risks early. For example, during periods of economic uncertainty, Aladdin's stress tests provide insights into portfolio vulnerabilities, enabling proactive adjustments.
- Operational Efficiency: Aladdin's AI-driven automation has streamlined many back-office operations, from compliance monitoring to transaction processing. This reduces employees' manual workloads, cuts operational costs, and enhances accuracy, allowing BlackRock to focus on strategy and client relations.
- Setting Industry Standards: Aladdin's success has influenced the broader investment management industry, leading other firms to invest in AI and machine learning technologies for similar capabilities. As a result, Aladdin's model has set new standards for AI applications in investment management.

C.5.3 Conclusion

BlackRock's Aladdin platform exemplifies how AI can transform investment management by enhancing portfolio optimisation, risk assessment, client personalisation, and operational efficiency. Through machine learning, NLP, and predictive analytics, Aladdin offers a comprehensive toolset for managing complex investment portfolios, benefiting both BlackRock and its clients while shaping the future of AI in the investment industry.

Executive Order on Advancing US Leadership in Artificial Intelligence Infrastructure

D

On 14 January 2025, then US President Joe Biden signed an Executive Order on Advancing US Leadership in AI Infrastructure [137]. The order "instructs the Departments of Defence (DOD) and Energy (DOE), in coordination with other federal agencies, to lease federal sites to private-sector entities for constructing large-scale AI data centres and associated clean energy facilities". Thus, the executive order seeks to expedite the development of large-scale AI infrastructure within the United States, putting a premium on sustainable energy ambitions and national security concerns. The declaration directs agencies to take radical steps to accelerate large-scale AI infrastructure development at federal sites. More significantly, it imposes key requirements and safeguards for developers building in these locations.

D.1 Accelerating AI Infrastructure Development

The executive order directs certain agencies to make federal sites available for AI data centres and new clean power facilities, facilitate this infrastructure's interconnection to the electric grid, fulfil permitting obligations expeditiously, and advance transmission development around federal sites. According to the executive order, in order to accelerate AI infrastructure development, agencies will leverage their authorities to achieve the following:

- Lease federal sites owned by the DOD and DOE to host gigawatt-scale AI data centres. These departments will select sites where the private sector can build AI data centres and clean power facilities.
- Catalyse deployment of new clean energy generation to support AI infrastructure. Developers selected to build on DOD and DOE sites will be required to bring online sufficient clean energy generation resources to match the entire electricity needs of their data centres, consistent with applicable law.
- Prioritise full and expeditious permitting of AI infrastructure on federal sites. Agencies will prioritise and dedicate staff towards permitting this infrastructure in a timely manner.
- Accelerate transmission development around federal sites. To help ensure the timely operation of AI infrastructure on federal sites, the DOE will take appropriate steps to coordinate with developers in constructing, financing, facilitating, and planning the upgrade and development of transmission lines around those sites.
- Facilitate interconnection of AI infrastructure to the electric grid. The DOE will engage utilities on their requirements for interconnecting AI infrastructure on federal sites and on opportunities

A. G. O. Mutambara, *Deploying Artificial Intelligence to Achieve the UN Sustainable Development Goals*, Sustainable Development Goals Series,
https://doi.org/10.1007/978-3-031-88423-8

to accelerate interconnection via grid-enhancing technologies, operational changes, and other measures.

- Ensure low electricity prices for consumers. Developers selected to build on DOD and DOE sites will be required, consistent with applicable law, to pay all costs of building and operating AI infrastructure so that this development does not raise electricity prices for consumers.
- Advance allies and partners' development of AI infrastructure. The Department of State will engage allies and partners on steps to build trusted AI infrastructure worldwide.
- Bolster domestic manufacturing and increase investment in US semiconductor chip production by mandating that companies building on federal land are to utilise an appropriate share of US-made semiconductors, with specific procurement requirements determined on a case-by-case basis.

D.2 Lease Obligations for Developers on Federal Sites

As already intimated, the AI efforts must support US clean energy leadership and a competitive technology ecosystem while empowering involved workers and affected communities. These US obligations, as expressed in the executive order, include:

- Paying the full cost of building, operating, and maintaining AI infrastructure—including the costs of building new data centres and clean power facilities, transmission development, and upgrades.
- Procuring new clean energy generation resources that can be delivered to the data centre and that accurately match their electricity and capacity needs. This buildout will prevent electricity price increases and advance US energy leadership.
- Strengthening lab-security requirements and evaluating the national security implications of AI models developed on federal sites. The evaluations will assess models for risks to safety and security and their potential to advance national security objectives.
- Adhering to high labour standards, promoting positive labour-management relations, and paying workers prevailing wages.
- Purchasing an appropriate share of domestically manufactured semiconductors to help ensure a robust domestic semiconductor supply chain.

Interview with DeepSeek Founder: We're Done Following

E

E.1 Enter DeepSeek-R1

DeepSeek-R1 is shaking Silicon Valley. In an interview in February 2025, the founder Liang Wenfeng said: "We're done following. It's time to lead". What a poignant message—an instructive lesson to the least industrialised and emerging economies. A seismic shift in AI dominance is underway, and all eyes are on China and the Global South. DeepSeek-R1—an open-source inference model from China—has sent shockwaves through the tech world by matching OpenAI and Google's top-tier LLM performance at 3.3% of the application programming interface (API) cost, the price a company charges for developers to access and use their API, while embracing full openness. With a meagre $6 million, China built one of the world's finest AI models, contrasting with the billions spent by companies such as OpenAI, Meta, Google, and Microsoft. Across the world, users—especially individuals and SMEs—are switching to DeepSeek-R1, adopting it as their foundational model. Clearly, a Chinese startup dared to leapfrog industry giants and redefine the rules of AI innovation. Other Chinese AI giants, including ByteDance, Alibaba, Baidu, and Tencent, have joined the race, ignited by DeepSeek [149].

INTERVIEW Excerpts

"We never intended to be a disruptor; it just happened by accident. We didn't expect pricing to be such a sensitive issue. We were simply following our own pace, calculating costs, and setting prices accordingly. Our principle is neither to sell at a loss nor to seek excessive profits. The current pricing allows for a modest profit margin above our costs.

If the goal is to develop applications, adopting Llama's structure to quickly launch a product is a reasonable choice. However, our goal is AGI (Artificial General Intelligence), which requires us to explore new model structures to achieve superior capabilities within limited resources. This is foundational research for scaling up. Beyond architecture, we've studied data curation and human-like reasoning—all reflected in our models. Also, Llama's training efficiency and inference costs lag behind cutting-edge global standards by about two generations.

We believe the most important thing right now is to participate in global innovation. For years, Chinese companies have been accustomed to leveraging technological innovations developed elsewhere and monetising them through applications. But this isn't sustainable. This time, our goal isn't quick profits but advancing the technological frontier to drive ecosystem growth.

Innovation is undoubtedly costly, and our past tendency to adopt existing technologies was tied to China's earlier developmental stage. But today, China's economic scale and the profits of giants like

A. G. O. Mutambara, *Deploying Artificial Intelligence to Achieve the UN Sustainable Development Goals*, Sustainable Development Goals Series,
https://doi.org/10.1007/978-3-031-88423-8

ByteDance and Tencent are globally significant. What we lack isn't capital but confidence and the ability to organise high-calibre talent for effective innovation.

For three decades, we've emphasised profit over innovation. Innovation isn't purely business-driven; it requires curiosity and creative ambition. We're shackled by old habits, but this is a phase.

More investment doesn't always fuel more innovation.

We won't go closed-source. We believe that establishing a robust technology ecosystem matters more. I believe innovation is, first and foremost, a matter of belief. Why is Silicon Valley so innovative? Because they dare to try. When ChatGPT debuted, China lacked confidence in frontier research. From investors to major tech firms, many felt the gap was too wide and focused instead on applications. But innovation requires confidence, and young people tend to have more of it.

Achieving AGI could take two years, five years, or 10 years—but it will happen within our lifetime. As for our roadmap, there's no consensus even within our company. However, we are placing our bets on three directions:

- Mathematics and code, which serve as a natural testbed for AGI—much like Go, they are enclosed, verifiable systems where self-learning could lead to high intelligence.
- Multimodality, where the AI engages with the real world to learn.
- Natural language itself, which is fundamental to human-like intelligence.

All strategies are products of the past.

We are open to all possibilities. All strategies are products of the past generation and may not hold true in the future. Discussing AI's future profitability using the commercial logic of the internet era is like comparing Tencent's early days to General Electric or Coca-Cola. It's essentially carving a boat to mark a sword's position, an outdated approach." [149]

Bibliography

1. Abboud, L., & Heikkilä, M. (2025). US and UK refuse to sign summit declaration on AI. *Financial Times*, 11 February 2025.
2. African Union. (2015). *Agenda 2063: The Africa We Want.* African Union. https://au.int/en/agenda2063/overview
3. African Union (2024). *Continental Artificial Intelligence strategy: Harnessing AI for Africa's development and prosperity*. African Union.
4. Ahuja, V., & Ahuja, N. (2020). A framework for implementing Artificial Intelligence in addressing the sustainable development goals. *Journal of Cleaner Production, 253*, 119958. https://doi.org/10.1016/j.jclepro.2019.119958
5. Allen, C., & O'Connell, D. (2019). The SDGs as a tool for integration: A comparative analysis of policy coherence in national SDG strategies. *Sustainable Development, 27*(2), 273–284. https://doi.org/10.1002/sd.1895
6. Alzahrani, A., & Ameen, A. (2020). Applying Artificial Intelligence to achieve sustainable development: A systematic review. *Sustainable Development, 28*(6), 1630–1642. https://doi.org/10.1002/sd.2082
7. Anderson, K., & Parthasarathy, P. (2021). Evolution of control systems with Artificial Intelligence. *Control Engineering*, 24:51–59.
8. Bäckstrand, K., & Lövbrand, E. (2019). Evolving global governance: The SDGs and the role of non-state actors. *Environmental Politics, 28*(2), 221–240. https://doi.org/10.1080/09644016.2018.1532058
9. Banerjee, A. V., & Duflo, E. (2019). *Good economics for hard times: Better answers to our biggest problems.* PublicAffairs.
10. Bansal, G., & Kumar, A. (2020). Artificial Intelligence for sustainable development: A review on applications and challenges. *IEEE Access, 8*, 35567–35587. https://doi.org/10.1109/ACCESS.2020.2979566
11. Bennis, W., & Nanus, B. (1985). *Leaders: Strategies for taking charge.* Harper & Row.
12. Binns, R. (2018). Fair attribution: Mitigating discrimination in AI. In *Proceedings of the 2018 CHI Conference on Human Factors in Computing Systems,* pp. 1–14. https://doi.org/10.1145/3173574.3173575
13. Bostrom, N. (2014). *Superintelligence: Paths, dangers, strategies.* Oxford University Press.
14. Bressers, H., & Huitema, D. (2020). The role of national governments in achieving the sustainable development goals: Integrating policy and practice in the Netherlands. *Sustainability, 12*(4), 1502. https://doi.org/10.3390/su12041502
15. Brundtland, G. H. (1987). *Our common future: World Commission on Environment and Development.* Oxford University Press.
16. Bughin, J., Seong, J., Manyika, J., Chui, M., & Joshi, R. (2018). *Notes from the AI frontier: Modeling the impact of AI on the world economy*. McKinsey Global Institute.
17. Bughin, J., Hazan, E., Ramaswamy, S., Chui, M., Allas, T., Dahlström, P., Henke, N., & Trench, M. (2017). *Artificial Intelligence: The next digital frontier?* McKinsey Global Institute.
18. Calo, R. (2017). Artificial Intelligence policy: A primer and roadmap. *UC Davis Law Review, 51*(2), 399–435. https://doi.org/10.2139/ssrn.2942706
19. Cath, C., & Mayer, P. (2018). AI ethics: A systematic literature review of the technical and ethical perspectives. In *Proceedings of the 2018 International Conference on AI & Ethics,* pp. 1–7. https://doi.org/10.1145/3208769.3209367
20. Cernev, T., & Fenner, R. (2024). Beyond 2030: Structures for achieving sustainable development. *Frontiers in Climate*. https://www.frontiersin.org/journals/climate/articles/10.3389/fclim.2024.1453366/full
21. Chee, F. Y. (2025). EU's AI push to get 50 billion euro boost, says von der Leyen. *Reuters*, 11 February 2025.
22. Ciaburro, G. (2017). *MATLAB for machine learning*. Packt Publishing

A. G. O. Mutambara, *Deploying Artificial Intelligence to Achieve the UN Sustainable Development Goals*, Sustainable Development Goals Series,
https://doi.org/10.1007/978-3-031-88423-8

23. Cloete, F., & De Coning, C. (2019). The role of the South African government in implementing the sustainable development goals: Assessing national strategies and performance. *Administration Publica, 31*(1), 134–156. https://doi.org/10.10520/EJC-1c5e15d2
24. Collins, J. C. (2001). *Good to great: Why some companies make the leap . . . and others don't.* Harper Business.
25. Corke, P. (2022) *Robotics, vision and control: Fundamental algorithms in MATLAB*. Springer Nature.
26. Cowls, J., Tsamados, A., Taddeo, M., & Floridi, L. (2021). AI and its impact on the SDGs. In *AI Ethics for Sustainable Development Goals.* IEEE. Available at: IEEE Xplore
27. Dados, N., & Connell, R. (2012). The Global South. *Contexts, 11*(1), 12–13.
28. D'Ignazio, C., & Klein, L. F. (2020). *Data feminism.* MIT Press.
29. Dowling, T. (2025). Key Takeaways from Keir Starmer's Action Plan for AI. *The Guardian (UK)*, 13 January 2025.
30. Elkington, J. (2018). 25 years of triple-bottom-line thinking. *Journal of Corporate Citizenship, 100*, 61–83. https://doi.org/10.9774/GLEAF.4700.2018.ju.00009
31. ERM. (2023). *AI & the sustainable development goals: The state of play.* ERM.
32. Eubanks, V. (2018). *Automating inequality: How high-tech tools profile, police, and punish the poor.* St. Martin's Press.
33. European Commission. (2021). *Proposal for a regulation on a European approach for Artificial Intelligence.* COM(2021) 206 final. https://ec.europa.eu/info/sites/default/files/%proposal_for_a_regulation_on_a_european_approach_for_artificial_intelligence.pdf
34. European Union Commission. (2024). *European Union AI Act.* European Union.
35. Folke, C., Jansson, Å., Rockström, J., Olsson, P., Carpenter, S. R., Chapin III, F. S., . . . & Westley, F. (2011). Reconnecting to the biosphere. *AMBIO: A Journal of the Human Environment, 40*(7), 719–738. https://doi.org/10.1007/s13280-011-0184-y
36. Fukuda-Parr, S., & Hulme, D. (2011). International norm dynamics and the "end of poverty": Understanding the Millennium Development Goals (MDGs). *Global Governance, 17*(1), 17–36.
37. Ghaffari, A., & Karami, N. (2020). How AI could help achieve the Sustainable Development Goals. *AI & Society, 35*(3), 481–493. https://doi.org/10.1007/s00146-020-00997-y
38. Ghaffari, A., & Karami, N. (2021). Strategic frameworks for using AI in sustainable development: A multi-dimensional perspective. *Sustainable Development, 29*(1), 23–34. https://doi.org/10.1002/sd.2046
39. González, A. G., & Fath, B. D. (2020). Advancing sustainable development goals in the United Arab Emirates: Opportunities and challenges. *Sustainable Development, 28*(3), 515–525. https://doi.org/10.1002/sd.2020
40. Goodfellow, I., Pouget-Abadie, J., Mirza, M., Xu, B., Warde-Farley, D., Ozair, S., . . . & Bengio, Y. (2014). Generative adversarial nets. *Advances in Neural Information Processing Systems, 27.* https://doi.org/10.5555/2969239.2969255
41. Goodfellow, I., Bengio, Y., & Courville, A. (2016). *Deep learning.* MIT Press. https://doi.org/10.5555/3298482
42. Griggs, D., Stafford-Smith, M., Gaffney, O., Rockström, J., Öhman, M. C., Shyamsundar, P., . . . & Noble, I. (2013). Policy: sustainable development goals for people and planet. *Nature, 495*(7441), 305–307. https://doi.org/10.1038/495305a
43. Herrmann, A., & Brenner, W. (2018) *Autonomous driving: How the driverless revolution will change the world.* Emerald Publishing.
44. Holden, M. H., & Bhatia, R. E. (2020). Sustainability in the global context: How the 2030 agenda can guide public health research. *International Journal of Public Health, 65*(2), 55–66. https://doi.org/10.1007/s00038-020-01327-5
45. Holland, S. (2025). *Trump announces private-sector US$500 billion investment in AI infrastructure*. Reuters. https://www.reuters.com/technology/. Cited 22 January 2025.
46. Holstein, K., Wortman, J., Daumé III, H., & Dudik, M. (2019). Improving fairness in machine learning systems: What do industry practitioners need? In *Proceedings of the 2019 CHI Conference on Human Factors in Computing Systems.* https://doi.org/10.1145/3290605.3300834
47. Hossain, M. A., & Hossain, M. I. (2021). AI-based frameworks for smart sustainable cities: A systematic review. *Sustainable Cities and Society, 64*, 102510. https://doi.org/10.1016/j.scs.2020.102510
48. Hulme, D. (2009). The Millennium Development Goals (MDGs): A short history of the world's biggest promise. *The European Journal of Development Research, 21*(1), 25–31. https://doi.org/10.1057/ejdr.2008.77
49. Inspire, A (2021). *How DR Congo $80 billion Grand Inga Dam will provide 40% of energy supply in Africa.* Youtube. https://ar.inspiredpencil.com/pictures-2023/inga-falls-map. Cited 10 Dec 2021.
50. International Science Council. (2017). *A guide to SDG interactions: From science to implementation.* International Council for Science. https://council.science/publications/a-guide-to-sdg-interactions-from-science-to-implementation/
51. Jansen, L. J. M., & Voss, J. (2019). Monitoring progress toward the sustainable development goals: Issues and challenges. *Environmental Science & Policy, 94*, 107–113. https://doi.org/10.1016/j.envsci.2018.12.005

52. Jitendra, R. R., & Ajith. K. G. (2013). *Mobile intelligent autonomous systems*. CRC Press.
53. Jitendra, R. R., & Ramakalyan, A. (2020). *Control systems: Classical, modern, and AI-based approaches*. CRC Press.
54. Jobin, A., Ienca, M., & Andorno, R. (2019). Artificial Intelligence: The global landscape of AI ethics guidelines. *Nature Machine Intelligence, 1*(9), 389–399. https://doi.org/10.1038/s42256-019-0088-2
55. Jordan, M. I., & Mitchell, T. M. (2015). Machine learning: Trends, perspectives, and prospects. *Science, 349*(6245), 255–260.
56. Kabir, A. (2024). *Top 10 LLMs in play in 2024*. https://www.threads.net/@altiamkabir/post/DD59Xzao4Gi. Cited 23 December 2024.
57. Kanyandekwe, J., & Mugisha, J. (2018). Assessing Rwanda's progress towards the sustainable development goals: Opportunities and challenges. *Sustainable Development, 26*(2), 149–158. https://doi.org/10.1002/sd.1664
58. Kaplan, J. (2016). *Artificial Intelligence: What everyone needs to know*. Oxford University Press.
59. Karp, R., Tzeng, Y., & Zhang, J. (2019). The role of advanced semiconductors in Artificial Intelligence: Opportunities and challenges. *Journal of Applied Physics, 125*(6), 060901. https://doi.org/10.1063/1.5023444
60. Kataria, J., & Kumar, A. (2023). *Connecting AI and SDGs for a sustainable world with smart solutions*. Lambert Academic Publishing.
61. Kates, R. W., Clark, W. C., Corell, R., Hall, J. M., Jaeger, C. C., Lowe, I., . . . & Svedin, U. (2001). Sustainability science. *Science, 292*(5517), 641–642. https://doi.org/10.1126/science.1059386
62. Kenya Ministry of ICT and the Digital Economy. (2025). *Kenya National AI Strategy 2025–2030*. Kenya Ministry of ICT and Digital Economy. https://ict.go.ke/. Cited 14 January 2025.
63. Kissinger, H. A., Schmidt, E., et al. (2024). *Genesis: Artificial Intelligence, hope, and the human spirit*. Hachette Book Group.
64. Kissinger, H. A. (2022). *Leadership: Six studies in world strategy*. Penguin Press.
65. Korkalo, E., & Vainio, P. (2020). Implementing the sustainable development goals in Finland: A national framework and practices. *Sustainability, 12*(9), 3940. https://doi.org/10.3390/su12093940
66. Kouzes, J. M., & Posner, B. Z. (2017). *The leadership challenge: How to make extraordinary things happen in organisations*. 6th ed. Jossey-Bass.
67. Krech, R., & Padmanabhan, M. (2019). Are the SDGs achievable? Assessing the challenges for sustainable development. *Environmental Science & Policy, 98*, 1–5. https://doi.org/10.1016/j.envsci.2019.05.019
68. Kroll, C., & Joseph, R. (2017). Tracking the sustainable development goals: An analysis of the inclusiveness of national SDG strategies in Asia. *Journal of International Development, 29*(6), 815–831. https://doi.org/10.1002/jid.3206
69. Kroll, C. (2019). Implementing the sustainable development goals in developing countries: Challenges and opportunities. *Sustainable Development, 27*(6), 1470–1474. https://doi.org/10.1002/sd.1952
70. Kroll, C. (2018). Measuring progress towards the sustainable development goals: Limitations and new approaches. *Sustainable Development, 26*(6), 522–532. https://doi.org/10.1002/sd.1741
71. Kroll, C., & Joseph, R. (2017). Tracking the sustainable development goals: An analysis of the inclusiveness of national SDG strategies in the United States. *Journal of International Development, 29*(6), 815–831. https://doi.org/10.1002/jid.3206
72. Lamarre, E., Smaje, K., & R. Zemmel, R. (2023). *Rewired: The McKinsey guide to outcompeting in the age of digital and AI*. Wiley.
73. Leach, M., Scoones, I., & Stirling, A. (2010). Dynamic sustainabilities: Technology, environment, social justice. *Earthscan*. https://doi.org/10.4324/9781849775069
74. Le Blanc, D. (2015). Towards integration at last? The sustainable development goals as a network of targets. *Sustainable Development, 23*(3), 176–187. https://doi.org/10.1002/sd.1582
75. Lo, S. C., & Wang, T. C. (2020). The AI semiconductor market: A comprehensive overview and future outlook. *Journal of Semiconductor Technology and Science, 20*(4), 305–312. https://doi.org/10.5573/JSTS.2020.20.4.305
76. Mahler, A. G. (2018). *Global South*. Oxford Bibliographies in International Relations.
77. Manyika, J., & Spence, M. (2023). The coming AI economic revolution: Can Artificial Intelligence reverse the productivity slowdown? *Foreign Affairs, 102*(6), 55–57.
78. Marr, B. (2018). *Artificial Intelligence in practice: How 50 successful companies used AI and Machine Learning to solve problems*. Wiley.
79. Marwala, T. (2024). *The balancing problem in the governance of Artificial Intelligence*. Springer.
80. McArthur, J. W., & Rasmussen, K. (2019). The SDG index and dashboards report 2019: Global goals, local actions. *Sustainable Development Solutions Network*. https://doi.org/10.1108/SAMPJ-11-2019-0451
81. McKinsey & Company. (2023). *What are AI guardrails?* McKinsey & Company website. www.mckinsey.com
82. Mignolo, W. D. (2007). Delinking: The rhetoric of modernity, the logic of coloniality and the grammar of de-coloniality. *Cultural Studies, 21*(2), 449–514. https://doi.org/10.1080/09502380601162647
83. Mitchell, T. M. (2007). *Machine learning*, vol. 1. McGraw-Hill.

84. Morales, A. M., & Gallego, J. (2019). Leveraging AI for sustainable development. *International Journal of Information Systems for Crisis Response and Management, 11*(3), 16–30. https://doi.org/10.4018/IJISCRAM.2019070102
85. Mutambara, A. G. O. (2024). *Design and analysis of control systems: Driving the Fourth Industrial Revolution*. Taylor & Francis Group.
86. Mutambara, A. G. O. (2025). *Artificial Intelligence: A driver of inclusive development and shared prosperity for the Global South*. Taylor & Francis Group.
87. Mutambara, A. G. O. (1998). *Decentralized estimation and control for multisensor systems*. CRC Press.
88. Mutambara, A. G. O. (1999). *Design and analysis of control systems*. CRC Press.
89. Nanus, B. (1992). *Visionary leadership: Creating a compelling sense of direction for your organisation*. Jossey-Bass.
90. Negnevitsky, M. (2005). *Artificial Intelligence: A Guide to intelligent systems*. Pearson Education.
91. Nilsson, M., Griggs, D., & Visbeck, M. (2016). Map the interactions between Sustainable Development Goals. *Nature, 534*(7607), 320–322. https://doi.org/10.1038/534320a
92. Noble, S. U. (2018). *Algorithms of oppression: How search engines reinforce racism*. NYU Press.
93. Nussbaum, M. C. (2011). *Creating capabilities: The human development approach*. Harvard University Press.
94. O'Neil, C. (2016). *Weapons of math destruction: How big data increases inequality and threatens democracy*. Crown Publishing Group.
95. OpenGov Asia. (2023). *Aadhaar: A global benchmark in digital identity and governance*. OpenGov Asia.
96. Paluszek, M., & Thomas, S. (2020). *Practical MATLAB Deep Learning: A project-based approach*. Apress.
97. Parris, T. M., & Kates, R. W. (2003). Characterising a sustainability transition: Goals, targets, and indicators. *Environment: Science and Policy for Sustainable Development, 45*(8), 6–19. https://doi.org/10.1080/00139157.2003.10524405
98. Pitschas, C., & Krech, A. (2020). The role of local governments in Germany's implementation of the sustainable development goals. *Sustainable Development, 28*(6), 1573–1584. https://doi.org/10.1002/sd.2000
99. Porter, M. J. (1996). *What is strategy?* Harvard Business Review.
100. Quartly, N., & Lopes, C. (2019). Ethical implications of Artificial Intelligence in the context of sustainable development goals. *AI & Society, 34*(4), 669–676. https://doi.org/10.1007/s00146-018-0870-2
101. Quijano, A. (2000). Coloniality of power and Eurocentrism in Latin America. *Nepantla: Views from South, 1*(3), 533–580.
102. Radford, A., Wu, J., Child, R., Luan, D., Amodei, D., & Sutskever, I. (2019). Language models are unsupervised multitask learners. *OpenAI*. Retrieved from https://openai.com/research/language-unsupervised
103. Ranjan, A., & Kumar, P. (2018). Sustainable development goals: The challenges of implementation in developing countries. *Sustainable Development, 26*(6), 513–523. https://doi.org/10.1002/sd.1740
104. Ranjan, P., & Fox, I. (2020). Sustainable development goals and the role of Artificial Intelligence: A review. *AI & Society, 35*(2), 389–401. https://doi.org/10.1007/s00146-019-00903-z
105. Rask, M. (2019). The enchained state: A resource-based view on sustainable development. *Sustainability, 11*(1), 142. https://doi.org/10.3390/su11010142
106. Raworth, K. (2017). *Doughnut economics: Seven ways to think like a 21st-century economist*. Random House Business.
107. Rigg, J. (2019). More than the soil: Rethinking development in the Global South. *Progress in Human Geography, 43*(6), 1111–1131.
108. Rockström, J., Steffen, W., Noone, K., Persson, Å., Chapin III, F. S., Lambin, E., . . . & Foley, J. A. (2009). A safe operating space for humanity. *Nature, 461*(7263), 472–475. https://doi.org/10.1038/461472a
109. Russell, S., & Norvig, P. (2016). *Artificial Intelligence: A modern approach* (3rd ed.). Pearson.
110. Sachs, J. D. (2012). From Millennium development goals to sustainable development goals. *The Lancet, 379*(9832), 2206–2211. https://doi.org/10.1016/S0140-6736(12)60685-0
111. Sachs, J. D. (2015). *The age of sustainable development*. Columbia University Press.
112. Saetra, H. S. (2022). *AI for the sustainable development goals*. CRC Press.
113. Sashkin, M., & Sashkin, M. G. (2003). *Leadership that matters: The critical factors for making a difference in people's lives and organisations' success*. Berrett-Koehler Publishers.
114. Schoenmaker, D., & Schramade, W. (2019). Investing for long-term value creation: The intersection of financial and ESG performance. *Journal of Sustainable Finance & Investment, 9*(4), 356–377. https://doi.org/10.1080/20430795.2019.1501191
115. Schwab, K. (2016). *The fourth industrial revolution*. Crown Business.
116. Schwab, K., & World Economic Forum. (2021). *COVID-19: The great reset*. World Economic Forum. https://www.weforum.org/reports/the-great-reset
117. Shakya, S., & Kumar, V. (2021). A strategic framework for the deployment of Artificial Intelligence for sustainable development: Evidence from India. *Artificial Intelligence for Sustainable Development, 1*(2), 125–139. https://doi.org/10.1007/s43681-021-00015-9

118. Sharma, S., & de Vries, J. (2020). Decolonizing AI: Actions for equitable design. *Feminist Review, 126*(1), 115–133. https://doi.org/10.1177/0141778920909219
119. Siddique, N. (2016). *Intelligent control: A hybrid approach based on fuzzy logic, neural networks and genetic algorithms.* Springer International Publishing.
120. Skilton, M., & Hovsepian, F. (2017). *The 4th industrial revolution: responding to the impact of Artificial Intelligence on business.* Springer Nature.
121. Steffen, W., Richardson, K., Rockström, J., Cornell, S. E., Fetzer, I., Bennett, E. M., ... & Sörlin, S. (2015). Planetary boundaries: Guiding human development on a changing planet. *Science, 347*(6223). https://doi.org/10.1126/science.1259855
122. Sunder, V., Modukuri, S., & Srivastava, R. (2021). *Aadhar: The digital multiplier of the Indian economy.* Indian School of Business.
123. Tiwari, R. (2017). Integrating the Sustainable Development Goals into national policy frameworks: Lessons from developing countries. *Journal of Cleaner Production, 179*, 105–116. https://doi.org/10.1016/j.jclepro.2017.12.217
124. Tsanni, A. (2024). *Will Artificial Intelligence help or hinder progress on the SDGs.* Nature. https://www.nature.com/articles/d41586-024-03959-1. Cited Dec 2024.
125. United Nations. (2024). *The global goals.* United Nations SDG Website. https://www.globalgoals.org/
126. United Nations. (2015). *Transforming our world: The 2030 agenda for sustainable development.* United Nations. https://doi.org/10.1007/978-3-319-71690-3-14
127. United Nations. (2019). *The age of digital interdependence: Report of the UN Secretary-General's High-Level Panel on Digital Cooperation.* United Nations. https://www.un.org/en/pdfs/DigitalCooperation-report-for%20web.pdf
128. United Nations Statistics Division. (2016). *The 2030 Agenda for Sustainable Development: A new framework for global reporting.* United Nations. https://unstats.un.org/unsd/publications/SDGreport2016.pdf
129. United Nations Statistics Division. (2024). *Revised list of global sustainable development goal indicators.* United Nations. https://unstats.un.org/sdgs/. Cited 10 November 2024.
130. United Nations. (2023). *Secretary-General's remarks to launch the special edition of the SDG progress report.* United Nations. https://www.un.org/. Cited 25 April 2023.
131. United Nations. (2023). *The sustainable development goals report 2023: Special edition.* United Nations. https://desapublications.un.org/publications/. Cited July 2023.
132. United Nations. (2024). *The sustainable development goals report 2024.* United Nations. https://unstats.un.org/sdgs/report/2024/. Cited June 2024.
133. United Nations. (2024). *Rankings: The overall performance of all 193 UN Member States 2024.* United Nations. https://dashboards.sdgindex.org/rankings. Cited June 2024.
134. United Nations Secretary-General's High-level Advisory Body on AI. (2023). *Interim UN Report on governing AI for humanity.* United Nations.
135. United Nations Secretary-General's High-level Advisory Body on AI. (2024). *Final UN Report on governing AI for humanity: Global framework for AI oversight.* United Nations.
136. United States Government. (2023). *Executive order on the safe, secure and trustworthy development and use of Artificial Intelligence.* United States Government. https://www.whitehouse.gov/. Cited 30 October 2023.
137. United States Government. (2025). *Executive order on advancing United States leadership in Artificial Intelligence infrastructure.* United States Government. https://www.whitehouse.gov/. Cited 14 January 2025.
138. van Zanten, J. A., & van Tulder, R. (2018). Multinational enterprises and the Sustainable Development Goals: An institutional approach to corporate engagement. *Journal of International Business Policy, 1*(3-4), 208–233. https://doi.org/10.1057/s42214-018-0008-x
139. Vinuesa, R., Azizpour, H., Leitschuh, C., Beyan, O., & Guntuku, S. C. (2020). The role of Artificial Intelligence in achieving the Sustainable Development Goals. *Nature Sustainability, 3*(8), 501–515. https://doi.org/10.1038/s41893-020-0505-1
140. Waisman, H., & Ryser, J. (2019). Sustainable development in the European Union: Integrating sustainable development goals into national and local policies in Finland. *Sustainability, 11*(10), 2741. https://doi.org/10.3390/su11102741
141. Walsh, C. E. (2018). *Decoloniality: Ideas for a world to come.* In C. E. Walsh (Ed.), *The decolonial turn: An introduction* (pp. 1–27). Zed Books.
142. Wang, Y., Kung, L. A., & Byrd, T. A. (2018). Big data in health care: A systematic review. *Health Information Science and Systems, 6*(1), 1–10. https://doi.org/10.1007/s13755-018-0275-0
143. Wang, Z., & Xiong, Q. (2018). Achieving the sustainable development goals in China: Current status and a pathway. *Sustainable Development, 26*(1), 2–11. https://doi.org/10.1002/sd.1586
144. World Bank. (2024). https://data.worldbank.org/. World Bank Website.
145. World Economic Forum. (2025). *Future of jobs report 2025: Insight report.* World Economic Forum.

146. World Economic Forum. (2025). *Davos 2025: Special address by Donald J. Trump, President of the United States of America*. World Economic Forum. https://www.weforum.org/. Cited 23 January 2025.
147. Weidmann, N. B., & Schutte, S. (2020). The role of Machine Learning in supporting the achievement of the SDGs. *Data Science for Sustainable Development, 1*(1), 5–19. https://doi.org/10.1007/s43545-020-00002-6
148. Yap, M. H., & Yang, Y. (2018). The implementation of the sustainable development goals in Singapore: A multi-stakeholder perspective. *Sustainable Development, 26*(4), 309–320. https://doi.org/10.1002/sd.1673
149. Yong, A. (2025). *Interview with Deepseek founder: We're done following*. The China Academy. 27 January 2025.
150. Zeng, S. X., Xu, X. J., Dong, Y., & Zeng, Z. (2018). The roles of government and collaboration in achieving sustainable development goals in China. *Sustainability, 10*(10), 3381. https://doi.org/10.3390/su10103381
151. Zeng, J., Lu, E. G., & Huang, J. (2021). Governance of Artificial Intelligence for sustainable development: A framework for policy making. *Sustainable Development, 29*(1), 134–143. https://doi.org/10.1002/sd.2062
152. Zeng, Y., et al. (2021). AI Ethics by Design: Implementing Customizable Guardrails for Responsible AI Development. *arXiv*. Available: ar5iv.org.
153. Zhang, Y., & Huang, J. (2019). Sustainable development goals in China: Progress and challenges. *Journal of Cleaner Production, 233*, 697–711. https://doi.org/10.1016/j.jclepro.2019.06.176
154. Zounmenou, D. (2018). Africa's prospects for peace and security under Agenda 2063: Opportunities and challenges. *African Security Review, 27*(2), 118–130. https://doi.org/10.1080/10246029.2018.1487556

Index

A. G. O. Mutambara, *Deploying Artificial Intelligence to Achieve the UN Sustainable Development Goals*, Sustainable Development Goals Series,
https://doi.org/10.1007/978-3-031-88423-8